Bioinformatics Methods and Protocols

METHODS IN MOLECULAR BIOLOGY™

John M. Walker, SERIES EDITOR

145. **Bacterial Toxins: *Methods and Protocols***: edited by *Otto Holst, 2000*
144. **Calpain Methods and Protocols**, edited by *John S. Elce, 2000*
143. **Protein Structure Prediction: *Methods and Protocols***, edited by *David Webster, 2000*
142. **Transforming Growth Factor-Beta Protocols**, edited by *Philip H. Howe, 2000*
141. **Plant Hormone Protocols**, edited by *Jeremy A. Roberts and Gregory A. Tucker, 2000*
140. **Chaperonin Protocols**, edited by *Christine Schneider, 2000*
139. **Extracellular Matrix Protocols**, edited by *Charles Streuli and Michael Grant, 2000*
138. **Chemokine Protocols**, edited by *Amanda E. I. Proudfoot, Timothy N. C. Wells, and Christine Power, 2000*
137. **Developmental Biology Protocols, Volume III**, edited by *Rocky S. Tuan and Cecilia W. Lo, 2000*
136. **Developmental Biology Protocols, Volume II**, edited by *Rocky S. Tuan and Cecilia W. Lo, 2000*
135. **Developmental Biology Protocols, Volume I**, edited by *Rocky S. Tuan and Cecilia W. Lo, 2000*
134. **T Cell Protocols:** *Development and Activation*, edited by *Kelly P. Kearse, 2000*
133. **Gene Targeting Protocols**, edited by *Eric B. Kmiec, 1999*
132. **Bioinformatics Methods and Protocols**, edited by *Stephen Misener and Stephen A. Krawetz, 2000*
131. **Flavoprotein Protocols**, edited by *S. K. Chapman and G. A. Reid, 1999*
130. **Transcription Factor Protocols**, edited by *Martin J. Tymms, 1999*
129. **Integrin Protocols**, edited by *Anthony Howlett, 1999*
128. **NMDA Protocols**, edited by *Min Li, 1999*
127. **Molecular Methods in Developmental Biology:** Xenopus *and Zebrafish*, edited by *Matthew Guille, 1999*
126. **Adrenergic Receptor Protocols**, edited by *Curtis A. Machida, 2000*
125. **Glycoprotein Methods and Protocols: *The Mucins***, edited by *Anthony P. Corfield, 2000*
124. **Protein Kinase Protocols**, edited by *Alastair Reith, 1999*
123. ***In Situ* Hybridization Protocols (2nd ed.)**, edited by *Ian A. Darby, 2000*
122. **Confocal Microscopy Methods and Protocols**, edited by *Stephen W. Paddock, 1999*
121. **Natural Killer Cell Protocols:** *Cellular and Molecular Methods*, edited by *Kerry S. Campbell and Marco Colonna, 1999*
120. **Eicosanoid Protocols**, edited by *Elias A. Lianos, 1999*
119. **Chromatin Protocols**, edited by *Peter B. Becker, 1999*
118. **RNA–Protein Interaction Protocols**, edited by *Susan R. Haynes, 1999*
117. **Electron Microscopy Methods and Protocols**, edited by *M. A. Nasser Hajibagheri, 1999*
116. **Protein Lipidation Protocols**, edited by *Michael H. Gelb, 1999*
115. **Immunocytochemical Methods and Protocols (2nd ed.)**, edited by *Lorette C. Javois, 1999*
114. **Calcium Signaling Protocols**, edited by *David G. Lambert, 1999*
113. **DNA Repair Protocols:** *Eukaryotic Systems*, edited by *Daryl S. Henderson, 1999*
112. **2-D Proteome Analysis Protocols**, edited by *Andrew J. Link 1999*
111. **Plant Cell Culture Protocols**, edited by *Robert D. Hall, 1999*
110. **Lipoprotein Protocols**, edited by *Jose M. Ordovas, 1998*
109. **Lipase and Phospholipase Protocols**, edited by *Mark H. Doolittle and Karen Reue, 1999*
108. **Free Radical and Antioxidant Protocols**, edited by *Donald Armstrong, 1998*
107. **Cytochrome P450 Protocols**, edited by *Ian R. Phillips and Elizabeth A. Shephard, 1998*
106. **Receptor Binding Techniques**, edited by *Mary Keen, 1999*
105. **Phospholipid Signaling Protocols**, edited by *Ian M. Bird, 1998*
104. **Mycoplasma Protocols**, edited by *Roger J. Miles and Robin A. J. Nicholas, 1998*
103. ***Pichia* Protocols**, edited by *David R. Higgins and James M. Cregg, 1998*
102. **Bioluminescence Methods and Protocols**, edited by *Robert A. LaRossa, 1998*
101. **Mycobacteria Protocols**, edited by *Tanya Parish and Neil G. Stoker, 1998*
100. **Nitric Oxide Protocols**, edited by *Michael A. Titheradge, 1998*
99. **Human Cytokines and Cytokine Receptors**, edited by *Reno Debets and Huub Savelkoul, 1999*
98. **Forensic DNA Profiling Protocols**, edited by *Patrick J. Lincoln and James M. Thomson, 1998*
97. **Molecular Embryology:** *Methods and Protocols*, edited by *Paul T. Sharpe and Ivor Mason, 1999*
96. **Adhesion Protein Protocols**, edited by *Elisabetta Dejana and Monica Corada, 1999*
95. **DNA Topoisomerases Protocols:** *II. Enzymology and Drugs*, edited by *Mary-Ann Bjornsti and Neil Osheroff, 1999*
94. **DNA Topoisomerases Protocols:** *I. DNA Topology and Enzymes*, edited by *Mary-Ann Bjornsti and Neil Osheroff, 1999*
93. **Protein Phosphatase Protocols**, edited by *John W. Ludlow, 1998*
92. **PCR in Bioanalysis**, edited by *Stephen J. Meltzer, 1998*
91. **Flow Cytometry Protocols**, edited by *Mark J. Jaroszeski, Richard Heller, and Richard Gilbert, 1998*
90. **Drug–DNA Interaction Protocols**, edited by *Keith R. Fox, 1998*
89. **Retinoid Protocols**, edited by *Christopher Redfern, 1998*
88. **Protein Targeting Protocols**, edited by *Roger A. Clegg, 1998*
87. **Combinatorial Peptide Library Protocols**, edited by *Shmuel Cabilly, 1998*
86. **RNA Isolation and Characterization Protocols**, edited by *Ralph Rapley and David L. Manning, 1998*
85. **Differential Display Methods and Protocols**, edited by *Peng Liang and Arthur B. Pardee, 1997*
84. **Transmembrane Signaling Protocols**, edited by *Dafna Bar-Sagi, 1998*
83. **Receptor Signal Transduction Protocols**, edited by *R. A. John Challiss, 1997*
82. **Arabidopsis Protocols**, edited by *José M Martinez-Zapater and Julio Salinas, 1998*

METHODS IN MOLECULAR BIOLOGY™

Bioinformatics
Methods and Protocols

Edited by

Stephen Misener

Queen's University, Kingston, Ontario, Canada

and

Stephen A. Krawetz

Wayne State University, Detroit, MI

Humana Press Totowa, New Jersey

999 Riverview Drive, Suite 208
Totowa, New Jersey 07512

This publication is printed on acid-free paper. ∞
ANSI Z39.48-1984 (American Standards Institute) Permanence of Paper for Printed Library Materials.

Cover design by Patricia F. Cleary.

For additional copies, pricing for bulk purchases, and/or information about other Humana titles, contact Humana at the above address or at any of the following numbers: Tel: 973-256-1699; Fax: 973-256-8341; E-mail: humana@humanapr.com, or visit our Website at www.humanapress.com

Printed in the United States of America. 10 9 8 7 6 5 4 3

Library of Congress Cataloging-in-Publication Data

Bioinformatics methods and protocols/ edited by Stephen Misener and Stephen A. Krawetz
p. cm. -- (Methods in molecular biology ; v. 132)
Includes index.
ISBN 0-89603-732-0 (alk. paper)
1. Molecular biology--Computer network resources. 2. Internet (Computer network) 3. World Wide Web (Information retrieval systems) 4. Molecular biology--Study and teaching. I. Misener, Stephen. II. Krawetz, Stephen A. III. Series: Methods in molecular biology (Totowa, N.J.) ; v. 132.
QH506.B535 2000
572.8'0285--DC21
99-17054
CIP

Preface

Computers have become an essential component of modern biology. They help to manage the vast and increasing amount of biological data and continue to play an integral role in the discovery of new biological relationships. This *in silico* approach to biology has helped to reshape the modern biological sciences. With the biological revolution now among us, it is imperative that each scientist develop and hone today's bioinformatics skills, if only at a rudimentary level. *Bioinformatics Methods and Protocols* was conceived as part of the *Methods in Molecular Biology* series to meet this challenge and to provide the experienced user with useful tips and an up-to-date overview of current developments. It builds upon the foundation that was provided in the two-volume set published in 1994 entitled *Computer Analysis of Sequence Data*. We divided *Bioinformatics Methods and Protocols* into five parts, including a thorough survey of the basic sequence analysis software packages that are available at most institutions, as well as the design and implementation of an essential introductory Bioinformatics course. In addition, we included sections describing specialized noncommercial software, databases, and other resources available as part of the World Wide Web and a stimulating discussion of some of the computational challenges biologists now face and likely future solutions.

Part 1, Sequence Analysis Packages, provides a resource guide to some of the analysis packages that are currently available, including the client-server GCG package found at most institutions, and several PC- and Macintosh-based packages suitable for standalone computing. The Staden Package is also featured because it is a very widely used and fully integrated set of sequence analysis and assembly software tools that is available free to academics. We have also covered the use of freeware that can be used to build an analysis suite to meet specific needs.

Part 2, Molecular Biology Software, provides a collection of software resources used to address some of the basic tasks of bioinformatics. This section begins with an overview of the currently available freeware for the various computer platforms. Specific examples then follow that include sequence-similarity searching using *FASTA, CLUSTAL* multiple-sequence alignment, and phylogenetic analysis. This is followed by a discussion of *Genotator,* a very powerful sequence annotation and presentation suite that integrates the output of multiple and varied analysis into a format suitable for publication. This part concludes with a discussion of common image analysis techniques.

Part 3, Web-Based Resources, highlights the essential primary sequence databases and the varied analysis tools that are available. This part also includes a unique description of the clinical resources that are quickly becoming an integral part of the emerging field of Molecular Medicine. Primary sequence analysis methods, including the means to identify transcriptional control regions using MatInspector, and a review of the current approaches to gene identification are included. Part 3 concludes with a discussion of oligonucleotide and PCR-primer design and the very practical manner in which laboratory methods and reagents can be distributed through the WWW.

In Part 4, Computers and Molecular Biology, the authors directly address the limitations of *in silico* analysis and present possible solutions. This section closes by addressing the still unanswered question of how to detect biologically meaningful patterns from a sequence string of As, Cs, Gs, and Ts.

Teaching Bioinformatics is fast becoming an integral part of the core curriculum at most universities. Part 5 concludes this volume with the authors' recommendations on the delivery of an introductory bioinformatics course. An in-depth examination of how to keep current with the ever increasing body of literature is also included in this section. In short, this section offers essential, practical answers to the day-to-day problems encountered in a successful career in modern biology.

Bioinformatics has helped to make possible the current revolution in modern biology. It is only by understanding and wisely using this resource that we will be able to push the frontier forward. In *Bioinformatics Methods and Protocols*, a broad overview of each subject area was emphasized to help orient those just beginning to use computational tools to address biological problems. We believe that even the novice can quickly tackle each computational problem and arrive at a satisfying result when guided by this unique collection of software and illustrative examples.

Stephen A. Krawetz
Stephen Misener

Contents

Contributors

Ashok Aiyar • *University of Wisconsin-Madison, Madison, WI*
Roger Anderson • *Anderson Unicom Group, Inc., Yorba Linda, CA*
Kathryn F. Beal • *MRC Laboratory of Molecular Biology, Cambridge, UK*
James K. Bonfield • *MRC Laboratory of Molecular Biology, Cambridge, UK*
Timothy G. Burland • *DNASTAR, Madison, WI*
Brian Fristensky • *University of Manitoba, Winnipeg, Manitoba, Canada*
Don Gilbert • *Indiana University, Bloomington, IN*
Nomi L. Harris • *Lawrence Berkeley National Laboratory, Berkeley, CA*
Jack P. Jenuth • *Base4 Bioinformatics, Mississauga, Ontario, Canada*
Lila Kari • *University of Western Ontario, London, Ontario, Canada*
Jeffrey A. Kramer • *Monsanto Life Science Company, St. Louis, MO*
Stephen A. Krawetz • *Wayne State University School of Medicine, Detroit, MI*
MaryAnn Labant • *Anderson Unicom Group, Inc., Yorba Linda, CA*
Laura F. Landweber • *Princeton University, Princeton, NJ*
Avi Orr-Urtreger • *Genetic Institute, Tel Aviv, Israel*
William R. Pearson • *University of Virginia, Charlottesville, VA*
Promila A. Rastogi • *Oxford Molecular Group, Campbell, CA*
Keir Reavie • *Wayne State University, Detroit, MI*
Jeffry A. Reidler • *Scion Corporation, Frederick, MD*
Jacques D. Retief • *University of Virginia, Charlottesville, VA*
Patricia Rodriguez-Tomé • *EMBL European Bioinformatics Institute, Hinxton, Cambridge, UK*
Steve Rozen • *Whitehead Institute for Biomedical Research, Cambridge, MA*
Helen Skaletsky • *Whitehead Institute for Biomedical Research, Cambridge, MA*
Gautam B. Singh • *Oakland University, Rochester, MI*
Rodger Staden • *MRC Laboratory of Molecular Biology, Cambridge, UK*
Paul Stothard • *University of Alberta, Edmonton, Alberta, Canada*
Gary H. Van Domselaar • *University of Alberta, Edmonton, Alberta, Canada*
Thomas Werner • *Institute of Mammalian Genetics, Neuherberg, Germany*
David S. Wishart • *University of Alberta, Edmonton, Alberta, Canada*
David D. Womble • *Wayne State University School of Medicine, Detroit, MI*
Yuval Yaron • *Genetic Institute, Tel Aviv, Israel*

1

Sequence Analysis Packages

1

GCG:

The Wisconsin Package of Sequence Analysis Programs

David D. Womble

1. Introduction

The GCG programs, also called the "Wisconsin Package," comprise a powerful suite of tools for manipulating, analyzing, and comparing nucleotide and protein sequences *(1)*. The initials GCG stand for Genetics Computer Group, which is a subsidiary of Oxford Molecular Group (Campbell, CA). The Wisconsin Package includes more than 130 programs, each of which functions as a tool for performing a specific task, such as translating a nucleotide coding sequence or determining restriction enzyme cutting sites. Most GCG programs use one file as input and write the results to another file. The output files from many GCG programs are suitable as input to other GCG (or other) programs. In many cases, complex problems can be solved by using several GCG programs in succession.

The Wisconsin Package is installed commonly on a shared computer on a network, such as a UNIX server, so that individuals may access the programs and use them from remote locations, such as from their own personal computers (PCs) or other kinds of terminals. There are several different methods for operating the GCG programs. Two methods included with the package are the command line interface, which is the traditional method in which users type the name of a GCG program to initiate an interactive program session, and a graphical user interface (GUI) called *SeqLab,* in which users open a set of windows to the GCG programs and interact graphically to select sequences and program functions. *SeqLab* also includes a powerful color-coded graphical multiple sequence editor. With either interface, all programs operate similarly to each other. Once an individual is familiar with how to run any given pro-

From: *Methods in Molecular Biology, vol. 132: Bioinformatics Methods and Protocols*
Edited by: S. Misener and S. A. Krawetz © Humana Press Inc., Totowa, NJ

gram in the suite, all the other programs will operate in a similar fashion. In the author's experience, students who are newly introduced to the GCG programs often prefer the easy-to-use *SeqLab* graphical interface, whereas experienced GCG users often prefer the command line, because it is inherently faster, especially over a network from a remote workstation. Both interfaces work well. A recently introduced Web-based interface, called *SeqWeb*, is also available from the GCG and allows users to run GCG programs and manipulate sequence files through a Web browser such as *Netscape Communicator* or *Internet Explorer*. Web-based interfaces for GCG are described in Chapter 2.

The complete GCG package also includes a full set of nucleic acid and protein sequence databases. The sequences in the databases are suitable for direct submission into the GCG programs for analysis, manipulation, or comparison. Also included are complete sets of user manuals, in both printed and online Web-based versions. The program manuals provide complete descriptions of all the programs, including examples of how to use each one. A user's guide provides a complete introduction to all of the general features of the Wisconsin Package, including helpful information about useful, simple UNIX commands that will help one manage GCG sequence files. A system support manual describes how to install and maintain the Wisconsin Package and its databases. It is possible to print additional copies of most of the manuals from within the GCG programs if that is desired.

2. Materials

The methods described here are for the GCG program package version 9.1, installed on a shared computer with a UNIX operating system connected to a TCP/IP network *(2)*. The package can be installed on several different kinds of computers, including Digital Alpha machines running *Digital UNIX 4.0*, Silicon Graphics RISC-based machines running *IRIX* versions 6.2, 6.3, or 6.4, and Sun SPARC-based machines running *Solaris* versions 2.51 or 2.6. The package will also run on a Digital Alpha machine running *OpenVMS* versions 6.2 or 7.1. A minimum of 15 gigabytes of hard disk space is needed to install and maintain the Wisconsin Package with its entire set of databases. The disk space requirement is increasing rapidly with the expansion of the databases. Additional disk space for individual users' files is also needed. It is suggested that a minimum of 128 megabytes of core memory be provided with 200 megabytes of virtual memory. The programs are generally run from the C shell in the UNIX environment. The package can be obtained from the Genetics Computer Group, 3575 Science Drive, Madison, WI 53711, (608) 231-5200, Fax: (608) 231-5202, e-mail: `info@gcg.com`, URL: `http://www.gcg.com`.

The GCG programs can be operated directly on the console of the UNIX computer or from remote workstations (e.g. PCs running Windows or MacOS). To operate the programs from the command line, a terminal or PC running telnet software with VT100 terminal emulation is suggested. To operate *SeqLab*, *X-Windows* terminals or personal computers running *X-Windo*ws server software are needed.

Most GCG program results are written to ordinary text (ASCII) files. The text files can be imported into any text or word processor for further manipulation. In addition, many GCG programs result in graphical output, such as restriction maps or RNA secondary-structure predictions. Included among the choices for handling the graphical output are: displaying it on the terminal screen; printing it to a printer or plotter attached to the terminal; or saving it to a file to view or print later. For displaying the graphics on screen, a graphics terminal or emulator is needed. *X-Windo*ws emulation works well for displaying onscreen graphics as well as for using the *SeqLab* graphical interface. Documentation that accompanies the GCG programs suggests different kinds of terminal and graphics software suitable for use with the package. The author's personal suggestions are included in **Subheading 4.**

For printing GCG graphics, printers that can handle either PostScript or HPGL graphics languages are useful. For printing graphics directly from a remote server to a printer attached to the user's terminal, a terminal program capable of printing in transparent mode is required, so that the files are sent directly to the printer without processing by the personal computer.

3. Methods

3.1. Program Descriptions

There are more than 130 programs included in the Wisconsin Package. Although the programs can be used as independent tools, for the purposes of description they can be grouped by related program functions. Here are some of the general program functions included in the package, with short descriptions of some of the programs included as examples. Although a complete description of all of the GCG programs is beyond the scope of this chapter, these examples should provide enough information to give the reader a general idea of the comprehensive nature of the tool kit included with this package.

3.1.1. Comparison

3.1.1.1. Pairwise Comparison

These programs compare one sequence with a second sequence. The choices available include creating the best overall, i.e., global, alignment of the two sequences (*Gap*), finding the best segment of similarity between two sequences (*BestFit*), or creating a X/Y plot of sequence similarity (*Compare/DotPlot*).

3.1.1.2. Multiple Comparison

The *PileUp* program creates multiple sequence alignments from groups of related sequences using progressive, pairwise alignments. Other programs in this group allow manual editing of the aligned sequences (*SeqLab*), display various attributes of the aligned sequences or create profiles from the aligned sequences that can be use for database searching.

3.1.2. Database Searching

3.1.2.1. Reference Searching

These programs (*LookUp*, *StringSearch*) can identify sequences by name, accession number, author, and other kinds of key words.

3.1.2.2. Sequence Analysis

The programs in this group (*BLAST*, *NetBLAST*, *FASTA*, and so on) allow searches for similarity of a query sequence to those in a database. *NetBLAST* directly searches the databases at the National Center for Biotechnology Information (NCBI). The others search the locally installed databases.

3.1.3. Editing and Publication

Programs in this group allow editing of single (*SeqEd*) or multiple (*LineUp*, *SeqLab*) sequence files, as well as preparation of sequence data for publishing or preparation of plasmid maps.

3.1.4. Evolution

The programs *PAUPSearch*, *PAUPDisplay*, *Distances*, *GrowTree*, and *Diverge* allow comparison of multiply aligned sequences for sequence similarity and phylogenetic relatedness.

3.1.5. Fragment Assembly

The GCG fragment assembly system is a set of programs that allow entry of sequence data from a sequencing project and assembly of those data into a contiguous sequence.

3.1.6. Gene Finding and Pattern Recognition

More than a dozen programs are included in this group (*TestCode*, *Frames*, *Motifs*, and so on), which assist in identifying protein-coding regions, protein-binding motifs, direct repeats and other patterns, and other similar tasks.

3.1.7. Importing / Exporting

Fifteen programs in this group assist in entering sequence data and converting the data between the various sequence file formats, including formats for GCG, *Staden*, *EMBL*, *GenBank*, *IntelliGenetics*, *PIR*, and *FASTA*.

3.1.8. Mapping

The mapping programs (*Map*, *MapPlot*, *MapSort*, and so on) can create and display restriction maps, open reading frame maps, peptide digestions maps, T1 ribonuclease digestions maps, plasmid maps, and so on.

3.1.9. Primer Selection

The *Prime* program selects oligonucleotide primers for polymerase chain reaction (PCR) experiments and for DNA sequencing.

3.1.10. Protein Analysis

Programs included in the protein analysis group (*PeptideMap*, *PepPlot*, *PeptideStructure*, and so on) assist in determining information about protein amino acid sequences, such as plotting the isoelectric point, location of functional motifs, and predictions of various aspects of protein secondary structure, including antigenicity and secretory signals.

3.1.11. RNA Secondary Structure

Programs in this group (*Mfold*, *StemLoop*, and so on) can predict and display in multiple formats information about RNA secondary structure, based on the method of Zuker ***(3)***, as well as locate inverted repeat sequences.

3.1.12. Translation

The translation programs (*Translate*, *BackTranslate*, *PepData*, and so on) translate nucleotide sequences into peptide sequences or vice versa.

3.1.13. Utilities

3.1.13.1. Sequence Utilities

These include several useful programs (*Reverse*, *Shuffle*, *Simplify*, and so on) for reversing a nucleotide sequence, randomizing sequences, or replacing low complexity regions with X characters, among others.

3.1.13.2. Database Utilities

With these programs, one can create a GCG personal database from any set of sequences in GCG format, combine any set of GCG sequences into a database that can be searched with *BLAST*, or extract sequence fragments randomly from sequence(s).

3.1.13.3. Printing/Plotting Utilities

These programs (*Lprint*, *ListFile*, *Figure*, and so on) are used for displaying, printing, or plotting GCG results files, either text or graphics files, to various kinds of display, printing, or plotting devices. See **Subheading 3.5.** for more information on displaying or printing GCG results files.

3.1.13.4. File and Miscellaneous Utilities

A number of other utility programs (*ChopUp*, *Replace*, *Reformat*, and so on) assist in manipulating text files, printing GCG documentation, and other tasks.

3.2. Databases

A comprehensive set of sequence databases is included with the Wisconsin Package. These include the *GenBank* and *EMBL* nucleotide sequence databases (with the *EMBL* databases abridged to avoid duplication with sequences in *GenBank*), and the *PIR* and *SwissProt* protein sequence databases. The sequences in the databases are in GCG file format so that they can be used directly as input for the GCG programs. Because most sequences are present in the databases, it is not necessary for individual users to collect their own copies of those sequences. Referring to the copies in the databases is sufficient. Also included are various kinds of databases used by the GCG programs, including restriction enzymes, scoring matrices, proteolytic enzymes, and reagents, protein analysis data files, transcription factor database (TFD), codon frequency tables, translation tables, and the *PROSITE* dictionary of protein sites and patterns. Those data are stored in text files that individuals may retrieve and edit for customization and special purposes.

3.3. Interfaces

Two interfaces are included with the Wisconsin Package: the command-line interface and a graphical user interface known as *SeqLab*. With the command-line interface, a user types the name of a GCG program to initiate an interactive program session. The program then prompts the user for the information needed to run the program, such as the name of the input sequence file, and allows the user to choose from a menu of various options for how the program should operate. Finally, after pressing the return key for the last time, the program runs and, typically, saves the results to a file. All GCG programs operate similarly from the command line, and once one is familiar with the procedure for one program, operation of the other programs will also seem familiar. Operation of the programs from the command line can also be scripted, including command-line switches, which can be a powerful method for running GCG programs multiple times with multiple input files. To use the command-line interface from a remote location, one needs to connect to the GCG server computer by telnet with a terminal emulation program that uses the VT100 set of terminal functions. An example of the command-line interface is shown in **Fig. 1**.

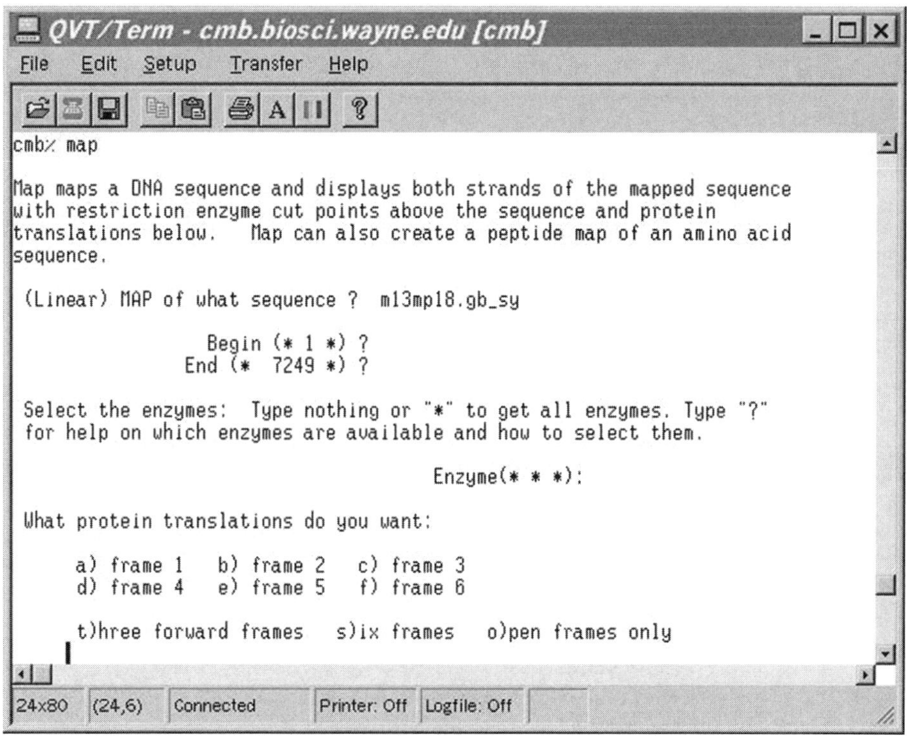

Fig. 1. The GCG command-line interface.

SeqLab is a graphical user interface for GCG that provides an easy-to-use method for operating Wisconsin Package programs. *SeqLab*'s pull-down menus allow one to choose programs to manipulate sequences. When a GCG program is selected from a pull-down menu, a separate window specific for that program appears. Options are then clicked with the mouse, including which sequences are to be analyzed, followed by clicking the **Run** button. Results from the selected GCG programs are then listed in another window called the **Output Manager Window**. The functionality of *SeqLab*, however, goes beyond the command-line interface to include a rich visual display of sequences by individual bases or residues or by known sequence features. This visual display makes it easy to hand edit sequences or create and manipulate multiple sequence alignments. The GUI used by *SeqLab* is called *X-Windows*, which is a windowing system used by computers with a UNIX operating system. To use *SeqLab*, one needs an *X-Windows* display, such as an *X* server program running on a Windows PC or Macintosh, or a workstation that runs *X-Windows*. An example of the *SeqLab* interface is shown in **Fig. 2**.

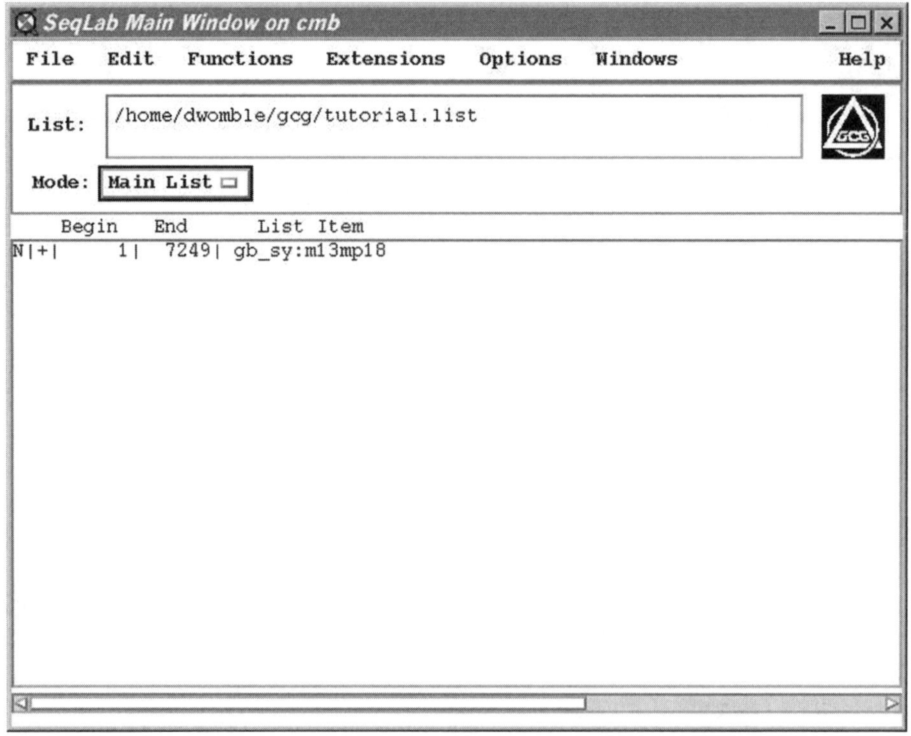

Fig. 2. The GCG *SeqLab X-Windows* interface.

A program called *HYGCGmenu* (hypertext menus for GCG) can be used as an enhancement to the command-line interface. *HYGCGmenu* creates a set of menu screens for the GCG programs and allows a "point and shoot" method of operating the GCG programs. The arrow cursor keys are used to select sequences and to initiate interactive GCG program sessions. The GCG programs are organized in the menus by program function within *HYGCGmenu*, which makes it easy to select programs without having to remember the names of the individual GCG programs. *HYGCGmenu* also has a suite of directory browsing and file management tools, such as for copying, renaming, editing, and so on, which enhances its usefulness. An advantage of using *HYGCGmenu* is that it requires no additional software on the user's end, operating through the VT100 emulation of a telnet terminal. *HYGCGmenu* is not produced by GCG or distributed with the Wisconsin Package, but can be downloaded free of charge by individual educational users or system managers (*see* **Subheading 4.**). An example of the *HYGCGmenu* interface is shown in **Fig. 3**.

Another interface for operating GCG programs is called *SeqWeb*. *SeqWeb* is

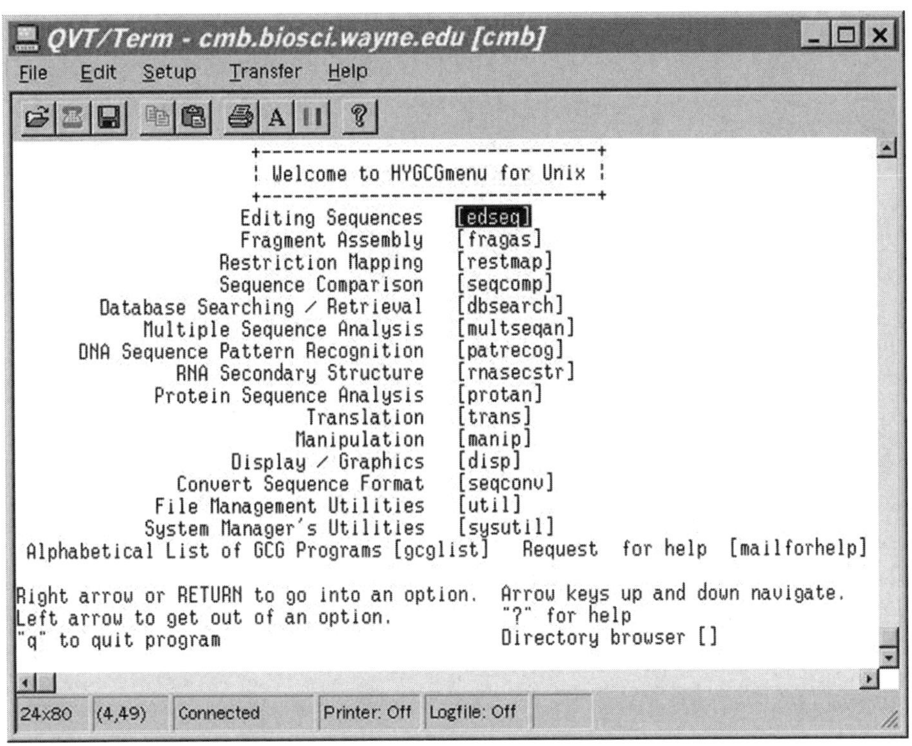

Fig. 3. The *HYGCGmenu* hypertext menu interface for GCG.

sold as a separate product by GCG. As might be expected from the name, *SeqWeb* is a Web-based interface for the Wisconsin Package that operates through a Web browser, such as *Netscape Communicator* or *Internet Explorer*. Web-based interfaces for GCG are described in Chapter 2.

3.4. Tutorials

The tutorials in this section illustrate the basic use of GCG programs, both from the command line and from *SeqLab*. Because all GCG programs behave similarly, familiarity with the operation of one program will be useful for the operation of other GCG programs. The step-by-step instructions in these tutorials demonstrate how to use the *Map* program to create a restriction map and determine the location of open reading frames in a DNA sequence retrieved from the GCG nucleic acid databases. These steps should work with little modification on most GCG servers running with a UNIX operating system and connected to a TCP/IP network. The user's login environment should be set to use the C shell.

3.4.1. Basic Tools Needed

To complete these tutorials, in addition to a login account on a GCG server, the reader will need to have several basic tools installed on a personal computer (Windows or MacOS) or a UNIX workstation connected to the network. The required tools are a telnet terminal program (for connecting to and controlling the GCG server) and an *X-Windows* server program (for displaying *X-Windows* graphics and the *SeqLab* windows on the PC screen). Additional helpful tools include a Web browser (for reading the online GCG manuals), an FTP client (for transferring files to and from the GCG server), a text editor, such as the Windows *Notepad* program (for editing text files on the PC), and printer utilities (for sending PostScript and HPGL files to a printer attached to the PC). See **Subheading 4.** of this chapter for information on where to obtain the tools used by the author. Additional information about recommended software tools is available at the GCG Web site, also listed in **Subheading 4.**

3.4.2. Command Line Tutorial

Use the telnet terminal program on the PC to open a connection to the GCG server and login with the pre-assigned **userid** and **password**. The UNIX command prompt will appear, looking something like this:

```
UNIX%
```

Commands are typed at the prompt. Use UNIX commands to create a folder to contain the GCG working files by typing:

```
UNIX% mkdir gcg
```

and pressing the return key (type only the part after UNIX%). Change into the gcg directory by typing:

```
UNIX% cd gcg
```

and pressing the return key. Startup the GCG programs by typing:

```
UNIX% gcg
```

and pressing the return key. The GCG welcome banner will scroll across the screen, listing the available databases, and, after a moment, the command prompt will return.

Use the GCG fetch command to bring back a copy of the DNA sequence of the M13mp18 cloning vector by typing:

```
UNIX% fetch m13mp18
```

and pressing the return key. A file named m13mp18.gb_sy will now reside in the gcg folder. Its presence can be verified with the UNIX list files command by typing:

```
UNIX% ls
```

and pressing the return key. The contents of the m13mp18.gb_sy file can be examined with the UNIX more command by typing:

```
UNIX% more m13mp18.gb_sy
```

and pressing the return key. Press the spacebar to advance through the file one page at a time. The extension on the file name, gb_sy, indicates that the sequence is from the *GenBank Synthetic* sequences database.

Run the GCG *Map* program on the m13mp18 sequence by typing:

```
UNIX% map m13mp18.gb_sy
```

and pressing the **return** key. The *Map* program begins to run, presents a short description of the program on the screen, and then prompts the user to make some choices about how the program will operate. For this tutorial, rather than making choices at each step, all the default options will be accepted by simply pressing the **return** key at each prompt. That will cause the *Map* program to use the entire sequence (beginning at base number 1 through the last base number 7249), to search for all known restriction enzyme cutting sites (***), to translate the three forward reading frames into single-letter amino acid code (* t *), and to save the results into a text file named m13mp18.map. Dots will scroll across the screen as the program is working, followed by displaying a summary of the results when the program has finished running. The UNIX list files command, **ls**, can be used as above to verify the presence of the new results file m13mp18.map, and the UNIX more command, **more**, can be used as above to examine the contents of the results file. The results file will contain both strands of the DNA sequence, with positions of the cutting sites of the known restriction enzymes indicated above the DNA sequence, and with the amino acid single-letter codes for the translated reading frames listed below the DNA sequence, similar to **Fig. 4**. The results file can be printed (*see* **Subheading 3.5.**) or transferred by FTP to the PC for importing into a word processor or text editor program, if desired.

If the *Map* program were started without giving the name of the input file on the command line, then the first prompt from the program would be for the name of the input sequence file. That input sequence could be a file in the user's folder, as in this example, or a file in the GCG databases. To run the *Map* program directly on a sequence in a database, add the database specifier to the front of the sequence name, like this: gb_sy:m13mp18. That can be done either on the command line, as used in this tutorial, or after starting the program when answering the prompt for input sequence (**Fig. 1**). During the interactive program sessions, the user may choose among the various menu options to control how the program operates. The other GCG programs operate similarly.

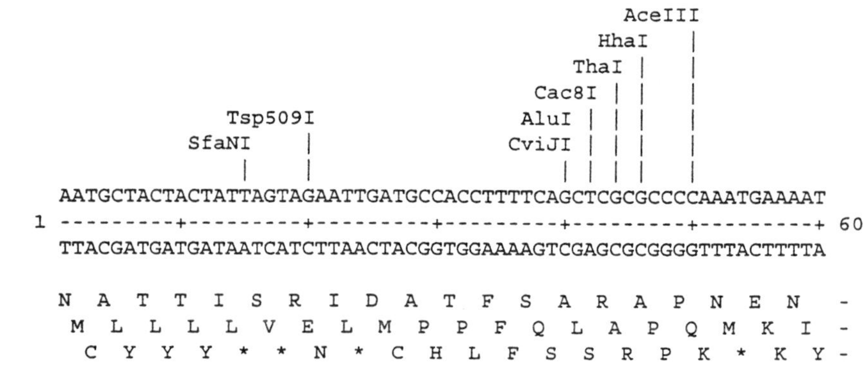

Fig. 4. Map program output.

HYGCGmenu could also be used to create the gcg folder, retrieve the m13mp18 sequence file from the databases, and run the *Map* program, with similar results. Instead of typing the commands as described above, one uses the arrow cursor keys to select the *Map* program, points it to the input sequence file, and runs the program. The results file can then be examined from within *HYGCGmenu* as well.

3.4.3. SeqLab *Tutorial*

X-Windows emulation is used in the *SeqLab* tutorial. The steps listed below are not the only possible way to set up *X-Windows* and *Seqlab*, but they are generic and should be applicable in most situations.

Start the *X-Windows* server program on the PC and leave it running in the background. Follow the instructions in the first part of the command line tutorial (**Subheading 3.4.2.**) to use telnet to connect to and login to the GCG server, change into the gcg folder, and startup the GCG programs. The IP (Internet) address of the PC is needed to set the **DISPLAY** environment for *X-Windows*. If it is not already known, it can be determined on the GCG server by typing:

```
UNIX% who|grep userid
```

and pressing the return key, substituting the actual userid on the command line. The IP address from which the user is logged in will be listed at the right end of the line on the screen beside your login ID. Set the **DISPLAY** environment by typing:

```
setenv DISPLAY my.ip.address:0
```

and pressing the **return** key. Substitute the actual IP address for my.ip.address, and do not forget the :0 on the end. That will allow the *X-Windows* from the GCG server to display on the user's PC screen, assuming the *X-Windows* progam on the PC was started as requested in paragraph two, this subheading.

Start *SeqLab*, the graphical interface to GCG, by typing:

```
UNIX% seqlab &
```

and pressing the **return** key. The "&" causes the program to run in the background, so that other commands can be typed, if desired. Two windows will open on the screen, "SeqLab Main Window" (**Fig. 2**) and "About SeqLab." Click the mouse on the **OK** button on the **About SeqLab** window to dismiss it.

Files created during a *SeqLab* session are stored in a "Working Directory." Set the working directory to the gcg folder (created in the command line tutorial in **Subheading 3.4.2.**) by following these steps: On the SeqLab Main Window, click **Options**, click **Preferences**. On the new User Preferences window that appears, click **Working Dir. . .** , double click the **gcg folder**, then click **OK**. Click **Apply**, then click **OK** again.

SeqLab works by using a list of sequence files. A different list file may be created for each project. Here, a list file for this tutorial will be created that contains an entry for the DNA sequence of the M13mp18 cloning vector. On the SeqLab Main Window, click **File**, click **New List,** type tutorial.list and click **OK**. Click **File**, click **Add Sequences From**, click **Databases**. In Database Specification, type m13mp18. Click **Show Matching Entries**. Under Entries, click m13mp18 to select it, then click **Add to Main Window**, and click **Close**. An entry for gb_sy:m13mp18 should now be in the list. Click **File,** click **Save List** to save the tutorial.list file that was created. The working list can contain entries for sequences from the databases, as in this example, as well as sequence files stored in the user's working directory or other folders.

The next steps in this tutorial run the *Map* program on the m13mp18 DNA sequence. On the SeqLab Main Window, click the m13mp18 sequence in the list to highlight it, if not already selected. Click **Functions**, click **Mapping**, click **Map**. A new window for the *Map* program opens. On the *Map* program window, there are buttons for selecting various options. For this tutorial, just click the **Run** button to start the *Map* program and run it with all default parameters. On the SeqLab Main Window, click **Windows**, click **Job Manager**. The Job Manager window opens. After the *Map* job has completed successfully, as reported in the Job Manager Window, click **Open Output Mgr**. The Output Manager window opens. In the Output Manager window click on the results file, **m13mp18_nn.map**, to select it, then click **Display**. The results of the *Map* program are displayed on the screen. The results appear similar to the results obtained by running the *Map* program from the command line in **Subheading 3.4.2.**, except that all six reading frames have been translated. The results file can be printed (*see* **Subheading 3.5.**) or transferred by FTP to the PC for importing into a word processor or text editor program, if desired. Other GCG programs operate similarly in *SeqLab*. The various windows can be closed by clicking the **Close** buttons. The SeqLab Main Window can be closed by clicking **File**, then clicking **Exit**.

3.5. Displaying and Printing GCG Graphics and Other Files

Many GCG programs have graphic output, such as restriction maps or RNA secondary structure predictions. In the author's experience, printing GCG graphics results is an important skill to master. However, because there are so many different kinds of printing devices, terminal programs, and personal computers used to operate the GCG programs, it is not possible to cover all possible combinations. This discussion will use examples familiar to the author to illustrate some of the important techniques that can be used to print GCG graphics results as well as other files. These examples are for connecting a personal computer (Windows or MacOS) via telnet to a GCG server, with GCG installed on a shared computer with a UNIX operating system on a TCP/IP network. The printer is a PostScript-capable printer connected directly to the PC, although having the printer connected via the network is also feasible.

There are several choices for how to handle graphic output from GCG programs. Those described here are: display it on the screen; print it directly to a printer attached to a PC terminal; save the output to a file and then print it later. Most GCG graphics programs can output their data in either PostScript or HPGL graphics format. Printing those results requires a printer capable of PostScript or HPGL output. For printing PostScript or HPGL files through the network directly to a printer attached to a personal computer, it is necessary to use a terminal program capable of printing in "transparent mode," so that the files are sent directly to the printer without processing by the PC. Note that these methods can be used for printing any files, not just files created by GCG. If a PostScript-capable printer is not available, one can use the freeware programs *GhostScript* and *GhostView* for viewing and printing PostScript files on a PC (see **Subheading 4.**). The first examples presented here are for printing from the command line or from within *HYGCGmenu*, followed by examples for printing from *SeqLab*.

3.5.1. The GCG SetPlot *Program*

Before running a GCG program with graphics output, use the GCG *SetPlot* program to tell GCG how to display, print, or save the graphics output. This program presents a menu of choices. The choices depend on how the GCG systems administrator has set up the GCG programs. Typical choices include the following.

3.5.1.1. X-WINDOWS

The *X-Windows* choice will display the graphic in a window on the screen. This works very nicely if an *X-Windows* server program is installed on a PC, or if using a UNIX workstation with *X-Windows* built in. For this to work, one must first start up the *X-Windows* program on the terminal, then tell GCG where to display its *X-Windows* by setting the **DISPLAY** environment variable (*see* the tutorial in **Subheading 3.4.3.**).

3.5.1.2. PSFile and HPFile

These choices save the graphic to either a PostScript or HPGL file, respectively. They can be printed later on any PostScript or HPGL printer. To print the file later, download it to a PC and send it to the printer. During downloading and copying to the printer, remember that PostScript files are simple ASCII text files, whereas HPGL files are binary files. Printing instructions are included in **Subheading 3.5.3.**

3.5.1.3. LW

With this option, the graphic is sent directly to a PostScript printer attached to a PC. It can be any PostScript printer, not just a LaserWriter. Be sure to set the terminal to print in transparent mode, and it may be necessary to set up page protection from the printer's control panel to get it to print the entire graphic on one page.

3.5.1.4. Laser

With this option, the graphic is sent directly to a HPGL printer attached to a PC. It can be any HPGL printer, not just a LaserJet. Be sure to set the terminal to print in transparent mode, and it may be necessary to set up page protection from the printer's control panel to get it to print the entire graphic on one page.

3.5.1.5. *Tek*

Some telnet programs have built-in Tektronix emulation. With this choice, the graphic is displayed in a *Tek* window on the screen, similar to the *X-Windows* example in **Subheading 3.5.1.1.** However, it is not necessary to set the **DISPLAY** variable with this choice.

3.5.2. Testing the Graphics Set Up

After using *SetPlot* to make the choice for graphics output, the graphics set up can be listed with the *showplot* program. The graphics set up can be tested with the plottest program, which plots a GCG test pattern.

3.5.3. Printing GCG Output and Other Files

Most GCG results files are plain text that can be printed on any printer or imported into any word processor. Some of the programs store graphics output in HPGL or PostScript files. Here are some examples that illustrate some of the options for printing these files. Note also that these methods can be used for printing any files, not just files created by GCG.

3.5.3.1. Printing a Simple Text File from the GCG Server Directly to a Printer Connected to a PC

Use the GCG **listfile** command to print any text file to the printer attached to a PC or Macintosh. Example:

```
UNIX% listfile -noheading filename.txt
```

The `-noheading` command line option will prevent the header (name of the file, date) from printing at the top of the first page.

3.5.3.2. Printing a Simple Text File from the GCG Server Directly to a PostScript Printer Attached to a PC

Use the GCG **lprint** command to print any text file to a PostScript printer attached to a PC or Macintosh. Be sure to set the terminal to print in transparent mode before printing. Example:

```
UNIX% lprint -noheading filename.txt
```

The `-noheading` command line option will prevent the header (name of the file, date, page number) from printing at the top of each page.

3.5.3.3. Printing a PostScript Graphic File from the GCG Server Directly to a PostScript Printer Attached to a PC

Use the GCG listfile command with the `-noheading` option to print a PostScript graphic file to a PostScript printer attached to a PC or Macintosh. Be sure to set the terminal to print in transparent mode before printing. Also, it may be necessary to set up page protection from the printer's control panel to get it to print the entire graphic on one page. Example:

```
UNIX% listfile -noheading graphics.ps
```

3.5.3.3. Downloading and then Printing a PostScript Graphics File on a PC

First, download the file to the PC with ftp. Use ASCII text mode to transfer a PostScript file. Then, on a PC, go to a DOS prompt and copy the file to the printer attached to the appropriate printer port, such as:

```
C:\ copy graphics.ps lpt1
```

Windows users may prefer to use the freeware drag-and-drop *PrFile* utility (*see* **Subheading 4.**). On a Macintosh, use the LaserWriter font utilities to send the file to the printer.

3.5.3.4. Downloading and then Printing a HPGL Graphics File on a PC

First, download the file to the PC with FTP. Use **BINARY** mode to transfer a HPGL file. Then, on a PC, go to a DOS prompt and copy the file to the printer attached to the appropriate printer port, such as:

```
C:\ copy /b plot.hp lpt1
```

Be sure to use the /b switch to send the binary file to the printer. Windows users may prefer to use the freeware drag-and-drop *PrFile* utility.

3.5.4. *Printing GCG Output and Other Files from* SeqLab

When GCG programs are run from *SeqLab*, the *X-Windows* interface, the output files are either text files, such as the results of a *BLAST* search, or are .figure files, which are files that the GCG *figure* program or *SeqLab* can display as graphics on an *X-Windo*ws display screen. Both kinds of output can be printed from *SeqLab*. The methods are similar to those for printing from the command line described above, except that the print jobs are controlled from within the GUI.

An important element for printing from GCG is to use a telnet program with good printing capabilities, such as the shareware *QVT/Net Terminal* program (*see* **Subheading 4.**). Be sure to start *SeqLab* from *QVT/Net Terminal* (*see* the tutorial in **Subheading 3.4.3.** for starting *SeqLab*). Note also that these methods can be used for printing any files, not just files created by GCG.

3.5.4.1. Printing a Simple Text File from *SeqLab* Directly to a Printer Attached to a PC

Any text files listed in the **SeqLab Output Manager Window** can be printed. The text file can be the output of any GCG program, or any text file that is added to the **Output Manager Window** with the **Add Text File . . .** button. To print the text file, click the name of the file in the list to highlight it, then click the **Print** button. Under **Output Format**: choose **ASCII**. Under **ASCII print command:** choose or type:

```
listfile -noheading
```

then click **OK**. That will use the GCG listfile command to print the text file, and the `-noheading` switch will prevent the header (name of the file, date) from printing at the top of the first page.

3.5.4.2. Saving a Graphic from *SeqLab* to a PostScript File

Any .figure file in the **SeqLab Output Manager Window** can be saved to a PostScript file. If the .figure file is not listed in the **Output Manager Window**, it can be added to the list with the **Add Graphics File . . .** button. To save the .figure graphic to a file, click the name of the file in the list to highlight it, then click the **Print** button. Under **Output Device:**, choose **PostScript output saved as file graphics.ps**, and under **Port or File** choose or type:

```
graphics.ps
```

or type any name under which the file is to be saved. If making multiple runs, save each file with a different filename. Next, click **Proceed**. The file will be saved in the folder on the GCG server's hard disk from which *SeqLab* was started. The file can then be downloaded and printed as described for PostScript files above (**Subheading 3.5.3.3.**).

3.5.4.3. Printing a PostScript Graphic File from *SeqLab* to a Printer Attached to a PC

First save the graphic (i.e., .figure file) to a PostScript file as described just above. Then add the graphics.ps file to your **Output Manager Window** by clicking the **Add Text File . . .**" button (remember that PostScript files are just text files that a PostScript printer can interpret and print as a graphic). Then print the file as described above (**Subheading 3.5.4.1.**) for a text file using the listfile `-noheading` command, except for the following: Set the terminal program to print in transparent mode; choose a PostScript-capable printer on which to print it.

3.5.4.4. Saving a Graphic from *SeqLab* to a HPGL File

Any .figure file in the **SeqLab Output Manager Window** can be saved to a HPGL file. If the .figure file is not listed in the **Output Manager Window**, it can be added to the list with the **Add Graphics File . . .** button. To save the .figure graphic to a file, click the name of file in the list to highlight it, then click the **Print** button. Under **Output Device:**, choose **HPGL output saved as file plot.hp**, and under **Port or File** choose or type:

```
plot.hp
```

or type any other name under which the file is to be saved. If making multiple runs, save each file with a different filename. Next, click **Proceed**. The file will be saved in the folder on the GCG server's hard disk from which *SeqLab* was started. The file can then be downloaded and printed as described for HPGL files above (**Subheading 3.5.3.4.**). Remember that HPGL files are binary files.

4. Notes

1. Further information about the Wisconsin Package is available at the Genetics Computer Group Web site at `http://www.gcg.com/`.
2. The *HYGCGmenu* hypertext menus for GCG, and the companion program *HYBROW*, can be obtained at `ftp://biomed.nus.sg/pub/biocomp/`.
3. The author uses the *QVT/Net Terminal* (telnet) program because of its ability to print GCG graphics in transparent mode (passthrough printing). This works well for printing GCG graphics from either the command line or from within *SeqLab* (if *SeqLab* is started from a *QVT/Net Terminal* session). Further information about the shareware *QVT/Net Terminal* program is available at `http://www.frontiernet.net/~qpcsoft/`.
4. The freeware terminal program *TeraTerm* can display Tektronix graphics on the screen. The quality is not as high as using *X-Windows* or printing PostScript or HPGL files, but is sufficient for getting an idea of what the graphic will look like. The latest version of the *TeraTerm* software is available at `http://hp.vector.co.jp/authors/VA002416/teraterm.html`.
5. The author uses the drag-and-drop PrFile printer utility for sending PostScript, HPGL, binary (e.g. *.prn files), and text files to the printer (after downloading the files from the server to his PC). The freeware *PrFile* printer utility for Windows can be obtained at `http://hem1.passagen.se/ptlerup/prfile.html`.
6. The freeware programs *Ghostscript*, *Ghostview*, and *Gsview* can be used for viewing and printing PostScript files even when a PostScript-capable printer is not available. *Ghostscript*, *Ghostview*, and *Gsview* can be downloaded from `http://www.cs.wisc.edu/~ghost/`.
7. The *Micro X-Win32 X* server program works well with the *SeqLab* GUI for GCG and with other UNIX *X* programs such as *File Manager* or *ClustalX*. Further information about the *Micro X-Win32 X-Windows* server program for Windows, including free demo versions, is available at `http://www.starnet.com/docs/xwin32.html`.
8. The latest version of the shareware *WS_FTP FTP* client software is available at `http://www.ipswitch.com/`.
9. The latest version of the *Netscape Communicator* web browser software is available at `http://home.netscape.com/computing/download/index.html`.
10. The latest version of the *Internet Explorer* web browser software is available at `http://www.microsoft.com/ie/ie40/`.
11. The UseNet discussion group for GCG, *INFO-GCG/bionet.software.gcg*, is located at `http://www.bio.net/hypermail/INFO-GCG/`.
12. The author's home page can be found, after a little browsing, at `http://cmmg.biosci.wayne.edu/`.
13. The author has no affiliation with any of the software sources listed above.

References

1. Devereux, J., Haeberli, P., and Smithies, O. (1984) A comprehensive set of sequence analysis programs for the VAX. *Nucleic Acids Res.* **12,** 387–395.
2. *Wisconsin Package Version 9.1.* Genetics Computer Group (GCG), Madison, WI.
3. Zuker, M. (1989) Computer prediction of RNA structure, in *Methods in Enzymology*, vol. 18. (J.E. Dahlberg and J.N. Abelson, eds.), Academic, San Diego, CA, pp. 262–288.

2

Web-Based Interfaces for the GCG Sequence Analysis Programs

David D. Womble

1. Introduction

The Genetics Computer Group (GCG) programs, also called the Wisconsin Package, comprise a powerful suite of tools for manipulating, analyzing, and comparing nucleotide and protein sequences. More than 130 programs are included in the Wisconsin Package, each of which functions as a tool for performing a specific task, such as translating a nucleotide-coding sequence or determining restriction enzyme cutting sites. The Wisconsin Package is commonly installed on a shared computer on a network. Traditionally, users login to the GCG server with a terminal program to operate the GCG programs. Recent efforts have resulted in development of Web-based interfaces for the Wisconsin Package, which allow the GCG programs to be run from Web browsers such as *Netscape Communicator* (*see* **Note 6**) or *Internet Explorer* (*see* **Note 7**). Because Web browsers are now ubiquitous and relatively easy to use, they offer a way to operate the GCG programs that will be familiar and comfortable to most people. GCG has recently introduced its version of a Web-based GCG interface called *SeqWeb. SeqWeb* is distributed separately from the Wisconsin Package as an add-on for an existing GCG installation (**Note 3**). Another Web-based interface for GCG is called *BioPortal* (**Note 4**), developed at the National University of Singapore. *BioPortal* is also an add-on for an existing GCG installation. Other Web-based interfaces for GCG are also in development, but are not addressed here.

The presentation in this chapter assumes the reader is already familiar with the Wisconsin Package. A discussion about the GCG programs themselves is included in Chapter 1.

From: *Methods in Molecular Biology, vol. 132: Bioinformatics Methods and Protocols*
Edited by: S. Misener and S. A. Krawetz © Humana Press Inc., Totowa, NJ

2. Materials

The methods reported here are for the GCG program package version 9.1, installed on a shared computer with a UNIX operating system connected to a TCP/IP network. The package can be installed on several different kinds of computers, including Digital Alpha machines running Digital UNIX 4.0, Silicon Graphics RISC-based machines running IRIX versions 6.2, 6.3, or 6.4, and Sun SPARC-based machines running Solaris versions 2.51 or 2.6. A minimum of 15 gigabytes of hard disk space is needed to install and maintain the Wisconsin Package with its entire set of databases. The disk space requirement is increasing rapidly with the expansion of the databases. Additional disk space for individual users' files is also needed. It is suggested that a minimum of 128 mb of core memory be provided with 200 mb of virtual memory.

SeqWeb is a Web-based interface for operating the GCG programs. The *SeqWeb* product includes Web server software and runs only on UNIX-based computers, as listed above. *SeqWeb* must be installed on a server on which the Wisconsin Package is already installed and running. The *SeqWeb* version described here was a prerelease version. The final product has since been released. Installation of *SeqWeb* requires approx 16 megabytes of hard disk space. The Wisconsin Package and *SeqWeb* can be obtained from the Genetics Computer Group, 575 Science Drive, Madison, WI 53711; (608) 231-5200, Fax: (608) 231-5202; e-mail: `info@gcg.com`, URL: `http://www.gcg.com`.

BioPortal, developed at the National University of Singapore, is another Web-based interface for operating the GCG programs. Like *SeqWeb*, it includes Web server software and must be installed on the UNIX server on which GCG is already installed and running. Installation of *BioPortal* requires approx 17 megabytes of hard disk space. Further information about *BioPortal* is available at the *BioPortal* Web site, URL: `http://bic.nus.edu.sg:8888/`, or by sending e-mail to: `meena@bic.nus.edu.sg`.

To operate the Web-based GCG interfaces, one needs a personal computer (Windows or MacOS) or a UNIX workstation connected to a TCP/IP network, with either *Netscape Communicator 4.0* or higher or *Internet Explorer 4.0* or higher installed. The Web-based GCG interfaces use java, which is the reason for requiring the most recent browsers. An access account with a userid and password must also be setup for each user on the GCG server.

3. Methods

3.1. The SeqWeb *GCG Interface*

3.1.1. SeqWeb *General Description*

SeqWeb is an add-on product for the Wisconsin Package of sequence analysis tools that provides a Web-based interface to a core set of GCG programs.

Access to the Wisconsin Package is made available by simply logging in to the GCG server via *Netscape* or *Internet Explorer*. GCG programs are selected from a menu, and options for running a program are selected with check boxes, pull-down menus, text boxes, and so forth.

Sequences can be loaded into the **SeqWeb Work Area** for analysis by several methods. They can be uploaded from the user's PC to the server by browsing the files on the PC hard disk, by pasting in sequence data from the clipboard, or by selecting sequences from the GCG databases on the server. Once loaded into *SeqWeb*, the sequence files are stored in the user's assigned space on the server's hard disk. Sequence data need not be in GCG file format. *SeqWeb* translates other common file formats automatically, such as *FASTA*, *EMBL*, and *GenBank*, for processing by the GCG programs. Access to a set of *SeqWeb Sequence Manager* functions allows adding, viewing, editing, renaming, copying, or deleting of sequence files.

Results from running the GCG programs are stored in files in the user's assigned space on the server's hard disk. Text and graphic results appear on screen in color, and can either be printed or saved on the hard disk of the user's PC. Text output may be saved in HTML format, which can be particularly useful when the results contain links to internal databases as well as external data on the Web. When program results contain a list of database sequences, selecting those sequences of interest will cause *SeqWeb* to load them into the **Work Area**. The results files can be managed from a *SeqWeb Results Manager*, with functions for viewing, editing, or deleting the results files. This kind of ease of access and use, combined with the power of the GCG programs, results in a very effective overall package.

3.1.2. SeqWeb *Appearance*

The *SeqWeb* GCG interface uses frames, with a **Contents** frame on the left side of the browser screen, a **Work Area** frame on the right, and a **Utilities** frame at the bottom (**Fig. 1**). The **Contents** frame can be toggled to show either a menu-based program function, or an index of all programs available. Clicking on either a program function, such as **Mapping**, or a program name, such as *Map*, opens up new contents in the **Work Area** frame. The **Work Area** frame contains such items as buttons, check boxes, pull-down menus, text frames, a list of sequence files, and so on, that are used to operate a given GCG program and choose which sequence(s) upon which to run it. The contents of the **Utilities** frame at the bottom change according to the context of the **Work Area** frame. The utilities include access to a *Sequence Manager*, a *Project Manager*, a *Results Manager*, user preference settings, and a **Help** button to open the online GCG documentation. Buttons for **Run** and **Reset** are also

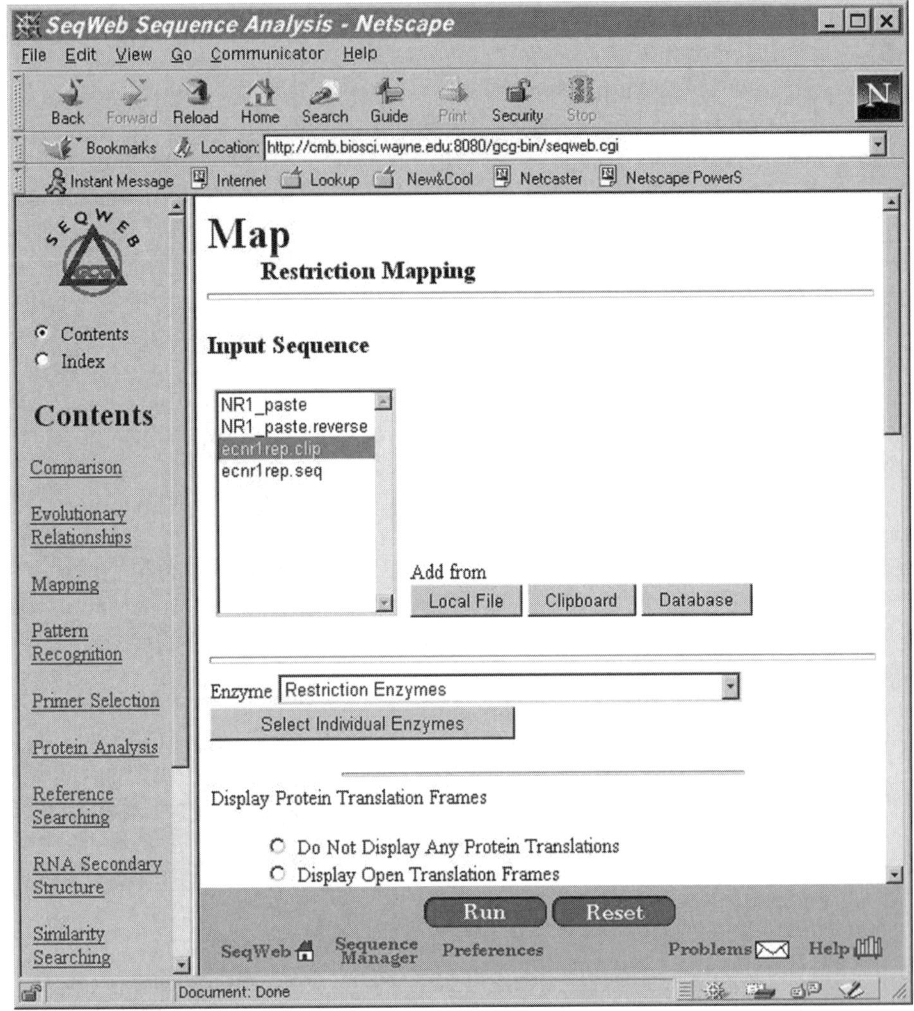

Fig. 1. The *SeqWeb* interface to the GCG programs.

located in the **Utilities** frame when a program is loaded into the **Work Area** frame.

3.1.3. GCG Programs Available in SeqWeb

SeqWeb provides access to some of the most frequently used Wisconsin Package programs. These programs fall into the following functional categories:

Comparison Pairwise comparisons can be either alignments or dot-plots. Multiple sequence comparisons are displayed as multiple sequence alignments.

Database searching Sequence databases can be searched using either sequences or text queries.

Evolution These programs allow investigation of the relationships within a group of pre-aligned sequences. One can compute the pairwise distances between sequences in an alignment, reconstruct phylogenetic trees using distance methods, and calculate the degree of divergence of two protein-coding regions.

Gene finding and pattern recognition These programs help in recognition of coding regions, terminators, repeats, and consensus patterns. Several of the programs help to analyze sequence composition.

Mapping These programs calculate and display restriction maps and peptide cleavage patterns.

Primer selection This program helps to select oligonucleotide primers for a template DNA sequence.

Protein analysis These programs do analyses specific to protein sequences. They can identify sequence motifs as well as predict secondary structure, hydrophobicity, and antigenicity.

RNA secondary structure These programs predict RNA secondary structures, which can then be plotted graphically in several ways. Inverted repeats can also be identified.

Translation These programs translate nucleic acids into proteins and vice versa.

3.2. The BioPortal *GCG Interface*

The *BioPortal* Web-based interface to the GCG programs is similar overall to *SeqWeb*, although the appearance and many of the details are different. *BioPortal* also offers access to a core set of some of the most frequently used GCG programs. The java-based interface uses frames, with a **Contents** frame at the left of the browser screen and a **Work Area** screen at the right (**Fig. 2**). The **Contents** are organized by program function, such as **Mapping**, and the function menus may be expanded to show lists of programs, such as *Map*. Individual programs can then be selected to load the selected program into the **Work Area** at the right.

Sequence files may be uploaded from the user's PC with a browse function, pasted into a text box from the clipboard, or called in from the GCG databases on the server. Sequence data need not be in GCG file format, but can be translated on the fly from most of the common file formats, including *FASTA*, *EMBL*, and *GenBank*. Unlike *SeqWeb*, *BioPortal* stores uploaded files only temporarily on the server. However, the input files remain available for submission to other GCG programs for 24 h.

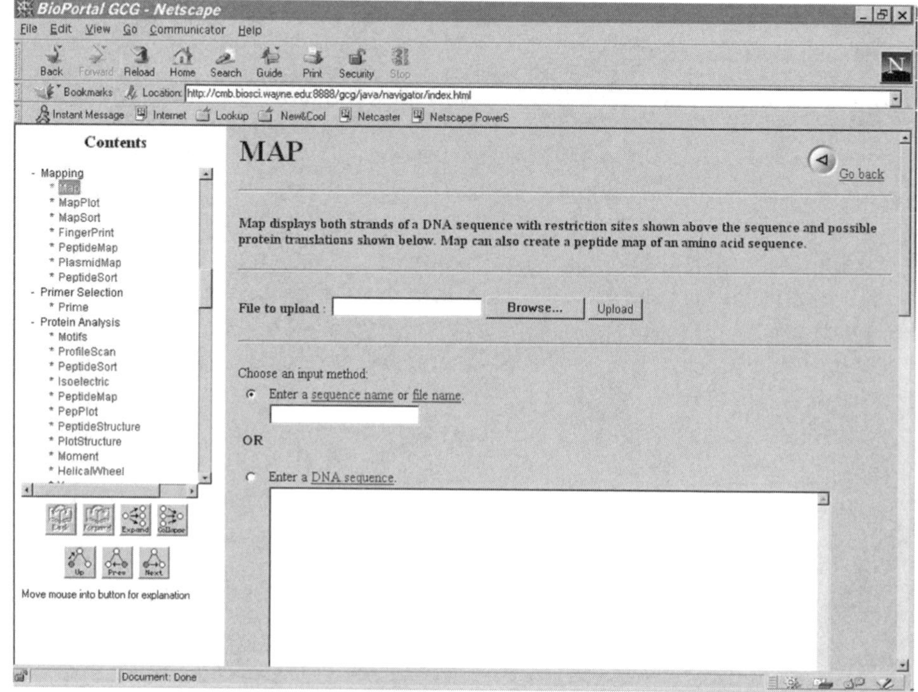

Fig. 2. The *BioPortal* interface to the GCG programs.

Once a sequence is loaded into a program, program options can be selected with check boxes, radio buttons, text boxes, and so on, and the program is then started by clicking a **Submit** button. The results of the program then appear on screen. The results can be viewed, printed, or saved to the user's PC hard disk. The results files are also stored temporarily on the server so that they can be submitted to other GCG programs for further analysis, when appropriate.

In addition to the interface for GCG programs, *BioPortal* also comes with a suite of other useful sequence analysis programs. These include *CLUSTAL W*, *PHYLIP*, *Primer*, and *ReadSeq*, among others. Each of those programs also has a Web-based interface included within *BioPortal*. Together with the interface to the GCG programs, this provides an impressive suite of sequence-analysis tools together in one convenient location, with easy-to-use Web access for them all.

3.3. Other Web-Based Interfaces for GCG

Although not covered in this chapter, there are other Web-based interfaces available for the GCG programs. Two examples are *WWW2GCG*, developed at the Belgian *EMBNet Node*, and *W2H*, developed as a collaborative project between European Bioinformatics Institute (EMBL-EBI), Hinxton, UK and German Cancer Research Center (DKFZ), Heidelberg, Germany. Notes on how to obtain further information about these interfaces are included in **Subheading 4.**

3.4. SeqWeb *Tutorial*

The tutorial in this section illustrates the basic use of GCG programs through a Web interface. This tutorial is for *SeqWeb*, but the procedures for using *BioPortal* are quite similar. The step-by-step instructions demonstrate how to use the *Map* program to create a restriction map and determine the location of open reading frames in a DNA sequence retrieved from the GCG nucleic acid databases. These steps should work with little modification on most GCG servers running a UNIX operating system and connected to a TCP/IP network, on which *SeqWeb* is installed and running.

3.4.1. Basic Tools Needed

To complete this tutorial, in addition to a login account on a *SeqWeb*/GCG server, the reader will need access to a personal computer (Windows or MacOS) or a UNIX workstation connected to the network. A Web browser (either *Netscape Communicator 4.0* or higher, or *Internet Explorer 4.0* or higher) must be installed on the PC or workstation in order to connect to the *SeqWeb* server.

3.4.2. Using the SeqWeb *Interface*

To open the *SeqWeb* interface to GCG, the browser is pointed to the *SeqWeb* server's URL. A login screen prompts for a userid and password. Once logged in, the main *SeqWeb* screen appears. Click the **Index** button on the **Contents** frame at the left of the screen to reveal all GCG programs available. Scroll down the **Index** until **Map** is shown, then click on the **Nucleic** button, which opens the *Map* program in the **Work Area** at the right (**Fig. 1**).

Click the **Databases** button, which opens the **Add** from **Databases** screen. On the **Search** set pull-down menu, select **nucleic: All DNA Databases**. In the **Sequence locus name or accession number** text box, type m13mp18, then click on **OK**. On the screen that comes back, select the gb_sy:m13mp18 entry by clicking the check box next to it, then click on **Add Selected**. Click on **Close** to dismiss the window. The gb_sy:m13mp18 sequence will have been added to the **Input Sequence** list in the *Map* program frame.

If it is not already highlighted, click the gb_sy:m13mp18 sequence in the list to select it. Various options for running the program are available, but for this tutorial, just click on the **Run** button to submit the job with all default parameters. The results appear on the screen, and will contain both strands of the DNA sequence, with positions of the cutting sites of the known restriction

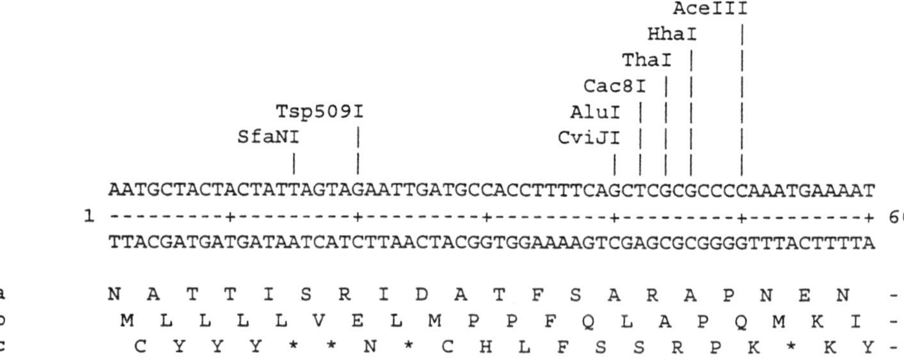

Fig. 3. Output from *Map*.

enzymes indicated above the DNA sequence, and with the amino acid single letter codes for the translated reading frames listed below the DNA sequence, similar to **Fig. 3.**

The results can be printed, or they can be saved onto the user's PC hard disk as either HTML or plain text. The results are also saved on the server's hard disk in a file with a name similar to m13mp18_map_14885.htm, which can be retrieved later in the *SeqWeb Results Manager*. The input sequence file, gb_sy:m13mp18, can be viewed by clicking on the link near the top of the results screen, or by scrolling through the sequence box near the bottom of the results screen. Other GCG programs operate similarly.

4. Notes

1. Further information about the Wisconsin Package and SeqWeb is available at the Genetics Computer Group Web site at `http://www.gcg.com/`.
2. Further information about *BioPortal* is available at `http://bic.nus.edu.sg:8888/`.
3. Information about the *W2H* Web-based interface for GCG is available at `http://industry.ebi.ac.uk/w2h/`.
4. Information about the *WWW2GCG* Web-based interface for GCG is available at `ftp://alize.ulb.ac.be/pub/www2gcg/`.
5. The UseNet discussion group for GCG, INFO-GCG/bionet.software.gcg, is located at `http://www.bio.net/hypermail/INFO-GCG/`.
6. The latest version of the *Netscape Communicator* web browser software is available at `http://home.netscape.com/computing/download/index.html`.
7. The latest version of the *Internet Explorer* web browser software is available at `http://www.microsoft.com/ie/ie40/`.
8. The author's home page can be found, after a little browsing, at `http://cmmg.biosci.wayne.edu/`.
9. The author has no affiliation with any of the software sources listed above.

3

Omiga

A PC-Based Sequence Analysis Tool

Jeffrey A. Kramer

1. Introduction

Computer-based sequence analysis, notation, and manipulation are a necessity for all molecular biologists working with any but the most simple DNA sequences. As sequence data becomes increasingly available, tools that can be used to manipulate and annotate individual sequences and sequence elements will become an even more vital tool in the molecular biologist's arsenal. The *Omiga DNA and Protein Sequence Analysis Software* tool, version 1.1 provides an effective and comprehensive tool for the analysis of both nucleic acid and protein sequences and runs on the ubiquitous standard PC. *Omiga* allows the import of sequences in several common formats. Upon importing sequences and assigning them to various projects, *Omiga* allows the user to produce, analyze, and edit sequence alignments. Sequences may also be queried for the presence of restriction sites, sequence motifs, and other sequence features, all of which can be added into the notations accompanying each sequence. Finally, *Omiga* allows rapid searches for putative coding regions as well as PCR and sequencing primers.

In this chapter, attention is given mainly to the capabilities provided by *Omiga*, as well as the type of output generated and the manner in which this output is generated. It would be difficult to go into great detail on how to perform individual tasks in the space of a single chapter. In fact, *Omiga 1.1* comes with a manual containing more than 350 pages *(1)*. The manual is quite useful, and it describes how to perform tasks in great detail. It is written in an easy-to-follow style, and includes several step-by-step tutorials. These tutorials are

From: *Methods in Molecular Biology, vol. 132: Bioinformatics Methods and Protocols*
Edited by: S. Misener and S. A. Krawetz © Humana Press Inc., Totowa, NJ

valuable for learning the *Omiga* interface, and familiarizing oneself with the individual tasks that can be performed in *Omiga*.

2. Materials

Omiga version 1.1 is designed to run in Windows 95 or Windows NT and requires a CD-ROM drive and 17 mb of disk space. An additional approx 15 mb is at least required to load all of the peripheral utilities, including *VecBank* sequence files, *RasMol,* and *Adobe Acrobat.* It is provided with an exhaustive user guide, as well as several online tutorials that require an additional 1 mb of hard drive space *(1)*. For testing purposes, the software was loaded and run on a 233 MHz PC equipped with an AMD K6 processor, however a lesser machine would suffice.

3. The General Program Interface

3.1. Projects

Omiga uses a project concept to organize sequences, their individual notations, and any additional results and information generated for those sequences by the program. A project represents a simple way to organize an individual investigator's data. Different projects can be localized onto different sectors of the hard drive, or even on a remote storage device such as a Zip drive. Multiple users can also use several separate projects to organize unrelated sequences, and sequence from unrelated lab projects to reduce the amount of superfluous data in each individual interface. Occasional users will most likely not require more than a single project.

3.2. The Project View

3.2.1. A General Overview

The first step in utilizing *Omiga* involves setting up a project. This is done easily by selecting the **new project** button on the toolbar upon opening *Omiga.* The **new project** folder is assigned a name and a file location on the computer. Once the new project is defined, a **project view** is initiated. The **project view** is the main window used in *Omiga* (**Fig 1**). It provides an overview of all of the sequences and related data in the project. Within the **project view,** the user can create additional folders and subdirectories to organize data and information as desired. For example, it may be desirable to store peptide and nucleic acid sequences in separate folders, and store alignments generated by *Omiga* in still another folder.

3.2.2. The **Tree Panel**

The **project view** allows the user to visualize large amounts of data quickly. It contains three windows, referred to as panels. The **tree panel**, which appears

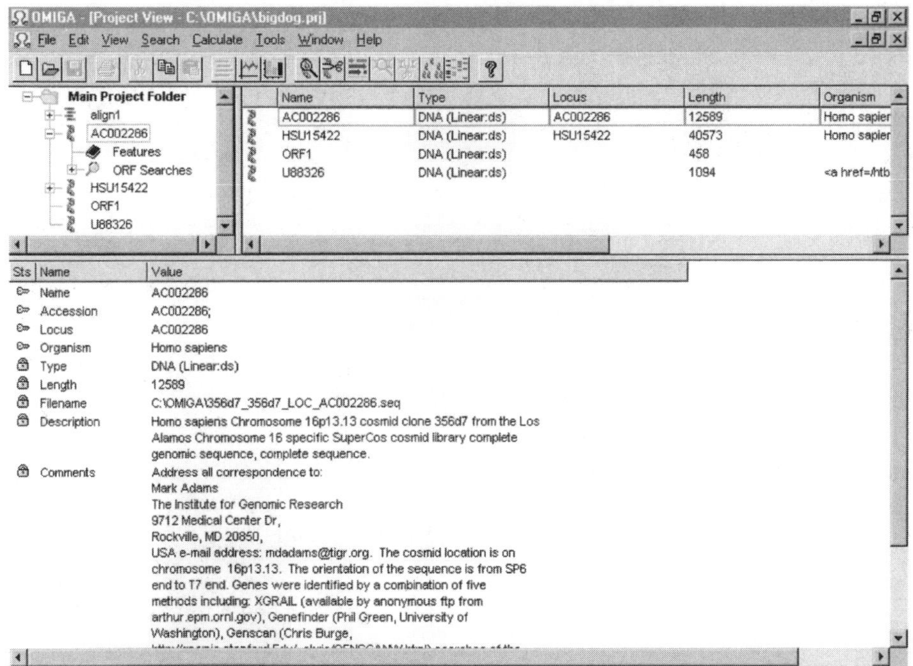

Fig. 1. The **project view** provides an overview of all the sequences and other data in the project.

on the left side of the *Omiga* **project view** window gives a hierarchical view of the individual objects within a project. Clicking on a "+" sign to the immediate left of a folder will expand the folder, showing its contents. Each type of object is indicated by a name as well as a characteristic icon. For example, the main project as well as all subdirectories are indicated by folder icons, whereas DNA and RNA sequences are shown as double and single helices, respectively. The icons even indicate circular or linear DNA sequences, which affects how restriction site search results are derived. The icons allow rapid identification of the content of each object within the entire project.

3.2.3. The Summary and Attributes Panels

Upon expanding the contents of a folder in the **tree panel**, a list of the items within the folder appears in the summary panel. The **summary panel** is positioned on the right side of the *Omiga* **project view** window. Expanding the contents of a folder in the **tree panel** causes a more detailed list of the contents of the folder to be shown within the **summary panel**. For example, expanding a folder containing DNA sequences reveals a list in the **summary panel** detailing the name, size, and accession number (if available) of the sequence, as well as additional information such as the source of the sequence, where available. By selecting an item listed in the **summary panel** (this is done by clicking once on the item in the **tree panel**), the attributes associated with that item are indicated in the third panel. The third panel in the **project view** window is the **attributes panel**. It appears at the bottom of the project view, and lists any attributes for the selected object. This view looks very much like the view commonly seen in a *GenBank* sequence.

The three panels within the **project view** can be resized as desired simply by clicking and dragging the boundaries between them. As discussed above, the contents of folders in the **tree panel** can be visualized by clicking on the "+" sign to the immediate left of a folder's icon. Similarly, additional information available for individual sequences can also be seen. Sequences for which additional information is available will have a "+" sign to the immediate left of their icon, just as folder icons do. Clicking on this expands an additional branch within the tree panel, and shows a list of the additional data available. This includes restriction maps and functional maps, typically generated by the user. Generating and viewing such features will be discussed in later sections.

4. Project Management and Program Organization

4.1. The Sequence View

Managing projects and all of the elements within the project is made easy by the various views available. As discussed above in **Subheading 3.**, the **project view** is the main view in *Omiga.* However several other views are available, all of which can be accessed directly from the **project view**. By double clicking on an individual sequence within the **tree panel** of the **project view,** a **sequence view** is opened for the selected sequence. The **sequence view** can be used to edit a sequence by hand. This feature can be used to reflect site-specific mutations introduced into the sequence in the laboratory as well as sequence polymorphisms. The edited sequence can then be used to identify new restriction sites generated by the specific alterations. Editing can also be used to add a large piece of one sequence to another sequence to reflect cloning experiments in the laboratory. Thus, an experiment involving the cloning of an insert isolated from a library into a plasmid vector can be simulated using *Omiga,* provided the sequence of the vector and the insert are known. The new sequence can then be queried for restriction patterns characteristic of the construct that can be used to verify the actual experiment performed in the laboratory. This new construct and all of the analyses performed on it can then be saved as a new sequence within the project. Additionally, all features such as origins of replication and ampicillin-resistance genes associated with the original sequence fragments used to construct the new sequence will then be part of that new sequence. The **sequence view** also allows the user to translate and reverse-translate sequences, generate complementary strands, and generate the reverse complement of the sequence strand provided.

4.2. The Alignment View

Another view available in *Omiga* is the **alignment view** (*see* **Subheading 6.**). As indicated by its name, the **alignment view** is utilized when performing or viewing alignments for multiple sequences. To perform an alignment, two or more sequences may be selected in the **tree panel** of the **project view**. The **align sequences** option is then selected from the pull-down menu displayed upon selecting the **calculate** option from the tool bar. **Subheading 6.** provides additional information regarding the generation of sequence alignments. The **alignment view** contains the **summary panel**, which displays sequence names, and the **display panel**, which shows the actual alignment. Alignments for up to 500 sequences of 10,000 or fewer residues can be aligned, however aligning an extremely large number of sequences is a CPU-intensive process, and may cause an unstable operating system to crash. The **alignment view** can also be used to group sequences, add or edit gap positions, and add additional notations to alignments. Finally, the consensus alignment can be saved as a data object associated with individual projects.

4.3. The Search Results View

The **search results view** is used to display in a tabular form the results of the various searches that can be performed using *Omiga* (*see* **Subheadings 7.** and **8.**). A **search results view** is generated each time a search is performed (*see* **Subheadings 7.** and **8.** for additional information on performing searches in *Omiga*). The **search results view** allows results to be tabulated according to a number of criteria, and can be optimized to suit an individual user's preference. Thus, a list of restriction sites can be visualized in decreasing order from the most frequent to the least frequent, or increasing order from the least to the most frequent. Sites may also be listed according to position within the sequence being queried, and other search results can be listed in other orders as preferred by each individual user. Searches can also be filtered to reduce unnecessary information. For example, certain restriction sites may be filtered so that only sites of interest are indicated. Data displayed in a **search results view** may be saved as a data object within a project, or printed directly.

4.4. The Features View

Perhaps the most useful view, along with the project view, is the **features view**. The **features view** displays in a graphical fashion the results of the vari-

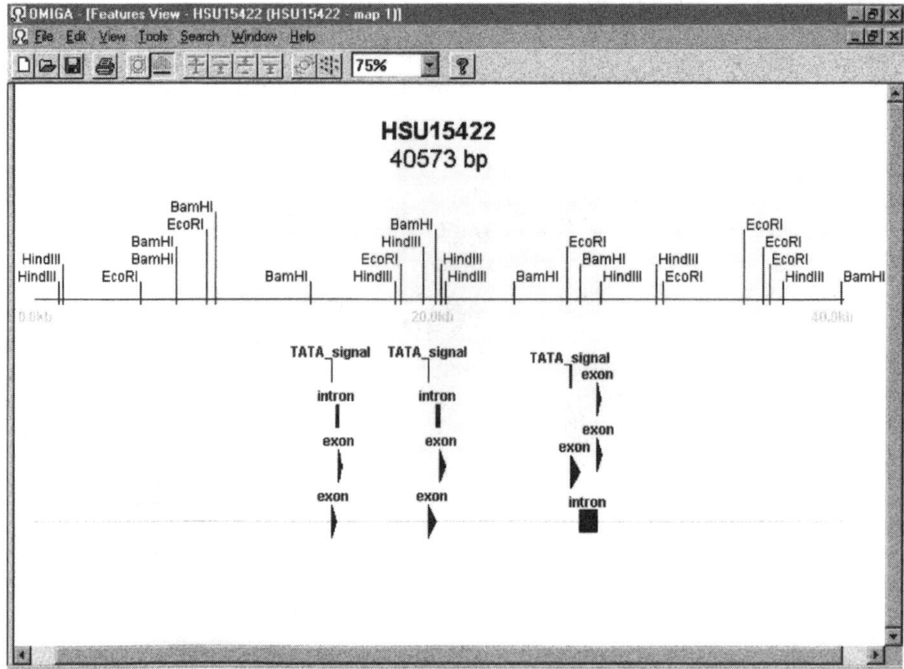

Fig. 2. The **features view** gives a visual representation of a sequence that becomes increasingly informative as various features are toggled on.

ous searches that can be performed using *Omiga* (*see* **Subheadings 7.** and **8.**). It provides a visual representation of a sequence, on which the various sites and features identified for that sequence can be overlaid. For example, displaying the approx 53 kb human protamine locus (accession nos. U15422 plus AC002286) in the features view displays initially a simple horizontal line representing the linear sequence. However, as the various types of features are toggled on, the on-screen representation becomes increasingly informative (**Fig. 2**). The genes identified within the locus can be displayed in varying levels of complexity, such as representing the entire gene as a single arrow, or breaking the gene down into its individual exons. Levels of complexity can be added by displaying coding regions, restriction fragments, and other important sequence elements. Whereas it may be arduous to identify repetitive sequence elements using *Omiga,* these may be added manually to the individual sequences within a project, provided that their location is known. Other analysis tools, such as the *Censor* server (`http://charon.lpi.org/~server/`) can identify repetitive elements, and provides their exact location within the sequence queried. Alternatively, such elements can be imported into *Omiga* if they are already present in the sequence imported from one of the various online databases. Like the **project view,** the **features view** consists of multiple panels, including a **tree panel** and a **display panel**. The **tree panel** is used to toggle the various overlaying groups of features on and off, whereas the **display panel** contains the actual visual representation of the elements toggled on in the **tree panel**.

4.5. The Profile and Composition Analysis Views

The final two views available in *Omiga* are the **profile view** and the **composition analysis view.** These views are displayed for a sequence or sequences by selecting **property profile** or **composition analysis**, respectively, from the pull-down menu displayed when the **calculate** option is selected from the tool bar at the top of the *Omiga* window. The **profile view** allows visualization of property profile searches, such as a plot of the GC content of a sequence. The **profile view** can be used to compare several different plots from one sequence, or identical plots from several different sequences. The **composition analysis view** presents a tabulation of the number and percent composition of nucleotide or amino acid residues within a sequence. Utilizing the views described above, all of which are accessible via the **project view**, allows numerous possibilities for analysis. Several of these are discussed in more detail in the following sections.

5. Importing and Exporting Sequences

5.1. Importing Sequences Into a Project

A primary task in *Omiga* is that of importing sequences. Obviously, importing a sequence is far easier and less prone to error than inputting a sequence by hand. *Omiga* supports several formats, including *ASCII, EMBL, FASTA, GCG, GenBank, PC-Gene,* and *SwissProt.* Imported sequences are converted to the *Omiga* format. The *Omiga* format includes any additional features and information that was in the original sequence file, such as coding regions, transcription start sites, termination codons, polyadenylation signals, and so on. Upon importing new sequences, many such features may be identified based on primary sequence data alone. However, it is often both useful and timely to have these items already identified. Further, many of the features identified in *GenBank* sequences, for example, are based not on primary sequence data but rather on experimental data. Thus, a transcription start site, or an intron–exon boundary may be determined experimentally by comparing genomic and cDNA sequences. The presence of such elements can not always be predicted with certainty simply by inspecting the sequence data alone, and thus these elements may not be identifiable using *Omiga.* For example, importing a large nucleotide sequence from *GenBank* such as the human protamine gene cluster would also import the additional sequence features identified in the original

GenBank accession, including experimentally defined TATA and CAAT boxes, transcription start sites, exons, start and termination codons, polyadenylation signals, and exon–intron boundaries. Whereas *Omiga* could be used to identify ATG or TATA motifs, only experimental evidence can determine if these are the actual start codons and transcription start sites utilized for a particular gene. *Omiga* also imports data on repetitive elements, when they are provided with the original sequence being imported. Clearly, this is an important functionality that adds considerably to the usefulness of *Omiga*.

5.2. Exporting Sequences

As well as importing sequences, *Omiga* can also be used to export sequences, with their dependent feature information where appropriate, into several formats including *ASCII, EMBL, FASTA,* and *GenBank* formats. This functionality will prove useful when submitting sequences to the various databases. Thus, sequence data generated in a laboratory can be mined for information using *Omiga*. Then, upon submission of a manuscript, the sequence data, and all of the characteristic features identified using *Omiga* can be exported to a *GenBank* or *EMBL* format for submission to one of these databases. Presently *GenBank* provides the opportunity to submit sequences online using the *Bank-it* interface. Simply exporting sequences from *Omiga* may speed up the sequence submission task even more. Finally, sequences can also be transferred, complete with all auxiliary features, between two or more projects. Thus, identifying desirable features within a frequently used vector sequence need only be done once, and the information can then be transferred to other projects and other users.

6. Sequence Alignments

6.1. Performing Alignments

Omiga version 1 utilizes the *Clustal W* algorithm for multiple sequence alignments *(2)*. To perform alignments using *Omiga,* two or more sequences in the **tree panel** of the **project view** are selected. The **align sequences** option is then selected from the pull-down menu displayed by selecting the **calculate** option on the tool bar. *Omiga* allows sequences of up to 10,000 nucleotides or amino acids in length, and can accommodate up to 500 sequences. As discussed above, working with too large a number of sequences can be taxing on the CPU, and may even cause the operating system of some computers to lock up. Thus, it is advisable to use far fewer than the 500 sequence maximum.

Alignments can be created in *Omiga* by two methods. Either two or more sequences selected from the project view can be aligned, or sequences can be added to existing alignments within the alignment view. This is similar to the *CLUSTAL W* interface, which allows both possibilities. Although adding sequences to pre-existing alignments is typically quicker, re-aligning all of the component sequences, including the new sequence(s), is often a more accurate means of providing the best alignment. This is because adding a sequence to an existent alignment treats all of the sequences in the alignment as a single sequence (that of the consensus sequence), and thus information may be lost. However, for very similar sequences, such as orthologous genes from closely related species, adding sequences to previous alignments is often sufficient, and can yield identical results to the slower method.

Omiga has a number of preset parameters for performing alignments, which may be changed at the user's discretion. Groups of specific parameters may be saved as alignment protocols. Thus gap penalties, weighted mis-matches, and divergent sequence delays can be selected as desired, and saved as a protocol for use when aligning numerous groups of sequences without customizing the alignment parameters each time. *Omiga* also offers a choice of scoring matrix, including *BLOSUM, MD, PAM,* and *Identity* for protein sequence alignments. Additional information specific to *CLUSTAL W* and the various scoring matrices mentioned above can be found elsewhere in the literature (**ref.** *2,* and references therein), and is beyond the scope of this chapter. (*See* Chapter 11.)

6.2. Editing Alignments

In addition to performing sequence alignments, the **alignment view** allows alignments to be edited, and additional notations to be added. Sequences within alignments can be grouped such that changes to one sequence are performed automatically to the other sequences within the group. This function is useful when introducing a gap at the same location in more than one sequence in an alignment that is suboptimal. It is also possible to pin a sequence, causing it to remain fixed at the top of the alignment view window regardless of scrolling. This function is useful when visually comparing an individual sequence in an alignment with each other sequence in turn. At the bottom of each line of an alignment, *Omiga* can display either an individual sequence of particular interest, an identity/similarity line, or a consensus line. The identity/similarity line marks positions where all sequences are identical with a colon, whereas positions with very few, or very conservative deviations from the consensus are indicated with a period. Thus, in a pairwise alignment of peptide sequences, a lysine and an arginine will be marked by a period, because they are both basic residues, even if the codons used in the nucleotide sequence were divergent, such as AAG and CGC. Similarly, a tyrosine and an aspartate residue (codons TAY and GAY, respectively) though chemically quite different will be indicated by a period, as they represent conservative nucleotide changes. Finally,

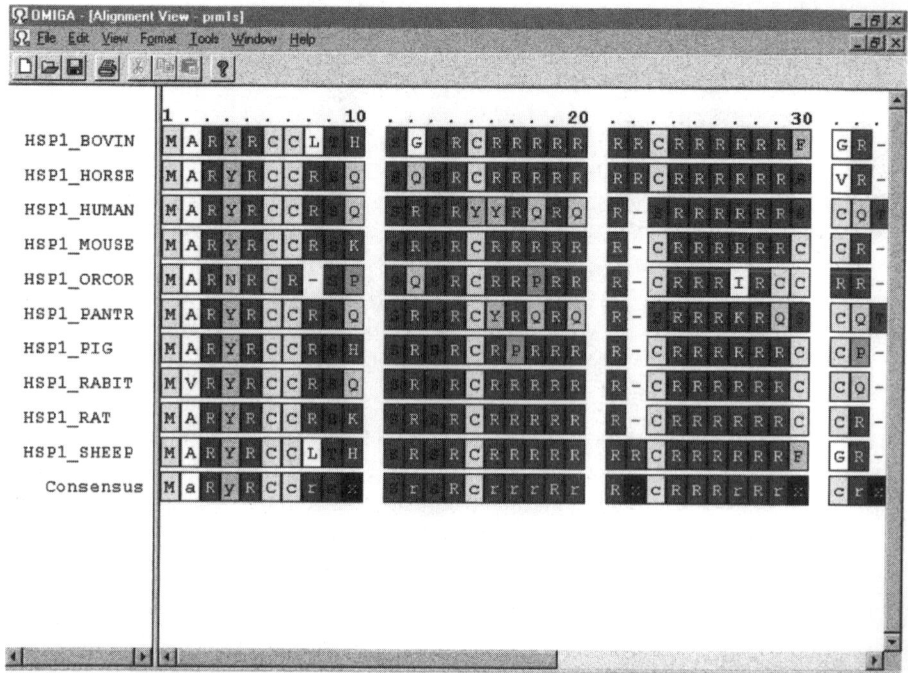

Fig. 3. In the **alignment view,** alignments can be visualized with color, here indicated by shades of gray.

in the alignment view the way in which alignments are visualized can be modified, including the use of user-selected color schemes as well as boxing and/or shading of regions of sequence conservation. This function, along with a color printer, provides for the generation of visually informative documentation for both nucleotide and peptide sequence alignments (**Fig. 3**).

7. Searching for Sequence Motifs

7.1. Identifying Nuclease and Protease Sites

Whereas sequence features and notations provided with sequences from databases are imported into *Omiga* along with the actual sequence itself, it is often desirable to search for and add additional features and notations. This is perhaps the primary functionality of analysis tools such as *Omiga. Omiga* allows the user to identify restriction sites and proteolytic sites in nucleotide and peptide sequences, respectively. When searching for restriction endonuclease cleavage sites, as with searches for proteolytic sites, the user may search for each type of site in turn by inputting the actual sequence recognized by the endonuclease or protease in question. Alternatively, *Omiga* contains the

REBASE database of all common restriction endonuclease sites as well as a *PABASE* database of proteolytic sites that can be used to identify all sites within a sequence. This is done by selecting either the **restriction sites** or **proteolytic sites** option from the pull-down menu generated by selecting the search feature from the tool bar. The output thus generated and displayed in the **search results view** can then be filtered to identify sites ideal for subcloning reactions. In this way, *Omiga* may serve as a valuable tool for planning experiments. An example of this is provided in the tutorials that are included with *Omiga*. In these tutorials, the novice user is led through a process of importing the human *β-globin* domain sequence, and identifying restriction endonuclease sites that will allow subcloning of the *β-globin* gene into the vector pUC 19 in a single ligation. The new pUC–*β-globin* construct can then be saved as a new sequence and searched for restriction endonuclease sites that will give a characteristic restriction pattern to verify the success of the ligation and subsequent cloning reactions in the laboratory. Especially with large sequences, such as the approx 75 kb human *β-globin* domain (*GenBank* locus HUMHBB), this is far preferable to identifying such sites simply by inspection with the unaided eye.

7.2. Searching for User-Defined Sequence Motifs

Omiga also performs searches for user-defined sequence motifs, and allows users to store motifs and search parameters in protocols that can be used for additional queries on other sequences. Searches for user-defined motifs are performed in a similar fashion to searches for restriction and proteolytic sites, except that the **nucleic acid motifs** or **protein motifs** selections are chosen from the **search** pull-down menu. User-defined motifs may be as simple as searching for all ATG trinucleotides independent of reading frame, or as complex as identifying segments with a percentage identity to a complex hormone-receptor-binding site that lie within a predetermined distance from a previously identified transcription start site. Upon initiating the motif search, the user is prompted for information regarding the search parameters using the nucleic acid or protein motif search parameters box. Individual user-defined search protocols can be saved and stored in databases for later use, without the need to redefine the search parameters each time. In this way, segments with a high degree of identity to reported promoters, regulatory sequences, and other common *cis* elements can be identified. Theoretically, it would be possible to identify alu repeats or other repetitive elements, provided a consensus sequence is known. This would require the identification of regions with a similarity above a user-defined cut-off point to the repetitive element consensus sequence of interest. All of these elements can then be labeled and included as features of that sequence. Another potential use for motif identification involves a search for restriction endonuclease sites within PCR products. By searching for and identifying the previously designed primer sequences, the user could delimit the search for restriction fragments to show only those sites that occur between the primers. This would be useful for identifying restriction patterns indicative of the successful amplification of a particular amplicon. Other applications for this functionality will become obvious as users gain familiarity with the *Omiga* interface and its capabilities.

8. Searching for Coding Regions

Another function provided by *Omiga 1* is the identification of putative coding regions within nucleic acid sequences. Searching for open reading frames is carried out by selecting the **open reading frames** selection from the **search** pull-down menu. This function is particularly useful when large pieces of genomic sequence data are being analyzed. As with the sequence alignments discussed above, individual protocols can be generated for use when searching sequences for potential coding regions. *Omiga 1* comes with an internal copy of the *GENMOTIFS* database. In addition, sequences characteristic of genes (gene-associated motifs) can be added by the user. For example, it has been suggested that uneven positional base preference can be used as an accurate indicator of expressed segments within large genomic sequences *(3)*. Whereas this may be difficult to fashion into a gene-associated motif, it is conceivable that other motifs may be identified that are characteristic of coding regions. It appears that the search for coding regions in *Omiga* represents a more complex version of this search protocol, utilizing a search for multiple weighted motifs. That is, *Omiga* appears to search for the occurrence of statistically significant clusters of gene-associated motifs within the sequences being queried. This method is likely to be inferior to the neural-network approach utilized in search tools such as *GRAIL (4)*.

9. Primer Identification

A welcome addition to *Omiga* version 1.1 is the primer-design capability. Both sequencing and PCR primers can be designed using user-defined criteria. This is carried out by selecting a sequence in the **tree panel**, then selecting **PCR Primer Pairs** or **Sequencing Primers** from the **Search** pull-down menu. Default protocols are provided, however the user may also define search parameters of his or her own and has the option of saving these as additional protocols. When searching for PCR primers, the parameters over which the user has control include primer length, GC content, melting temperature, salt concentration, and primer concentration. The user may specify specific regions of the template sequence to be searched or omitted from the search. The user may specify a 3′ clamp, as well as a number of ambiguous nucleotides for degenerate primers. Another feature available for primer design is the ability to omit duplicate end points. That is, if one 5′ primer is compatible with numerous potential 3′ primers, the user has the option to show only that compatible 3′ primer that produces the shortest amplification product, thus reducing the clutter of acceptable pairs that can result from identifying primer pairs in a very large template sequence. However, there does not seem to be a feature that allows the user to design primers compatible with an already defined primer. For example, it may be desirable to design a primer pair that will amplify across an intron (to differentiate genomic from cDNA), and then design an additional 3′ primer that is compatible with the 5′ primer of the original pair. The additional 3′ primer could be designed within an intron, to identify pre-mRNA, or within a different exon to show alternative splicing. The ability to identify new primers compatible with a preselected primer for PCR amplification is a common and useful feature of several other primer-design tools.

When identifying PCR primer pairs, *Omiga* will identify three types of primer-to-primer interactions, including primer–primer annealing, primer self-annealing, and primer–template annealing. The way in which *Omiga* handles these three types of annealing can be specified by the user. Primer–primer interactions include the dreaded primer–dimer formation. If very stable primer–dimers are formed, particularly at the 3′ end of the primers, little or no amplification of template will occur. Primer self-annealing deals with hairpin formation within each individual primer. As with primer–dimers, if too large a fraction of an individual primer is in the form of hairpin structures, little or no amplification will occur, as much of the one primer will be sequestered in the hairpin form, and will be unavailable for annealing to the template. Finally, *Omiga* will also search for nonspecific primer annealing within a template sequence. This is important to avoid primers that contain low-complexity sequence elements, and those primers that may be directed towards repetitive regions within the template. An additional feature that is missing from *Omiga*'s primer design capabilities is the ability to search primers against additional sequence databases. Whereas it can often be time consuming to compare primer sequences against large databases, it is also often a very valuable functionality. For example, when designing primers for the amplification of regions within the genome of a very simple organism for which the entire genome is known, it might be desirable to compare your primers to the entire genome of that organism, to avoid spurious priming. In the age of whole genomes, such a search will become increasingly possible. Similarly, if primers are designed within a short sequence from the public databases, it may be desirable to scan them against a database of known repetitive elements. Whereas such sequences are rare within coding sequence, and thus not likely to be identified by the primer–

template annealing comparisons performed by *Omiga,* the presence of even a slight similarity to an Alu or Line elements will result in frequent inappropriate priming and poor amplification of the desired amplicon.

Searching for sequencing primers in *Omiga* is similar to searching for PCR primer pairs, except that primer–primer interactions need not be considered. Of course the individual characteristics of a sequencing primer are different from that of a PCR primer, and thus a different set of default parameters are used by *Omiga.* As with the search for PCR primers, these values can be edited by the user, and then saved as protocols. Thus, this feature may be used for the identification of individual primers for uses other than sequencing. For example, it is conceivable that a user could design single primers for the generation of cDNA nearer the 5′ end of extremely long messages where poly-T primers are inadequate. This function will be useful for designing primers for both 5′ and 3′ RACE experiments.

Upon identification of primers and primer pairs, the results can be displayed in either a graphical format using the **map** button or tabular format using the **table** button. Within the graphical format, it is possible to display a map of all optimal primers, or of all compatible primer pairs. Furthermore, if no primer pairs are identified, *Omiga* gives the user the ability to analyze the reasons for this failure. In this way, it is possible to alter individual search parameters accordingly, until optimal primers can be identified. Finally, as with many features in *Omiga,* it is possible to make an individual user's primer identification protocols available across multiple projects, and for multiple users. Thus, optimal primer identification protocols can be shared by all members of a laboratory.

10. Notes and Comments

Omiga 1.1 can be used as a valuable aid in planning common experiments that are performed in most molecular biology laboratories. The interface is relatively easy to learn and use. Perhaps the greatest shortcoming of *Omiga* is the clutter than can be created by certain analyses. For some analyses, such as the one detailed in the third tutorial provided with *Omiga 1,* a confusing number of windows must be open at one time, which could quickly become unwieldy. This is partly unavoidable because of the great functionality available to *Omiga* users. It is quite conceivable to plan an entire experiment using *Omiga;* from amplifying a sequence from a cDNA library through cloning of a restriction digested amplification product to the verification of successful isolation of the desired construct. Understandably, the amount of information necessary for such computations will make for an occasionally busy computer screen. Despite this, *Omiga 1* proves to be a valuable tool, and should fulfill the computer-based sequence analysis needs of most molecular biology laboratories.

A useful feature that ought to be included in future versions of *Omiga* would be the ability to import sequence alignments produced independently by stand-alone versions of *Clustal W* or performed by other alignment tools. Whereas *Clustal W* is a useful and popular alignment tool, other alignment tools exist, and some of these are preferred by some investigators. Another feature that might improve the utility of *Omiga* would be a sequence compiler/contig identification functionality. Clearly genome sequencing centers use highly sophisticated software on large mainframe machines, but other PC and Mac software tools exist for compiling sequences on a more modest basis. The Oxford Molecular Group, the producers of *Omiga,* have recently acquired GCG, the most widely used and comprehensive sequence analysis tool presently available, and will be supporting it in the future. It would be ideal if these two tools were made fully compatible. Then, a researcher could compile information from sequencing experiments using the powerful mainframe computer-based GCG program, yet retaining the ability to perform more simple and common analyses on a more easy to use PC-based application such as *Omiga*. The primer identification protocol is a welcome addition to *Omiga 1*, but as discussed above, additional functionality similar to other commonly available primer design software tools is required.

References

1. Oxford Molecular Group. (1997) *Omiga 1.0 User Guide.* Oxford Molecular Ltd., Oxford, England.
2. Thompson, J. D., Higgins, D. G., and Gibson, T. J. (1994) CLUSTAL W: improving the sensitivity of progressive multiple sequence alignment through sequence weighting, position-specific gap penalties and weight matrix choice. *Nucl. Acids Res.* **22,** 4673–4680.
3. Kramer, J. A., Adams, M. D., Singh, G. B., Doggett, N. A., and Krawetz, S. A. (1998) Extended analysis of the region encompassing the PRM1 → PRM2 → TNP2 domain: genomic organization, evolution and gene identification. *J. Exp. Zool.* **282,** 1–2, 245–253.
4. Uberbacher, E. C. and Mural, R. J. (1991) Locating protein-coding regions in human DNA sequences by a multiple sensor-neural network approach. *Proc. Natl. Acad. Sci. USA* **88,** 11,261–11,265.

4

MacVector

Integrated Sequence Analysis for the Macintosh

Promila A. Rastogi

1. Introduction

Whether a researcher is working on a genome project, or is cloning and characterizing a gene of interest, the ability to manipulate, analyze, and annotate sequence data is becoming increasingly important. It is evident by now that efficiently performing the above operations on thousands or millions of base pairs by hand is so difficult as to be impossible, and computer programs that do sequence analysis are becoming more and more ubiquitous in laboratories practicing molecular biology. *MacVector*™, from Oxford Molecular Group, Campbell CA, is one such computer package. It is an integrated comprehensive sequence analysis program that runs on the Macintosh. It provides all of the most commonly used nucleic acid and protein analysis tools and also provides access via the Internet to the public *Entrez* databases at the National Center for Biotechnology Information (NCBI). At the time of writing, the version of *MacVector* available was 6.5. *MacVector* also comes with a module for contig assembly, called *AssemblyLIGN. AssemblyLIGN* will not be discussed here.

2. Materials

MacVector 6.5 is designed to run on MacOS 7.1 or higher and requires a CD-ROM drive and 18 mb of disk space. Minimum memory required is 4 mb of RAM, though at least 6 mb of RAM is recommended. Connection to the Internet is required for access to the NCBI databases. For greatest utility a color monitor and a Macintosh-compatible printer are also recommended. The

From: *Methods in Molecular Biology, vol. 132: Bioinformatics Methods and Protocols*
Edited by: S. Misener and S. A. Krawetz © Humana Press Inc., Totowa, NJ

user guide *(1)* is provided in book format and as a PDF file on the *MacVector 6.5* CD.

Being a commercial program, *MacVector* is copy protected. The copy protection can either be via hardware or software. The hardware copy-protection device is called the EvE key, which connects to the Macintosh's Apple Desktop Bus (ADB). The device can be daisy chained to the computer's keyboard or mouse. The computer must be switched off before connecting or disconnecting the EvE key to the ADB. The software copy-protection solution is *KeyServer*™ from Sassafras Software™ (Hanover, NH). *KeyServer* is used to copy protect multiple copies within the same software license, and is installed on one machine in the network. *MacVector* is installed on each of the clients, which connect via *AppleTalk* to the *KeyServer* for copy protection. The *KeyServer* is programmed with the number of users (x) allowed by the license. When the (x+1)th user tries to launch his/her copy of *MacVector,* the *KeyServer* denies copy protection until one of the other users exits the *MacVector* program.

3. Methods

3.1. The General Program Interface

MacVector adheres to the Macintosh Human Interface Guidelines, and provides an interface that is easy to use and familiar to any Macintosh user. The program menu bar has six items: **File**, **Edit**, **Options**, **Analyze**, **Database**, and **Windows**. The **File**, **Edit**, and **Windows** menus are similar to these menus in other Macintosh programs. The **Options** menu offers three commands that can be used to modify the way information is presented in analysis results windows (**Format Annotated Display**, **Format Aligned Display**, and **Default Symbols**). The user can modify or create the genetic code assignments to be used in all translations by selecting the **Modify Genetic Codes** command, or create a codon bias table for an organism by using the **Make Codon Bias Table** command in conjunction with the *Entrez* databases at the NCBI. The **Analyze** menu lists the various analysis functions that *MacVector* can perform on nucleic acid and protein sequences. The **Database** menu includes commands pertaining to database searching, such as BLAST searches, *Entrez* browsing and pairwise alignments of a query sequence to a user database of sequences saved locally.

The sequence is the central object in *MacVector*, and a sequence window must be active before an analysis can be performed on it. The program is highly interactive. The commands in the **Analyze** menu are available depending on what sequence type is active. When a nucleic acid sequence is active, all protein-analysis functions are grayed out, and vice versa. Most of the analyses store intermediate results in memory and allow the user to apply various filters to view different subsets of the results over and over without having to redo the entire analysis. Many analyses also offer a number of different ways in which to view the results: in tabular format, graphical format, and on the annotated sequence. The user can zoom in to view local regions of the graphical results in more detail. Results may be saved as text (tabular results) or PICT (graphical results) files, which may be reviewed or edited in word processing or drawing programs.

When a user initiates an analysis, a dialog box specific to that analysis appears wherein the user can either enter parameters of their choice or opt to perform the analysis with the default parameters. The user submits the analysis and once it is complete, *MacVector* presents the user with the results dialog box. The results of most analyses in *MacVector* can be presented in more than one form, and the results dialog box presents the user with the choice of result formats, as well as any possible options to filter the results before they are presented. Once the choice of result formats is made, *MacVector* presents them.

3.2. Entering and Editing Data

MacVector supports most of the common sequence formats. The native *MacVector* sequence file type is a binary file. The other file formats are in text format, and include *GenBank, GCG, IG-Suite, CODATA* including *NBRF PIR, EMBL, SWISS-PROT, FASTA,* and line (sequence only). Sequence files in these formats can be opened in *MacVector* using **File|Open**, and then saved in *MacVector* format by using the **File|Save** command. Sequences can also be entered by hand: A new sequence window is opened by selecting the sequence type from the pull-down menu next to the **File|New** command. Entering the actual residues is like word processing; the keyboard or the numerical keypad can be used. Once sequence is entered, the proofreader can check it. Protein sequences can also be created by translating portions of or entire nucleic acid sequences, via the **Analyze|Translation** command. Nucleic acid sequences can similarly be created by reverse translating protein sequences in part or in full, via the **Analyze|Reverse Translation** command.

The sequence window has buttons linking it to the **Features Table**, the **Annotations Table**, the **Annotated Sequence View** and the **Features Map** for the sequence (**Fig. 1**). The **Features** and **Annotations Tables** may either be populated with information that was contained in the original *GenBank* file for the sequence or with information entered by the user. The **Annotated Sequence View** is a text representation of features annotated on the sequence. The **Features Map** is a customizable graphical representation of the information contained in the **Features Table**.

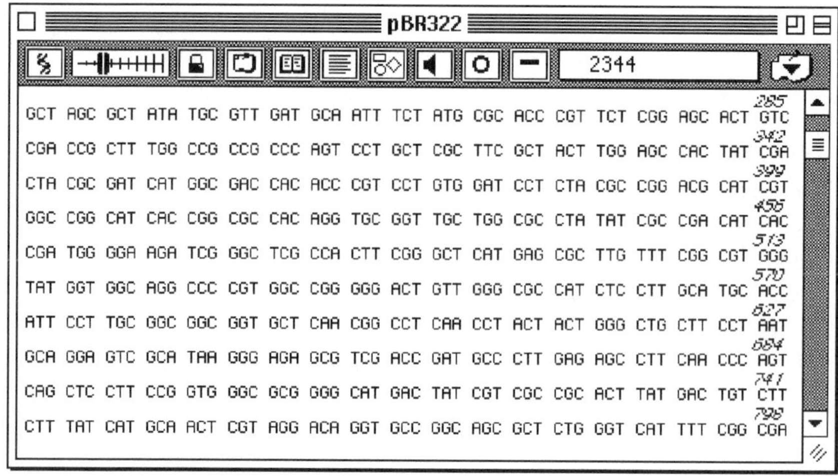

Fig. 1. The sequence window for pBR322.

MacVector comes with a variety of different data file types, which are used in the many analyses that *MacVector* offers. These are in binary format and may be edited from within *MacVector*. These files include restriction enzymes, proteolytic agents, nucleic acid and protein subsequences (motifs), and scoring matrices for nucleic acid and protein sequence comparisons. New windows for any of these data file types can be opened by selecting the file type from the pull-down menu next to the **File|New** command. Codon bias tables for a number of organisms are provided with the program. They can also be created by *MacVector* but cannot be viewed or edited by the user.

3.3. Exporting Sequences

Sequences can be exported from *MacVector* in the following formats: *GenBank, GCG, FASTA, IG Suite, Staden,* and text (ASCII). The **File|Save As** command is used to export a sequence from *MacVector* in any one of these formats. The resulting file is saved as a text file in the location selected by the user. Double clicking on a file created in this fashion launches *MacVector,* opening the sequence in *MacVector* format. Exporting a sequence in *GenBank* or *GCG format* ensures that the features and annotations associated with the sequence are transferred along with it.

3.4. Customizing Features Maps and Results Maps

MacVector 6.5 has considerably better graphics than previous versions of the software. **Features** and **Results Maps** are customizable: the user can select

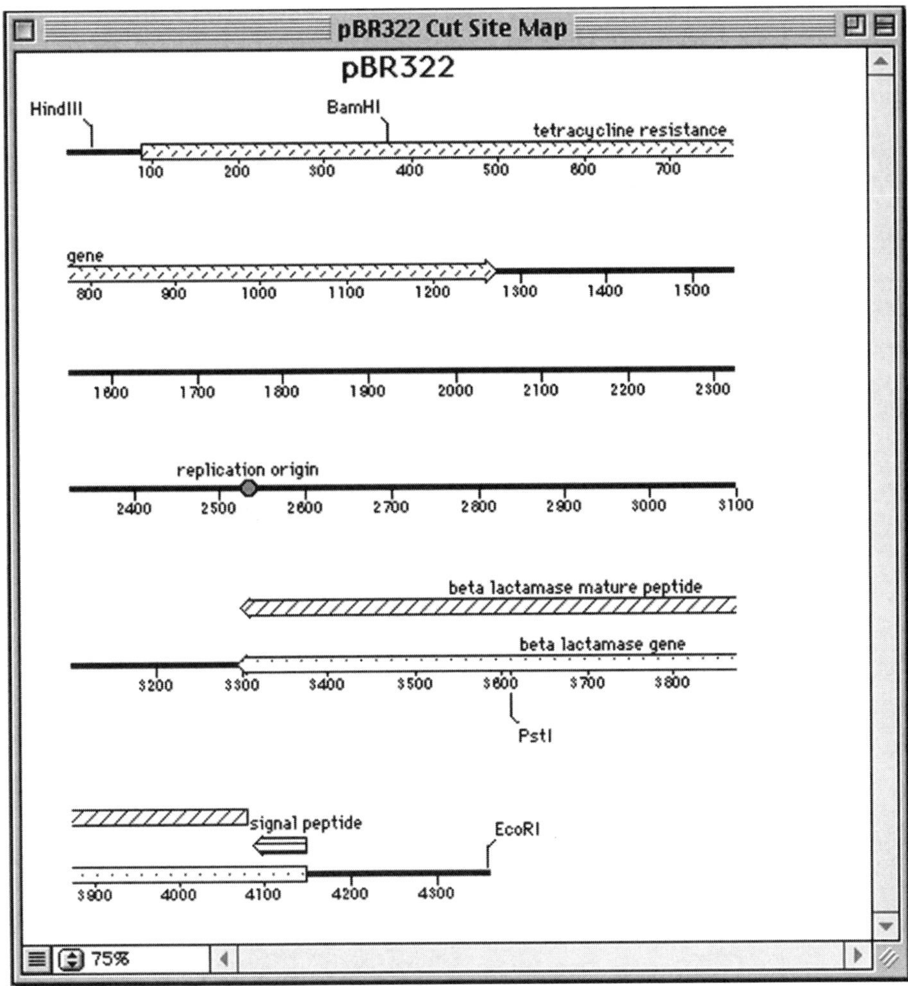

Fig. 2. pBR322 restriction map. The user has customized the appearance of the features in this map.

the color, pattern, and style of individual features and results, as well as the font and style of text used in the maps for features and results labels and the title. The user can also decide which features and/or results will be displayed in the maps (**Fig. 2**).

The **Options|Default Symbols** command allows the user to set global symbols for **Features Maps** and **Results Maps.** This menu option is available only if a sequence window or results window is not active on the desktop; if a

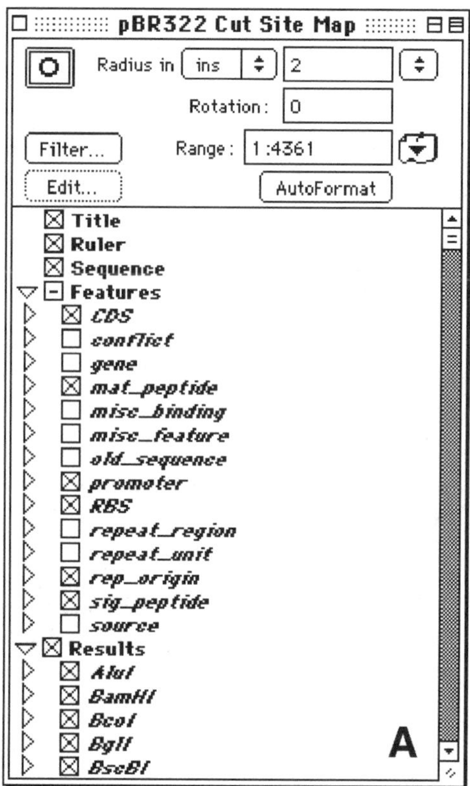

Fig. 3A–B. The graphics palette for the circular and linear forms of the pBR322 restriction map. The check boxes adjacent to the individual map elements enable the user to switch the elements on or off in the map.

sequence window or results window is active, the **option** key must be pressed before making the menu selection. The **Default Symbols** window has a pull-down menu in the upper left corner that offers the selection of **Title**, **Features**, **Result**, **Ruler**, and **Sequence**. The user can select the font, style, color, and position of the ruler, sequence, and title as shown in the maps. The user can set default symbols for individual features types and individual result types. This includes style (shape), color, fill pattern, and position of the feature or result, the position of the feature or result label and the font, style, and color of the label text. Symbols for individual sequences can be set in a similar fashion by selecting the **Options|Symbols for <sequence_name>** when the sequence window is active.

When the **Features Map** or **Results Map** of a sequence is the active win-

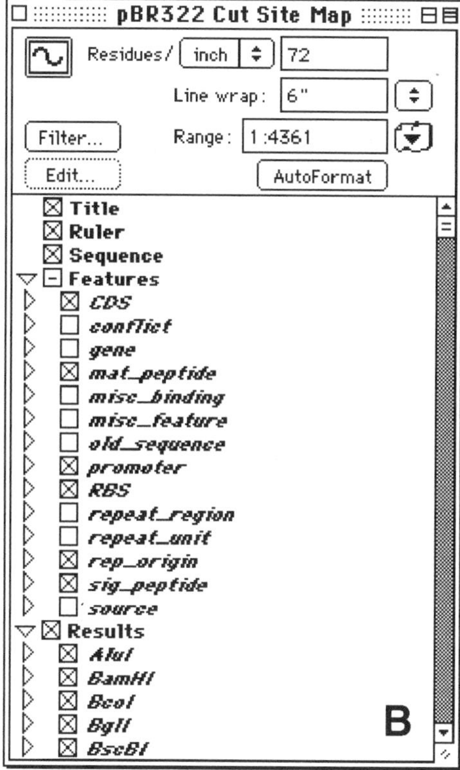

Fig. 3B.

dow, the **Graphics Palette** dialog box is also displayed. The scrolling list in this dialog box lists the individual map elements whose appearance can be edited. In case of a **Features Map**, the list includes **Title**, **Ruler**, **Sequence**, and **Features**, whereas for a **Results Map** the list includes these items and **Results**. The **Graphics Palette** (**Fig. 3**) allows the user to switch the individual map elements on or off, e.g., the user may decide that a certain feature should not be shown on the map or that only a particular restriction site should be shown. For DNA sequences, the user can toggle between linear and circular display by selecting the linear/circular button. The **Graphics Palette** also allows control of the display density of residues (number of residues per inch, centimeter, or line for linear maps or radius of circle for circular maps), line length for linear displays, and sequence range displayed in the map. The **Filter** button in the **Graphics Palette** for the **Results Map** can also be used to filter the results.

3.5. DNA Analyses

3.5.1. Restriction and Motif Analyses

MacVector comes with the *REBASE* restriction enzyme database, which consists of several restriction enzyme files. The restriction enzyme files are lists of enzymes carried by different manufacturers. The **File|Open** command is used to open the restriction enzyme files: Each row has the name, cut site, and any known isoschizomers of a restriction enzyme (**Fig. 4**). Clicking on an enzyme puts a check mark at the left of its name, and selects this enzyme for use in restriction analysis. Selecting the **Analyze|Restriction Enzyme** command brings up the restriction analysis dialog box. The user clicks on the **Enzyme File** button to select a restriction enzyme file for the analysis. The pull-down menu next to **Search using** offers the user the choice of using all enzymes in the restriction enzyme file or only the enzymes that were selected earlier.

Once the analysis is complete, the results dialog box lists the output options: a list of restriction enzymes that cut, a list of those that do not, the restriction map, the cut site annotated sequence, and finally the fragment prediction. The restriction map can be customized to include features and the results can be filtered so that only a subset of the cut sites is shown. The map can be toggled between circular and linear by clicking on the **circular/linear** button in the **Graphics Palette**. Fragments can be predicted using single or double digests.

In *MacVector,* motifs are referred to as *subsequences. MacVector* comes with six nucleic acid subsequence files, each of which can be opened using the **File|Open** command. As in the restriction enzyme files, each row of a subsequence file consists of the name, the recognition site, and any additional information (such as references) of a nucleic acid subsequence. Select subsequences can be highlighted by Shift-clicking, copied to the clipboard and pasted into a new nucleic acid subsequence file (opened by the **File|New** command and selecting **Nucleic subsequence** from the pull-down menu). The new subsequence file can be saved and then used in nucleic acid subsequence analysis. The analysis itself is done in a fashion similar to restriction analysis, by selecting the **Analyze|Nucleic Acid Subsequence** command.

The results dialog box for nucleic acid subsequence analysis is similar to that of restriction analysis. The output options include lists of sites found and sites not found, the subsequence map, the subsequence annotated sequence, and the fragment predictions. The last term implies that the size of the regions that lie between occurrences of given subsequences is calculated.

New restriction enzymes (or subsequences) can be added to a restriction enzyme (or subsequence) file by clicking on the + button, which brings up the enzyme (or subsequence) editor. This editor requires the entry of the name and

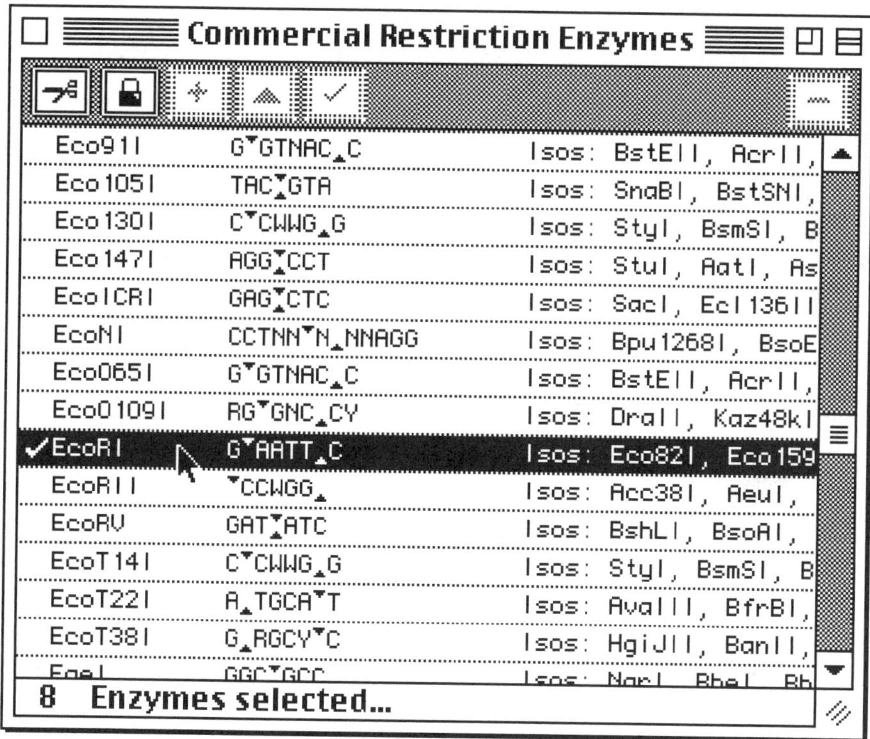

Fig. 4. A restriction enzyme file in *MacVector*. *Eco*RI is one of the eight enzymes selected, as evidenced by the check mark adjacent to its name.

cut site (or recognition pattern) for the enzyme (or subsequence). There is an optional comment box as well. In case of subsequences, motifs of up to three parts can be entered and either **AND** or **OR** logic specified. If the **AND** logic is used, the user also enters the distance between the individual parts.

3.5.2. Gene Prediction

Unlike many other sequence-analysis programs, *MacVector* allows the user to define which codons are to be used as start and stop codons. The **Options|Modify Genetic Codes** command opens a dialog box containing a chart of the 64 codons. The user can designate a codon as a start or stop codon by clicking on the cell containing the codon to highlight it; clicking on the rightward pointing arrow designates it a start codon and clicking on the leftward pointing arrow designates the codon as a stop codon.

The open reading frame (ORF) analysis is initiated by selecting the **Analyze|Open Reading Frames** command. *MacVector* allows the user to designate the 5′ and 3′ ends as start and/or stop codons. This ensures that an open reading frame that extends beyond the ends of the sequence is not overlooked because of the absence of a start or stop codon. Once the analysis is complete, the results dialog box appears, with the choice of output options: list of ORFs, ORF map, and ORF-annotated sequence. The ORF map is a graphical representation of the results, and the map can be customized to show some or all of the features that may be present in the **Features Table** of the sequence. The ORF-annotated sequence shows the ORFs annotated along the sequence, in text format.

MacVector provides two methods that could help determine if a predicted open reading frame has the characteristics of a protein-coding region: Fickett's *TESTCODE* algorithm *(2)* and codon preference analysis *(3)*. Fickett's method is based on a statistical analysis of known coding and noncoding regions, and can be used to predict coding regions in DNA sequences of at least 200 base pairs. To use Fickett's method to predict open reading frames, the **Analyze|Open Reading Frames** command is selected and the checkbox next to **Fickett's method** is checked.

The codon preference analysis is based on the fact that some organisms exhibit a preferential use of codons according to the availability of the corresponding tRNAs. There is no codon preference in noncoding regions; regions coding for proteins with low expression levels show codon distribution similar to the occurrence of the corresponding tRNAs; and highly expressed proteins show a higher preference for codons corresponding to the more abundant tRNAs. Before the codon preference analysis is done, the codon bias table for the organism must be available. If it is not provided with *MacVector,* it can be created using the **Options|Make Codon Bias Table** command, using the *Entrez* database at the NCBI. The actual codon preference analysis is initiated by the **Analyze|Codon Preference Plot** and the results are plotted for each reading frame. Prominent peaks in the codon preference plot are indicative of regions exhibiting a codon bias (possibly coding for highly expressed proteins).

3.5.3. Primer Analysis

MacVector can predict PCR primer pairs for a given template, based on parameters selected by the user *(4)*. Here the user provides *MacVector* with criteria for an acceptable primer, the program scans the sequence and presents the user with a list of primers, after eliminating those primers that don't match the user's criteria. *MacVector* can also test primers against a template sequence for usability in PCR. In this case, the program analyzes the primer input by the user, and provides information about the primer. The user can then decide, based on the results of the analysis, whether the primer should be used in an experiment. Similarly, *MacVector* can also predict primers to be used in sequencing a template and test a user-defined primer against a template for its usability as a sequencing primer. Whereas the PCR primer-analysis functions predict or test primer pairs, the sequencing primer-analysis functions predict or test single primers.

PCR primer-pair prediction is initiated by the **Analyze|Primers|PCR Primer Pairs** command; sequencing primer prediction by **Analyze|Primers|Sequencing Primers/Probes.** The initial dialog allows the user to specify the region of interest; for PCR primer pairs it is based on either product size or flanking regions. Other parameters that are user-specified include range of melting temperature for the primer-template duplex, G+C content of the primers, range of desired primer length, and whether a G-C anchor is required in the primer. To eliminate primers with too much secondary structure, the user specifies the maximum number of consecutive bonds the primer is allowed to have when it forms a hairpin with itself or a duplex with another primer. Primers that bind to other sites on the sequence are also eliminated. In the case of PCR primer pairs, the program compares the 3′ end of each primer with the product that will be amplified in the reaction. For sequencing primers, the 3′ end of each primer is compared with the entire sequence, or a specified region of the sequence.

The results dialog box presents a list of statistics, including the number of primers tested, number of primers accepted and reasons why primers were rejected, and the choice of viewing the results as a list or a graphical map. The list of PCR primer pairs provides T_m and %GC information for each primer of a pair, the length of the product, the product's T_m and %GC, and the optimum temperature to be used for the reaction.

User-defined primer testing is initiated by the command **Analyze|Primers|Test PCR Primer Pairs** (or **Analyze|Primers|Test Sequencing Primers/Probes**). A dialog box appears, in which one or two primers can be entered (for testing sequencing primers there is space for only one primer). Once the user clicks on the **Apply** button, statistics for the primers appear (**Fig. 5**). Information provided includes T_m and %GC for the primers, whether the primers actually anneal to the template, and whether the primers show any excessive secondary structure. In the case of user-defined PCR primer pairs, additional information pertinent to PCR is provided, including the size of the product formed, if any, whether the 3′ end of either primer binds within the product and the T_m difference of the primer pair. Any information that would be detrimental to the success of the reaction is highlighted in red.

Hybridization probes for a nucleic acid sequence can be predicted in the same fashion as sequencing primers. To obtain a hybridization probe for a gene whose exact sequence is not known, but whose amino acid sequence is known, the reverse-translation function can be used. *MacVector* then scans the result-

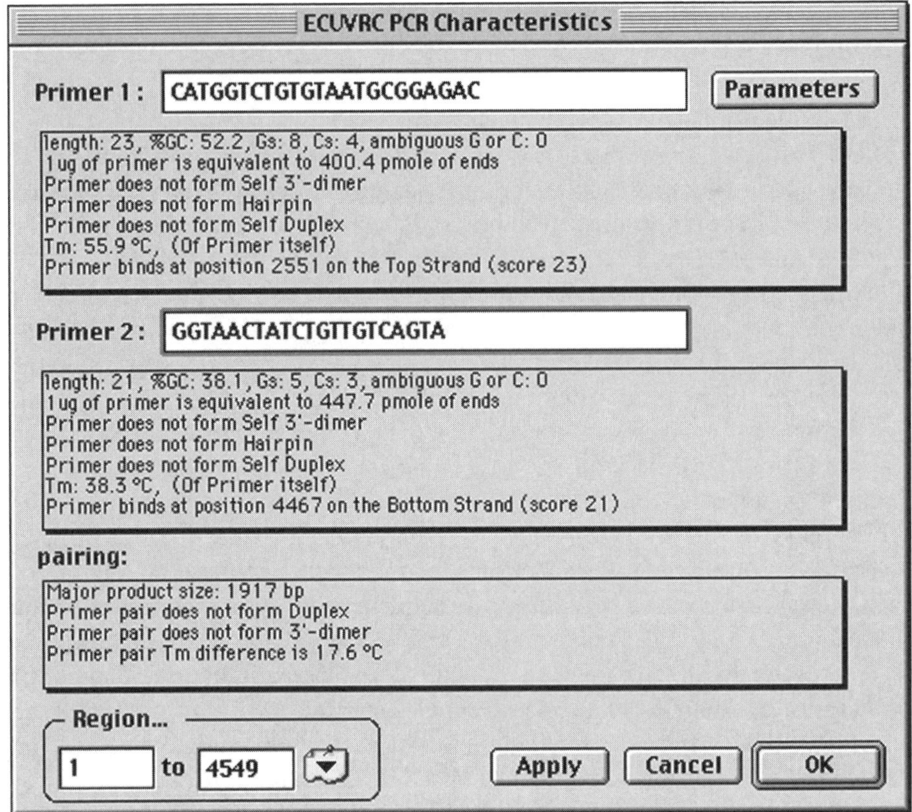

Fig. 5. The dialog box for user-defined primer testing.

ing DNA sequence to find a region that is not too degenerate and outputs a list of oligos along with their dissociation temperatures, G+C content and the number of oligos that would have to be synthesized to cover all possible sequences represented by the degenerate oligo. The user can modify the genetic code used in the reverse-translation by reducing the number of codons each amino acid has (based on knowledge of the organism's codon bias). This will reduce the degeneracy in the hybridization probes predicted by *MacVector.*

3.6. Protein Analyses

3.6.1. Protease and Motif Analyses

Protein pattern analysis is analogous to nucleic acid pattern analysis. Proteolytic enzyme analysis utilizes proteolytic enzyme files and protein subsequence (motif) analysis uses protein subsequence files. Subsequences can be selected by opening a protein subsequence file and clicking on their names, causing a check mark to appear on the left of each name. Proteases can be similarly selected by opening the proteolytic enzyme file provided with *MacVector*. As with restriction enzyme or nucleic acid subsequence files, new proteolytic agents (or protein subsequences) can be added by clicking on the + sign in a proteolytic enzyme (or protein subsequence) file. Entries in these files can be removed by clicking on the – sign or changed by clicking on the Δ sign.

The proteolytic enzyme file includes both chemical and enzymatic proteolytic agents and includes cut site and reference information. The analysis is initiated by the **Analyze|Proteolytic Enzyme** command. The analysis dialog box allows the user to select the proteolytic enzyme file and whether *MacVector* should use all proteolytic agents in the file, or only the agents selected by the user. The user can also select the region of the sequence to be analyzed. The results dialog box allows the user to select the output options: lists of proteolytic agents that cut or do not cut, a proteolytic map, proteolytic cut site annotated sequence in text, and fragment predictions from single and/or double digests.

Protein subsequence analysis can be done using either selected or all the subsequences included in the subsequence file of choice. The subsequences themselves may be composed of up to three parts. The **Analyze|Protein Subsequence** command initiates the analysis. The results can be viewed as a list of subsequences found in the sequence, a list of subsequences not found in the sequence, the subsequence map, and the subsequence annotated sequence in text format. The subsequence map can be customized to show the features of the protein alongside subsequences, and the results can be filtered to show only a subset of subsequences in the map.

3.6.2. Protein Analysis Toolbox

The **Protein Analysis Toolbox** is used to analyze a given protein sequence for a large number of properties, all of which could be used to deduce some information about the three-dimensional structure and, possibly, the function of the protein. The **Toolbox** offers more than one algorithm for each of the following properties: secondary structure, hydrophilicity, hydrophobicity, antigenicity, flexibility, surface probability, amphiphilicity, and prediction of transmembrane helices. Other analyses include amino acid composition, molecular weight and p*I*. These methods are based on data derived from proteins whose structure and function are known. In analyzing a protein of unknown structure or function, the user should first determine whether the new protein is similar to any known protein. The algorithms included in the **Protein Analysis Toolbox** should be used in conjunction with biochemical and biophysical data before making any conclusions about the structure and/or function of the new protein.

Secondary structure prediction is based on the two methods: Chou-Fasman *(5)* and Robson-Garnier *(6)*. *MacVector* does not resolve any conflicts between these two prediction methods. The user should also realize that each method has about a 60% likelihood of being correct. Thus, even if both methods agree on the conformation of a particular region of the sequence, the consensus of both methods does not necessarily give the right prediction.

The **Protein Analysis Toolbox** is accessed by the **Analyze|Protein Analysis Toolbox** command. The dialog box has four panels (**Fig. 6**). The first panel contains a list of all the algorithms included in the **Toolbox**: The user has the option of listing and/or plotting the results of each. It also offers the option to list the p*I*, molecular weight, and amino acid composition. The second panel allows the user to select the region of the sequence on which the analysis is done. The third panel allows the user to select the window size for each algorithm. When a particular algorithm is highlighted in the first panel, a brief description of the algorithm appears in the fourth panel.

The list shows the results of the analyses in tabular format. Each column lists the results of a particular algorithm and the rows represent the amino acids in the sequence. The graphical plots from the algorithms are displayed in a single window. The user can zoom into the graph by highlighting the region using the cursor. Double-clicking the mouse button returns the graph to full size. Amino acid composition, molecular weight, and p*I* are shown in a third window.

3.7. Sequence Comparisons

Comparisons of sequences can range from determining regions of similarity between two sequences to aligning a single sequence to an entire database. *MacVector* offers four different sequence alignment functions: multiple sequence alignment using *CLUSTAL W,* BLAST searches against the *Entrez* databases via the Internet, the Pustell matrix analysis and the **Align to Folder** function. The *CLUSTAL W* algorithm aligns two or more sequences simultaneously while the **Align to Folder** function individually aligns a query sequence to sequences in a folder. The BLAST algorithm searches databases at the NCBI for sequences that are similar to a query sequence. The Pustell matrix analysis searches for similarities between two given sequences graphically.

3.7.1. Multiple Sequence Alignment using CLUSTAL W

CLUSTAL W (7) aligns multiple sequences such that regions of identical and similar residues are lined up. This type of alignment is useful in comparing sequences from different sources; for example: studying differences and similarities between sequences of the same type from different organisms, resulting in the identification of regions that are conserved from species to species.

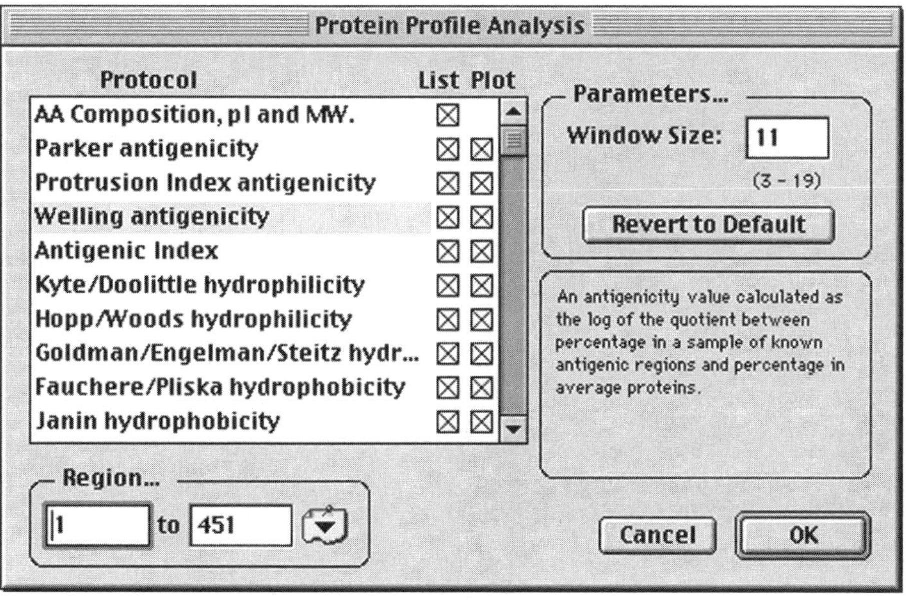

Fig. 6. The protein analysis toolbox dialog box. A brief description of an algorithm appears in the fourth panel when its name is highlighted in the first panel.

CLUSTAL W does a pairwise comparison of every sequence first and then starts the multiple alignment with the pair of sequences that is most similar. Sequences are added one by one to the alignment based on their similarities to the starting pair.

At least two sequences of the same type (nucleic acid or protein) must be open on the desktop before the **Analyze|Clustal W Alignment** command is available. The *CLUSTAL W* alignment dialog box has four panels: **Pairwise Alignment, Multiple Alignment, Sequences to Align,** and **Protein Gap Parameters** (**Fig. 7**). The **pairwise alignment** panel includes parameters that control the speed (and hence the sensitivity) of the initial pairwise alignments. These parameters are scoring matrix (the choice is between **BLOSUM 30**, **PAM 350**, **MD 350**, or **Identity**), alignment speed (**Slow** vs. **Fast**), and gap penalties for inserting and extending gaps in the alignment. The **multiple alignment** parameters control the gaps in the final multiple alignment and include scoring matrix, open and extend gap penalties, and the delay divergent (the alignment of a sequence is delayed if its % identity to any other sequence is less than the value of this parameter). The **Transitions** parameter is displayed in the multiple alignment panel only for nucleic acid alignments and toggles between **Weighted** (transitions A ⇔ G and T ⇔ C are weighted more strongly than transversions A ⇔ T, A ⇔ C, G ⇔ T and G ⇔ C) and **Unweighted**. The

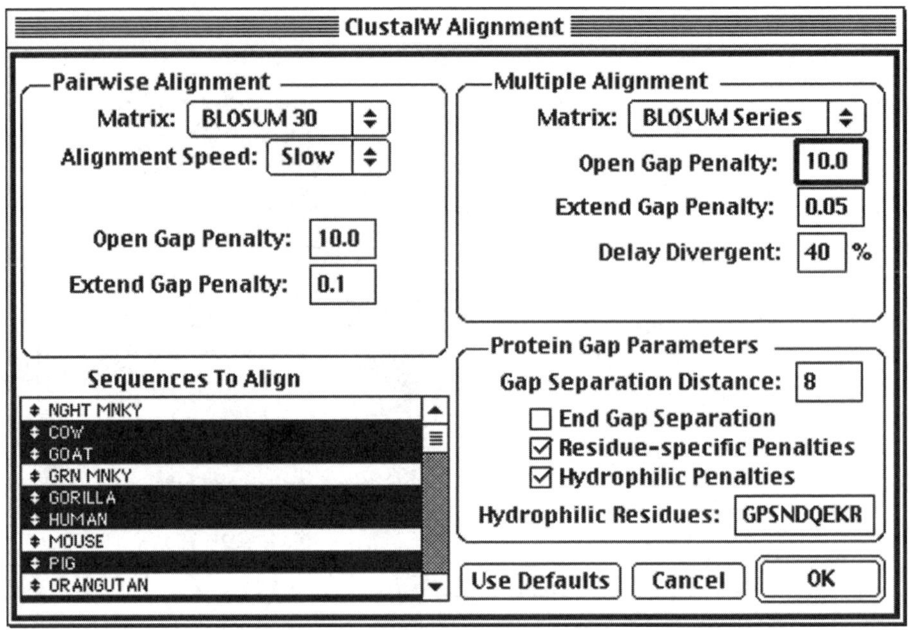

Fig. 7. The *CLUSTAL W* multiple sequence alignment dialog box.

sequences to align panel consists of a scrolling list containing the names of all sequences open on the desktop that are of the same type as the active sequence. If only a subset of these sequences is to be aligned, they can be selected by shift-clicking their names. The **protein gap parameters** are available only when protein sequences are being aligned. These parameters are gap separation distance (the minimum allowed distance between two gaps), end gap separation (if this is enabled, end gaps are treated as internal gaps for the purpose of the gap separation distance parameter), residue-specific penalties (if this is enabled, amino acid specific penalties decrease or increase the gap opening penalties at each position in the alignment), and hydrophilic penalties (if this parameter is enabled, these penalties increase the possibility of a gap opening within a hydrophilic stretch: hydrophilic residues are defined by the user in the **Hydrophilic Residues** box).

The multiple sequence alignment editor is displayed at the same time as the results dialog box. The editor allows the user to view and modify the alignment. Individual residues in the alignment are color coded, and the user can change the colors. *MacVector* offers a number of color groups for amino acids, ranging from chemical type to steric bulk of the side chain. The user can choose the color group from the list that *MacVector* provides, or create new color groups. In the alignment editor reference sequences can be pinned so that they

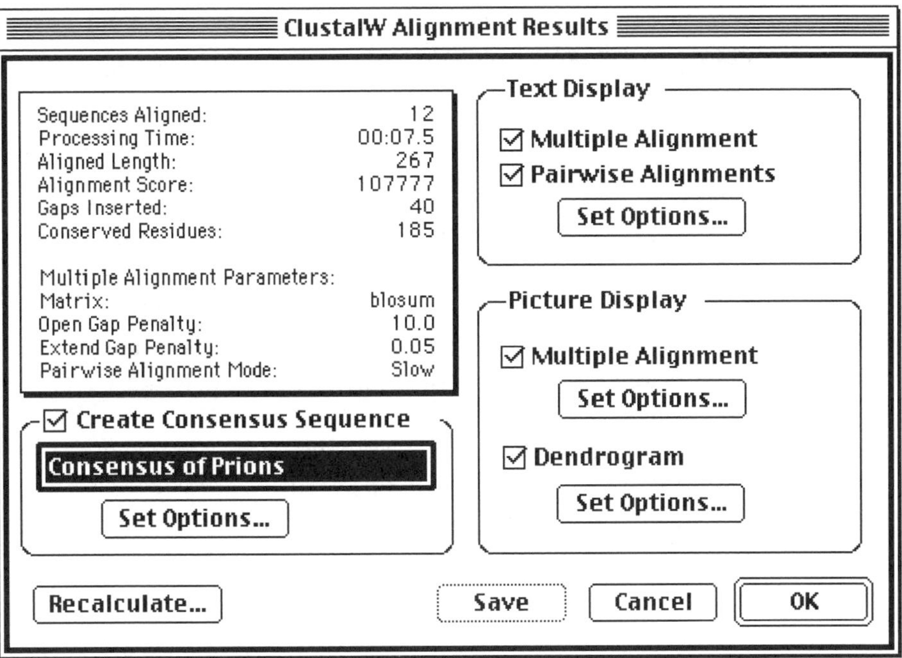

Fig. 8. The *CLUSTAL W* multiple sequence alignment results dialog box.

remain stationary at the top of the list while the other sequences are scrolled under them. The alignment itself can be edited to either add or remove gaps and moving sequences along the vertical axis can change the order of the sequences in the alignment.

The results dialog box lists various statistics from the alignment, such as number of sequences aligned, processing time taken by the alignment, length of the alignment, gaps inserted, and so on (**Fig. 8**). The alignment itself may be viewed in text or *PICT* formats. There are two text results windows: the multiple alignment, which shows the alignment in *FASTA* format, and the pairwise alignment that displays all pairwise combinations of the sequences aligned with each other. The user can select characters to represent conserved and dissimilar residues in the consensus line. The *PICT* format alignment view is a publication-quality output, which can be customized. Identical and/or similar residues can be shown in bold or italics and they can be boxed and shaded. The user also has the option of changing the font and style of the residues and the title. The alignment can also be viewed as a dendrogram in which the sequences are arranged in tree format, showing the relationship between them. The dendrogram may be viewed either in cladogram or phenogram format, and its appearance can be formatted. The consensus sequence from the multiple sequence alignment can be saved as a *MacVector* sequence file by checking the checkbox next to **Create Consensus Sequence.**

3.7.2. BLAST *Searches*

The BLAST (*Basic Local Alignment Search Tool*) ***(8)*** heuristic search algorithm can be used to search the sequence databases at the NCBI for sequences that are similar to the query sequence. *MacVector* accesses the Internet BLAST function at the NCBI for this purpose: the query sequence is submitted to the BLAST server at the NCBI, where the search occurs and the results are sent back, via the Internet, to *MacVector.* There are five BLAST programs that can be used to find homologues to the query sequence: *blastn* (compares a nucleic acid query sequence to a database of nucleic acid sequences), *blastp* (compares an amino acid query sequence to a database of amino acid sequences), *blastx* (compares the six-frame conceptual translation products of a nucleic acid query sequence to a protein sequence database), *tblastn* (compares a protein query sequence to a nucleic acid sequence database dynamically translated in all six frames), and *tblastx* (compares the six-frame translations of a nucleic acid query sequence to the six-frame translations of a nucleic acid sequence database).

The analysis is initiated with the **Database|Internet BLAST Search** command. If the computer is connected to the Internet, the BLAST search dialog box appears with the choices of program, database and **expect** value. The choice of databases is based on the choice of program (*blastn, blastx,* or *tblastx* for nucleic acid query sequences and *blastp* or *tblastn* for amino acid query sequences) and depends on what databases are available at the NCBI at the time of the analysis. The **expect value** is a measure of the statistical significance threshold for reporting matches against sequences in the database. This is the number of matches expected to be found in the database merely by chance.

Once the search is complete, the results dialog box presents the statistics of the search, including the number of entries in the database searched and the number of matches observed, saved, and aligned. The results can be viewed as a list of matching sequences and as pairwise alignments of the query sequence with the matching sequences. Matching sequences can be retrieved from the database by highlighting them in the list and then using the **Database|Retrieve to Desktop** command. Upon receiving this command, *MacVector* opens a connection to the NCBI *Entrez* server and individual sequence windows containing each of the selected sequences appear on the desktop. Alternately, the **Database|Retrieve to Disk** command can be used to save the sequences to the hard disk.

3.7.3. Pustell Matrix Analysis

A matrix comparison plots the residues of one sequence along the X-axis and the residues of the other along the Y-axis. A dot is placed at each (x, y) coordinate where the residues of the two sequences are identical. The similarities between the two sequences are best shown by a matrix comparison, because a matrix analysis can detect features that other methods of comparison may miss. By displaying the results graphically, this method allows the human eye to detect even weak similarities between the two sequences being compared. Because the sequences are compared globally, weak regions of similarity such as duplications and rearrangements, which could be missed by computational alignment methods, are revealed.

In *MacVector* the Pustell matrix analysis ***(9)*** can be used to compare DNA to DNA, protein to protein or DNA to protein. These analyses can be initiated by selecting **Analyze|Pustell DNA Matrix**, **Analyze|Pustell Protein Matrix**, or **Analyze|Pustell Protein & DNA**. There are a number of parameters that can be used to customize the comparison and they are **scoring matrix, hash size, window size, minimum percent score,** and **jump parameter**. The scoring matrix lists the weights given to different types of matches. (The **DNA Identity Matrix,** which is provided with *MacVector,* is recommended for DNA–DNA comparisons. Any of the protein scoring matrices provided with *MacVector* may be used for protein–protein comparisons or protein–DNA comparisons). The scoring matrix can be modified such that certain mismatches can be designated as partial matches according to criteria such as hydrophilicity, charge, or structure, and certain mismatches can be considered worse than others. The **hash size** (also known as the k-tuple size or word size) determines the sensitivity and speed of the analysis. It is a measure of the length of an exact match between two sequences that must be found before *MacVector* will attempt to score and align the matching region. The **jump parameter** is used in conjunction with the hash size. When the **jump** setting is 1, the **hash** value represents the number of residues in a row that must match perfectly; when the **jump** is set at 3, the **hash** value represents the number of triplets in a row whose first residue must match perfectly. Once a hit is found in the initial hashing step, the second step of the comparison is the scoring step, which utilizes the scoring matrix, the window size and the minimum percent score. When *MacVector* finds a match in accordance with the **hash** and **jump** settings, it examines the segment of sequence of the aligned sequences that surrounds the matching region, the size of this segment being determined by the **window size** parameter. The window is scored using the values for the residue pairs found in the scoring matrix. The window is saved if the percent score (actual score divided by score that would occur if all residues were identical) is greater than or equal to the **minimum % score** parameter.

The results can be viewed as a matrix map and as aligned sequences. The matrix map is a dot-matrix plot of the results of the comparison. The user can zoom in to view regions of the plot at higher magnification by dragging the cursor over the region of interest. The aligned sequence display shows the X-axis sequence aligned with all of the regions of the Y-axis sequence that meet or exceed the minimum percent score parameter.

3.7.4. Folder Searches

MacVector offers the **Align to Folder** function that can also be referred to as personal database searching. It is accessed through the **Database|Align to Folder** command. A query sequence is compared to a folder containing one or more sequences using the Lipman-Pearson DNA alignment algorithm or the Wilbur-Lipman protein alignment algorithm ***(10–12)***. These methods use an initial hashing step to screen the sequences in the folder against the query sequence, a scoring step, and an alignment step. There are three parameters involved in the alignment step and they can be changed by opening the scoring matrix: **cut-off score, deletion penalty** (gap-opening penalty), and **gap penalty** (gap-extension penalty). Only those sequences are aligned whose initial score is greater than or equal to the **cut-off score.** The alignment step inserts gaps in order to improve the final (optimized) score. The **deletion** and **gap penalties** are subtracted from the score of the alignment, so gaps are introduced only if they improve the alignment. The **processing** parameter determines whether the alignments are carried out on-the-fly or at the end of the search.

The results can be viewed as a list of saved matches, a horizontal map of the alignments, and as aligned sequences. The information in the list of matches includes the locus name of the sequence and the initial and optimized scores from its alignment with the query sequence. The horizontal map is a graphical representation of the alignment showing which parts of the folder sequences match the query sequence. As with any graphical result, the user can zoom in for higher magnification by dragging the cursor over the region of interest. The aligned sequences window displays the optimized alignments between the query sequence and the matching sequences.

3.8. *Browsing* Entrez

The NCBI produces the *Entrez* database, which contains nucleic acid and protein sequences and related bibliographic information. The sequence data include complete nucleotide and protein sequence data from the *GenBank, EMBL, DDBJ, PIR, PRF, PDB, SWISS-PROT, dbEST,* and *dbSTS* databases, and data from U.S. and European patents. *Entrez* also includes a subset of the *MEDLINE* database: references and abstracts that are cited in the sequence databases and other related *MEDLINE* records. *MacVector* can browse the *Entrez* database via the Internet. The *Entrez* browser feature is advantageous because it allows the user to extract sequences from the *Entrez* database to the desktop directly in MacVector format.

The **Database|Browse Entrez|via Internet** command accesses the *Entrez* browsing function. There are icons for the DNA, protein, and *MEDLINE* portions of the database, one of which must be selected first. The three pull-down menus on the top allow the user to define a single or combined (Boolean) annotation search. The first and third pull-down menus list all the possible search categories. For a single annotation search, the user selects a category from the first pull-down menu and enters the search string in the box below the category menu before initiating the search. For a combined annotation search, the user selects search categories from the first and third pull-down menus and either **and** or **not** from the middle pull-down menu (**Fig. 9**). Once the search strings are entered, the search can be started.

If there are any matches to the search string, the results are displayed in the two panels below the search string. The left **Matches** panel contains a list of matches to the search string. Highlighting a match makes it active. The right **Document ID** panel contains a list of document ID numbers of the sequences corresponding to the match highlighted in the matches panel. Shift-clicking can be used to highlight more than one match or document ID. Clicking on the **Browse** button brings the annotations and features information of the highlighted sequences into the large panel in the bottom half of the window. Selected sequences can be extracted to individual sequence files on the desktop by clicking on the **To Desk** button. Clicking on the **To Disk** button extracts the highlighted sequences to an existing or new folder on the hard drive.

4. Notes

MacVector 6.5 is a comprehensive, easy-to-use sequence-analysis software package written for the Macintosh. Along with its module *AssemblyLIGN,* it offers the full gamut of sequence-analysis functions for the Macintosh desktop. These functions include restriction and protease analysis; motif analysis for nucleic acid and protein sequences; PCR and sequencing primer prediction and analysis; a large variety of protein-analysis algorithms; browsing the *Entrez* databases via the Internet; and sequence comparisons including BLAST searches, multiple sequence alignment using *CLUSTAL W,* pairwise sequence comparisons using the Pustell matrix dotplot method, and pairwise comparisons of a query sequence and sequences saved in a user database (folder searches).

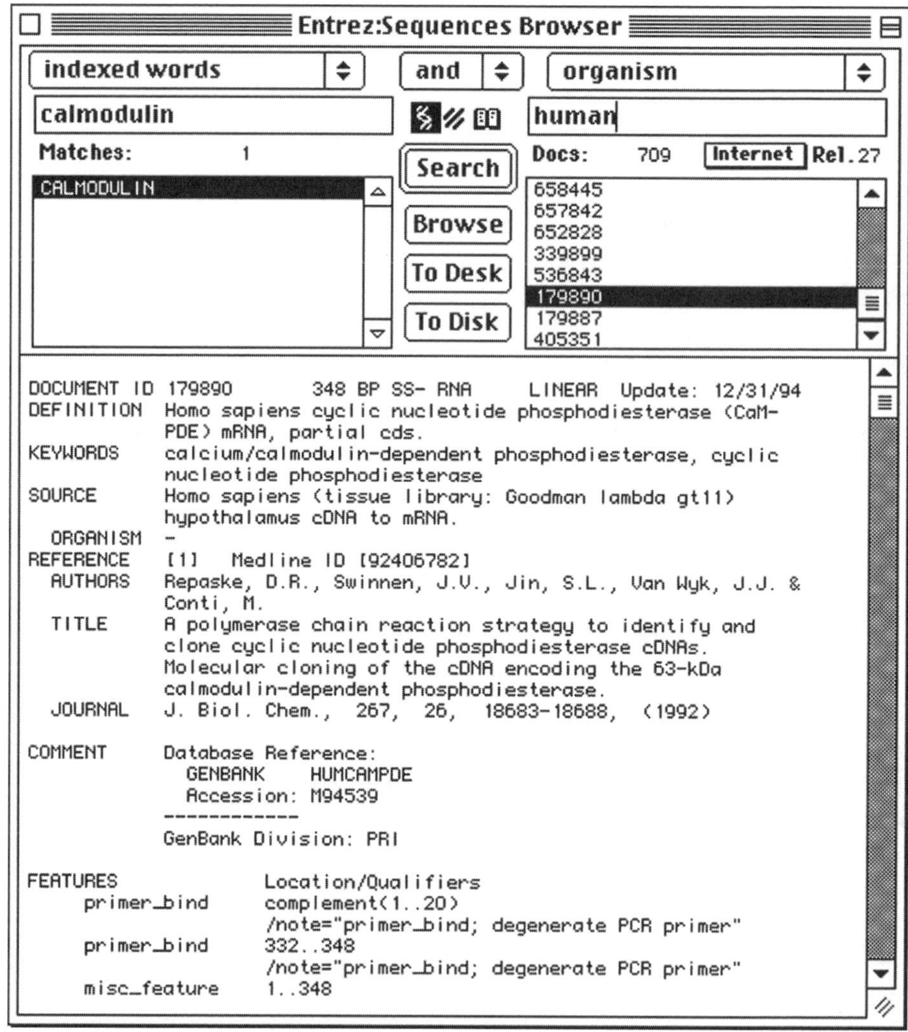

Fig. 9. The *Entrez* browser window in *MacVector.*

The producers of *MacVector,* Oxford Molecular Group, also produce *Omiga*™, a sequence analysis software package for Windows™. (*Omiga* is reviewed in Chapter 3). Though *MacVector* and *Omiga* are developed at the same company, their user interfaces are quite different, which may make it somewhat difficult for users of both programs to move between them. However, the developers of *MacVector* have chosen to continue to adhere to the Human Interface Guidelines from Apple Computer so that the users of the Macintosh platform remain completely at ease while using the program. Also,

Omiga is able to import and export sequences in *MacVector* format, allowing for the easy flow of information between users of both programs.

References

1. Oxford Molecular Group (1998) *MacVector 6.5 User Guide.* Oxford Molecular, Oxford, England.
2. Fickett, J. W. (1982) Recognition of protein coding regions in DNA sequences. *Nucl. Acids Res.* **10,** 5303–5318.
3. Gribskov, M., Devereux, J., and Burgess, R. R. (1984) The codon preference plot graphic analysis of protein coding sequences and prediction of gene expression. *Nucleic Acids Res.* **12,** 539–549.
4. Rychlik, W. and Rhoads, R. E. (1989) A computer program for choosing optimal oligonucleotides for filter hybridization, sequencing and in vitro amplification of DNA. *Nucl. Acids Res.* **17,** 8543–8551.
5. Chou, P. Y. and Fasman, G. D. (1974) Conformational parameters for amino acids in helical, beta-sheet, and random coil regions calculated from proteins. *Biochemistry* **13,** 211–22.
6. Garnier, J., Osguthorpe, D. J., and Robson, B. (1978) Analysis of the accuracy and implications of simple methods for predicting the secondary structure of globular proteins. *J. Mol. Biol.* **120,** 97–120.
7. Thompson, J. D., Higgins, D. G., and Gibson, T. J. (1994) CLUSTAL W: improving the sensitivity of progressive multiple sequence alignment through sequence weighting, positions-specific gap penalties and weight matrix choice. *Nucleic Acids Res.* **22,** 4673–4680.
8. Altschul, S. F., Gish, W., Miller, W., Myers, E. W., and Lipman, D. J. (1994) Issues in searching molecular sequence databases. *Nat. Genet.* **6,** 119–129.
9. Pustell, J. and Kafatos, F. C. (1982) A high speed, high capacity homology matrix zooming through SV40 and polyoma. *Nucleic Acids Res.* **10,** 4765–4782.
10. Lipman, D. J. and Pearson, W. R. (1985) Rapid and sensitive protein similarity searches. *Science* **227,** 1435–1440.
11. Pearson, W. R. (1990) Rapid and sensitive sequence comparison with FASTP and FASTA. *Meth. Enzymol.* (ed. Doolittle, R. F.) **183,** 63–98.
12. Wilbur, W. J. and Lipman, D. J. (1983) Rapid similarity searches of nucleic acid and protein databanks. P*roc. Natl. Acad. Sci. USA* **80,** 726–730.

5

DNASTAR's Lasergene Sequence Analysis Software

Timothy G. Burland

1. Introduction

Lasergene comprises eight applications, organized into functional units. A user with the full *Lasergene* system might employ the software as follows:

SeqManII: Trim and assemble sequence data, and determine the consensus sequence.
GeneQuest: Discover and annotate genes, patterns and other features among small, BAC-sized or larger sequences.
Protean: Predict protein secondary structure and identify antigenic regions.
MegAlign: Align sequences in pairwise or multiple configurations, and build phylogenetic trees.
GeneMan: Search sequence data with Boolean queries constructed from sequence similarity, consensus sequence and text terms.
PrimerSelect: Design primers for PCR, sequencing, hybridization, and transcription.
MapDraw: Create restriction maps displaying sites, translations, and features.
EditSeq: Import and manipulate sequences from other applications for analysis in *Lasergene.*

Examples of how the *Lasergene* applications may be used follow. The procedures are the same whether the software is run on *Windows 95/98/NT* or *Macintosh* computers. Examples are given for *Lasergene99* v4.0.

2. Sequence Assembly with *SeqManII*

SeqManII is the assembly software of choice for the *Escherichia coli* genome sequencing project *(1)*, among others. It can assemble sequencing projects

From: *Methods in Molecular Biology, vol. 132: Bioinformatics Methods and Protocols*
Edited by: S. Misener and S. A. Krawetz © Humana Press Inc., Totowa, NJ

ranging from less than a kilobase to as large as a bacterial genome. As an example, this section describes how almost 1200 ABI®-format sequence trace files from the *E. coli* genome project were assembled into one 93-kb contig.

2.1. Sequence Entry

The simplest way to add large numbers of sequence files to a *SeqManII* project is by drag-and-drop from the *Windows Explorer* or *Macintosh Finder.* Multiple folders may be added, and files may be ABI and SCF3 trace files, or *DNASTAR* sequence files. *SeqManII* accepts up to 64,000 sequences, ample for >5x shotgun sequence coverage of most bacterial genomes. Once sequences are added, a file of filenames (fof) can be created so that adding the same set of sequences to future assemblies can be done by adding the fof rather than the individual files.

2.2. Trimming Vector and Host Data

Sequence data may be contaminated with sequences from the host used to grow clones, and from the vector used to clone the target sequences. Such contaminants can compromise sequence assembly seriously. *SeqManII* removes contaminating vector or host sequences prior to assembly. To trim vector from the *E. coli* sequence reads, the sequences are sorted into two groups then entered into *SeqManII.* The forward reads are marked as cloned in the Janus vector *(2)* and the reverse reads are marked with the InvJanus vector from the vector catalog. For trimming, the recommended default stringency is used for sequence similarity to the vectors.

2.3. Quality Trimming

Sequence reads vary in quality over their length, with poor quality data typically occurring near 5′ and 3′ ends of each read. Poor-quality regions contain frequent basecalling errors, which compromise sequence assembly. *SeqManII* evaluates peak quality directly in fluorescent trace data, and trims data that fall below a specified quality threshold *(3)*. This retains good data and removes poor data without the need for human editing. In this *E. coli* example, quality trimming stringency was set to the recommended medium value.

After trimming, another fof may be saved, which stores the trimming information as well as the filenames. Thus, trimming need not be repeated for additional assemblies involving the same sequences.

2.4. Sequence Assembly

If all sequence reads originate from one contiguous piece of DNA, and there are reads for every segment of DNA, it should be possible to assemble the reads into one contiguous piece—or contig—corresponding to the original piece of DNA. For sequence assembly, several parameters may be adjusted, including the extent of match needed to join sequence reads into contigs, and the penalties applied for introducing gaps in alignments of overlapping sequence reads. For most data, default values give the best assemblies—they provide the fewest number of contigs and minimize the possibility of false joins.

Once adjustable parameters are chosen, *SeqManII* can execute vector and quality trimming and sequence assembly with one click of the mouse. *SeqManII* trimmed and assembled the 1200 *E. coli* sequence reads into one 93-kb contig (**Fig. 1**) in 15–20 min on a 200-MHz PentiumPro™ Windows PC or Macintosh G3.

2.5. Consensus Calling

Once sequences are aligned and assembled (**Fig. 1**), the consensus sequence is called. *SeqManII,* like other assembly programs, can choose the base occurring most frequently at a given position for the consensus—a majority calling system. However, where fluorescence trace data are available, as in this project, *SeqManII* determines the consensus sequence based on evaluation of peak quality. This system is much more accurate than majority-based consensus calling *(3)*. Unlike majority-based systems, *SeqManII* can call the consensus correctly when a majority of the original basecalls are in error (**Fig. 1**). Such functionality is a prerequisite of assembly software suitable for the >99.99% accuracy goal of the human genome project.

Changes to the consensus calling criteria may be made without repeating the assembly. For trace data, the **Evidence Percentage** parameter controls the stringency used to make unambiguous consensus calls—i.e., specific base calls an International Union of Biochemistry (IUB) ambiguity code representing a heterozygous or questionable position. Setting **Evidence Percentage** to a higher value increases the likelihood that ambiguous or heterozygous calls will be made for the consensus. Setting **Evidence Percentage** too high might result in spurious heterozygous or ambiguous calls in the consensus. Conversely, reducing **Evidence Percentage** too far increases the risk of making an unambiguous call in the consensus when the evidence for that call is equivocal, or when the position may be heterozygous.

2.6. Editing

Although data trimming, assembly, and consensus calling may be run automatically, *SeqManII* provides a graphical interface for manual editing of the sequence reads and consensus (**Fig. 1**). Sequence reads, traces, and six-frame translation are all aligned with the consensus sequence. Open reading frames and stop codons in the translation direct attention to potential frameshifts or other sequencing problems.

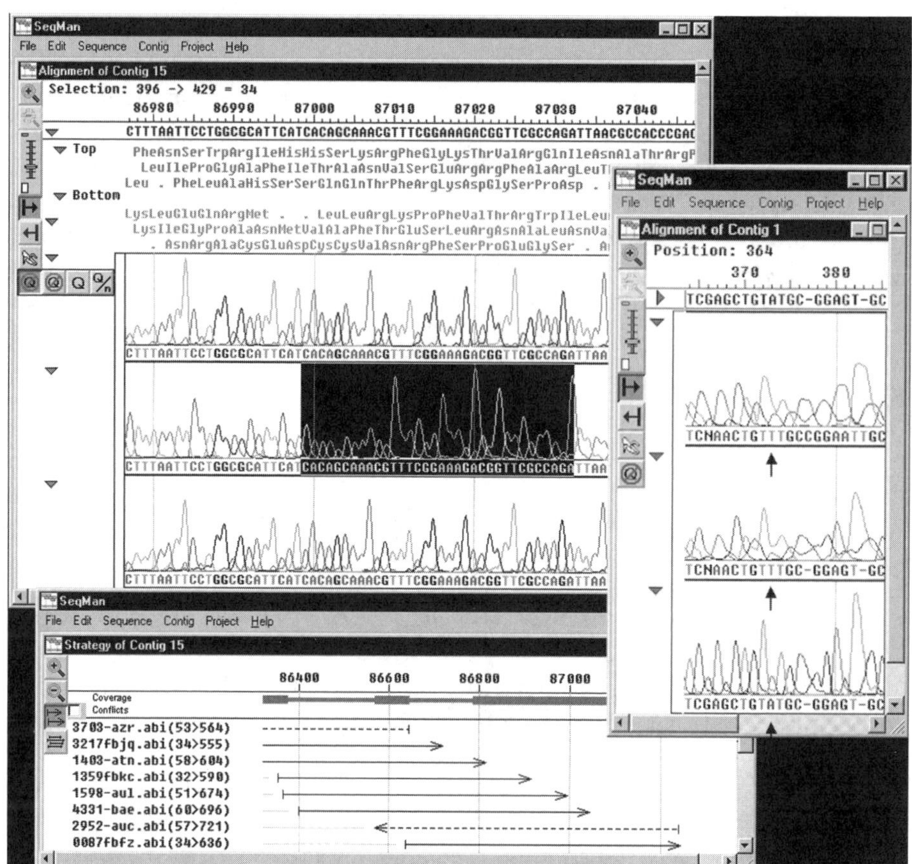

Fig. 1. The left side shows the 93.8-kb *SeqManII* project. Upper left is the Alignment Window with traces (four-color when viewed on color monitors) and consensus sequence aligned with six-frame translations. Dots in the translations are stop codons. Bottom left is the strategy view, showing where each sequence read covers the contig. The alignment view on the right shows a small project where *SeqManII*'s trace-quality-based consensus caller has correctly called an A in the consensus sequence when the basecalls in the corresponding column were T, T, and A (arrows). A majority-based consensus caller would incorrectly call the consensus a T at this position.

The **Strategy View (Fig. 1)** displays depth and orientation of sequence coverage throughout the contig. Thresholds for "complete" and "partial" coverage can be chosen by the user. In this *E. coli* project, complete was defined as fourfold coverage, with at least two reads in each direction. Different colors and thicknesses for the contig indicate where coverage is only a single read or on only one strand. In this example, a region around 87,000 nucleotides has full coverage but regions close by do not (**Fig. 1**). The **Strategy Viewer** simplifies decisions as to whether additional experiments are needed to complete the project.

2.7. Further Analysis

Once satisfied with a contig, any or all of the sequence may be selected as a query for *BLAST* searching over the internet using NCBI's *BLAST* server. Results are returned detailing public sequences with matches to the contig. This is a useful preliminary indication of the nature of the sequence, and if segments of the sequence are closely related to published ones, it can serve as an independent way to assess the assembly.

The contig may be exported as a single file in *DNASTAR, GenBank*, or *FASTA* formats. For sequences that do not match public data closely over their whole length, a typical next step would be gene discovery using *GeneQuest.*

3. Gene Discovery using *GeneQuest*

GeneQuest identifies a wide range of features in DNA sequences, and provides tools for annotation and visualization. Once characterized, the information may be submitted to public databases.

3.1. Sequence Entry

GeneQuest accepts sequence files in *DNASTAR, Genbank, ABI* and *SCF3* formats. Projects may be any size up to whole bacterial genomes. As an example, this section describes how gene discovery might proceed for a 28.67-kb cosmid clone of *Caenorhabditis elegans* imported from *GenBank* (accession no. Z46240).

3.2. Finding Coding Regions

3.2.1. Repeat Sequences and Base Distribution

Highly repetitive regions are unlikely to encode proteins. Thus finding repeats can exclude segments of DNA from the search for genes. *GeneQuest* identifies direct, dyad, and inverted repeat sequences. The occurrence of direct repeats around 20–21.7 kb of the *C. elegans* sequence (**Fig. 2**) suggests the absence of coding potential in this region.

Determining base content can assist in gene identification in organisms that have distinct G+C contents in coding vs. noncoding regions. For this example, a window of 50 nucleotides was used to average G+C content (**Fig. 2**). G+C content varies nonrandomly along the length of this sequence, with regions of higher G+C interspersed with stretches of lower G+C. One might focus first on the higher G+C regions for gene discovery.

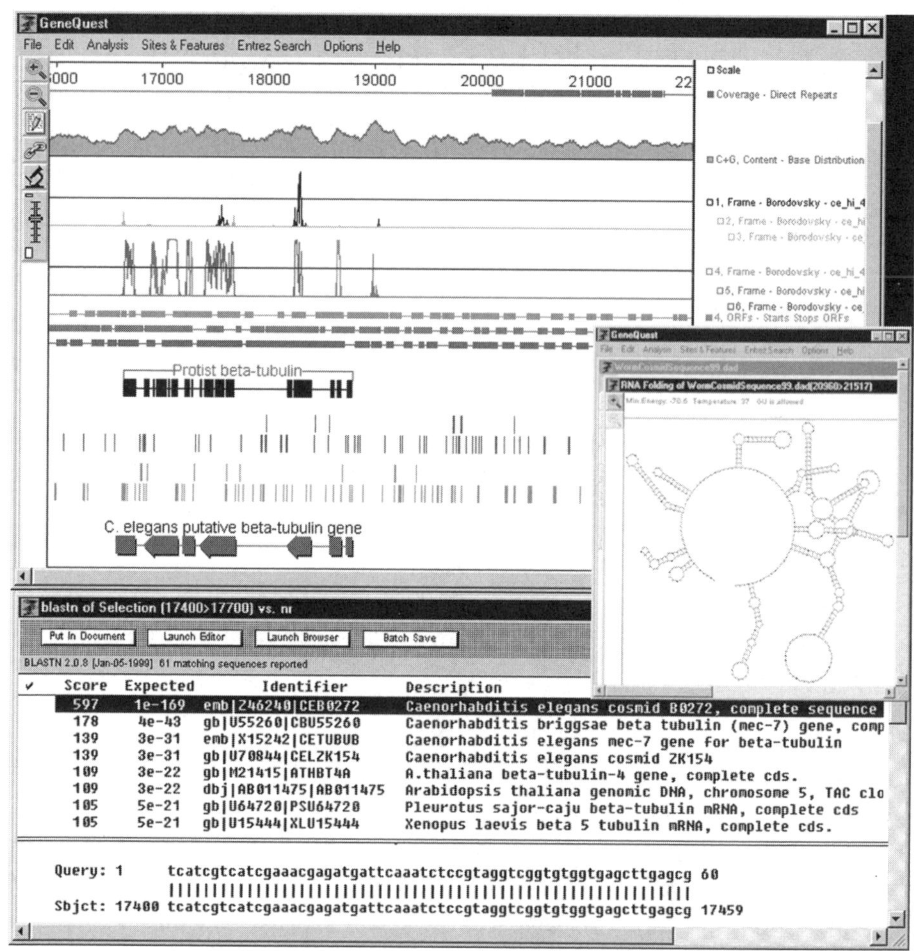

Fig. 2. The upper window shows *GeneQuest*'s main view, the "Assay Surface," which displays method results. Below the scale at top the first line shows direct repeats, and the graph below charts G+C content. The next plot is Borodovsky statistics, superimposed for frames 1–3, with little evidence of coding potential. The following bars indicate ORFs for frames 4–6, and immediately above them are the Borodovsky data for the same frames, consistent with the presence of a multi-exon gene. The *Gene Finder* result below shows that the protist β-tubulin coincides with the Borodovsky peaks. Viewing all these data in the context of the splice sites (next sets of hash marks) permits prediction of intron–exon boundaries. The authors of the sequence submission annotated a tubulin gene in the same location *GeneQuest* finds one. Bottom right shows predicted folding for RNA from the repeat sequence. Bottom left shows the results of a *BLAST* search using the ORF spanning coordinates 17,700–17,400 as query.

3.2.2. ORFs, Stops, and Starts

In prokaryotes, which lack introns, coding regions may be identified from sequence data by seeking long ORFs between start codons and stop codons, though this simple approach is rarely sufficient as a gene-finding tool. For eukaryotes, where introns are the norm and exons may comprise a minority of the sequence that makes up a gene, even genuine ORFs are typically much shorter than genes. **Figure 2** shows the ORFs and stops for the *C. elegans* sequence. *GeneQuest* displays plenty of candidate coding regions, but few ORFs are larger than 500 nucleotides long.

3.2.3. Coding Regions—Borodovsky Statistics

Statistical methods of identifying candidate coding regions offer considerable power, and are indispensable for finding coding regions in eukaryotes. Borodovsky's method *(4)* finds sequence patterns characteristic of coding regions. *C. elegans*-specific matrix files were used with this method to generate Borodovsky plots for all six reading frames (**Fig. 2**). There are compelling statistical indications of coding capacity for frames 4, 5, and 6 around sequence coordinates 19,000–16,500 nucleotides (**Fig. 2**). There is only marginal evidence for coding capacity in frames 1–3, and ≈2 kb on either side of this region there is little evidence at all. These results are consistent with the presence of a multi-exon gene in frames 4, 5, and 6.

3.2.4. Related Published Sequences—BLAST *searching*

For a fast indication of what a sequence might encode, a region with coding potential, such as the *C. elegans* segment 17,700–17,400 nucleotides, may be selected as query for a *blastn* search of *GenBank* over the Internet. In this case the sequence naturally found itself as the top match, as it was already published (**Fig. 2**). However, all of the top matches to the query are β-tubulins, and the matches are highly significant. This indicates that the segment used as query may be part of a *β-tubulin* gene.

3.2.5. Search for Specific Genes—Gene Finder

The coding region predicted in this example appears closely related to a known sequence. *GeneQuest*'s *Gene Finder* functions may be used to test explicitly for relatedness to β-tubulin. A file for a prototypical β-tubulin polypeptide (*GenBank* accession no. M58521) from a Protist was chosen as the protein to find. **Figure 2** shows where the regions of this tubulin match *C. elegans* codons. There is compelling overlap with the cluster of Borodovsky peaks, strongly indicating the presence of a *β-tubulin* gene.

3.2.6. Splice Sites—Statistical Patterns

GeneQuest provides statistical methods to locate intron–exon boundaries. In this example, *C. elegans*-specific matrix files were used to predict potential splice sites (**Fig. 2**).

The precise limits of the coding region and the intron–exon boundaries may be investigated by viewing the predicted splice sites in the context of the ORFs—ideally, the Borodovsky peak should coincide with an ORF in the same reading frame, bounded by donor and acceptor splice sites. If candidate splice sites are not found using the statistical pattern method, one can zoom in to magnify the *GeneQuest* display so that individual bases can be resolved and appropriate dinucleotides for splice sites may then be sought by eye.

Putting together all the information now gathered for this *C. elegans* example, the tubulin-coding region could be divided into seven exons (**Fig. 2**). As this example sequence was already published, one can examine the published annotations. Indeed, the authors of the *GenBank* submission had already annotated a putative 7-exon *β-tubulin* gene in the same location that *GeneQuest* predicts one (**Fig. 2**).

3.2.7. Other Discovery and Annotation Functions

In addition to the methods cited in the *C. elegans* example, *GeneQuest* can identify transcription factor binding sites, restriction sites, any pattern typed in by the user, codon usage for all or part of the sequence, and regions of DNA that are likely to bend *(5)*. *GeneQuest* can also simulate separation of restriction fragments by agarose gel electrophoresis, and predict how RNA corresponding to the selected DNA will fold (**Fig. 2**).

3.3. Text Searches to Find Related Sequences

GeneQuest provides access to the National Center for Biotechnology Information (NCBI) *Entrez* server, which processes text queries of sequence databases and returns the results over the internet. For example, to find other tubulins from *C. elegans,* one could build an *Entrez* query of the nucleic acid data for ["tubulin" in a text field] **AND** ["*Caenorhabditis*" in the organism field]. In mid 1998 this would match 68 database entries containing α, β- and γ-tubulins from *C. elegans*, tubulins from other members of the genus, tubulin tyrosine ligases from the genus, and other entries that refer to tubulins. However, the *Entrez* server does not support searches as sophisticated as the Boolean text/sequence searches that *GeneMan* provides (**Subheading 6.**).

3.4. Annotating and Displaying Features

GeneQuest provides tools for displaying and annotating features, whether they are found using *GeneQuest,* or in public data files imported into

GeneQuest. Once a region of interest is selected with the range selector tool, *GeneQuest* enters the sequence coordinates automatically into the annotation window when the **Features_New Feature** menu option is chosen. Information ranging from standard feature descriptions to free-form notes may be added. To annotate the *tubulin* gene in this example, the first exon was selected and annotated as a feature, then each of the other exons was selected and joined to the first exon. If conclusions change as to the boundaries of any features, the sequence coordinates may be adjusted in the annotation window, saving the effort of creating annotations anew.

Features may be displayed in a variety of forms, including graphs, boxes, arrows, bars, hash marks, and text, depending on the feature (**Fig. 2**). Colors and fill patterns may be selected from the *GeneQuest* tool kit, and graphic elements may be superimposed, rearranged, and juxtaposed by dragging them up and down. A useful approach is to specify, e.g., red, blue, and green for reading frames 1, 2, and 3, and use these colors consistently for frame-specific elements such as ORFs, stops, starts, and Borodovsky plots.

3.5. Creating Sets of Methods for Re-use

The range of analytical methods in *GeneQuest* is broad, and the scope for varying method parameters greatly multiplies the combinations of methods available for analyzing DNA sequences. Once a set of methods has been applied to a sequence with specific parameters, *GeneQuest* can save the **Method Outline,** which is analogous to a template used in word processing. The same methods and parameters can be quickly applied to another sequence simply by applying a saved Method Outline.

3.6. Further Analysis and Publication of the Results

To use external illustration functions, *GeneQuest*'s graphical views may be copied to the clipboard, then moved to other applications using the **Paste_Special** option and choosing to paste the picture (rather than the sequence). To edit the picture in *Microsoft Powerpoint*, for example, one may use the drawing tools to ungroup the pasted image. Each element—graphs, bars, arrows, labels, legends, and so on—may then be separately edited, resized, moved, or deleted.

GeneQuest saves projects as *GeneQuest* document files. For further analysis in other *Lasergene* applications, data may be saved as *DNASTAR* or *FASTA* files. For submission to public databases, the sequence and annotations may be saved as a *GenBank* flatfile.

4. Protein Structure Analysis with *Protean*

Protean works the same way *GeneQuest* does, but with polypeptides rather than DNA sequences. *Protean* accepts sequence files in *DNASTAR* format. For files in other formats, sequences may be converted to *DNASTAR* format using the *EditSeq* module (**Subheading 7.**).

4.1. Analyzing Sequences

Protean has over 20 analytical methods for predicting secondary structural and physicochemical properties of proteins. As with *GeneQuest* (**Subheading 3.**), most methods provide for customized graphical display of the results. For the example in this section, a human calmodulin protein was analyzed.

4.1.1. Predicting Alpha Helices, Beta Sheets, Coils, and Turns

Protean provides four methods for predicting secondary structures ***(6–9)***. In this example, the Garnier-Robson method ***(7)*** was used to predict helices and turns. Calmodulins are well-documented proteins, so it is possible to view how *Protean*'s *in silico* predictions of helices and turns compare with reality—very well in this case (**Fig. 3**). For proteins that are not well characterized, use of multiple methods to analyze secondary structures is recommended.

4.1.2. Predicting Hydropathy and Amphiphilicity

Three methods are provided for predicting hydropathic character ***(10–12)***. Hydrophobicity plots may also be predicted using the Eisenberg method ***(13)***, which predicts amphipathic character. For the calmodulin example, the Kyte-Doolittle method ***(11)*** predicts that calmodulin is hydrophilic over most of its length (**Fig. 3**), and thus is unlikely to be embedded in membranes—again, a good match with reality.

4.1.3. Finding Motifs and Sequence Similarities

Searching the *PROSITE* database ***(14)*** for published motifs that match one or more segments of a protein, *Protean* located the four known "EF-hand" calcium-binding sites in calmodulin protein. They are located in the predicted turn regions (**Fig. 3**)—independent evidence of structural similarity for the four sites.

Protean provides *BLAST* search capability. As with *GeneQuest,* the results of a *BLAST* search may vary substantially depending on whether the whole sequence or a specific segment is chosen as the query. The ability to choose a specific segment of the protein sequence increases the probability of finding matches to a motif or structural element in unrelated proteins compared with a query based on the entire sequence.

4.1.4. Predicting Antigenic Regions, Surface Probability, and Flexibility

Identifying antigenic regions *in silico* can aid generation of useful antibodies greatly. *Protean* provides four methods for predicting antigenicity ***(15–18)***. The Jameson-Wolf method ***(16)*** indicates five highly antigenic regions in this

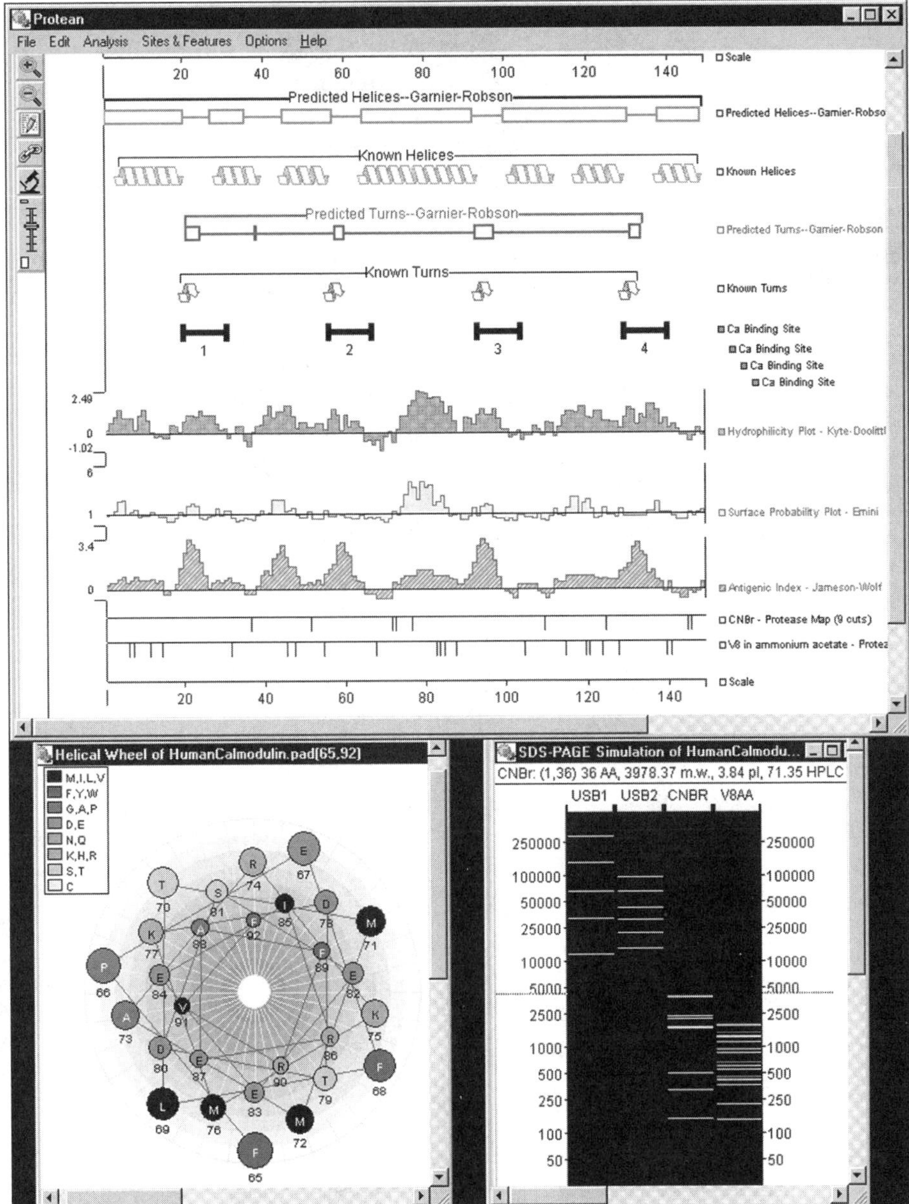

Fig. 3. The upper window shows *Protean*'s main view. Bottom right is a simulation of electrophoretic separation of fragments of the sequence that would be generated by CNBR and trypsin (TRYT), compared with known molecular size standards (USB1, USB2, BRL); the characteristics of the selected band are displayed above the gel image. Bottom left displays a schematic view of the alpha-helical segment of the protein from residues 65–92, as seen by an imaginary observer looking down the center of the helix.

small protein (**Fig. 3**). Four coincide with the calcium-binding sites. Antibodies generated from these may cross-react with EF-hand calcium-binding sites in other types of proteins. The *Protean* analysis suggests that the antigenic region around residues 40–50 may be a better choice of antigen if reaction with other calcium-binding proteins is to be avoided.

Surface regions are often good antigens for antibodies for immunocytochemistry—antibodies generated may yield better experimental results if the epitope lies on the surface of the native protein. For this calmodulin protein, the Emni method *(19)* suggests that surface probability is moderate for the region around residues 40–50, indicating that this is an acceptable target region for immunocytochemical experiments. However, the Emni method also points to the hydrophilic region around residues 75–85 as a prominent surface region (**Fig. 3**). This region is predicted to have moderate antigenic character, and does not overlap any calcium-binding site. It may be a good alternative antigen to the 40–50 region.

The search for antigenic sites may be combined with the built-in *BLAST* search, to limit candidate antigenic segments to those that have suitable sequence specificity.

4.1.5. Finding Proteolytic Sites

Protean has a database of 24 proteolytic activities, including enzymes and chemical agents. A built-in editor allows addition or modification of activities. For calmodulin, cleavage sites for cyanogen bromide (CNBR) and V8 protease are shown (**Fig. 3**). As with *GeneQuest* (**Subheading 3.2.7.**), simulations of electrophoretic separation of proteolytic fragments can be displayed (**Fig. 3**).

4.1.6. Modeling Structures

Protean displays model structures for all or part of a sequence. Models include helical wheel (**Fig. 3**), helical net, beta net, linear space fill, and chemical formula. These models provide clues as to the three-dimensional structure of the gene product.

4.2. Creating Sets of Methods for Re-Use

To make repeated analyses with similar groups of methods more efficient, *Protean* can save and re-use "Method Outlines" just like *GeneQuest* (**Subheading 3.5.**).

4.3. Publication of the Results

As with *GeneQuest* (**Subheading 3.6.**), the graphical views in *Protean* may be altered using the built-in graphing tools, or copied to the clipboard for export to other applications for further modification.

5. Sequence Alignment/Phylogenetic Tree Building with *MegAlign*

MegAlign makes pairwise or multiple alignments of DNA or protein sequences, and generates graphical views of sequence similarities and differences, phylogenetic trees, and tables of the numerical data underlying the comparisons.

5.1. Sequence Entry from Files or Public Databases

MegAlign accepts sequence directly from *DNASTAR* files, and from ABI and SCF3 trace files. When both DNA and protein sequences are entered, *MegAlign* translates the DNA sequences and aligns all sequences as proteins unless set to backtranslate proteins for alignment as DNA.

If a protein is new to the investigator, sending the protein sequence as query to the NCBI *BLAST* server usually returns a list of similar sequences. Any of these may then be added directly into *MegAlign*.

5.2. Aligning Multiple Sequences

Multiple alignments may be performed using *Jotun Hein* ***(23)*** and *ClustalV* ***(24)*** algorithms. The two methods typically yield slightly different results, both for the sequence alignment and for the phylogenetic tree generated. The ability to compare results using two methods is important, as it underscores the need for caution in interpreting the results.

For the example in this section, a set of bacterial RecA sequences was aligned. The *MegAlign Worktable* (**Fig. 4**) is where the user edits and arranges the sequence names, and adjusts the segment of each sequence for alignment. Following alignment, the Worktable provides tools for making small adjustments to the alignment. Two panels display the left and right ends of the alignment, so that when adjustments to one region of the alignment are made, the effects on another region can be viewed immediately. The consensus sequence may be displayed above the alignment in the Worktable (**Fig. 4**).

A useful strategy is to perform an alignment on a set of complete sequences, and use the results to choose a common subsegment for final alignment. **Figure 4** shows a preliminary alignment of the RecA sequences. Most of the sequence regions match quite well, but the beginning and end show major variation that could make these regions unsuitable for inclusion in phylogenetic tree data. At this point, the sequences could be trimmed and realigned, using either the original or modified alignment parameters.

Users may experiment with realigning the sequences using different alignment parameters. A phylogenetic tree may be viewed at any time during the alignment process (**Fig. 4**). This is important, as minor editing of the alignment and changes to alignment parameters each may have a measurable impact on the phylogenetic trees built from the alignments. Iteratively changing align-

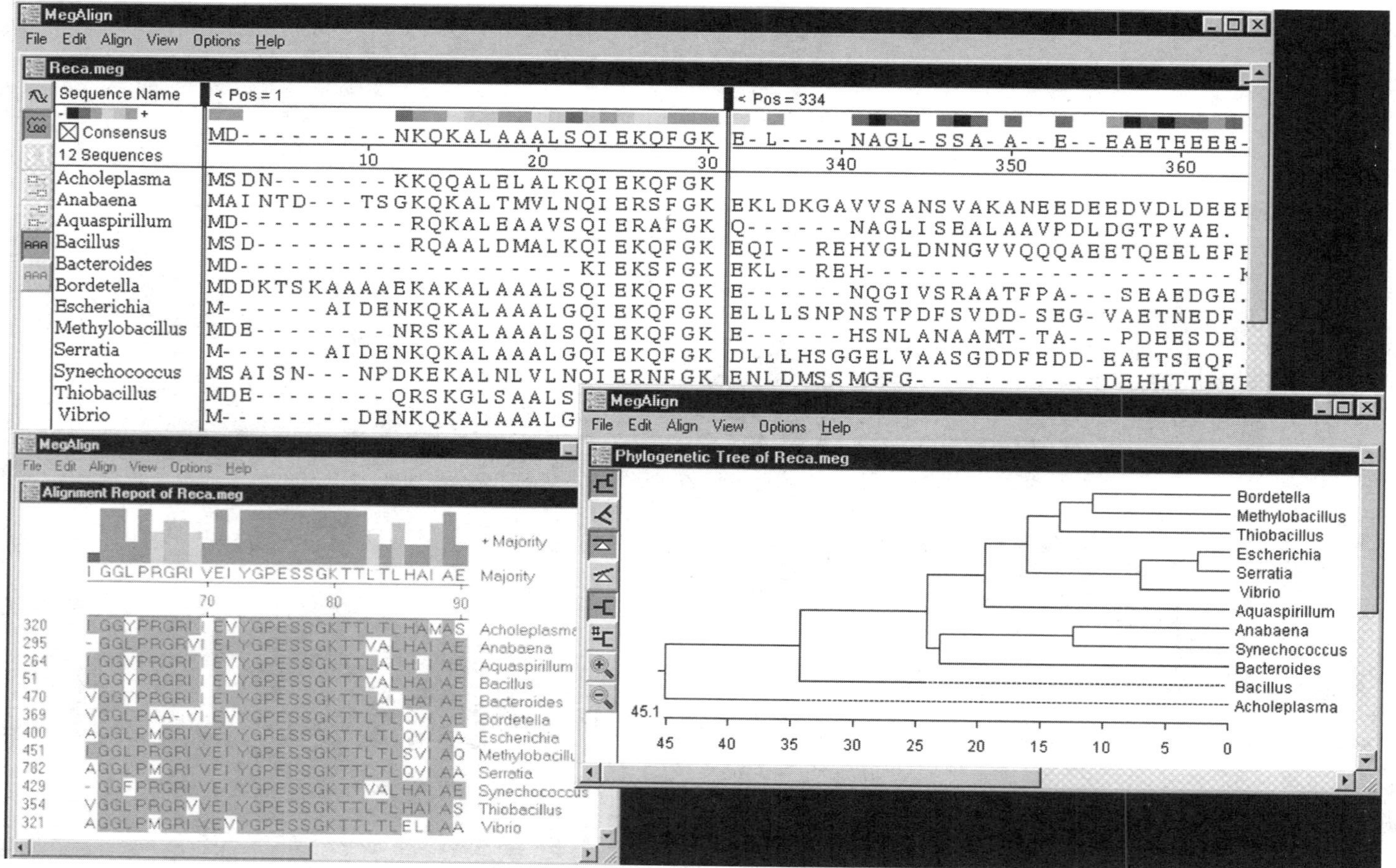
MegAlign
File Edit Align View Options Help
Reca.meg
Sequence Name
Consensus
12 Sequences
Acholeplasma
Anabaena
Aquaspirillum
Bacillus
Bacteroides
Bordetella
Escherichia
Methylobacillus
Serratia
Synechococcus
Thiobacillus
Vibrio
Pos = 1
Pos = 334
MegAlign
File Edit Align View Options Help
Alignment Report of Reca.meg
Majority
MegAlign
File Edit Align View Options Help
Phylogenetic Tree of Reca.meg
Bordetella
Methylobacillus
Thiobacillus
Escherichia
Serratia
Vibrio
Aquaspirillum
Anabaena
Synechococcus
Bacteroides
Bacillus
Acholeplasma
45.1
45 40 35 30 25 20 15 10 5 0

ment parameters and viewing the resulting trees can distinguish phylogenetic placements that are robust to parameter changes vs those that are sensitive. The latter class would warrant more attention in refining the phylogeny.

5.3. Pairwise Alignments

Four algorithms are available for pairwise alignments: Martinez Needleman-Wunsch *(20)* and Wilbur-Lipman *(21)* for DNA alignments, and Lipman-Pearson *(22)* for protein alignments. The fourth method, dot plot, may be applied to both DNA and protein.

Pairwise alignments may be done on any two sequences selected in the Worktable window—there is no need to remove other sequences. Parameters may be adjusted for each pairwise alignment, and for each the result is displayed in its own window.

If part of a pairwise alignment seems unsatisfactory, that segment may be selected and realigned, either using the original alignment parameters or after changing them. This is particularly useful for optimizing longer alignments.

5.4. Viewing and Publishing Results

The "Alignment Report" view (**Fig. 4**) gives a customizable display of the same data shown in the Worktable. Adjustments may be made to the number of residues per line, the typeface, whether to display the sequences, and/or the consensus, whether to display a graphical representation of the similarity across the alignment, and how to display the individual differences and similarities—for example, identical residues may be shaded or hidden to emphasize residues that differ from the consensus.

As with other *Lasergene* applications, the Worktable view, alignment report, and phylogenetic tree may be copied to the clipboard and pasted into other applications (**Subheading 3.6.**) for further illustration. If additional computational analyses on the numerical data are desired, the sequence distance and residue substitution tables may be pasted directly into a *Microsoft Excel*® spreadsheet.

To use the data in other applications, they may be saved in formats suitable for use in the PAUP and GCG *Pileup* programs. The sequences in the *MegAlign*

Fig. 4. *(opposite page) MegAlign*'s *Worktable* (upper window) is where sequences are added and aligned, and where manual adjustments to the alignment may be made. The two panels show different segments of an alignment of 12 RecA-like proteins. This alignment took about five seconds using the *Clustal* method on a 200-MHz *PentiumPro* PC running *Windows NT4.* Bottom left shows the customizable Alignment View. The bar-chart displays the extent of agreement for the consensus residue in each column. Bottom right shows the phylogenetic tree.

project may also be exported as a set of individual *DNASTAR* sequence files. This is useful if the data were originally obtained as a *MegAlign* document from a colleague, or from the results of a *BLAST* search. One can then analyze any of the individual sequences in *GeneQuest* or *Protean* to elucidate what the consequences may be of any sequence variations found among the sequences examined in *MegAlign*.

6. Finding Public Data using *GeneMan*

GeneMan searches public data that are stored locally on CD-ROMs. Databases include *GenBank/EMBL*, *GBTrans*, and *PIR/NBRF*. Users can search the multiple CD-ROMs directly, or load the data onto a local hard drive and search from a single location. The latter option is quite feasible since a 17-gb hard drive can be purchased in 1998 for approx $400.

6.1. Sequence Similarity

GeneMan searches public data based on sequence similarity for either protein or DNA, using a modified *FASTA* algorithm ***(25)***. *GeneMan* formulates the query from a sequence file, and provides options to change the sequence coordinates to use as query, the k-tuple (unit of comparison), the window size over which similarity is calculated, the percent similarity required for a match, and the penalty for introducing a gap between the query and the database match when the two are aligned (**Fig. 5**). For DNA searches, the "rapid screen" option screens the database before the *FASTA*-based search, reducing the number of sequences that FASTA subsequently searches dramatically, thereby accelerating the search.

Display of search results is initially as one-line summaries. Users may choose to expand the view to include more or all of the information available. Additional displays include a plot of the frequency of database hits vs the percent similarity, and the alignment between the query and the database hit (**Fig. 5**).

6.2. Consensus Sequences

GeneMan's consensus search function accepts a consensus sequence query up to 256 characters, and permits adjustment of the percent similarity for a match. Searching for a consensus is a powerful way to find known motifs. The syntax for defining the consensus is based on the conventions for *Prosite* ***(26)***. These conventions support IUB codes, and for any sequence position allow explicit specification of alternative residues, excluded residues, specific distances between residues, and whether a pattern must be located at the amino or carboxyl terminus.

An example of a consensus sequence is the tubulin GTP-binding consensus [SAG]-G-G-T-G-[SA]-G, which specifies S or A or G in position 1, followed by GGTG, followed by S or A, followed by G (all one-letter amino acid codes).

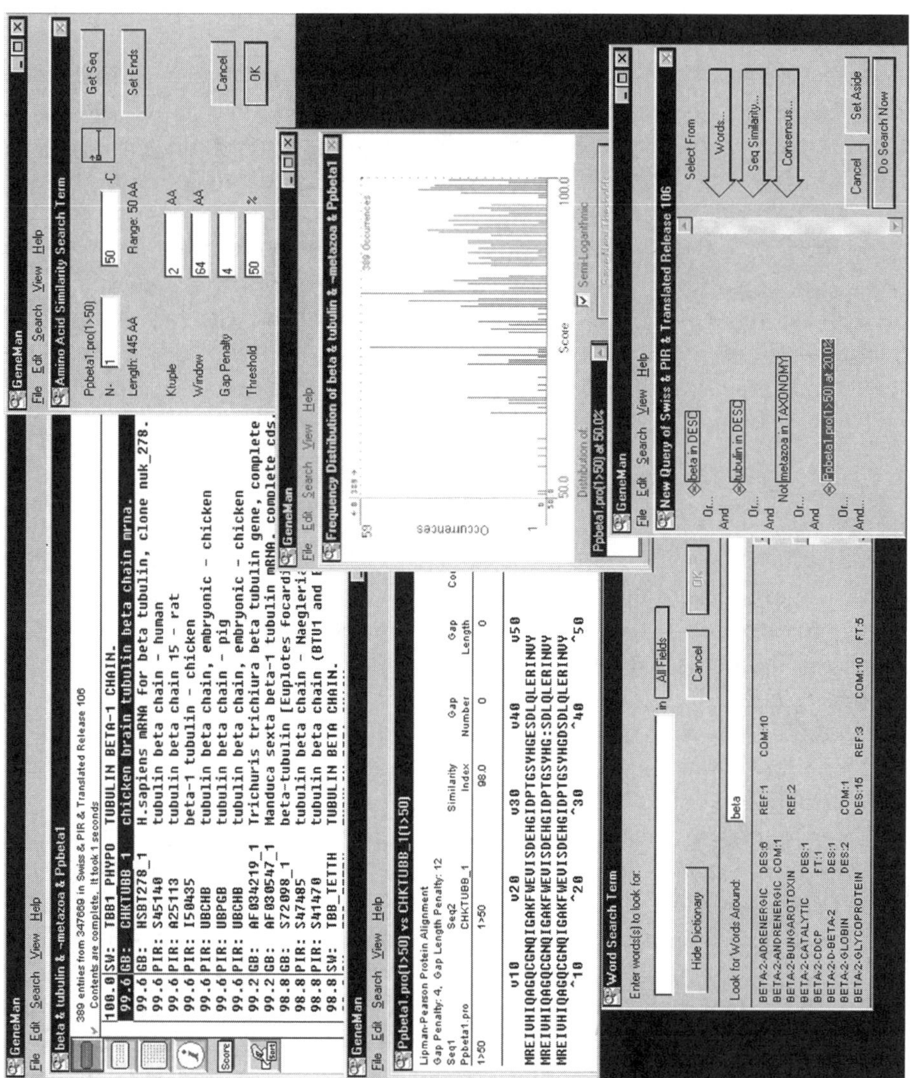

Fig. 5. Top left shows part of the match list for a sequence similarity search of the protein databases with a β-tubulin query. Immediately below is the alignment for the best match—alignments for any match may be displayed by selecting the match result with the mouse. Top and bottom right are dialogs for setting up queries and for adjusting parameters for sequence similarity terms. Bottom left is the word search dialog and associated dictionary listing the indexed words. Middle right shows a summary plot of the number of database matches with various levels of similarity to the sequence query.

6.3. Text Words

GeneMan provides tools to build a comprehensive text query before initiating a search, or to begin with a simple query and further query the results of the first search. In the latter case, the results for each stepwise query are kept in separate windows, allowing one to step back and perform multiple distinct searches on a prior set of results.

For each text word, a single field in the database may be specified, or all the fields may be searched. To check whether a word is indexed in the dictionary, the full dictionary may be viewed and any word selected from it (**Fig. 5**). Searching for multiple words, one can use Boolean operators AND, OR, NOT in combination with the field limitations. This allows quite specific queries to be built, increasing the focus on the data sought.

6.4. Complex Boolean Queries and Data Subsets

GeneMan can combine sequence similarity searching, text searching, and consensus searching in a single, comprehensive Boolean query. For example, imagine a user has found a potential GTP-binding site in a sequence. The next step might be identification of more gene products that have GTP-binding sites. However, the user knows this sequence is not a tubulin, and in searching for sequences encoding GTP-binding sites, wants to avoid perusing the thousand or so tubulin sequences in the public databases. One way to achieve this would be to combine with a text query for GTP binding the NOT operator for the tubulin GTP-binding consensus sequence [SAG]-G-G-T-G-[SA]-G.

Care is required in building the queries and interpreting the search results. As with any database search tool, *GeneMan* will not correct errors or omissions in the source data. Despite the best efforts of custodians, many entries in public databases contain errors, and annotations are not completely consistent. For example, to find all mammalian β-tubulin sequences, one could logically specify the text words "tubulin" **AND** "beta" in the **Definition** field, and the text word "mammal" in any field. This would find most mammalian β-tubulin sequence entries, but not those few β-tubulin sequence entries that do not have the word "tubulin" in the **Definition** field. However, searching for sequence similarity to β-tubulin, **AND** the text word "mammal" in any field, would find mammalian β-tubulins even if they lack "tubulin" in the **Definition** field, as β-tubulin sequences are highly conserved.

Similarly, imagine trying to formulate a query that eliminates Metazoan sequences from the hit list. Logically, this could be accomplished by choosing the Boolean operator **NOT** for the text query "Metazoa" restricted to the database field **Source.** But this query will produce a list of database entries that includes some sequences from the Metazoan *Drosophila,* simply because the term “Metazoa” is not in the **Source** field in some *Drosophila* database entries. *GeneMan*’s ability to combine sequence similarity searches with text searches provides the power to find relevant sequences even when database entries contain errors or omissions.

7. Primer Design, Restriction Maps, and Sequence Editing

The *Lasergene* system includes three more applications. These are described briefly.

PrimerSelect provides tools for design and analysis of oligonucleotides, including primers for PCR, sequencing, probe hybridization, and transcription. Using DNA, RNA, or backtranslated proteins as templates, *PrimerSelect* details thermodynamic properties for annealing reactions, identifies all possible primers, and ranks them in order of suitability for specified conditions. *PrimerSelect* also highlights potential pitfalls in both standard and multiplex PCR experiments.

MapDraw generates restriction maps and displays sites, translations, and features of sequences in six different formats, including circular and linear maps. The sequence entered may be as small as an oligonucleotide or as large as the largest BAC insert. From the database of nearly 500 restriction sites, any subset may be selected for mapping. Sets of restriction sites may be combined using Boolean operators, providing powerful site selection tools to assist cloning strategies.

EditSeq is provided with every *Lasergene* system to facilitate work on nucleic acid and protein sequences of all sizes from a variety of formats, including *GeneMan, GenBank, FASTA,* text, ABI, ALF, *Staden*, clipboard, GCG, *MacVector*™, and the efficient Lasergene sequence file format *DNASTAR* established in 1982. In addition, sequences may be obtained by accession number from NCBI’s databases over the internet, and sequences related to a sequence in *EditSeq* may be identified using the integrated *BLAST* search tool. *EditSeq* provides basic analytical tools, including editing, reverse complementing, translation, back translation, ORF identification, and simple annotation.

8. Summary

Lasergene’s eight modules provide tools that enable users to accomplish each step of sequence analysis, from trimming and assembly of sequence data, to gene discovery, annotation, gene product analysis, sequence similarity searches, sequence alignment, phylogenetic analysis, oligonucleotide primer design, cloning strategies, and publication of the results. The *Lasergene* software suite provides the functions and customization tools needed so that users can perform analyses the software writers never imagined.

Acknowledgments

I thank my colleagues Carolyn Allex and Sharon Savage for comments on the manuscript and Jeff Briganti for his patient explanations of all the myriad *Lasergene* features.

References

1. Blattner, F. R., Plunket III, G., Bloch, C. A., Perna, N. T., Burland, V., Riley, M., Collado-Vides, J., Glasner, J. D., Rode, C. K., Mayhew, G. F., Gregor, J., Davis, N. W., Kirkpatrick, H. A., Goeden, M. A., Rose, D. J., Mau, B., and Shao, Y. (1997) The complete genome sequence of *Escherichia coli* K12. *Science* **277,** 1453–1462.
2. Burland, V., Daniels, D. L., Plunkett III, G., and Blattner, F. R. (1993) Genome sequencing on both strands: the Janus strategy. *Nucleic Acids Res.* **21,** 3385–3390.
3. Allex, C. F., Baldwin, S. F., Shavlik, J. W., and Blattner, F. R. (1997) Increasing consensus accuracy in DNA fragment assemblies by incorporating fluorescent trace representations, in *Proceedings of the Fifth International Conference on Intelligent Systems for Molecular Biology* (Gaasterland, T., Karp, P., Karplus, K. Ouzounis, C., Sander, C., Valencia, A.), AAAI Press, Menlo Park, pp. 3–14.
4. Borodovsky, M. and McIninch, J. (1993) GeneMark: parallel gene recognition for both DNA strands. *Comp. Chem.* **17,** 123–133.
5. Trifonov, E. N. and Sussman, J. L. (1980) The pitch of chromatin DNA is reflected in its nucleotide sequence. *Proc. Natl. Acad. Sci. USA* **77,** 3816–3820.
6. Chou, P. Y. (1990) Prediction of protein structural classes from amino acid composition, in *Prediction of Protein Structure and the Principles of Protein Conformation* (Fasman, G. D., ed.), Plenum, New York, NY, pp. 549–586.
7. Garnier, J., Osguthorpe, D. J., and Robson, B. (1978) Analysis of the accuracy and implications of simple methods for predicting the secondary structure of globular proteins. *J. Mol. Biol.* **120,** 97–120.
8. Deléage, G. and Roux, B. (1987) An algorithm for protein secondary structure prediction based on class prediction. *Protein Eng.* **1,** 289–294.
9. Parry, D. A. (1982) Coiled coils in α-helix-containing proteins: analysis of the residue types in the heptad repeat and the use of these data in the prediction of coiled coils in other proteins. *Biosci. Rep.* **2,** 1017–1024.
10. Engelman, D. M., Steitz, T. A., and Goldman, A. (1986) Identifying nonpolar transbilayer helices in amino acid sequences of membrane proteins. *Annu. Rev. Biophys. Biophys. Chem* **15,** 321–54.
11. Kyte, J. and Doolittle, R. F. (1982) A simple method for displaying the hydropathic character of a protein. *J. Mol. Biol.* **157,** 105–132.
12. Hopp, T. P. and Woods, K.R. (1981) Prediction of protein antigenic determinants from amino acid sequences. *Proc. Natl. Acad. Sci. USA* **78,** 3824–3828.
13. Eisenberg, D., Weiss, R. M., and Terwilliger, T. C. (1984) The hydrophobic moment detects periodicity in protein hydrophobicity. *Proc. Natl. Acad. Sci. USA* **81,** 140–144.

14. Bairoch, A., Bucher, P., and Hofmann, K. (1997) The PROSITE database, its status in 1997. *Nucleic Acids Res.* **25,** 217–221.
15. Margalit, H., Spouge, J. L., Cornette, J. L., Cease, K. B., Delisi, C., and Berzofsky, J. A. (1987) Prediction of immunodominant helper T cell antigenic sites from the primary sequence. *J. Immunol.* **138,** 2213–2229.
16. Jameson, B. A. and Wolf, H. (1988) The antigenic index: a novel algorithm for predicting antigenic determinants. *Comp. Appl. Biosci.* (now *Bioinformatics*) **4,** 181–186.
17. Sette, A., Buus, S., Appella, E., Smith, J. A., Chesnut, R., Miles, C., Colon, S. M., and Grey, H. M. (1989) Prediction of major histocompatibility complex binding regions of protein antigens by sequence pattern analysis. *Proc. Natl. Acad. Sci. USA* **86,** 3296–3300.
18. Rothbard, J. B. and Taylor, W. R. (1988) A sequence pattern common to T cell epitopes. *EMBO J.* **7,** 93–100.
19. Emini, E. A., Hughes, J., Perlow, D., and Boger, J. (1985) Induction of hepatitis A virus-neutralizing antibody by a virus-specific synthetic peptide. *J. Virol.* **55,** 836–839.
20. Martinez, H. M. (1983) An efficient method for finding repeats in molecular sequences. *Nucleic Acids Res.* **11,** 4629–4634.
21. Wilbur, W. J. and Lipman, D. J. (1983) Rapid similarity searches of nucleic acid and protein data banks. *Proc. Natl. Acad. Sci. USA* **80,** 726–730.
22. Lipman, D. J. and Pearson, W. R. (1985) Rapid and sensitive protein similarity searches. *Science* **227,** 1435–1441.
23. Hein, J. (1990) Unified approach to alignment and phylogenies. *Meth. Enzymol.* **183,** 626–645.
24. Higgins, D. G. and Sharp, P. M. (1989) Fast and sensitive multiple sequence alignments on a microcomputer. *Comp. Appl. Biosci.* (now *Bioinformatics*) **5,** 151–153.
25. Pearson, W. R. (1990) Rapid and sensitive sequence comparison with FASTP and FASTA. *Meth. Enzymol.* **183,** 63–98.
26. Bucher, P. and Bairoch, A. (1994) A generalized profile syntax for biomolecular sequence motifs and its function in automatic sequence interpretation, in *Proceedings 2nd International Conference on Intelligent Systems for Molecular Biology* (Altman R., Brutlag D., Karp P., Lathrop R., Searls D., eds.), AAAI Press, Menlo Park, CA, pp. 53–61.

6

PepTool™ and GeneTool™:

Platform-Independent Tools for Biological Sequence Analysis

David S. Wishart, Paul Stothard, and Gary H. Van Domselaar

1. Introduction

PepTool™ and *GeneTool*™ are two new bioinformatics software packages currently being offered by BioTools Inc. (`www.biotools.com`). As the names might imply, *PepTool* is designed for protein sequence analysis and *GeneTool* is designed for DNA sequence analysis. The combined package is typically priced at $1500 for academic users and $1875 for commercial users. *PepTool* is actually based on two public domain programs originally developed at the University of Alberta—*SEQSEE (1)* and *XALIGN (2)*. These two UNIX-specific programs were later adapted to other platforms, given a graphical user interface *(3)* and subsequently licensed to BioTools as a commercial package called *PepTool*. *GeneTool* was developed independently by BioTools, although it uses some key concepts and algorithms originally found in *PepTool*. *PepTool* (version 1.0) was released in December 1997 and *GeneTool* (version 1.0) was released in December 1998.

Both *PepTool* and *GeneTool* are comprehensive, integrated programs that offer the full range of analytical and graphical features typically found in many advanced commercial bioinformatics products. *PepTool* and *GeneTool* also bring some much-needed advances into the bioinformatics arena in algorithm design, graphical-interface implementation, data compression, networked parallelism, and internet communication. However, probably the most eye-catching innovation lies in the fact that *PepTool* and *GeneTool* are both platform-independent software packages. This means that these two programs can run on just about any computer or just about any operating system (MacOS,

From: *Methods in Molecular Biology, vol. 132: Bioinformatics Methods and Protocols*
Edited by: S. Misener and S. A. Krawetz © Humana Press Inc., Totowa, NJ

Windows, UNIX) without any noticeable change to the programs' overall look and feel. BioTools was able to achieve this by developing the *PepTool/GeneTool* graphical-user interface (GUI) using a special language called *Smalltalk*. *Smalltalk*, which was developed by Xerox's Palo Alto Research Center in the early 1970s, is essentially a more sophisticated version of the better known platform-independent GUI programming language called Java. *Smalltalk* allows sophisticated GUIs to be prepared without the usual concerns of platform compatibility and back-end design or the constraints of having to work with a slow program interpreter.

In the following pages we will attempt to highlight some of the more useful features offered by *PepTool* and *GeneTool*. Particular attention will be paid to the unique or unusual components of each program. Space limitations prevent us from giving a complete overview of both packages. However, it is hoped that this short introduction may offer readers some insight into the design, intent, and general utility of these powerful new tools for biological sequence analysis.

2. Methods and System Requirements

Essentially all of *PepTool*'s and *GeneTool*'s complex analytical functions (i.e., the back-end) were written in ANSI C. The GUI and certain simple analytical functions were written entirely in VisualWorks *Smalltalk* (version 2.5.2, ObjectShare). *PepTool* and *GeneTool* are available for the Power Macintosh (OS version 7.5 and higher), Windows-compatible PC's (Win95, Win98 and WinNT), Silicon Graphics (Irix version 5.0 and higher) and Sun (Solaris version 2.0 and higher) platforms. Other versions for other operating systems may be specially ordered. Networked parallelism (*vide infra*) is available for SUN, SGI, and Windows machines and should be available for the Macintosh in late 1999. The combined package (*PepTool* plus *GeneTool*) without databases, requires 70 mb of disk space (25 mb for *PepTool* and 45 mb for *GeneTool*). Both packages come with their own sequence databases and users may arrange to purchase a variety of update options. The specially compressed *PepTool* protein database requires 60 mb, whereas the compressed *GenBank* database requires approx 3.5 gb. Computers running *PepTool* and *GeneTool* (plus databases) should have a minimum of 32 mb of RAM (64 mb is recommended) and 4 gb of free disk space. Both *PepTool* and *GeneTool* support WYSIWYG ("what-you-see-is-what-you-get") printing on PostScript-compatible printers, and on any printer running under Windows.

3. *PepTool*-Specific Program Features

Because of its requirement for universal platform compatibility, *PepTool*'s GUI does not strictly adhere to any single OS interface convention although, as

Fig. 1. The *PepTool* application *Launcher* as seen on a Win95 platform. A number of different applications or windows may be accessed from this program launcher.

a general rule, it tends to follow many MacOS stylistic tendencies. Depending on the platform being used, the program may be started from either the *Finder* or *Multifinder* (for MacOS), by clicking on the Windows Start button (for Win95/98/NT), or by typing *peptool* (for UNIX). After starting, an application "Launcher" (**Fig. 1**) appears at the top of the screen along with a Sequence Editor window at the center of the screen. The *PepTool Launcher* allows the user to launch additional windows, to access Help files, to change program preferences or to contact BioTools electronically. In fact, *PepTool* has at least a dozen different views or windows accessible through either the *PepTool Launcher* or the *Sequence Editor* including: a *Sequence Editor;* an *Alignment Editor;* a simple *Text Editor;* a *Graph Viewer/Editor;* a *DotPlot Viewer/ Editor;* a *Helical Wheel Viewer/Editor;* a *Structure Viewer/Editor;* a *Sequence Motif Viewer/Editor;* a *Sequence Statistics Viewer;* a *Help Viewer;* a *Preference Editor;* and a *Bug Reporter*. Text files, folders, or image files created with these different windows can be saved and are automatically marked (with an icon and a three-letter extension) in a format specific to that window. All of *PepTool*'s files, folders and directories can be searched or navigated with a File Chooser that typically resembles the user's system-specific file selector.

3.1. The Sequence Editor

The function of the *Sequence Editor* (**Fig. 2**) is to serve as a central workspace from which to enter, edit, retrieve, graph, or analyze protein sequences. As such, most of *PepTool*'s functionality is accessible through this particular window. The *Sequence Editor* contains a standard set of menu items including: **File** (for file handling and printing functions), **Edit** (for editing the viewed sequence), **Transfer** (for transferring the sequence or selected portions thereof to other applications or windows), **Search** (for finding or retrieving sequences in the database), **Analyze** (for performing statistical or structural

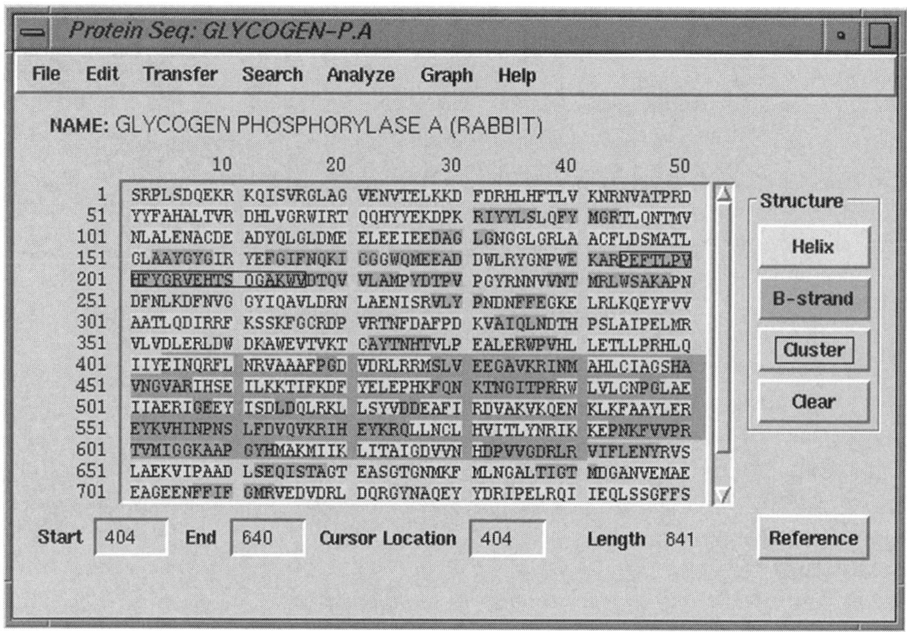

Fig. 2. The *PepTool Sequence Editor* as seen on a UNIX platform. Note how the secondary structure can be marked and viewed.

predictions), **Graph** (for plotting physicochemical properties or sequence similarities), and **Help** (for accessing the context-dependent hyperlinked Help system).

Sequences automatically loaded or manually entered into the *Sequence Editor* can be saved in either *Swiss-Prot, PIR, PepTool,* or *ASCII* format. The Editor also has the capacity to read "Foreign Format" files including GCG, *IntelliGenetics, FASTA, Swiss-Prot,* and NBRF-PIR as well as other common file types. The Foreign Format reader is both intelligent and general, meaning it does not require the user to know or to predesignate a given sequence file format. Similarly, if the Foreign Format reader encounters a file format it has not seen before, it is usually capable of making a reasonable choice about how to parse the sequence from superfluous text.

As with most sequence editors, the *PepTool Sequence Editor* supports autospacing, autowrapping, and mouse-driven text selection for the usual cutting, pasting, copying, and segment-deletion operations. It also has a text entry filter (the screen flashes when non-IUPAC letters are entered from the keyboard), a sequence ruler, a real-time sequence-length monitor, and an editable cursor position box that is updated instantly when the cursor position is changed by a mouse-click or text-entry operation. Information about the sequence and

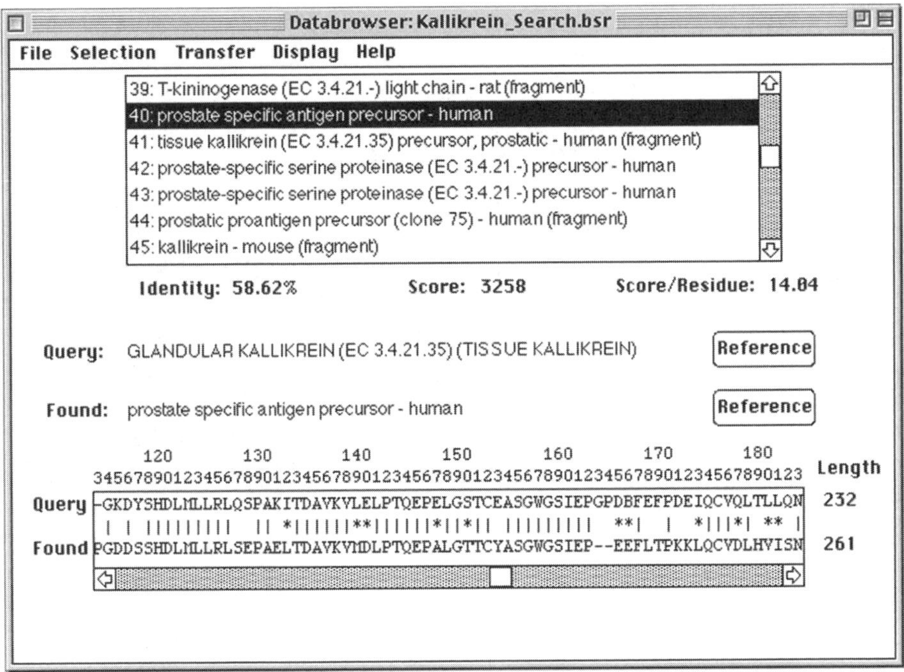

Fig. 3. A *PepTool Data Browser* as seen on a Power MacOS platform. Database hits may be selected in the upper window and the alignments viewed in the lower window.

the sequence file is displayed at the top of the window and additional data (such as the accession no., journal reference, date, and so on) can be read or entered from a pop-up sequence reference card (accessed by the **Reference** button on the lower right corner of the window).

A particularly useful feature of *PepTool's Sequence Editor* is its support of color-coded secondary structure display and editing. The buttons located on the right side of the window allow users to paint secondary structure (if known) directly on to a sequence or to precluster certain residues together when performing pairwise sequence alignments. These buttons also serve as a color-coded legend when viewing sequences loaded from *PepTool's Structure DB*—a database containing several hundred sequences with known secondary structures.

3.1.1. Database Searching

PepTool permits several kinds of sequence database searches from a variety of databases, all of which are launched from the *Sequence Editor* (under the **Search** menu item). Results from database searches can be viewed, saved, or transferred using a Data Browser (**Fig. 3**). *PepTool* supports database queries and sequence retrieval on the basis of keywords (such as organism, protein name, accession no., partial name, or logical combinations of the above); sequence patterns (simple sequence fragments or complex sequence patterns); subsequence similarity (short stretches of similar sequences); and, most importantly, global sequence homology. *PepTool* provides the option of conducting two kinds of global homology searches—a fast one and an exhaustive one.

The fast search (*FASTALIGN*) ***(1)***, which typically takes less than five minutes on a personal computer, is based on techniques similar to those described for *FASTDB, FASTA,* and *BLAST,* although it uses a specially developed scoring matrix and produces a global alignment instead of a partial local alignment (which is normally done by *BLAST*). Side-by-side comparisons of *FASTALIGN* to *FASTDB* have indicated that *FASTALIGN* is slightly faster and more sensitive than *FASTDB* ***(1)***.

The exhaustive search (*NWALIGN*) ***(1)***, which typically takes several hours on a personal computer (without networked parallelism), is based on the Needleman-Wunsch algorithm ***(4)***. Independent tests have shown this to be a very powerful algorithm for remote sequence identification with its performance easily exceeding that of *BLASTP, BLITZ, DFLASH* or *FASTA* ***(5)***. Interestingly, the same algorithm now used in *PepTool* played a key role in identifying a new class of poxvirus-encoded virulence proteins ***(6)*** and a novel uracil glycosylase ***(7)***.

3.2. *The* Alignment Editor

The *Alignment Editor* (**Fig. 4**) is an intuitive tool designed to permit the viewing, editing, and automatic generation of both pairwise and multiple sequence alignments. Typically data is transferred into this window from a Data Browser or Sequence Editor. Sequences may be transferred either individually or in groups. From the **Edit** menu, a user can easily add or delete specific sequences or change a given sequence or sequence name. Once the sequences have been loaded and/or edited, the alignment can be computed automatically by pressing the **Compute Alignment** button on the lower right corner. For this operation *PepTool* uses the *XALIGN* algorithm ***(2)*** which is capable of quickly aligning several hundred sequences using both sequence clustering and secondary structure information in the alignment process. A consensus sequence is automatically generated in the window above the alignment view using the threshold indicated in the Consensus Threshold box. Under the **Display** menu a user can select how the alignment should be displayed with options for coloring by structure (two colors), property (12 colors) or identity (one color). The pairwise comparison matrix can also be calculated and viewed from the **Display** menu item.

Manual alignment and manual editing of an automatically generated alignment can also be performed by selecting or painting over a sequence block.

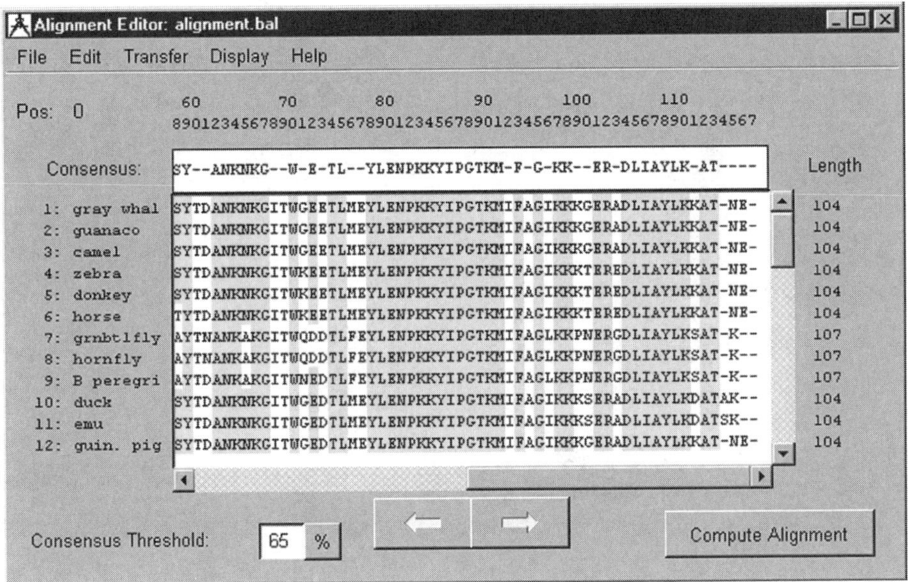

Fig. 4. The *PepTool Alignment Editor* as seen on a Win95 platform. Multiple alignments can be viewed and colored according to sequence identity as shown here.

Once highlighted, the entire sequence block (containing one or more partial sequences) can be moved right or left using the mouse-activated arrows at the bottom of the window.

3.3. *The* Structure Viewer

The Structure Viewer (**Fig. 5**) displays predicted secondary structure using specially shaded and color-coded helix and beta-sheet icons. Six different predictions, including the classic Chou-Fasman and Garnier-Osguthorpe-Robson (GOR) methods, are generated. A consensus result is produced based on the weighted average of all six predictions. The consensus result is typically 70% correct based on a simple three-state scoring system. The presence and location of membrane-spanning helices (colored in red) is also predicted using the technique of Klein et al. *(8)*. The order of the individual predictions can be rearranged by toggling a check-box at the bottom of the window and dragging the predicted structures to different locations. Under the **Display** menu it is also possible to selectively turn on or turn off certain predictions. At the top of the *Structure Viewer* the expected percent content of individual secondary structures is calculated (to allow comparisons with circular dichroism (CD) or Fourier transform infrared (FTIR) measurements) and the predicted folding class is identified (it is the one with the highest coefficient).

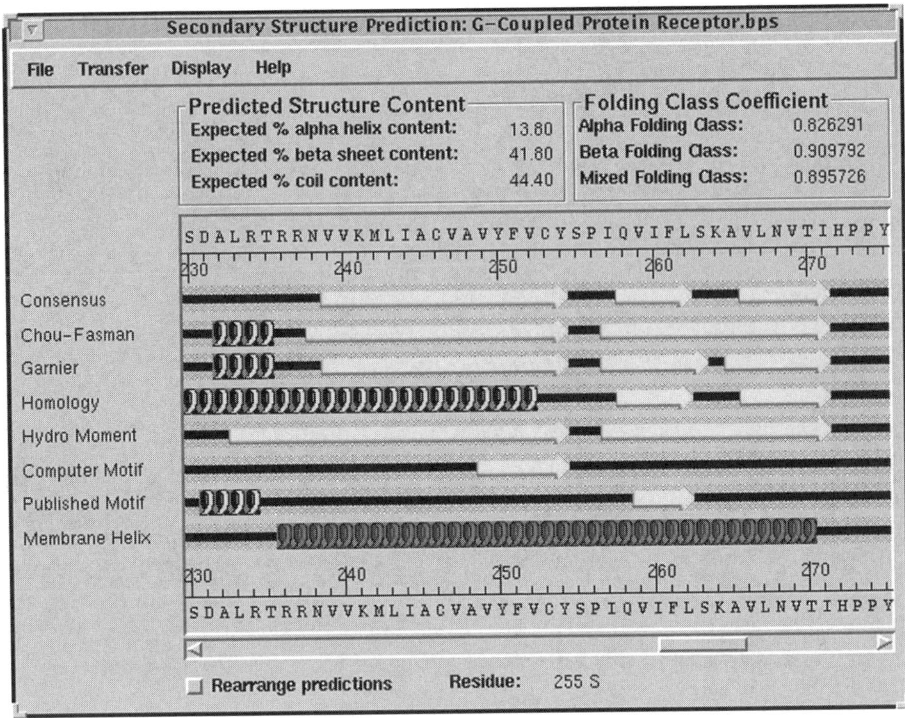

Fig. 5. The *PepTool Structure Viewer* as seen on a UNIX platform. The coils indicate helices and the arrows indicate beta-strands.

3.4. *The* Graph Viewer

This *Graph Viewer/Editor* (**Fig. 6**) shares many features with other windows including the *Helical Wheel Viewer* and the *DotPlot Viewer*. All three support fully scrollable displays, stepwise or regio-selective zooming and auto-scaling. Furthermore, all three permit the addition or deletion of text, lines, arrows, boxes, or circles to the displayed graph using a graphical palette located on the left side of the window. The *Graph Viewer* is specifically designed to display such functions as hydrophobicity, hydrophobic moments, and predicted flexibility. These protein property graphs may be further edited through the **Graph** menu, where the user may adjust the graph color, line width, graph title, and axis titles as well as turn on or turn off the grid lines and residue labels. Through the **Annotation** menu the color, linewidth and line style for any graphical annotation (except text) can also be interactively selected and adjusted.

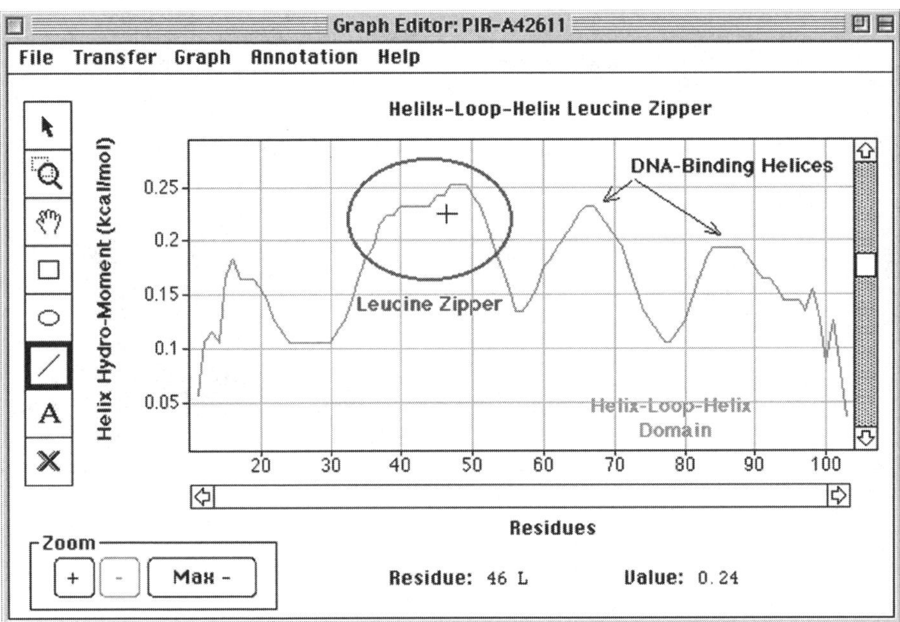

Fig. 6. A *PepTool Graph Viewer/Editor* as seen on a Power MacOS platform. An annotated plot of the helical hydrophobic moment is illustrated.

3.5. The DotPlot Viewer

Dot Matrix or Dot Plot sequence comparisons can be displayed, edited, annotated, and evaluated using *PepTool's DotPlot Viewer* (**Fig. 7**). Pairwise comparisons between two different sequences as well as simple self-sequence comparisons are possible. The number and length of plotted diagonals can be adjusted using the editable "Stringency," "Window Size," and "Diagonal Filter" boxes. Likewise the color of the plot (as well as the axis and graph titles) can be changed through options listed in the **Graph** menu. *PepTool's DotPlot* program is unique in that it displays the level of sequence similarity using a simplified color shading scheme, with identical matches appearing brightest and weak matches appearing progressively lighter. The *DotPlot Viewer* permits the usual zooming and annotation operations found in *PepTool*'s other graphical viewers although, unlike the others, it does allow the sequence for selected diagonals to be viewed in the lower sequence window. This is done by first clicking on the **ATGC** button on the annotation palette and then clicking on a specific diagonal line in the *DotPlot* window. The pairwise sequence alignment corresponding to that diagonal then appears

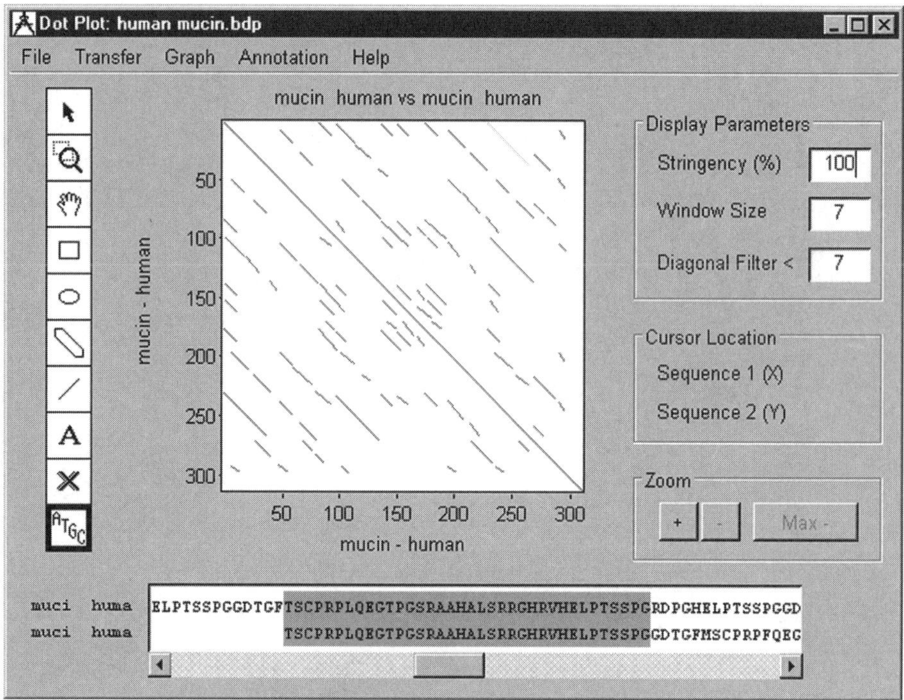

Fig. 7. The *PepTool DotPlot Viewer* as seen on a Win95 platform. The strong diagonal lines indicate the presence of multiple internal repeats in this protein.

highlighted in the lower window from which it can be viewed or inspected easily.

4. *GeneTool*-Specific Program Features

GeneTool shares many basic design and layout features with *PepTool*. However, it also has a number of important enhancements (many of which are expected to make their way into *PepTool, version 2.0*). In particular, *GeneTool* supports resizeable windows, resizeable fonts, multifeature display, multifeature editing, print-preview annotation, and audio playback. It also handles database searching, preference selection, reference information, window zooming, and window management in a more intuitive fashion. Just as with *PepTool, GeneTool* may be started from either the *Finder* or *Multifinder* (for MacOS), by clicking on the Windows **Start** button (for Win95/98/NT) or by typing *genetool* (for UNIX). After starting, the *GeneTool Launcher* appears at the top of the screen along with a *Sequence Editor* window at the center of the screen. *GeneTool* has over 20 different views or windows accessible

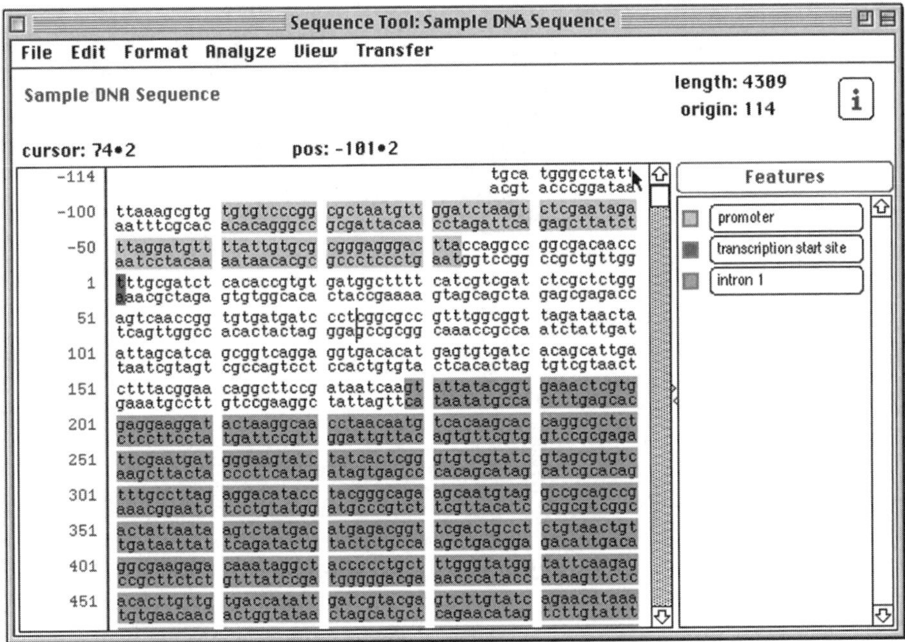

Fig. 8. The *GeneTool Sequence Editor* as seen on a Power Mac platform. Note the way in which sequence features can be marked and viewed.

through either the *GeneTool Launcher* or its *Sequence Editor* including: the *GeneTool Sequence Editor;* a *Translation Viewer/Editor;* a *Chromatogram Viewer/Editor;* an *Alignment Editor;* a *Contig Editor;* a simple *Text Editor;* a *Layout or Presentation Editor;* a *Graph Viewer/Editor;* a *Restriction Map Viewer/Editor;* a *Feature/Exon/Sequence Motif Viewer/Editor;* a *PCR Primer Designer;* a *Gel Simulation Viewer,* a *Sequence Statistics Viewer;* a *Help Viewer;* a *Preference Editor;* and a *Bug Reporter*. All of *GeneTool*'s files, folders, and directories can be searched or navigated with a File Chooser similar to the one in *PepTool*.

4.1. The Sequence Editor

Just like the *PepTool Editor,* the *GeneTool Sequence Editor* (**Fig. 8**) serves as *GeneTool*'s central operation window or central sequence worksheet. Consequently, most sequence-specific operations can be launched from this window. The *GeneTool Editor* maintains a similar arrangement of menu options (**File**, **Edit**, **Format**, **Analyze**, **View**, **Transfer**) and it permits the same wide choice of sequence formats to be read or saved (including *EMBL, GenBank,* and DNA Data Bank of Japan (*DDBJ*) formats) as the *PepTool Editor*. To limit the proliferation of file types found in *PepTool,* the designers of *GeneTool* have consolidated many of the multiple file types typically generated from a given sequence analysis into a single sequence file. The previously calculated graphs, plots, simulations, or other analysis functions associated with a given sequence file can be selected and viewed using the **View** menu. As with most other DNA sequence editors, the *GeneTool Editor* permits variable character grouping (1, 3, 5, 10, and so on), single- or double-strand display, DNA-to-RNA conversion, strand reversion, strand complementation, upper and lower case display, audio playback, autospacing, autowrapping, and mouse-driven text selection for cutting, pasting, copying, and segment deletion operations. It also supports the degenerate DNA alphabet (and flashes when non-IUPAC letters are entered from the keyboard), as well as continuously updated sequence length, reading-frame, and cursor position boxes.

There are a number of layout or design differences in the *GeneTool Editor* relative to the *PepTool Editor.* In particular, the sequence name, sequence length, cursor position, and reading-frame boxes now appear in a Sequence Status bar. Furthermore, the **reference information** button has been moved to the top and replaced with an "I" icon, which also permits more comprehensive annotation and reference display. Perhaps the most noticeable change is the fact that the *GeneTool Editor* supports a sophisticated feature display and markup system using an editable, scrollable "Feature Legend" box. With this system, *GenBank, EMBL,* or *DDBJ* sequences can be loaded and their feature tables automatically displayed using color-coded text selectors. The full name of the feature (as well as its corresponding color) can be viewed in this expandable Feature Legend box. Individual feature coloring in the text window can be toggled off and on using the colored radio button attached to each **Feature Name** button. By holding down the **shift** key and clicking on the **Feature Name** button (or alternately by clicking on the **Feature** button at the top of the Legend box), a dialog box containing additional information about that feature (and all other features) is displayed. This dialog box allows the user to add, reorder, edit, annotate, or prioritize overlapping features. A key advantage with this feature rendering method is that it allows users to add their own features to new sequence data (in a manner similar to the way *PepTool* permits secondary structure to be added or removed in its editor). This is simply done by: adding a new feature button to the Legend or editing an existing feature button to the desired feature name; highlighting the featured sequence in the text window; and clicking on the corresponding feature button to color the highlighted text.

4.2. The Chromatogram Viewer

Raw sequence data generated from automated DNA sequencers can be read, edited, and saved in a variety of formats using *GeneTool's Chromatogram*

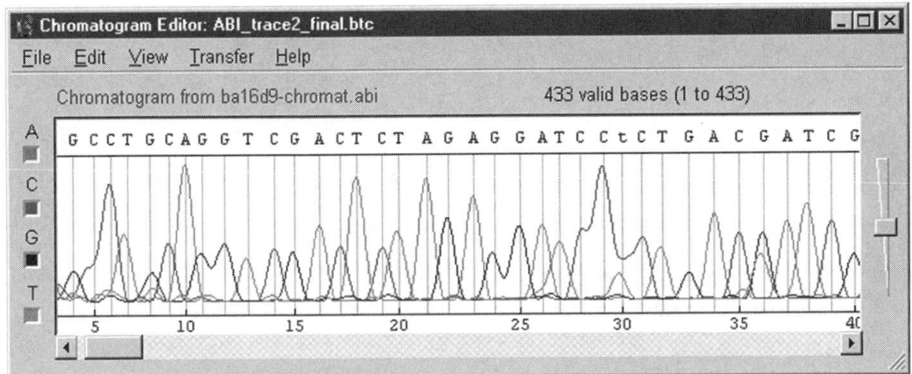

Fig. 9. *GeneTool's Chromatogram Viewer* as seen on a Win95 platform.

Viewer (**Fig. 9**). In particular, data can be read directly from ABI- or SCF-formatted chromatogram files as well as *GeneTool*'s own chromatogram format. Individual chromatogram traces can be toggled off or on using the colored **A,C,G,** and **T** buttons located on the left side of the window or, alternately, by checking off the **Base Trace** selections under the **Format** menu. Individual trace colors can be changed through a color palette presented in the Preferences window. Each trace can be selected and dragged up or down to help clarify base calls. A 5′/3′ trimming feature (located under the **Edit** menu) is also available to eliminate unwanted or unreadable data at the extreme ends of a chromatogram. The vertical scale of all four chromatogram traces can be adjusted using a scaling bar on the right side of the screen. Base calls made by the sequencer can be changed or deleted in a manner similar to most standard text editors. However, insertion of a base or bases must be done through a modal change in the **Edit** menu. The *Chromatogram Viewer* also supports two types of **Find** functions, one designed to locate ambiguous base calls (**Find Next Problem**) and the other to locate specific subsequences (**Find. . .**).

4.3. The Exon Finder

GeneTool uses a unique method for identifying exon/intron locations in eukaryotic DNA based on the reference point logistic (RPL) method developed by Peter Hooper at the University of Alberta. RPL is similar to a sophisticated neural network and can be trained to recognize very complex patterns and signals, such as those found at exon/intron boundaries. Performance evaluations using the test data (containing some 570 vertebrate genes) of Burset and Guigo *(9)* indicate that RPL can predict the location of exons and introns with a correlation coefficient of better than 0.92 (P. Hooper, pers. comm.). This is substantially better than most other gene-finding algorithms, including such

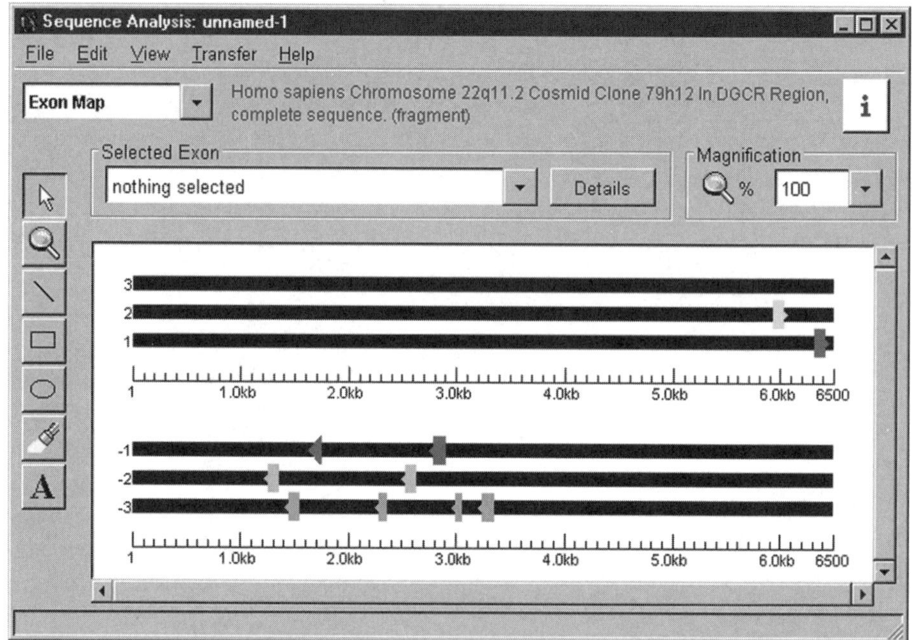

Fig. 10. An exon/intron map as generated by *GeneTool's Exon Finder* (Win95).

popular programs as *GRAIL* and *GRAIL 2* ***(9)***. Furthermore, the RPL prediction only takes a few seconds on a standard desktop machine. BioTools has enhanced this RPL technique by adding a database search method to fine-tune the initial exon/intron predictions. This typically improves the predictive performance by an additional 3 or 4%, although it adds another 4 min to the analysis time.

When **Find Exons/Introns** is selected from the **Analyze** menu, a dialog box is presented in which the user is asked to select either the fast search (which is the pure RPL method) or the exhaustive search (which combines RPL with a fast database scan). Graphical results are presented in a window like the one shown in **Fig. 10**. Individual exons or the complete set of exons may be selected by a mouse click and transferred to a *Sequence Editor*. Alternately, the displayed set of exons may be spliced together using the splice operation under the **Edit** menu. It is also worth noting that this window (and other graphical windows for motif or feature viewing) permits zooming all the way down to the sequence level so that the full gene sequence can be viewed and inspected.

4.4. *The* PCR Primer Designer

The *Primer Designer* (**Fig. 11**) is both an interactive and an automated tool for PCR primer selection and design. It may be launched either from within the

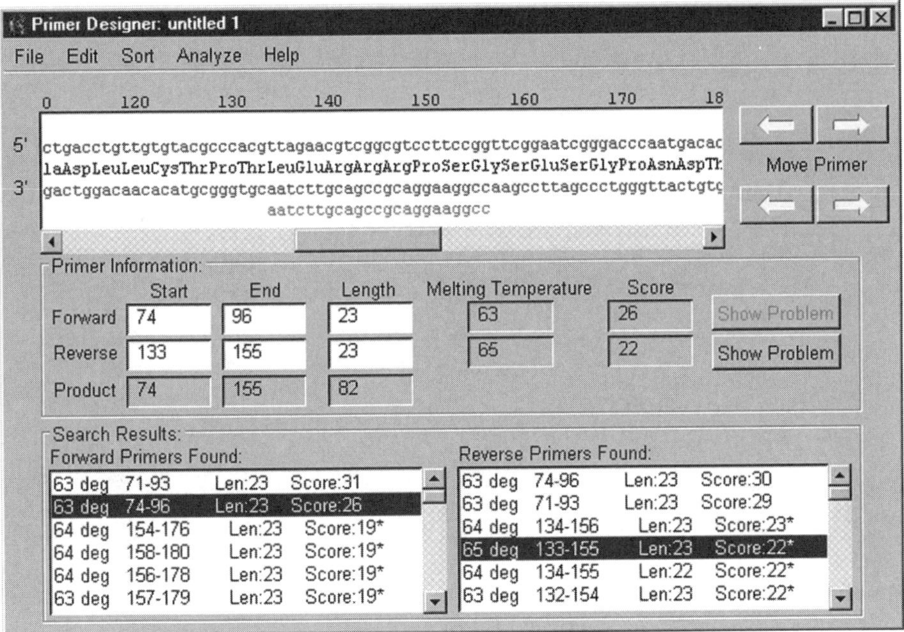

Fig. 11. The *PCR Primer Designer* as seen on a Win95 platform. Note the selectable list of primers on the lower half of the window.

Sequence Editor or from the *GeneTool Launcher*. To simplify primer analysis, sequence data is always presented in a double-stranded format, with an option to display the amino acid translation between the two strands. PCR primers may be created manually by clicking and dragging on the upper strand (for the forward primer) or the lower strand (for the reverse primer). During this operation, a primer sequence is automatically generated above (or below) the selected region while the primer length, product length, melting temperature, and primer score are calculated and updated in real time in the parameter boxes below. The primer score is an indication of the potential of the primer to form a good PCR oligo. High scores indicate a good primer, whereas low scores with asterisks indicate the presence of potential false-priming sites, hairpin turns, or incompatible melting temperatures. Primers generated through this interactive mode can be subsequently edited (to introduce point mutations) in the same manner one would edit characters in a standard text editor. Changes to a primer sequence automatically cause a corresponding change in the translated amino acid sequence (including a change in color) and an update to the primer's calculated melting temperature and PCR score. Note that the original DNA sequence and the translated sequence are not editable—only the PCR primer sequence is editable.

Automated PCR primer selection is also available under the **Analyze** menu. When the **Find Primers. . .** operation is selected, a series of compatible primers for both the upper and lower strands is calculated and presented in two data browser boxes that appear at the bottom half of the window. Forward (or upper) primers are shown on the left side and reverse (or lower) primers are shown on the right. These lists may be scrolled through and individual primers selected by a simple mouse click. Selecting a primer in this way brings the primer into the sequence view and updates the parameter boxes located in the center of the *Primer Designer* window. These primers may then be edited, lengthened, or shortened using the same primer editing techniques described earlier. Note that *PCR Primer* parameters for both the manual and automated mode may be set in the **Primer Parameters** dialog box.

GeneTool's PCR Primer Designer also supports functions to find sequences or subsequences in both the upper and lower strands; to sort identified primers by their length, position, melting temperature, or score; to check primers for specific problems; to rename primers and to save selected primers to a text file.

4.5. Restriction Map Viewer

Essentially every gene sequence analysis package has some kind of graphical restriction map viewer, and *GeneTool* is certainly no exception. Restriction digests are normally performed from the *Sequence Editor* (under the **Analyze** menu) although they may be initiated from the *Layout Editor* and the *Gel Simulation Viewer* as well. Both linear or circular DNA can be processed and presented. *GeneTool* comes with a database of some 400 restriction enzymes although it is possible for users to create their own sublibraries of enzymes, as well as add new enzymes. When the *Restriction Map* function is selected, a dialog box is presented which allows the user to select an enzyme library (the default is the full enzyme library) and to choose which enzymes in that library will be used. Specific enzyme selection may be done either on the basis of the overhang produced by the enzyme (5', 3', blunt end), the enzyme cut frequency within the DNA sequence being processed (single cutters, double cutters, and so on) or the enzyme name. Selecting enzymes by name is done using a scrollable check-box list located on the right side of the dialog box. This particular list allows any number of enzymes to be selected or deselected by name. It also indicates which enzymes have been chosen when the user has performed a selection-by-enzyme-type operation.

Once a restriction digest has been performed, a graphical map is generated as shown in **Fig. 12**. If sequence features have been identified previously, they are displayed as colored bars or semicircles. Clicking on any colored feature leads to that feature's information being displayed in a status bar at the top of the window. Once activated, that same feature may also be transferred to a

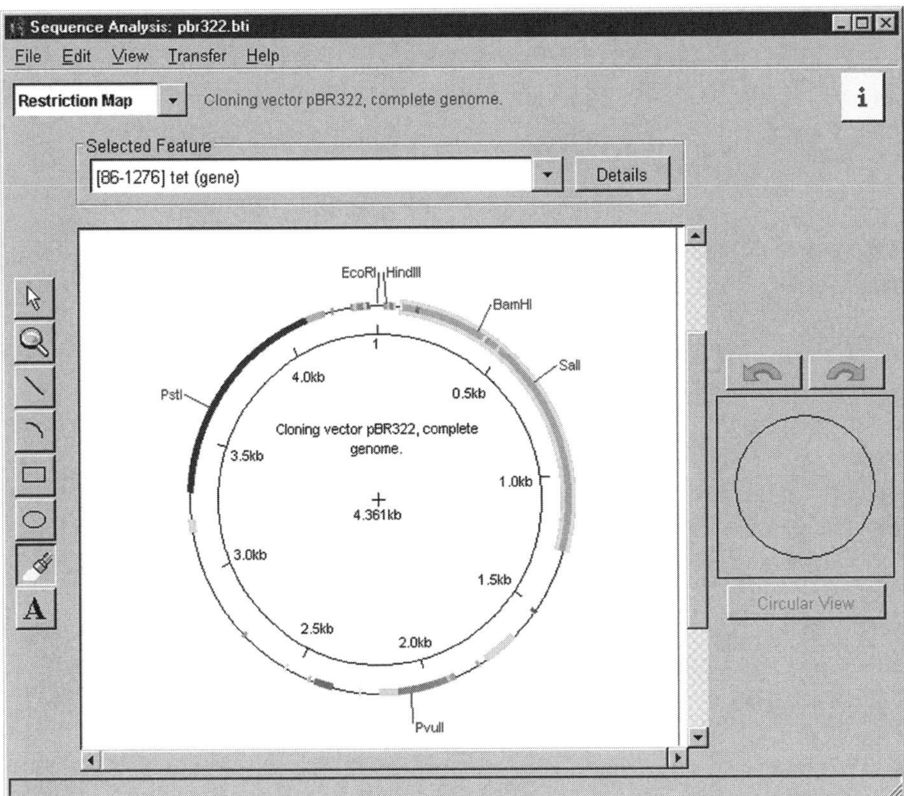

Fig. 12. A restriction map of the pBR322 plasmid prepared using *GeneTool's Restriction Mapper* (as seen on a Win95 platform).

Sequence Editor for further analysis. In addition to the sequence feature display, enzyme cut-sites are also displayed. Unique restriction sites are displayed in blue, whereas multiple restriction sites are displayed in black. Clicking on any restriction enzyme label leads to a pop-up box displaying a zoomed-in region of the sequence with the enzyme recognition sequence highlighted in red. Enzyme labels (with the attached site line) may be moved or dragged to any position on the screen to make for a more readable or symmetric presentation. Clicking on two enzyme names, while holding down the **shift** key, allows one to select the DNA sequence between the two cut sites. This graphical digest fragment may then be cut, copied, or pasted into another sequence or into another *Sequence Editor*.

Additional annotation (lines, circles, arcs, arrows, text, and so on) can be added to the map using the annotation icons on the left side of the window. Additional formatting or presentation changes can be performed through the

View menu where it is possible to selectively show or hide the sequence rulers, grid lines or enzyme labels. Under the **View** menu it is also possible to show a complete tabular summary of the restriction digest which includes the enzyme names, frequency of cuts, position of cuts and the recognition sequences. Under the **Help** menu, a user may view the complete *GeneTool* enzyme library with a full alphabetical listing of the enzyme names and recognition sequences.

4.6. *The* Layout Editor

The *Layout Editor* offers users the opportunity to create textually complex layouts or text figures (**Fig. 13**). These complex textual representations of DNA sequence data are commonly presented in published manuscripts, but typically require many tedious hours on a word processor. In an effort to reduce the difficulty associated with generating these kinds of text figures, BioTools has developed a specific *Layout Editor* to accelerate and simplify the editing process. As seen in **Fig. 13**, this editor essentially resembles the *GeneTool Sequence Editor* (minus the **Feature Legend** box) although it does have additional controls for adjusting the output. By selecting sections of the DNA sequence to be formatted (using the mouse) and then clicking either the **Grouping**, **CAPITALIZATION**, **Double Stranded**, **Translation**, or **Show Restriction Sites** buttons, it is possible to alter or annotate the highlighted sequence. The **Translation** button permits multiframe (one, three, or six) translation using either the single letter IUPAC amino acid code or the three-letter code. Likewise, the **Restriction Digest** button permits a textually annotated representation of restriction enzyme cut-site locations using the same dialog box and selection procedure found in the *Restriction Map Viewer*.

Additional formatting and annotation options are also available through *GeneTool's Print Previewer*. This particular window conveniently allows the user to view the text as it should appear on the printed page and to add or alter text, lines, arrows, boxes, or other useful annotations to selected regions of the PostScript image.

5. General Program Features—Networked Parallelism

Both *PepTool* and *GeneTool* offer a unique speed-up feature called networked parallelism. Networked parallelism allows a user to run a single program or a process simultaneously on several networked computers. The advantage to running a program on many computers as opposed to a single computer is that the program execution time can be accelerated by a factor roughly equal to the number of computers being used. Networked parallelism is actually a far cheaper alternative to purchasing a multimillion dollar supercomputer. Indeed, given that many laboratories, universities, and private

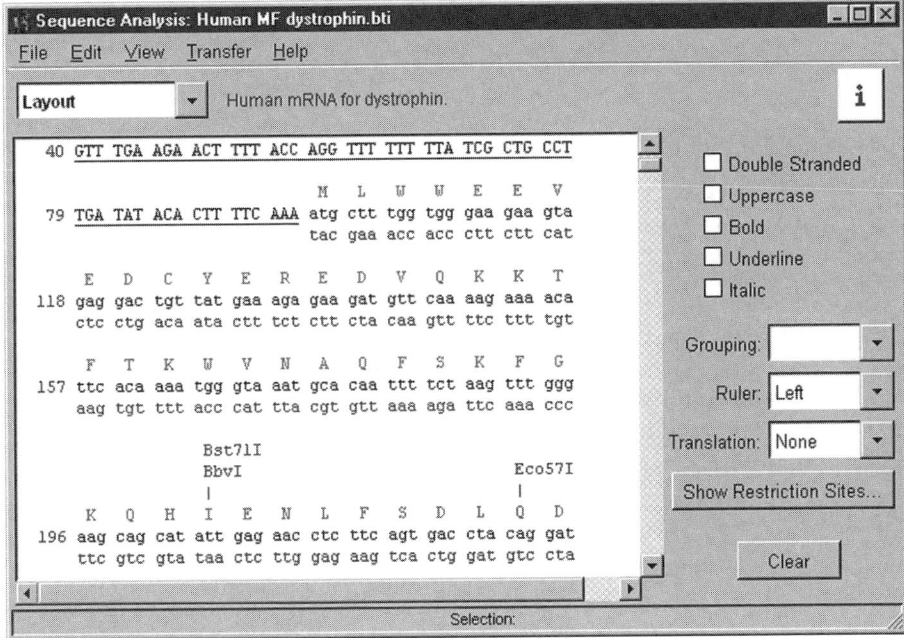

Fig. 13. An example of the textual figures that can be generated in *GeneTool's Layout Editor* (as seen on a Win95 platform).

companies already maintain networks of many personal computers, the use of networked parallelism means that it is relatively easy to get the power of 10s or 100s of networked computers for free (without disrupting the operation of other users in the network). BioTools has implemented networked parallelism (using PVM) in its most rigorous and time-consuming database searching routine (the Needleman-Wunsch algorithm). BioTools plans to extend this very useful option to other time-consuming operations such as multiple alignment, contig assembly, secondary structure prediction, and exon identification.

5.1. Database Compression

Protein and gene sequence databases are growing faster than hard drive capacity. The April 1998 release of *GenBank* (the last for which CDs were issued) required 12 CDs to hold all of the sequence data. Fortunately internet access to the *BLAST* servers at the NCBI or EBI now allows many researchers quick access to these huge databases without having to find a place to store >10 Gb of data or to read a dozen CDs at a time. However, these public servers are somewhat restricted in the types of searches that can be performed and the way that data can be saved, presented or downloaded. Furthermore, a growing number of university researchers and private companies are becoming increasingly concerned about internet security and firewall breaches that may occur when querying publicly accessible databases. The question is: How do you permit flexible database access and maintain security without the headache of purchasing a new hard drive every six months or a new CD every week?

One answer is to use data compression technology. BioTools has made use of the fact that most biological sequence data uses only a restricted alphabet of either four (for DNA) or 20 letters (for proteins). This means the size of the ASCII character set can be reduced from 8 bits per character to roughly 2.3 bits for DNA sequence data and 5 bits for protein sequence data. Further, by removing blanks, empty spaces, or redundant information from the database text fields and replacing common words with special characters, a good deal more compression can be achieved without significant loss of information. Finally, by combining multiple databases with duplicate entries (as there are for protein sequences) into a single nonredundant database it is possible to gain even more space savings. Using these and other data compression techniques, BioTools has reduced the size of the protein sequence databases from 300 mb to 60 mb and the *GenBank* database from 12 gb to 3.2 gb. This means that the complete set of databases can be delivered on 2 CDs (instead of 18) and easily stored on a regular 4 gb hard drive.

Although maintaining a local sequence database offers considerably more convenience, flexibility, and security than a remotely accessibly database, it is likely that researchers will continue to demand regular access to the NCBI's or EBI's super-fast facilities and highly integrated database features. To maintain this important database access route, BioTools also offers integrated WWW access to the NCBI server through its *GeneTool* package.

6. Summary

Although we are unable to discuss all of the functionality available in *PepTool* and *GeneTool*, it should be evident from this brief review that both packages offer a great deal in terms of functionality and ease-of-use. Furthermore, a number of useful innovations including platform-independent GUI design, networked parallelism, direct internet connectivity, database compression, and a variety of enhanced or improved algorithms should make these two programs particularly useful in the rapidly changing world of biological sequence analysis. More complete descriptions of the programs, algorithms and operation of *PepTool* and *GeneTool* are available on the BioTools web site (`www.biotools.com`), in the associated program user manuals and in the on-line Help pages.

Acknowledgments

The authors wish to thank Scott Fortin, Ann Leins, and Debby Waldman for their helpful comments and critical reading of the manuscript. We also thank the staff at BioTools for their assistance in preparing a number of the figures. G. H. Van D. is supported by a PMAC-MRC graduate scholarship and P. S. is supported by an NSERC postgraduate scholarship and an Alberta Heritage Foundation for Medical Research studentship stipend.

References

1. Wishart, D. S., Boyko, R. F., Willard, L., Richards, F. M., and Sykes, B. D. (1994) SEQSEE: a comprehensive program suite for protein sequence analysis. *Comput. Applic. Biosci.* (now *Bioinformatics*) **10,** 121–132.
2. Wishart, D. S., Boyko, R. F., and Sykes, B. D. (1994) Constrained multiple sequence alignment using XALIGN. *Comput. Applic. Biosci.* (now *Bioinformatics*) **10,** 687–688.
3. Wishart, D. S., Fortin, S., Woloschuk, D. R., Wong, W., Rosborough, T., Van Domselaar, G., Schaeffer, J., and Szafron, D. (1997) A platform-independent graphical user interface for SEQSEE and XALIGN. *Comput. Applic. Biosci.* (now *Bioinformatics*) **13,** 561–562.
4. Needleman, S. B. and Wunsch, C. D. (1970) A general method applicable to the search for similarities in the amino acid sequence of two proteins. *J. Mol. Biol.* **48,** 443–453.
5. Cattell, K., Koop, B., Olafson, R. S., Fellows, M., Bailey, I., Olafson, R. W., and Upton, C. (1996) Approaches to detection of distantly related proteins by database searches. *BioTechniques* **21,** 1118–1122.
6. Upton, C., Mossman, K., and McFadden, G. (1992) Encoding of a homolog of the IFN-γ receptor by myxoma virus. *Science* **258,** 1369–1372.
7. Upton, C., Stuart, D. T., and McFadden, G. (1993) Identification of a poxvirus gene encoding a uracil DNA glycosylase. *Proc. Natl. Acad. Sci. USA* **90,** 4518–4522.
8. Klein, P., Kanehisa, M., and DeLisi, C. (1985) The detection and classification of membrane-spanning proteins. *Biochim. Biophys. Acta* **815,** 468–476.
9. Burset, M. and Guigo, R. (1996) Evaluation of gene structure prediction programs. *Genomics* **34,** 353–367.

7

The Staden Package, 1998

Rodger Staden, Kathryn F. Beal, and James K. Bonfield

1. Introduction

For several years we have been concentrating on developing methods for large-scale sequencing projects and the resulting software is used in many major laboratories and genome centers. In the course of this work we devised a very powerful graphical user interface for use in our sequence assembly and editing program *GAP4* *(1)*, and recently we have started to write replacements for our old analysis programs *NIP* *(2)* and *SIP* *(3)*, and these entirely new programs have the same user interface as *GAP4*. The older programs were described in the previous edition of this book *(4)*, and are largely unchanged, but are still included in our package distribution. Here we give an overview of our methods for sequencing projects and also of our new analytical programs, all of which are fully documented in our 500-page manual which is available for printing or as an HTML document (`http://www.mrc-lmb.cam.ac.uk/pubseq`). This site also contains color versions of the figures used in this chapter and information about obtaining our package.

2. Methods for Managing Sequencing Projects

2.1. Introduction

Although much has been automated, sequencing projects still require human judgment for some of the more difficult contig joining and editing decisions. In our package all processing up to and including assembly is fully automated, and we provide tools that automatically analyze the contigs to suggest which experiments and templates will help to solve problems and join contigs. Manual checking and editing of contigs is kept to a minimum by directing the users' attention only to consensus bases that are not determined to the required level

From: *Methods in Molecular Biology, vol. 132: Bioinformatics Methods and Protocols*
Edited by: S. Misener and S. A. Krawetz © Humana Press Inc., Totowa, NJ

of accuracy. This is performed using *GAP4*'s powerful contig editor. Besides its own algorithms, *GAP4* provides a graphical user interface to the assembly engines *CAP2 (5), FAKII (6),* and *PHRAP (7)* (*see* **Subheading 4.**). We invented the SCF format *(8)* for storing trace data and accuracy values from fluorescence-based sequencing instruments, which in its latest form (J. K. Bonfield and R. Staden, unpubl.) reduces to one-tenth the storage space required for ABI-derived data; and for transferring sequence-related data between processing programs we use the **Experiment** file format *(9)*.

2.2. Preassembly Processing

Prior to assembly, sequence readings produced by automated sequencing instruments must be passed through several processes, which typically will include converting to the SCF format, calculation of base-calling accuracy or confidence values, quality clipping, sequencing vector clipping, cloning (e.g. cosmid) vector removal, and repeat tagging. Using our package, each of these individual steps is performed by a separate and specific program (**Fig. 1**), whereas the overall process is controlled by the program *PREGAP (9*; *see also* **Subheading 4.**), which can handle any number of readings in a single run.

The input to *PREGAP* is a file containing the names of all the sequencing instrument files to process. The output is generally an SCF file and an Experiment file for each of the input files processed, plus a new file of file names containing the names of all the newly created Experiment files, ready to be passed to *GAP4*. That is, *PREGAP* creates the initial Experiment file for each reading and then sends it, in turn, through each of the required processing steps. Note that the necessity to use file names to encode data about the readings they contain, imposed by external assembly engines such as *PHRAP,* is removed by the use of *PREGAP* and the Experiment files. *PREGAP* is modular and very flexible, and can be tailored for compatibility with local working practices. It can be configured to work completely automatically, or to be partially interactive.

2.3. Introduction to GAP4

GAP4 provides a comprehensive set of methods for assembly; checking assemblies; finding joins between contigs by using read-pair data and/or poor quality data at the contig ends; suggesting additional specific sequencing experiments to join contigs or to overcome other deficiencies in the data; to check the accuracy of the consensus; and to edit contigs to the required level of confidence. As all those who sequence know, if you try to deal with every disagreement between readings, it will take a long time, and it is the poor-quality data that causes assembly problems. For this reason we have advocated and employed *(10)* the use of base-accuracy estimates for use in consensus algorithms within *GAP4* that take these values into account. In this way, in

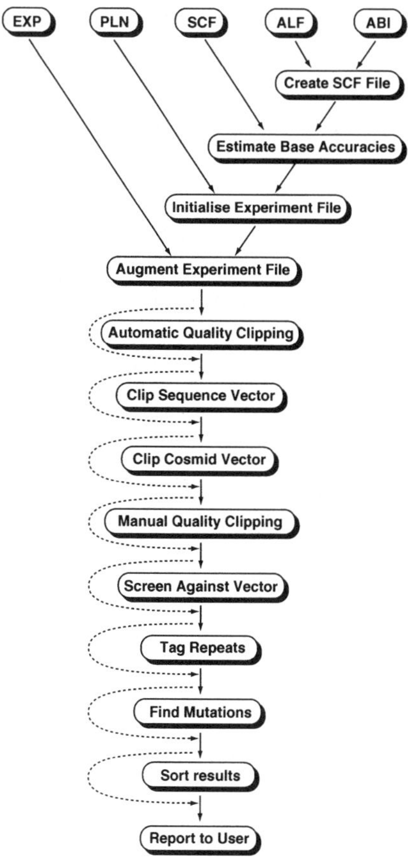

Fig. 1. The *PREGAP* processing steps.

places in which poor data disagrees with good data, no editing will be needed as long as the consensus calculation, using thresholds to ensure that the sequence meets required levels of confidence, produces A, C, G, or T. If the data is not sufficiently reliable a "-" or "N" will be generated. Throughout the whole of *GAP4* the consensus calculation is used as a basis for making decisions, it is continually updated to reflect any change, thus the user need only be concerned with those consensus locations that are not A, C, G, and T. Note that what we have just described is an ideal situation in which reliable base accuracy estimates exist, but we do not consider that the individual values calculated by our program (*EBA*) are sufficiently reliable (*EBA* was written simply to provide values to allow us to develop the associated methods; *see* **Subheading 4.**). We are aware of groups working to produce accuracy values (or confidence values), and hope that reliable methods are available soon.

The user interface enables users to view and manipulate their data at appropriate resolutions. It plays an important role in helping them to deal with difficult problems, and so simplifies and speeds up sequencing projects. The graphical display shown in **Figs. 1–6**, include the *Contig Selector*, the *Contig Comparator*, the *Template Display*, the *Quality Plot*, the *Restriction Enzyme Map*, and the *Stop Codon Map*. For editing aligned readings the *Contig Editor* and the *Join Editor* are used. From each of these the trace data for readings can be displayed and scrolled in register with the editing cursors.

The displays in *GAP4* communicate with one another and have linked cursors. For example, if the *Contig Editor* is being used on a contig that is also being viewed in a *Template Display* window, the position of the *Contig Editor*'s cursor will also be shown in the *Template Display*. Similarly, within the *Template Display* the user can use the mouse to drag the *Contig Editor*'s editing cursor. Also, if the *Contig Editor* is displaying traces, they too will scroll under the control of the *Template Display*. Note also that any number of displays of the same contig can be shown simultaneously, including several displays of the same type, such as several *Contig Editors*.

2.3.1. The GAP4 *Textual and Graphical Windows*

An example of the main window of *GAP4* (which, apart from the menus, is identical to those for *SIP4* and *NIP4*) is shown in **Fig. 2**. It consists of an **Output Window** for receiving textual results and below that an area for displaying error messages. Along the top of the window is a row of menus: the **File** menu includes database opening and copying functions, and consensus sequence file creation routines. The most important items in the other menus are shown in **Table 1**.

2.3.2. The Contig Selector *and* Contig Comparator

Once a database is loaded into *GAP4,* a graphical display called the *Contig Selector* will appear. When any analysis that compares contigs is performed, the *Contig Selector* is automatically transformed into the *Contig Comparator*. An example of which is shown in **Fig. 3**. At the top are three menus, and below that are buttons for stepping onto the next operation (*see* **Subheading 2.3.6.**), for unzooming the display, and for switching on the crosshairs. To their right, are boxes that display the positions in contigs. The panel below contains colinear horizontal lines, each terminated by short vertical bars, which represent the seven contigs in the database being viewed. When the *Selector* is transformed to make the *Comparator*, these lines are duplicated at right angles to create the square area in which to plot comparative results. Each type of analysis plots its results in a different color, e.g., to distinguish **Read Pairs** and **Find Internal Joins**. The former analysis finds templates that have been sequenced from both

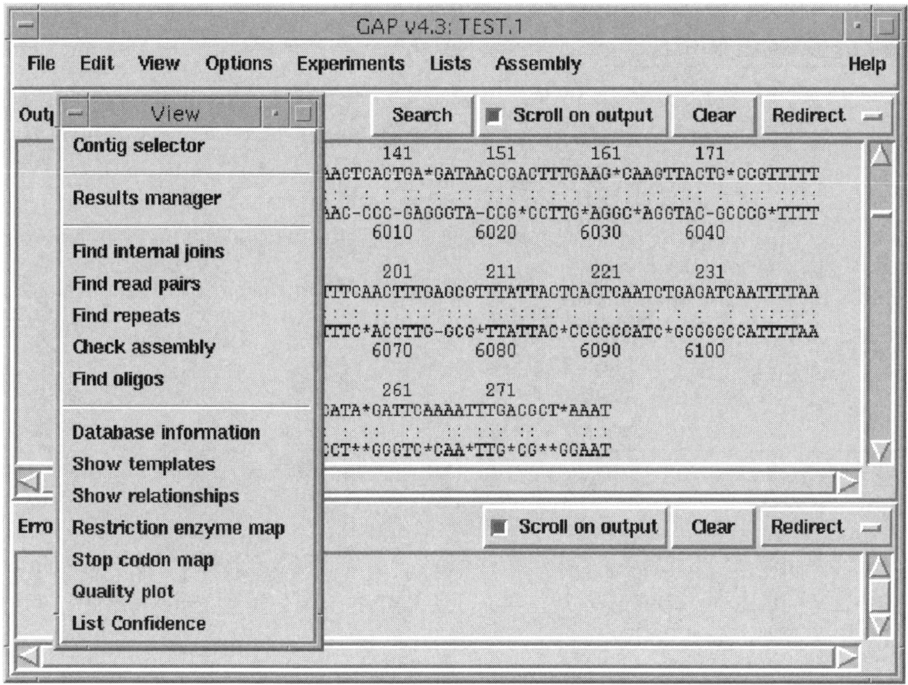

Fig. 2. The *GAP4* main window.

ends and that span contigs, and the latter finds matches between the sequences of contigs. Both plot their results as diagonal lines to show the pairs of contigs involved. If the lines are parallel to the main diagonal (shown here, top left to bottom right) the contigs involved are in the same orientation, if the lines are perpendicular to the main diagonal (as indicated by the white line), one of the contigs would need to be complemented before they could be joined. The *Contig Comparator* can, hence, show the results of completely independent forms of analysis that individually may not be convincing, but which together may, for example, be sufficient to confirm a join between a pair of contigs. Referring again to the figure: One of the diagonal lines appears lighter than the others, as do two of the contig lines. This simply denotes that the light diagonal line was the result last touched by the crosshair, and, as written in the **Information Line** at the base of the display, it is a **Read Pair** match between the two lighter contigs.

By clicking a mouse button on a line representing a contig the user can invoke a pull-down menu that contains a list of operations to perform on the contig. Similarly by clicking on a plotted result, a menu of relevant operations is revealed. For example if the user clicks on a **Find Internal Joins** result, the

Table 1
Menu Features in GAP4

Edit Menu	View Menu	Experiment Menu	Assembly Menu
Edit Contig	Contig Selector	Suggest Primers	Normal Shotgun
Order Contigs	Results manager	Suggest Long Reads	Directed Assembly
Join Contigs	Find Internal Joins	Compressions and Stops	Independent Assem
Break Contig	Find Read Pairs		CAP2 Assembly
Disassemble Readings	Find Repeats		FAKII Assembly
Complement Contig	Show Templates		PHRAP Assembly
	Restriction Enzyme map		Screen Only
	Stop Codon map		
	Quality Plot		
	Check Assembly		

Join Editor will start up with the two contigs aligned at the position of the reported match. The *Join Editor* is equivalent to two *Contig Editors,* (described below), but also shows the differences between the two aligned contigs, and can join them once editing is completed.

2.3.3. *The* Template Display

As shown in **Fig. 4**, the *Template Display* provides a schematic of a set of contigs or of a single contig. The information that can be displayed includes: readings, templates, tags, restriction enzyme sites, rulers, and consensus quality. As with the *Contig Selector*, if the mouse cursor is moved over any item in the display, textual data about it will appear in the Information Line at its base. The positions, both in the contig and in the set of contigs, of a vertical crosshair, under the control of the mouse, are continuously displayed in two boxes at the top of the window. The order of the contigs can be changed by dragging the lines that represent them horizontally in the display, whereupon all their associated data will be redrawn (this can also be done in the *Contig Comparator*). An example for five contigs is shown in **Fig. 4**. At the base of the display are lines representing the five contigs, which have been automatically spaced using Read Pair data. Two vertical lines are seen, one of which is the display crosshair, and the other marks the position of the editing cursor in an active *Contig Editor*. The long stacks of horizontal lines depict sequenced templates and the arrowed segments at their ends show the extent and direction of the sequence derived from them. All information is color coded, and so for example, templates with consistent Read Pairs are colored differently from those that are inconsistent.

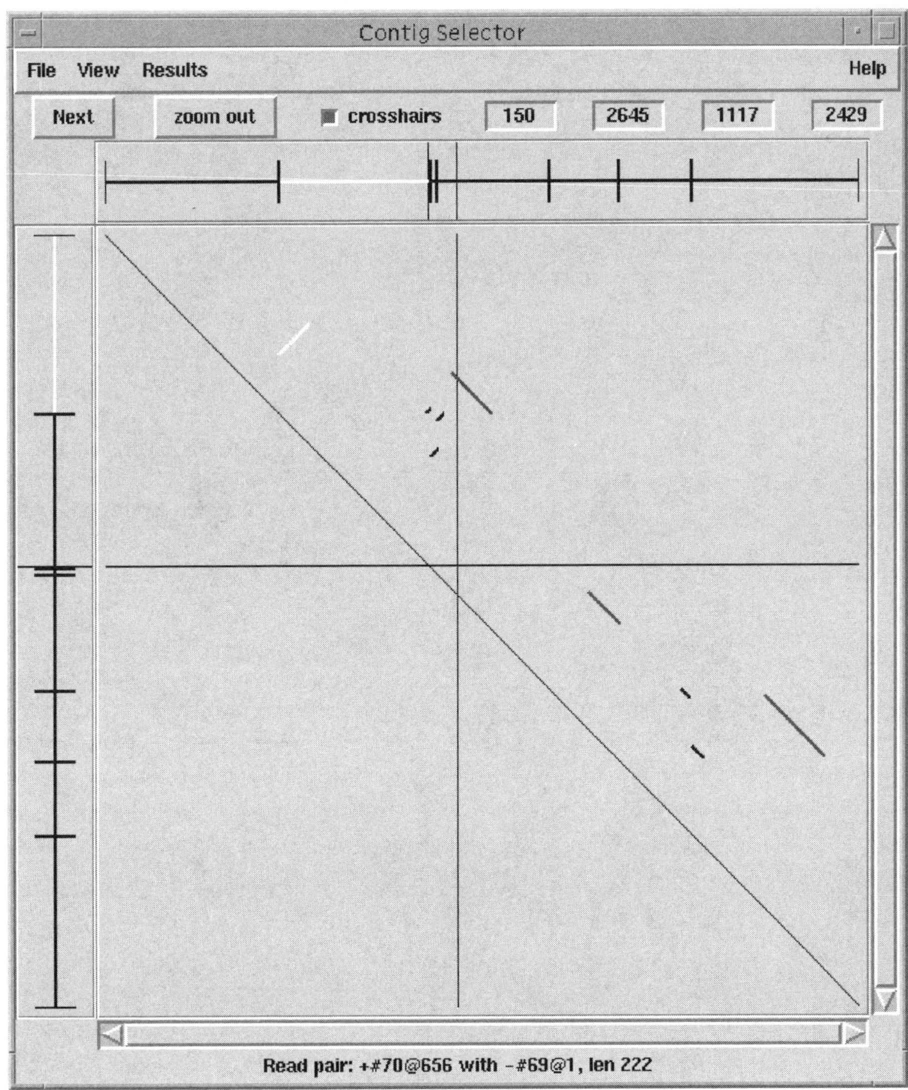

Fig. 3. The *GAP4 Contig Comparator*.

2.3.4. *The* Contig Editor

The *Contig Editor* (*see* **Fig. 5**) is one of the most important components of *GAP4*. It contains a range of sophisticated search functions that minimize the time taken to locate and deal with problems in contigs. In the ideal case, in which reliable base accuracy estimates or confidence values are available for

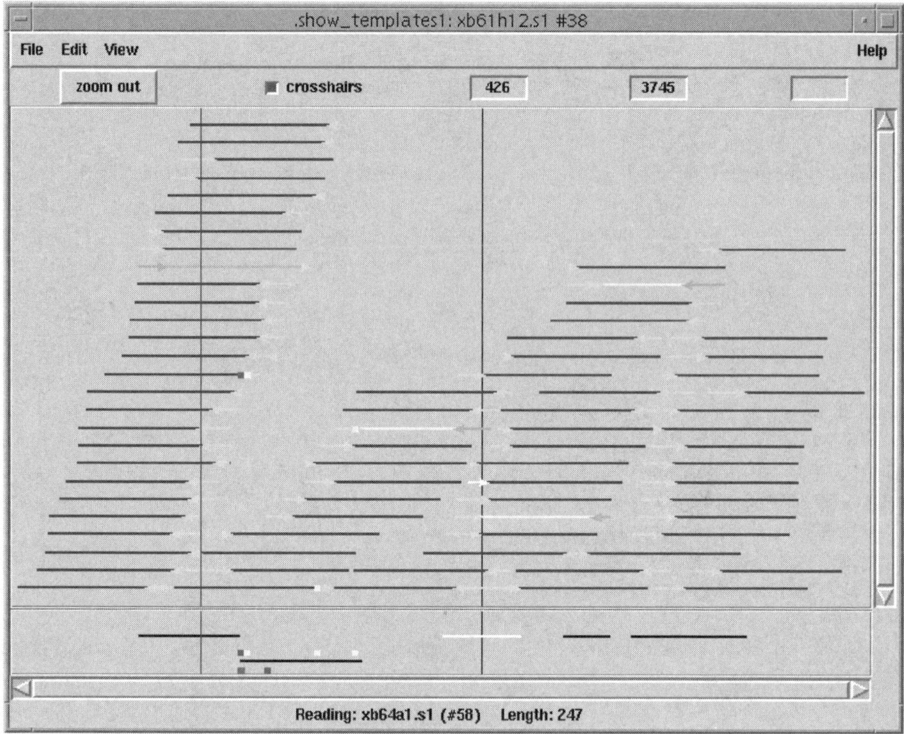

Fig. 4. The *GAP4 Template Display*.

the readings, searches can be made in which the consensus sequence is scanned for any positions that fail to reach a selected level of accuracy. This consensus is calculated on the fly and will update when any change to the readings is made. When confidence values are not available, different, on-the-fly consensus calculation functions are used (which can treat each strand separately, and hence ensure that the sequence is independently well determined on both strands), which again only direct the users' attention to where it is needed. Another type of search locates all edited positions where the consensus base does not appear in any of the original readings covering that location, and hence can be used to check the correctness of edits. Normally, when the search functions are used, the program is configured to automatically display the trace for the readings causing the problem, and also from a good reading from each strand.

Confidence values are shown using gray scales for each character in the readings and in the consensus. All edited positions are shown using a color scheme that identifies the type of edit performed. The dark segment of reading xb62c2.s1 is covered by a label known as a tag. A range of tag types are used for various purposes within the program, each type has a unique color, and

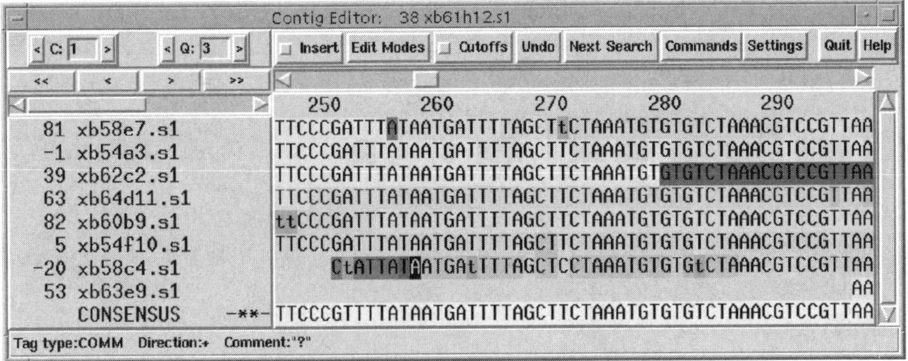

Fig. 5. The *GAP4 Contig Editor.*

each individual tagged segment has an associated and editable text string for recording notes. Tags can be made visible in all the *GAP4* displays.

2.3.5. GAP4 *Methods*

In this section we outline some of *GAP4*'s functions for aiding sequencing projects. As mentioned in **Subheading 2.1.**, *GAP4* contains several of our own assembly algorithms plus the global methods *CAP2* ***(5)***, *FAKII* ***(6)***, and *PHRAP (7)*. For processing large batches of shotgun data collected with the intention of producing almost complete coverage of the sequence, the global algorithms are best, especially if the sequence contains long repeats. Later, when adding new readings obtained to solve problems and fill gaps we would recommend our own shotgun assembly algorithm. Our *Directed Assembly* algorithm is very useful for projects in which the approximate location of the readings is known. Note that the *GAP4* database stores a great deal of extra information, other than just the readings. The main addition we have made to *CAP2, FAKII,* and *PHRAP* is to enable them to read and write Experiment files, and hence to be able to pass *GAP4* the data it requires to work to its full potential.

One way we use this extra information is in functions that can check on the relative order and orientation of readings derived from the same sequencing template. This can be used to find possible errors in the assembly and to find the most likely left-to-right order and orientation of contigs. The *Order Contigs* routine calculates the most likely order, and by clicking on the results, the user can invoke the *Template Display* (*see* **Fig. 4**).

Sometimes the data entered into the *GAP4* database may be incorrect (for example template names may be wrong) and this may cause an error in the contig-ordering process. If so, within the *Template Display*, the order can be changed by using the mouse to move the contigs to their appropriate positions.

As seen in **Fig. 3**, this type of Read Pair information can also be displayed in the *Contig Comparator*, and is particularly useful when combined with the **Find Internal Joins** function, which compares the ends of contigs to look for possible matches between contigs.

Using the same consensus algorithm as the *Contig Editor* and the routine that writes out the final finished sequence, other functions can analyse the contigs to find regions that are not sufficiently well determined, or which could be extended to help join contigs. These routines—known as the **Experiment Suggestion** methods—produce lists of template names and experiment types, and the results are written to disk, in an easily parsed format.

2.3.6. Using the Find Internal Joins *Function*

After assembly there may be potential joins between contigs that were too doubtful to be made automatically, but given the right information, a skilled user can judge which ones can be made. As an example of using *GAP4* we describe how the **Find Internal Joins** function is employed to locate these possible overlaps and to join them with the *Join Editor*.

1. Start *GAP4* by typing `gap4&`. The main window will appear as in **Fig. 2**.
2. From the File menu select **Open**, and use the browser to select and open the required project database. A summary of the current state of the project will be written to the Output Window and the *Contig Selector* will appear with lines representing all of the contigs.
3. From the **View** menu select **Find Internal Joins** and the dialog will appear as shown in **Fig. 6**.
4. The default parameters are set to compare every contig, in both orientations, against every other; to require a minimum overlap between the contigs of 20 bases, and a maximum percent mismatch after alignment of 30%, using a word length of 4. If necessary, change these settings, either by typing the new values in the text windows or using the adjacent sliders.
5. To extend the contigs with the best poor-quality data at their ends, leave the "Yes" button of **Use hidden data** active. Leave **Window size for good data scan**, and **Max number of dashes in scan window** set to 50 and 5, respectively. This means that the data added to the ends of the contigs will not contain any segments of length 50 that include more than 5 unknowns ("-" or "N").
6. Click on **Mask active tags**. This means that segments of sequence covered by selected tag types will not be used to initiate matches (but will be used in any alignments initiated elsewhere.) So for example, a segment tagged as an Alu repeat will only be included in a potential overlap if there is also a good match to sequence outside one of its edges. By default, the tag types ALUS, REPT, and MASK will be masked. To alter this click on **Select tags** and a dialog will appear containing the names of all the tag types. Unset or set the required types by clicking on their buttons. Hit **OK** to dismiss the tag selector window.

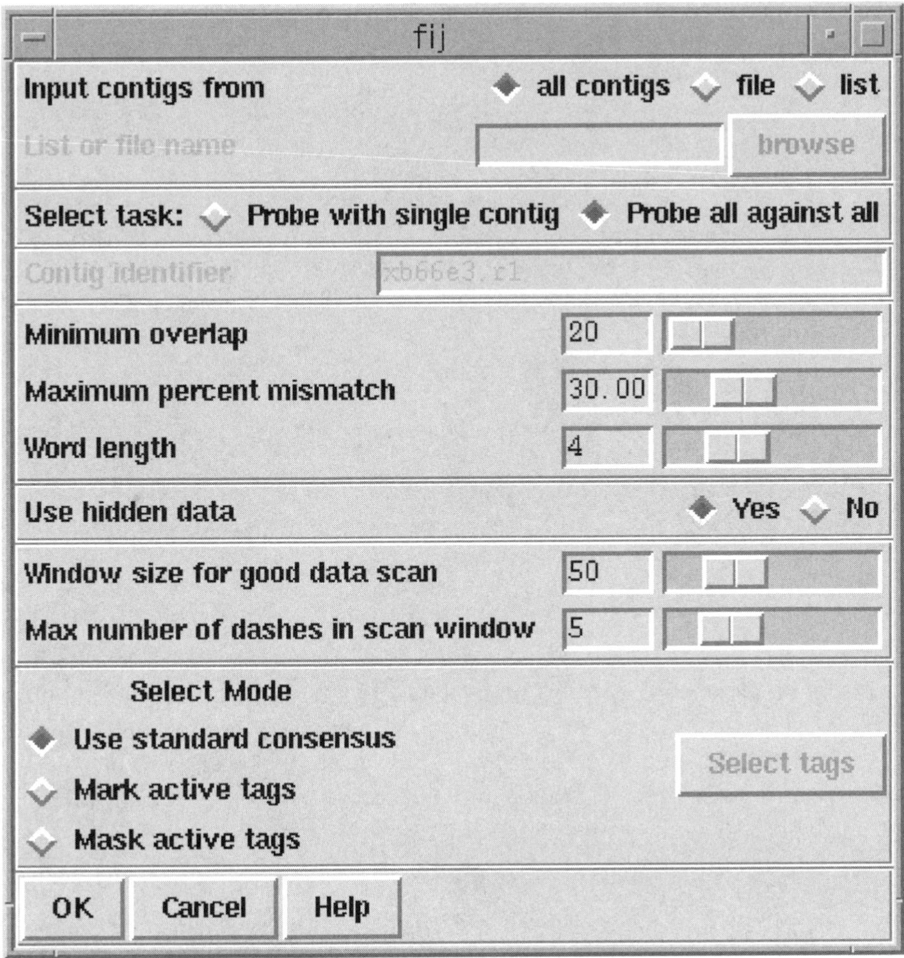

Fig. 6. The *GAP4* **Find Internal Joins** dialog.

7. Hit **OK** in the **Find Internal Joins** dialog and the search will commence. A busy cursor will appear while the search is performed, but it is usually very fast. If it has not already occurred, the *Contig Selector* window will transform itself into the *Contig Comparator* ready to receive any results produced. The results are plotted as described in **Subheading 2.3.2.** and the alignments from the overlaps are also written to the output window.
8. Click on the **Results** menu and drag down to the **Find Internal Joins** result. This reveals a pull-down menu. Drag down to the **Sort matches** button and release, and then repeat this with the **Use for Next** button. This causes the **Find Internal Joins** matches to be sorted into order on their percentage mismatch and to be passed to the **Next** command.

9. Click on the **Next** button which is at the top left of the *Contig Comparator* window. A *Join Editor* will appear containing the best alignment found from the **Find Internal Joins** search.
10. In the *Join Editor* window, click on the **Align** button. The two consensus sequences and their readings will be aligned using padding characters (*s).
11. Use the various movement methods to scroll along the full length of the overlap and inspect the traces for any disagreements (which is done by double-clicking on their sequence in the editor window).
12. Leave the Join Editor by clicking on the **Quit** button, and if satisfied that the join is genuine click on the **OK** button in the modal dialog that will appear. The plot will rearrange itself by joining the two contig lines and repositioning all the other results accordingly. This will also occur in any **Template Displays** currently showing the joined contigs.
13. Click on the **Next** button and repeat the process with the next best join. Continue this until all the results have been inspected. At any time, double clicking on any of the **Find Internal Joins** results plotted in the *Contig Comparator* will also bring up a *Join Editor*. The **Next** button is simply a shortcut to dealing with them in best-first order.

2.3.7. Further Notes on GAP4

Obviously *GAP4* contains many other functions and capabilities, such as those for breaking contigs or removing readings. It is worth noting that although here we have emphasized interactive operations using the program, most of *GAP4*'s functions can be used from scripts. A manual that explains how to write programs using the *GAP4* scripting language, written by James Bonfield, is available from our www site: `http://www.mrc-lmb.cam.ac.nk/pubseq`.

3. New Programs for Analyzing Finished Sequences

As stated in the **Subheading 1.**, we have started to build a new set of programs for analyzing finished sequences that benefit from the user interface we devised for *GAP4*. At the time of writing (May 1998) we regard *SIP4* (K. F. Beal, J. K. Bonfield, and R. Staden, unpubl.) as an excellent and well-featured sequence comparison program, but *NIP4* (K. F. Beal, J. K. Bonfield, and R. Staden, unpubl.), our nucleic acid investigation program, although having a good user interface and some useful functions, still needs work to reach the stage at which it can be a laboratory's primary program for this purpose. Our plan here, as we have done with *GAP4,* is to make it easy for others to add their algorithms to *NIP4*, and to provide routines for importing the results from external programs. Both programs have a powerful and easily used interface to the sequence libraries.

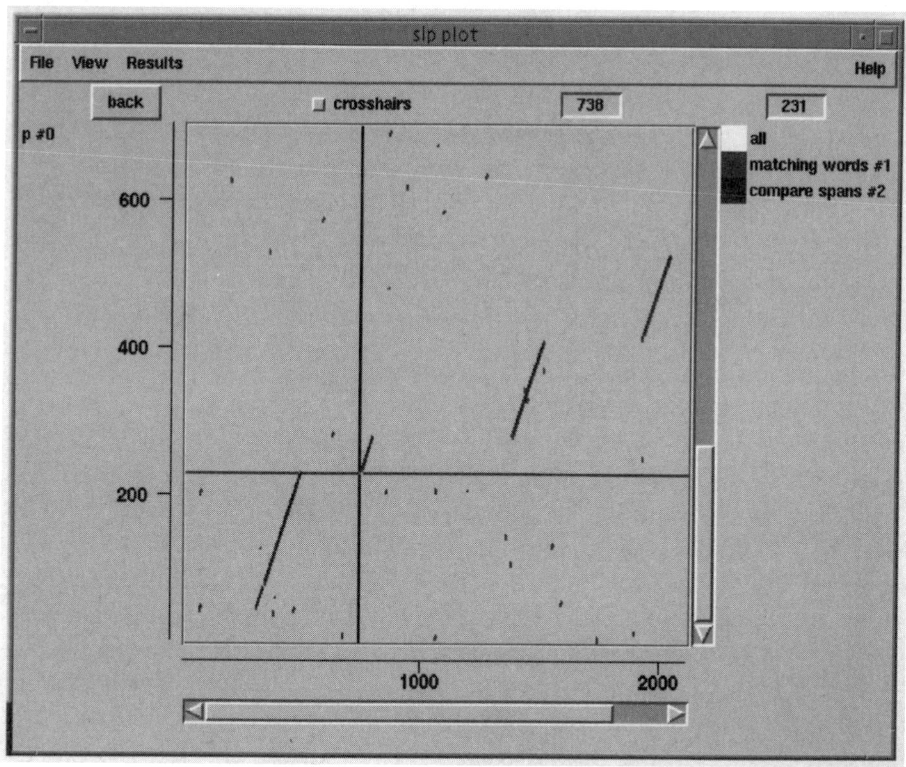

Fig. 7. The *SIP4 Graphics Display.*

3.1. The Sequence Comparison Program SIP4

SIP4 replaces the *SIP* program, which was published originally under the name *DIAGON* *(3)*, and is used for comparing pairs of sequences to find regions of similarity. It has a similar user interface to *GAP4,* but its main graphical window (*see* **Fig. 7**) is an interactive dot matrix display that is used for showing and analyzing the comparisons. The program contains several methods for comparing sequences, i.e., DNA against DNA, protein against protein, or DNA against protein, ranging from those that are extremely rapid but not very sensitive, to those that are slower but sensitive. All the comparison algorithms, including local *(11)* and global *(12)* alignment methods produce graphical results and alignments that can be viewed at the sequence level.

In **Fig. 7** we show a comparison between a region of genomic DNA and an mRNA derived from it. Any number of comparison methods or sets of thresholds can be applied and superimposed in a single plot, and each individual set

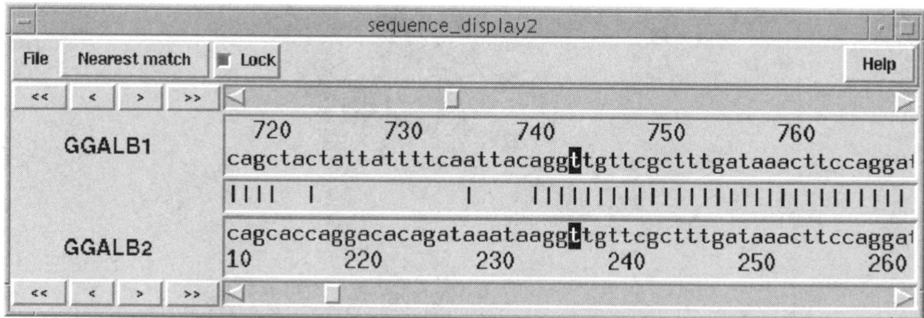

Fig. 8. The *SIP4 Sequence Display*.

of results will be shown in a different color. In addition, each set of results can be moved to its own display window, or selected ones superimposed. The plot can be zoomed and scrolled in both the x and y planes. The crosshair coordinates are displayed in the boxes at the top. For any set of results a scrolling sequence alignment window can be displayed as shown in **Fig. 8**. One sequence is shown in the top half and the other in the bottom. Each can be scrolled independently, or if the **Lock** button is set, they scroll in unison. Pressing the **nearest match** button will cause the sequences to be scrolled so that the nearest matching sequence block in the results list appears in the center of the window. In addition, the crosshair in the graphics window can be used to scroll the sequence display to enable matches to be examined in detail. A comparison of **Figs. 7** and **8** will reveal that the crosshair position and the sequence display are both positioned on an intron/exon boundary.

3.2. The Nucleic Acid Sequence Analysis Program NIP4

NIP4 replaces the old *NIP* program that was published originally under the name *ANALYSEQ (2)* and has an interface similar to *SIP4*. An example of its main graphics display is shown in **Fig 9**. The results from one of our gene-prediction methods that plots the likelihood of coding for each of the three reading frames, superimposed on plots of the stop codon positions are presented. Each result (e.g., the stop codons for one frame, or the likelihood of coding for one frame) is plotted in a unique color. Using the top left scale bars, the plots can be zoomed in the x and y planes. The data view can also be moved in x and y using the scroll bars around the plots. Individual results can be picked up and dropped in new locations to superimpose them, or can be put in separate windows. The crosshair is the light colored pair of lines in this figure and its coordinates are visible in the boxes at the top of the display. The darker vertical line bisecting the plots is the position of the cursor in the *Sequence Display* window, an example of which can be seen in **Fig. 10**. The *Sequence Display*

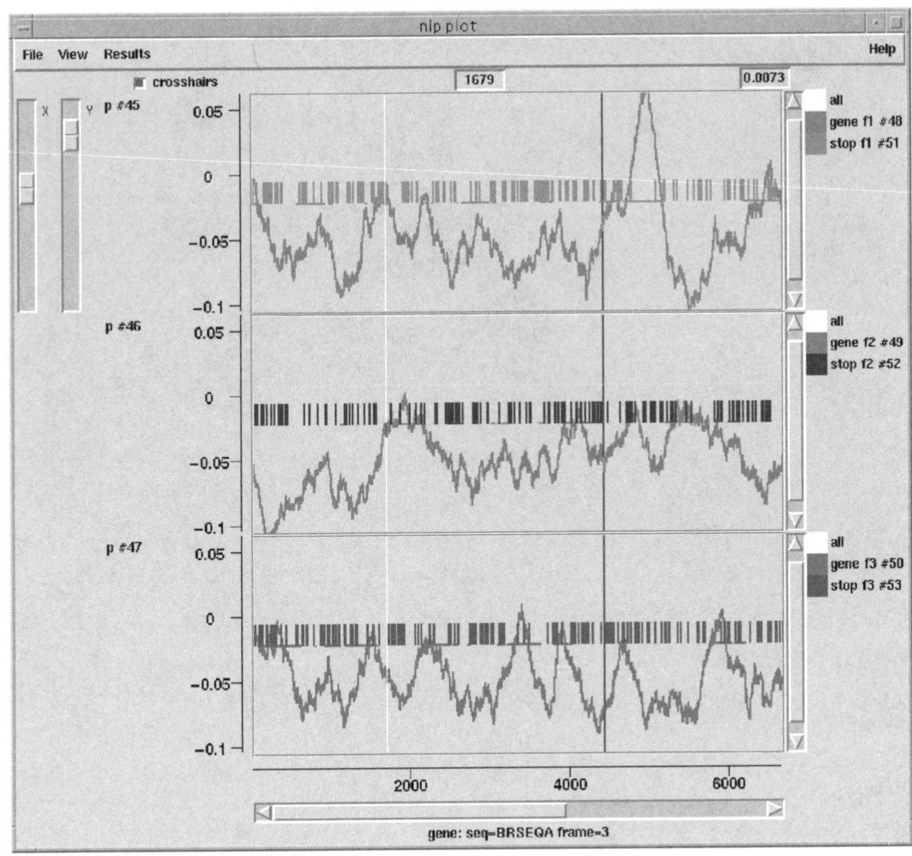

Fig. 9. The *NIP4 Graphics Display*.

can be scrolled using the cursor line in the graphics window, or by using its own internal methods. In the example presented, the display is showing both strands of the sequence, a three-frame translation, and the positions of restriction enzyme cutting sites. Each of these can be activated or switched off using the **Settings** menu. The **Search** button brings up a dialog to enable the user to perform a fuzzy search in either direction along the sequence, and the sequence can be scrolled to each match.

4. Obtaining the Package and Related Programs

Information about obtaining our package, which is free on UNIX to academics, can be obtained from our web site (`http://www.mrc-lmb.cam.ac.uk/pubseq`). To obtain the programs that can be used from within *GAP4*, e-mail the authors listed below requesting a copy of their program for use with

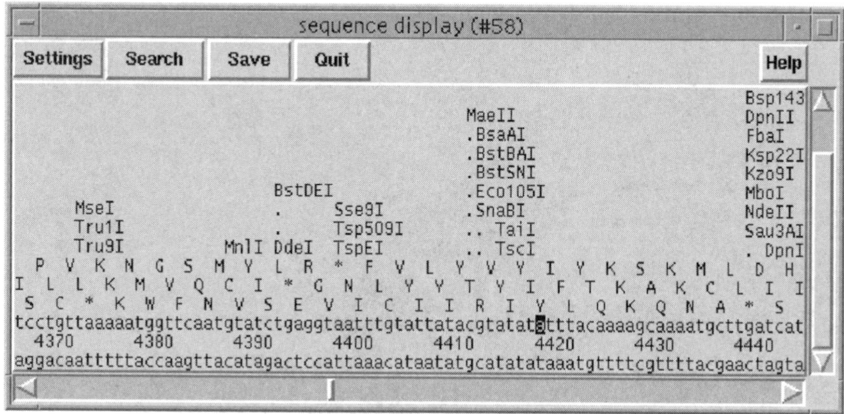

Fig. 10. The *NIP4 Sequence Display*.

our package. For *CAP2*: Xiaoqiu Huang, huang@cs.mtu.edu; for *FAKII:* Susan Miller, susanjo@cs.arizona.edu; for *PHRAP*: Phil Green, phg@u.washington.edu.

References

1. Bonfield, J. K., Smith, K. F., and Staden, R. (1995) A new DNA sequence assembly program. *Nucleic Acids Res.* **23,** 4992–4999.
2. Staden, R. (1984) Graphic methods to determine the function of nucleic acid sequences. *Nucleic Acids Res.* **12,** 521–538.
3. Staden, R. (1982) An interactive graphics program for comparing and aligning nucleic acid and amino acid sequences. *Nucleic Acids Res.* **10,** 2951–2961.
4. Staden, R. (1994), in Methods in Molecular Biology, vol. 25, *Computer Analysis of Sequence Data, Part II*. (Griffin, A. M. and Griffin, H. G., eds.) Humana Press, Totawa, NJ, pp. 9–170.
5. Huang, X. (1996). An improved sequence assembly program. *Genomics* **33,** 21–31.
6. Myers, E. W., Jain, M., and Larson, S. (1996) Internal report, University of Arizona.
7. Green, P. H. (1997) Pers. comm.
8. Dear, S. and Staden, R. (1992) A standard file format for data from DNA sequencing instruments. *DNA Sequence* **3,** 107–110.
9. Bonfield, J. K. and Staden, R. (1996) Experiment files and their application during large-scale sequencing projects. *DNA Sequence* **6,** 109–117.
10. Bonfield, J. K. and Staden, R. (1995) The application of numerical estimates of base calling accuracy to DNA sequencing projects. *Nucleic Acids Res.* **23,** 1406–1410.
11. Huang, X. Q. and Miller, W. (1991) A time-efficient, linear-space local similarity algorithm. *Adv. Appl. Math.* **12,** 337–357.
12. Huang, X. Q. (1994) On global sequence alignment. *Comp. Appl. Biosi.* (now *Bioinformatics*) **10,** 227–235.

8

Building a Multiuser Sequence Analysis Facility Using Freeware

Brian Fristensky

1. Introduction

Although many commercial packages exist for molecular sequence analysis, they are typically expensive. Whereas many Web-based applications are available for sequence analysis, the Web interface cannot store data on remote servers and is awkward to use. A good alternative is to build a sequence analysis facility on a local server. *BIRCH*, the *Biological Research Computer Hierarchy*, is an example of such a system (`http://home.cc.umanitoba.ca/~psgendb` and **ref. *1***). *BIRCH* is best thought of as a workbench containing tools for working with sequences, as well as software that minimizes the problems of putting tools together to perform a task. For example, in **Fig. 1**, several steps in phylogeny construction from an alignment were performed automatically. It is not possible to provide detailed instructions on installing all of the 300+ programs that currently reside in *BIRCH*. Rather, my purpose is to outline the strategies and tricks that make building and maintaining a sequence facility a smooth ongoing task.

2. Hardware, Software, and Know-How

2.1. Hardware and Operating System

BIRCH is currently implemented on a Sun workstation running Solaris 2.5. Since source code is available for most freeware programs used in BIRCH, it should be possible to recompile for other platforms. Almost all of the programs implemented in *BIRCH* have run under *LINUX,* and many have run on other UNIX platforms. If you are building *BIRCH* on an existing multiuser system,

From: *Methods in Molecular Biology, vol. 132: Bioinformatics Methods and Protocols*
Edited by: S. Misener and S. A. Krawetz © Humana Press Inc., Totowa, NJ

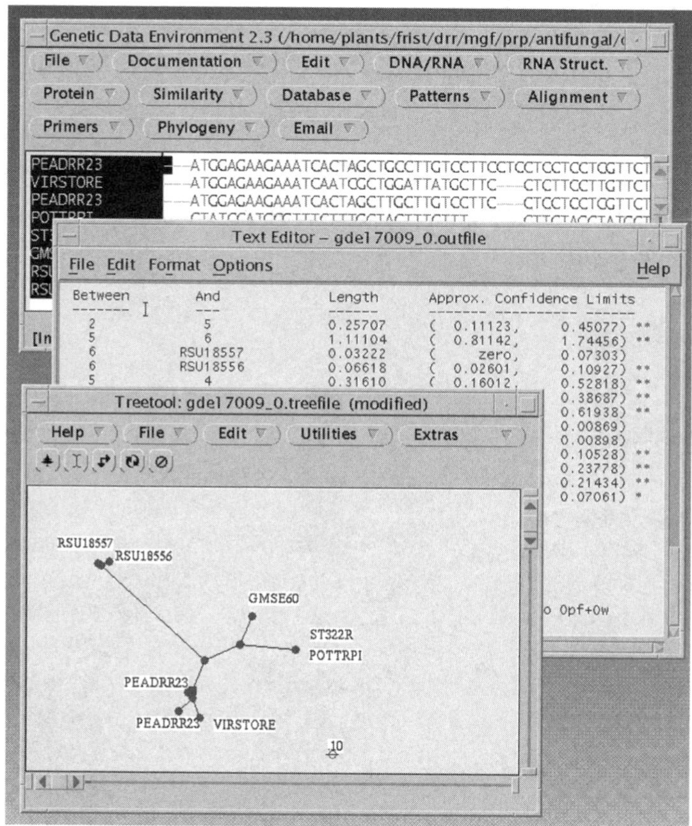

Fig. 1. Automated phylogeny. In *GDE* **(2)**, aligned sequences were selected, and *fastDNAml* **(3)** was called to produce a phylogeny. *GDE* automatically calls a text editor and *treetool* **(4)** to display results.

you will also need *X*-terminals or PC-based X-terminal emulators. When you log into your account, all programs run on the server, and *X-Windows* displays everything at the terminal. In comparison to a PC, where each machine has different software, data and hardware, you can log into the server from any terminal. *X-Windows* is one implementation of network computing, which is described in more detail in **ref. 5**.

2.2. Software

A C compiler, preferably *GNU C*, and a Web browser are necessary. *Netscape Communicator* is recommended because it comes with a visual HTML editor.

2.3. Know-How

You will need a working knowledge of UNIX *(6)*, some previous experience in programming (preferably C, C^{++}, or Java), and an ability to write HTML. Even if you don't know all of these things now, this is a good opportunity to gain these practical skills.

3. Setting up BIRCH

3.1. Create an Administrator's Account

Don't set up *BIRCH* on your own personal account. Get a separate account solely for this purpose. This keeps all of the *BIRCH* directories integrated as a separate unit. On your personal account, you are just another user, which is the only way to really test whether *any* user can run any program. As a corrolary, avoid doing preliminary installations on your personal account. Also, even if you have root privileges, avoid working as root except for systems level (i.e., non-*BIRCH*) activities.

3.2. Create a Directory Hierarchy

The directories needed to construct *BIRCH,* and their current sizes at our site, are as follows:

GenBank: *GenBank* DNA nucleic acid database (5.9 gb, release 106, April 1998).
PIR: 183 mb, release 55, February 1998.
admin: for administrative files and scripts. (0.4 mb).
bin: for executable files (60 mb).
dat: special data files for programs (e.g., scoring matricies) (5 mb).
doc: documentation files (18 mb).
install: working directory for software installation.
java: for Java classes.
manl: documentation in UNIX manual page format.
ncbi: directories for National Center for Biotechnology Information (NCBI) client/server programs (1.7 mb).
public_html: directory for the *BIRCH* web site (2.5 mb).

This directory hierarchy can be downloaded from `http://home.cc.umanitoba.ca/~psgendb/build/build.html`. Initially, all these directories contain is a series of shell scripts and datafiles for managing *BIRCH*. To ensure that the most recent versions of software are installed, it is best to download each program or package as you build your local *BIRCH* site.

These directories should be installed in *BIRCH* administrator's **$HOME** directory. On our system, the *BIRCH* **$HOME** directory is **/home/psgendb**.

We have set an environment variable, **$db**, to store this path. That is, when interpreting a command, the shell will replace **'$db'** with **/home/psgendb**. The administration directory **/home/psgendb/admin** can therefore be typed as **'$db/admin'** by any user on the system.

Whenever you install a program package, create specific subdirectories for that package in doc and dat, for their documentation and data (if any) respectively. In fact, it is a good working rule that when you start a new project of any kind, always create a directory specifically for that task, even if it is only temporary.

There are several important rules for these directories:

1. All directories and files, including the *BIRCH* **$HOME** directory, must be world-readable.
2. All directories must be world executable.
3. All programs must be world executable.

3.3. Configure the Administration Files

At our site, *BIRCH* contains over 300 programs and two major databases. The programs were written by different authors in different languages on different platforms using different file formats. In many cases they need to know the locations of datafiles, databases, or runtime libraries. If these things had to be set for each user, and changed by each user every time a new program was installed or updated, nothing would ever work. Fortunately, there is a clean solution to all these problems. All settings are read from **$db/admin**. Never deviate from this rule! The user should never have to configure his/her account for anything. (At the University of Manitoba, *BIRCH* has over 140 users. Imagine having to change settings for each user!)

Two files contain settings needed by all programs in *BIRCH*. **$db/admin/login.source** contains commands to be executed each time a user logs in. The most important command adds **$db/bin** to the user's **$PATH** environment variable. When the shell reads a command, the first nonblank string is interpreted as the name of a command. The shell searches for an executable file in every directory listed in **$PATH**. Thus, if all *BIRCH* programs are in **$db/bin,** all we have to do is to add **$db/bin** to **$PATH,** and the user can run any program. **login.source** also contains a command to print the contents of **$db/admin/Login_Message**, a file containing short announcements of interest to *BIRCH* users.

$db/admin/cshrc.source contains commands that need to be executed every time a new shell is started, e.g., when a window is opened, or a program is run.

Virtually all program settings are defined here. Most of this file contains commands to set environment variables. For example,

```
# Environment variables for sequence work.
# Upper and lowercase are supported.
setenv DB       /home/psgendb
setenv db       $DB
setenv DATA     $DB/dat
setenv data     $DATA
setenv DAT      $DATA
setenv dat      $DATA
setenv GENBANK  $DB/GenBank
setenv gb       $GENBANK
```

Here is where we define **$db,** and then use it to build other environment variables telling where data files are stored.

Each program or package may have specific settings as well. For example, the NCBI programs are configured as follows:

```
#NCBI
setenv NCBI $db/ncbi
alias entrez Nentrez
```

`setenv` tells the NCBI programs where to find necessary directories. The alias line tells the shell that when a user types "`entrez`", the network version of *entrez* (*Nentrez*) should be run.

To use *BIRCH,* the user must run **$db/admin/newuser**. This script adds a line reading

```
source /home/psgendb/admin/login.source
```

to the user's .login file, and

```
source /home/psgendb/admin/cshrc.source
```

to the user's **.cshrc** file. These two lines cause all commands in the **.source** files to be executed when the user logs in or starts a new shell, respectively. In this way, any change or addition to the .source files in **$db/admin** will automatically take effect for every user. The *BIRCH* administrator should never have to do anything to a user's account. This has worked very well, in practice.

login.source and **cshrc.source** will need to be modified to reflect local directory structures and installed software. For example, the **$db** environment variable will have to be changed to your local *BIRCH* **$HOME** directory. In **login.source** and **cshrc.source**, it is best to comment out all lines that refer to programs or databases that are not yet installed. These lines can be uncommented as *BIRCH* grows.

3.4. Create a Web Site

Consider the *BIRCH* web site to be your conceptual model of what you are building. Yes, it is also there to tell the user how to use the system and what is available, but the complexity of *BIRCH* demands a well-structured road map. Always have a copy of *Netscape* running on your screen, so that you can create web pages and modify them as you go. Because *BIRCH* is already documented on the University of Manitoba Web site *(1)*, you can often shortcut by downloading web pages and modifying them to meet your needs. The instructions in the following section assume that a web page exists called **programs.html** (*see* `http://home.cc.umanitoba.ca/~psgendb/programs.html`). This page contains links to all documentation, organized by category.

4. Building *BIRCH*

This section describes the overall process for installing several software packages, each chosen to illustrate some of the subtle problems that can be associated with getting programs to work for distributed users. The goal of this section is to provide a short path to getting a reasonably comprehensive suite of programs working quickly. This core of programs serves as the foundation for building a facility tailored to the needs of your local user base. For brevity, URLs from which programs can be downloaded are included in the references. Programs will usually include instructions for installation that are more detailed than what I can present here.

Whereas it is best to install programs on the administration account and test them on your personal account, it would be inconvenient to keep going back and forth between accounts. There are two ways around that problem. The ideal solution would be to have two *X*-terminals side by side, each logged into a different account. Because that is not always possible, the next best thing is to run an *X-Windows* session on your personal account, but log into the administration account in one or more windows. For example open up a command window (e.g., terminal window in *CDE*) and log into your administration account using *telnet*. For simple tasks, keep one or more *telnet* sessions logged into the administration account, one for each working directory. To get X11 programs to run on the administration account, but display on your personal account see the script **$db/admin/xdisplay**. It will have to be modified for your own site. If you are using the CDE desktop manager, it may prove less confusing to keep all windows from your personal account on one screen, and all windows from your administration account on a separate screen. Also, the *BIRCH* `newuser` script causes your UNIX prompt to display both the server name and the current directory, which should help you keep track of which window belongs to which account.

4.1. Install readseq *(7)*

The biggest single problem with sequence software is the plethora of file formats that must be used. *readseq* is a program that converts one format (e.g., *GenBank*) to another (e.g., GCG). The *readseq* source code and documentation are downloaded as a shell archive file, **readseq.shar**. To recreate the files in the archive type `sh readseq.shar`. You can compile readseq for your platform by typing `make`, which will create the executable file **readseq**. Make this file world readable

```
chmod a+rx readseq
```

and move it to the bin directory

```
mv readseq $db/bin
```

Also, create a directory to hold the help file

```
mkdir $doc/readseq
chmod a+rx $doc/readseq
```

and move the help file to this directory

```
mv readseq.help $doc/readseq/readseq.asc
```

Normally, I prefer not to rename files from other packages. However, because Web browsers often vary with regard to how they handle different file extensions, it is preferable to have a uniform file extension for all *ASCII* files. I use ".asc" for *ASCII* files. Finally, add a link for **readseq.asc** to **programs.html**.

Read the documentation for readseq and test the program on your personal account. For example, if you have a GenBank file called **PEADRRA.gen**, typing

```
readseq -p -oPEADRRA.wrp -fPearson <PEADRRA.gen
```

will create a file in Pearson/*FASTA* format called **PEADRRA.wrp**. *readseq* was originally developed under VMS, so the **"-p"** switch is necessary to pipe input to the program using the UNIX input redirection character **"<"**.

4.2. Install FSAP *(8,9)*

Many programs come in packages. The *FSAP* package includes programs for many common sequence tasks (e.g., printing sequences, translation, restriction site searches) all of which are run through interactive text-based menus. In this case, the package can be downloaded as a .tar archive, **fsap.tar.Z**. To recreate the directory hierarchy for *FSAP,* first uncompress the file

```
uncompress fsap.tar.Z
```

And then type

```
tar xvf fsap.tar
```

to create the **fsap** directory. If you type `ls -l` in the `fsap` directory, you should see the following:

```
drwx——  2 frist drr   512 Jun 4 1996 GDE/
-rw———  1 frist drr  7041 May 3 1996 INSTALL.doc
-rw———  1 frist drr   970 May 3 1996 RELEASE.NOTES
drwx——  2 frist drr   512 May 3 1996 bin/
drwx——  2 frist drr   512 May 3 1996 dat/
drwx——  2 frist drr   512 Mar 6 18:56 doc/
drwx——  4 frist drr   512 May 3 1996 src/
drwx——  2 frist drr   512 May 3 1996 src.c/
drwx——  2 frist drr  1024 May 3 1996 test/
```

INSTALL.doc contains step-by-step installation instructions. **src.c** contains C source code, which generates executable code. **doc** and **dat** contain, respectively, documentation and datafiles used by the programs. *Genetic Data Environment* (*GDE*) contains menu items and c-shell scripts that make it possible to run these programs through *GDE*. **test** is a directory in which you can run a script that will test all the programs to make sure that they function on your system. Many packages will have test scripts. When you have successfully tested the programs, installation is easy. Copy the contents of **fsap/bin** to **$db/bin**, **fsap/dat** to **$dat/fsap/dat**, and **doc** to **$doc/fsap/doc**. Make sure to add links for these documentation files in **programs.html**.

Again, log into your personal account and try out these programs. The first one to try is **numseq**, as described in **$doc/fsap/numseq.asc**. Any *GenBank* flat file will suffice for testing this program.

4.3. *Install* FASTA *(10)*

The *FASTA* package provides programs both for pairwise and database sequence comparisons. Compilation is done using the UNIX **'make'** command, and installation is as simple as copying executable files to **$db/bin**. This should be one of the easiest packages to install. One twist, though, is that the documentation is in the UNIX manual page format. *BIRCH* has a directory for manual pages called **$db/manl**. All files in this directory should be in the form 'name.l' (where 'l' stands for local). In **login.source**, the line

```
setenv MANPATH $MANPATH\:$DB
```

tells UNIX to look for manual pages in this directory, as well as in any other directory specified in the system's **$MANPATH**.

For example, to read the documentation for *align,* the user types `'man align'`, and the file **$db/manl/align.l** will be displayed.

It is also useful to create ASCII files from these manual pages for display by the web browser. To create an ASCII file for **align.l**, type `man align > $doc/fasta/align.asc`. Remember to make this file world readable, and create a link in **programs.html**.

4.4. *Install* GDE *(2)*

GDE, is a program that runs other programs. As illustrated in **Fig. 1,** GDE combines a multiple sequence alignment editor with a set of pull-down menus. For example, the **Similarity** menu contains calls to most of the *FASTA* similarity programs. The thing that makes *GDE* unique is its ability to have menus and menu items added with no reprogramming or recompiling. When *GDE* is launched, a file called **$GDE_HELP_DIR/.GDEmenus** is read, specifying the contents of each menu, and the commands to be executed to run each program. For example, the *lfasta* menu is shown in **Fig. 2**.

In **.GDEmenus**, the entry to create this menu begins like this:

```
#——————————— LFASTA ( 7/26/95) ——————-
item:LFASTA - Fast local alignment
itemmethod:(sed "s/[#%]/>/"<in1 >in1.tmp; readseq
in1.tmp -i1 -f8 > in1.seq1; readseq in1.tmp -i2
-f8 >in1.seq2; lfasta -w $RESPERLINE $MARKX -d
$NUMOFALN $MATRIX in1.seq1 in1.seq2 $KTUP >
in1.out;
fastaout.csh $MARKX in1.out; rm in1*) &
itemhelp: FASTA/fasta.asc
```

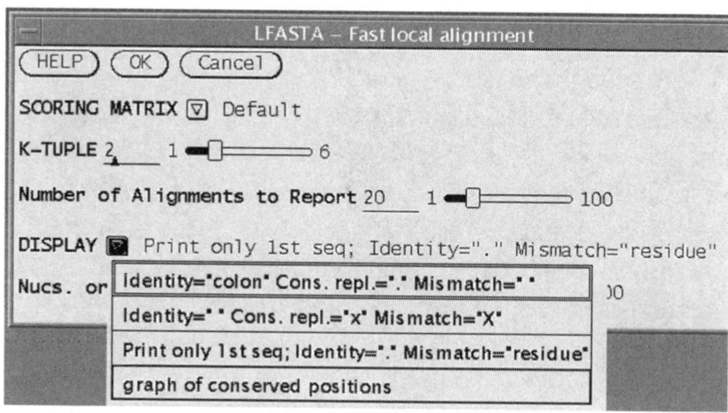

Fig. 2. Example of a *GDE* menu, illustrating pull down menus and sliders. This menu displays when **lfasta** is chosen from the **Similarity** menu.

The most important line is the `itemmethod`, which contains a string of commands to be run. For example, *readseq* is called to convert the selected sequence to *FASTA* format, and arguments are inserted into the command, each preceeded by a '$'. Each argument to *lfasta* can be specified in a few lines, such as that for the pull-down menu **DISPLAY**, shown in **Fig. 2**:

```
arg:MARKX
arglabel:DISPLAY
argtype:choice_menu
argchoice:Identity="colon" Cons. repl.="." Mismatch=" ":-m 0
argchoice:Identity=" " Cons. repl.="x" Mismatch="X":-m 1
argchoice:Print only 1st seq; Identity="." Mismatch="residue":-m 2
argchoice:graph of conserved positions:-m 4
argvalue:0
```

Whereas it is easiest to get *GDE* to run programs such as *readseq* that take all information from the command line, even interactive programs requiring user input can be called by *GDE*. For example, to run *numseq, GDE* sends the parameters set in the menu to a script called **numseq.csh**. **numseq.csh** reads the parameters and generates keystrokes that would normally be typed by the user in response to prompts by *numseq*. The ease with which new programs can thus be added to *GDE*'s menus makes *GDE* the foundation from which most of *BIRCH* is run. A **.GDEmenus** file and accompanying shell scripts necessary to run most of the programs in *BIRCH* can be downloaded from **ref. *1***.

4.5. Install NENTREZ *(11),* SEQUIN *(12),* BLASTCLI *(13), and* Cn3D *(14)*

The NCBI suite consists of networked client/server applications. *Nentrez* is a client that runs on your desktop, allowing text searches and sequence retrieval from the NCBI server. Its helper application, *Cn3D*, can download and display three-dimensional protein structures from structural databases at NCBI. *sequin* automates the process of annotating new sequences and submiting them to *GenBank*. *sequin* can also download sequences from NCBI for resubmission as updates. *blastcli* is a local client that submits sequences to the NCBI *BLAST* server.

Because these programs share a common directory containing configuration and datafiles, it makes sense to install them all at once. Generally, installation is as simple as copying the executables to **$db/bin** and running *netentcf*, the network client configuration program. All files and directories for these programs should be in a directory specified by **$NCBI** in **$db/admin/cshrc.source**. The first time you run *Nentrez*, a file called **$HOME/.ncbirc** will be created, containing configuration information. If you move this file to **$NCBI**, it will work for all users.

The workspace menu (**Fig. 3**) can be a valuable means of making it easy for users to know which programs are available on the system. All UNIX window managers have a configureable Workspace menu. The *CDE* manager is available on most UNIX platforms and is now the default on many. Therefore, I have created a **$db/.dt/dtwmrc** file to configure the *CDE* workspace menu for all *BIRCH* users. Programs are organized into submenus (e.g., Word Processing, Statistics, Molecular Biology). To add *sequin* to the menu,

```
"Sequin - submit seq. to GenBanl" f.exec /home/psgendb/bin/sequin
```

must be in **dtwmrc**. The setup script **$db/bin/menusetup** replaces the user's **dtwmrc** file with a symbolic link to **$db/.dt/dtwmrc**. Thus, as the *BIRCH* administrator updates this file, all users get the new menu.

GDE is conspicuously absent from this menu. This is a deliberate omission. If launched from the workspace menu, *GDE* will default to the **$HOME** directory for reading and writing files.

When working with *GDE,* it is best to create a separate directory for each project, e.g., phylogenetic analysis of a multigene family. If you **cd** to that directory and launch *GDE* from the command line, all file input and output

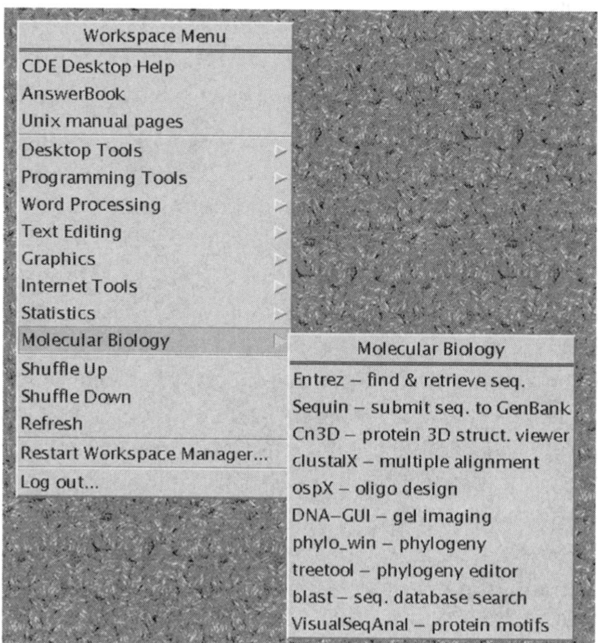

Fig. 3. Customized *CDE* workspace menu. Submenus, organized by category, make it easy for users to find and launch programs.

(including the creation of temporary files and directories) will occur in this directory.

4.6. Install XYLEM (15)

XYLEM is a set of tools for local database management. Although designed originally for creating subsets of *GenBank* or *PIR* files for projects such as phylogenetic studies, *XYLEM* can also be used for keyword searches and retrieving entries from these databases. **features** automates the process of extracting *GenBank* features (e.g., exon, intron, mRNA, CDS) from large sets of entries.

Installation for *XYLEM* is almost identical to installation for *FSAP*. The next section assumes that *XYLEM* is installed on your system.

4.7. Install GenBank and PIR

With programs like *Nentrez* and *blastcli,* it may not be necessary to have local versions of sequence databases. Since *GenBank* release 106 required 6 gb of space, this is an important consideration. However, a local copy of the database is useful for several reasons. First, networked *BLAST* programs do not give you the option of limiting your searches to specific *GenBank* divisions. Local database search programs such as *FASTA* can be tailored for specific needs. Finally *Nentrez* does not allow input of a group of accession numbers for retrieval of groups of sequences.

Automated downloading and formatting of *GenBank* and *PIR* is done through shell scripts called **gbupdate** and **pirupdate**. For example, the names of *GenBank* files to be downloaded are found in **$gb/master.filelist**. To begin after peak hours (e.g. after 7:00 P.M.) use the **at** command:

```
at 7pm
at>nice gbupdate master.filelist
at><ctrl>-D
```

The UNIX **nice** command runs the job at a low priority so that real time tasks (e.g., moving windows around the screen) will not be slowed down. **gbupdate** downloads each file, verifies that the downloaded file is the same size as the original, uncompresses the file, and for *GenBank* sequence files, runs *splitdb* ***(15)*** to split each *GenBank* division into annotation, sequence, and an index. The sequence files are written in *FASTA* format. Separating sequences and annotation into separate files speeds both *FASTA* searches of sequence and **findkey** searches of the annotation ***(15)***. *fetch* can retrieve sets of *GenBank* entries, rejoining annotation and sequence to recreate the original entries ***(15)***. Larger *GenBank* divisions are now split among several files. The EST division was split among 22 files, gbest1-gbest22 in release 106. **master.filelist** needs to be updated with each download to reflect these changes. However, the **fetch** and **findkey** automatically detect when divisions are split.

5. Training Users

Whereas it is vital to keep the documentation for *BIRCH* consistent and complete, human nature is such that generally people do not read it. Hands-on training sessions can be of great value to the user community, both in terms of teaching people how to use *BIRCH*, as well as in creation of a core of trained users who can help others.

Because most *BIRCH* users are also new to UNIX, it may seem a daunting prospect to cover both areas. Nonetheless, each year during the lab component of my course, Introductory Cytogenetics ***(16),*** students with no previous UNIX or bioinformatics background have learned enough to complete a simple sequence project over two hands-on sessions. At the end of the second session, each student is given a 300- to 400-bp unknown sequence, derived from the protein-coding sequence of a *GenBank* entry. Each student must be able to identify the parent sequence using *FASTA*, retrieve the parent sequence, identify the coding sequence from which the unknown was derived, print the entire coding sequence, with translation in the correct reading frame.

The sessions are run as follows:

1. **On screen demo** (30 min.): Starting with a demo gives students an idea of what things should look like. Using an *X*-terminal connected to a 1024 x 768 projector, I briefly explain how *X-Windows* works, and the basics of the *CDE* desktop. I also demonstrate examples of sequence analysis, using both command-line programs and programs run through *GDE.*
2. **Hands-on demo of UNIX, *CDE*, and simple sequence tasks** (2 h): Demos proceed step by step, making sure that everyone has successfully completed each step before the class moves on. It is valuable to have an assistant to help students when they encounter difficulties. First students run setup scripts **newuser** and **menusetup** found in **$db/admin**. Then students are introduced to the fundamentals of *CDE,* and use of a Web browser for reading documentation. Next, students learn about working with sequences by running *numseq* from the command line. *numseq* can be used to illustrate the ramifications of working with either one or both strands of a DNA sequence, the differences between linear and circular molecules, and how to translate sequences. Students then launch *GDE* and try repeating some of the same tasks, running *numseq* from a *GDE* menu. (Although graphic interfaces are good, it is still best to give people some exposure to the command line. Doing so provides important insights into what it is that the computer actually does.) The session ends with a quick discussion of *GenBank*. Students learn to search for sequence by keywords and to retrieve them.
3. **Similarity searches and Databases** (2 h): The second session opens with a short discussion of the theory behind both dot-matrix ***(9)*** and global ***(17,10)*** similarity searches. The concepts of look-up tables and optimal alignments are emphasized. Using *GDE,* students do pairwise comparisons of several related sequences, using *d4hom* ***(9)*** for dot-matrix searches and *align* ***(10)*** for global alignment. Finally, students run *fasta* to search for a DNA sequence in *GenBank*.

References

1. Fristensky, B. *BIRCH.* `http://home.cc.umanitoba.ca/~psgendb`.
2. Smith, S., Overbeek, R., Woese, C. R., Gilbert, W., and Gillevet, P. M. (1994) The tenetic data environment: an expandable GUI for multiple sequence analysis. *Comp. Appl. Biosci.* (now *Bioinormatics*) **10,** 671–675. `ftp://megasun.bch.umontreal.ca/pub/gde/`.
3. Olsen, G. J., Matsuda, H., Hagstrom, R., and Overbeek, R. (1994) FastDNAml: a tool for construction of phylogenetic trees of DNA sequences using maximum likelihood. *Comput. Applic. Biosci.* (now *Bioinformatics*) **10,** 41–48.
4. Maciukenas, M. (1994) Treetool. `ftp://rdp.life.uiuc.edu/rdp/programs/TreeTool/`.
5. Fristensky, B. (1999) Network computing: restructuring how scientists use computers, and what we get out of them, in *Methods in Molecular Biology* vol. 132.

Bioinformatics Methods and Protocols (Misener, S. and Krawetz, S. eds.), Chapter 22. Humana Press, Totowa, NJ.

6. Sobell, M. G. (1995) *A Practical Guide to the UNIX System.* Addison-Wesley Publishers.
7. Gilbert, D. (1993) `http://iubio.bio.indiana.edu/soft/molbio/readseq/.`
8. Fristensky, B., Lis, J. T., and Wu, R. (1982) Portable microcomputer software for nucleotide sequence analysis. *Nucl. Acids Res.* **10,** 6451–6463. `http://home.cc.umanitoba.ca/~psgendb/FSAP.html.`
9. Fristensky, B. (1986) Improving the efficiency of dot-matrix similarity searches through use of an oligomer table. *Nucl. Acids Res.* **14,** 597–610.
10. Pearson, W. R. (1990) Rapid and sensitive sequence comparison with FASTP and FASTA. *Meth. Enzymol.* **183,** 63–98. `ftp://ftp.virginia.edu/pub/fasta/.`
11. NCBI. *Nentrez.* `http://www.ncbi.nlm.nih.gov/Entrez/Network/nentrez.overview.html.`
12. NCBI. *Sequin.* `http://www.ncbi.nlm.nih.gov/Sequin/index.html.`
13. NCBI. *Blast client.* `ftp://ncbi.nlm.nih.gov/blast/network/.`
14. NCBI. *Cn3D.* `http://www.ncbi.nlm.nih.gov/Structure/cn3d.html.`
15. Fristensky, B. (1993) Feature expressions: creating and manipulating sequence datasets. *Nucl. Acids Res.* **21,** 5997–6003. `http://home.cc.umanitoba.ca/~psgendb/XYLEM.html.`
16. Fristensky, B. *Introductory Cytogenetics*, University of Manitoba. `http://www.umanitoba.ca/afs/plant_science/COURSES/CYTO/.`
17. Needleman, S. and Wunsch, C. D. (1970) A general method applicable to the search for similarities in the amino acid sequence of two proteins. *J. Mol. Biol.* **48,** 443–453.

2

Molecular Biology Software

9

Free Software in Molecular Biology for Macintosh and MS Windows Computers

Don Gilbert

1. Introduction

You will find a large collection of free applications for molecular biology and chemistry through Internet servers and from other sources. Most of these are written by biologists, chemists, and software developers, sometimes as part of a university- or government-funded project, sometimes as an unfunded project. Some of these offer a single function useful to you, one that may be found nowhere else.

What is free software for bioscientists, and especially where can you get it? How do you set it up and use it? What are some good programs, and what is available? These are some of the questions you can find answers to in this review, which concentrates on software for molecular biology for the commonly used *Macintosh* and *Microsoft* (Wintel) operating systems.

What you will not find covered here are the many free programs that now run as an Internet service through your Web browser. Neither are programs available mainly as source code (not ready to run) or UNIX programs, nor the choices you have with commercial software, nor software for related areas of chemistry, medicine, population biology, ecology, and others.

1.1. Free or Commercial Software?

Free data-analysis software is common in the sciences, as the scientists in need of new analyses develop algorithms for it, then crystallize the algorithms as software. Most of the basic biosequence analyses are scientist-developed, including *FASTA, BLAST, Clustal, MFOLD, PHYLIP, Paup, CAP,* to name a

From: *Methods in Molecular Biology, vol. 132: Bioinformatics Methods and Protocols*
Edited by: S. Misener and S. A. Krawetz © Humana Press Inc., Totowa, NJ

few. The source code of these is often shared freely. But often these programs lack ease of use and integration with other functions. Commercial software developers have incorporated such algorithms, along with their own, and added a much greater usability and integration, to allow you to analyze your data without spending a lot of time learning how to run the programs.

Besides adding integration and user interface, companies add the great value of good documentation and telephone and other support for their wares. When your funds permit, and a commercial package does what you need, it is usually a better choice than free software. Given market-place realities, companies charge you what seem like large fees, but these fees are needed to cover the costs of advertising, technical support, software developers, and so forth.

For scientists with limited budgets, students, and teachers, free software is often the only choice. You can also find unique programs that do things no commercial package does. Another great advantage of free software is that it often includes source code you can use to modify and extend an analysis.

Today with the rapid growth in the fields of bioinformatics and biocomputing, more good programmers are developing software, many making it freely available. You will find more sophistication in attention to user interfaces which eases your learning to use the software. Still there is no common means for funding development of free software: Government agencies do not generally fund projects from individual research/programmers, the main source of many of the free software packages. The shareware concept of users paying for software they use has never worked well. For most scientific software, the potential market is so small that we see a strong distinction between expensive commercial packages and free software.

Usability of free programs is variable, and depends in part on your computer and needs. Be aware that older programs may fail to work on new computers. Tolerance for program flaws and limitations is generally needed, and self-reliance in learning how they work. Often the authors either cannot be contacted, or do not have time to spare for supporting a product that they freely distribute.

1.2. Internet Sources

Access to and use of the Internet has become so ubiquitous in our society, especially among scientists, that it is assumed you have or can get such access. This was not true a few years ago, but since the 1980s, free software for biosciences has been available most widely through the Internet, because distribution that way is free (from cost and time) to software authors. There are a few widely used archives of biosciences software, and today many authors prefer to provide their own network server to distribute software, as that is very easy to do. Archives that collect such software still play an important role in providing these collections, and in archiving them for years past when an author may be able to.

Two commonly used archives of molecular biology software are *IUBio Archive,* at Indiana University, and the European Bioinformatics Institute (*EBI*) software archive. Internet resource locators (URLs) for these are:

EBI at `http://www.ebi.ac.uk/` or `ftp://ftp.ebi.ac.uk/`

The European Bioinformatics Institute is home to EMBL databank and others, and home to a large molecular biology software archive, including the very useful Bio Catalog of software (`http://www.ebi.ac.uk/biocat/biocat.html`).

IUBio at `http` or `ftp://iubio.bio.indiana.edu/`

IUBio Archive is home to a large collection of biology software, and also provides services for keyword searching of current *GenBank, SwissProt,* and *PIR* databanks, and the useful *Bionet* network news archive. It has been operating since 1989, as a user- and author-supported self-service archive. The molecular biology software collection at *IUBio* is also mirrored (copied) to sites around the world, including Finland, Sweden, Japan, United Kingdom, France, Spain, and Israel. See the following software listings section for URLs to these.

A few of many other Internet servers of note include:

`http://www.ncbi.nlm.nih.gov/`—home site for up-to-date access to *Entrez* software (databank lookups), *Sequin* for publishing your sequence in *GenBank, Macaw* (multiple alignment), and others.

`http://kiwi.imgen.bcm.tmc.edu:8088/search-launcher/launcher.html`—*Search Launcher* at Baylor College of Medicine, home to a selection of sequence analysis methods. Recommended is their *search launcher* program (a *Perl* script that is usable on Macintosh and Wintel systems) for batch access to this valuable resource.

`http://expasy.hcuge.ch/`—home of *SwissProt, PROSITE,* and home to a collection of useful database analyses functions, especially for protein data.

1.3. Fetching Software

File transfer protocol (ftp) is an Internet service designed specifically for file transfer. It is an antecedent of the popular hypertext transport protocol (http), or Web, Internet method. Generally ftp is still a better method for transferring large files such as software packages, than is a Web browser. One popular and easy to use FTP program for MacOS is the *Fetch* program by Jim Matthews of Dartmouth College. You can find more information at `http://www.dartmouth.edu/netsoftware/fetch.html`. The ubiquitous

Netscape or other Web browser also lets you fetch software either via http or ftp methods.

Generally software is stored and transferred to you in encoded formats from archives. Most current fetching software, like *Fetch* and *Netscape,* will decode this automatically for you. However, other software sometimes is used or needed for decoding. General Internet software packages and servers will get you started on what is needed. If in doubt, your local library or bookstore will have a few shelves full of books on using Internet services and general software archives.

The main trick in fetching software these days is knowing where to look. Besides this review, you can make use of Web links made by others, perhaps with better perspective than this author. See also the common Internet search services, like *Yahoo* (`http://www.yahoo.com/`), *AltaVista* (`http://www.altavista.digital.com`), *Lycos* (`http://www.lycos.com/`), or others.

Keep in mind that software is updated; if the version you try fails in some way, a newer version may be available now or soon. The author's preferred or home server is the best place to check for updates, as archive servers do not always have the current release.

1.4. Setup and Use

Once on your computer, the steps of installing and configuring free software range from easy to trying. Instructions are usually included, but are not always detailed enough, or cover the range of problems one can run into. Installing software is common with commercial software, but not with the free variety.

Report problems you have to the author, and suggestions for improving the software. Asking the frequently overworked authors for help with installation is not always a good solution though, so attention to their prewritten installation directions is essential.

Special kinds of software, especially those written for Java or Perl, will require that you also fetch and install other free, general software to use. In the case of Java, this is becoming less of a problem as new MacOS and UNIX systems are shipping with a Java runtime as part of the system.

1.5. Multiplatform Software

By 1998 counts of Internet browser contacts at *IUBio* and other biology Web servers, from 30 to 50% of biologists use Macintosh computers, 40 to 70% use Wintel systems, with somewhat less than 10% using *X-Windows* systems as their workstation (though many use UNIX or VMS for other things). Bioscientists remain diverse in their computing system choices and needs for software. Many use or have access to multiple operating systems, depending on what software runs on. Also, there are commercial emulator programs that allows one to run most MS Windows and MSDOS software on Macintoshes, and conversely to run some Macintosh programs on Wintel systems.

Some software is written to run on many operating systems. This may be a holy grail of those who develop scientific applications—one hopes a program can be used by anyone needing it, on any computer system. It is not as easy to write multiplatform software. Outside of the recent arrival of Java, there has been no easy and good way to do this inexpensively. Even large commercial developers expend effort that is not always successful (e.g., WordPerfect on UNIX pales compared to the MacOS or Wintel versions).

With the case of graphical interface software that most of us expect today, multiplatform programs may not look quite right on your particular system, even though they operate as intended. In the more common case of software with no graphical interface, you may need to devote extra time to learning its command-line- or menu-driven syntax.

The new Java development system born at Sun Microsystems (`http://www.javasoft.com`), is providing a means for developing useful software that works well on common systems. At present Java software is frequently slower than its counterpart written in C++, C or other languages. We may expect to see much more software in biocomputing written in Java in coming years. See for instance the new sequencing analysis package from Licor (`www.licor.com`).

1.6. Client-Server Biosequence Software

There are various developers working on the concept of separating the user interface from the analysis programs, and I think this is a useful approach to making such programs easier to use. This is the basis of a client-server design for software. Simple examples of this are Web interfaces that abound now for various data analyses.

This author's own work-in-progress *SeqPup* takes this approach: It allows you to use the analysis software you need, whether *Clustal, CAP, tacg, fastDNAml,* or others, running on your own computer or on a server computer. *SeqPup* provides a graphical interface and standard user interface methods for editing sequences, basic manipulations and alignments, and sophisticated display and output options. It also links to analysis engines in a way you can configure to your taste. These analysis programs encode the complex data-analysis algorithms but generally without a user-interface beyond command-line options. With a client program such as *SeqPup,* use of these programs is simplified and tied into a way for you to organize your sequence data.

Martin Senger at EMBL/EBI is working on a general CORBA interface to sequence analysis software called *AppLab* (`http://industry.ebi.ac.uk/applab/`), which is a similar approach.

Peter Rice's *EMBOSS* (`http://www.sanger.ac.uk/Software/EMBOSS/`) will be a freely distributable set of analysis programs. It is still under development at this writing. It will run on UNIX server computers, with a command-line interface. *EMBOSS* will include various sequence-analysis topics, and will include a major effort to provide easy integration of other public-domain packages. Analyses include rapid database searching with sequence patterns, and for sequence overlaps, simple and species-specific repeat identification, nucleotide sequence pattern analysis, codon usage analysis for small genomes, gene identification tools for genomic sequencing, rapid identification of sequence patterns in large scale sequence sets, protein motif identification, and presentation tools for publication.

These could be the basic analysis engines for sequence analysis, and some client program with a good user interface, perhaps like *SeqPup,* Java applets, or Web forms, can be the program you use to run the analyses.

2. Free Software Highlights

2.1. Clustal *Sequence Alignment*

Clustal provides automatic multiple sequence alignment. The current version is called *Clustal W,* and is available for MacOS, Wintel, UNIX, and VMS computers. The simultaneous alignment of many nucleotide or amino acid sequences is now an essential tool in molecular biology. Multiple alignments are used to find diagnostic patterns to characterize protein families; to detect or demonstrate homology between new sequences and existing families of sequences; to help predict the secondary and tertiary structures of new sequences; to suggest oligonucleotide primers for PCR; as an essential prelude to molecular evolutionary analysis. The *CpIt* program fits this need very well. It is available at `ftp://ftp-igbmc.u-strasbg.fr/pub/`-*Clustal W* and at *EBI* and *IUBio* archives. There is a companion program *Clustal X,* which provides a graphic interface to *Clustal,* and *Clustal* can be used from other sequence editors such as *SeqPup.*

2.2. Entrez *to Search Genome Data*

The *Entrez* program is used for keyword searches of gene sequence data and *MEDLINE* literature. It has been written by the programming staff at the National Center for Biotechnology Information (NCBI). It can be obtained from `http` or `ftp://ncbi.nlm.nih.gov/`. *Entrez* runs on a variety of computer systems. One big advantage of *Entrez* is the inclusion of a subset of *MEDLINE,* which covers the abstracts of entries submitted to the sequence databases. The Web service of NCBI also offers an *Entrez* type of capability through your Web browser. The program source for *Entrez* developed at NCBI has been instrumental in providing a software framework for other biosciences applications, including a version of *SeqPup* and *Clustal X.*

2.3. **NIH Image** *for Image Analysis*

A very useful Macintosh program for general image analysis is *NIH Image,* written by Wayne Rasband. *Image* can be used to measure the area, average density, center of gravity, and angle of orientation of a user-defined region of interest. It also performs automated particle analysis and can be used to measure path lengths and angles. Measurement results can be printed, exported to text files, or copied to the clipboard. Results can be calibrated to provide real world values. Find *Image* at `ftp://zippy.nimh.nih.gov/pub/nih-image/`, or `http://rsb.info.nih.gov/nih-image/`. There is now a version for MS Windows, from `http://www.scioncorp.com/`. The use of this program is descriped in Chapter 14.

2.4. **PHYLIP** *for Phylogeny Analyses*

The widely used *Phylogeny Inference Package, PHYLIP,* from Joseph Felsenstein, is a package of programs for inferring phylogenies (evolutionary trees), and written to work on as many different of computer systems as possible. It includes analyses of DNA and protein sequences, restriction sites, distance matrices and gene frequencies, quantitative and discrete characters, and plotting of evolutionary trees. Algorithms used include parsimony, maximum likelihood, neighbor joining, and several others. Many options for precise control of the analyses are available. The home for *PHYLIP* is at `http` or `ftp://evolution.genetics.washington.edu/`. The use of this program is discussed in Chapter 12.

2.5. **RasMol** *for Molecular Modeling*

RasMol is a widely used, free molecular graphics program for the visualization of proteins, nucleic acids, and small molecules. The program is aimed at display, teaching, and generation of publication-quality images. *RasMol* runs on all common computer systems. The program reads molecule co-ordinates and interactively displays the molecule in a variety of color schemes and molecule representations, including depth-cued wireframes, space-filling spheres, ball and stick, solid and strand biomolecular ribbons, atom labels, and dot surfaces. The home of *RasMol* is `ftp://ftp.dcs.ed.ac.uk/pub/rasmol/`.

2.6. **SeqPup** *for Sequence Editing*

SeqPup, and its predecessor *SeqApp,* are biological sequence editor and analysis programs. They includes links to network services and external analysis programs. *SeqPup* is usable on common computer systems, using the new Java language.

Features include multiple-sequence alignment and single-sequence editing, read and write several sequence file formats, pretty print of alignments and sequences with boxed and shaded regions, sequence feature editing, manipulation, and marking in prints, consensus, reverse-complement, distance/similarity, and translate DNA to/from protein. Print file formats include PICT, PostScript, and GIF.

User-definable links to external analysis programs, including *Clustal W* multiple alignment, *CAP* contig assembly, *tacg* restriction maps and *fastDNAml* phylogenetic analysis are included, and others can be added. One can use these running on your own computer or on an Internet server computer, using a new CORBA protocol (`www.corba.org`). Internet sequence analysis services include fetching sequences using SRS keyword search, and performing NCBI-*BLAST* similarity searches.

The home of *SeqPup* is `http://iubio.bio.indiana.edu/soft/molbio/seqpup/`. Note that this application is a work in progress; it has bugs. *SeqApp* is the Macintosh-only predecessor to *SeqPup*. Many folks find this currently a more useful program than *SeqPup*. It is faster, but lacks newer features of *SeqPup*.

3. Software Use Issues

3.1. Copyrighted vs Public Domain

Most free software is copyrighted by the author or sponsor, who retain all rights. They specifically grant you a right to use this software freely, perhaps only for noncommercial uses. Use of copyrighted software in commercial packages or other uses that make money require consent from the author. Many of the free software programs come with source code, so you can modify and extend it. This is a great boon to let sophisticated users do a needed analysis, but keep in mind use of such in a commercial product is not allowed. If the author explicitly places his work in the public domain, he or she retains no control, and it can be used in commercial applications.

3.2. Citing Software Publications

Publicly available software is a publication, and free software that you use should be treated with consideration that you give other publications used in your research. Whereas some free software has a companion paper publication to cite, some do not. It is usual practice to cite a software publication with its Internet URL in place of the journal/volume portion, e.g.,

> Felsenstein, J. 1993. *PHYLIP* (*Phylogeny Inference Package*) version 3.5c. Distributed by the author at `ftp://evolution.genetics.washington.edu/`. Department of Genetics, University of Washington, Seattle.

Gilbert, D.G., 1996. *SeqPup,* biosequence editor & analysis platform, version 0.6. Bionet.Software, July 1996. <news://4rb7hr$6rc@usenet.ucs.indiana.edu> See also ftp://iubio.bio.indiana.edu/molbio/seqpup/

Some of these programs have been available at the same location for 10 years or more, so there isn't a general problem of impermanence with Internet locators.

Acknowledgments

The many developers of free software for biosciences, some of them mentioned herein, are the real authors of this document. If you use their software, please let them know you find it useful. Often one program builds upon others. This author would like to thank Jonathan Kans, Joseph Felsenstein, Michael Zuker, Gary Olsen, Dan Davison, Rob Harper, Dave Kristofferson, Reinhard Doelz, Rainer Fuchs, Peter Markiewicz, Thure Etzold, Xiaoqiu Huang, Des Higgins, Harry Mangalam, Jim Brown, Bill Pearson, and many others who provided their sweat and ideas to help make his own works useful. The many users of software who offer suggestions, criticisms and insights on how software should work, also contribute enormously to making free software better for all to use. The former *GenBank* home at Intelligenetics in the 1980s held an archive for free molecular biology software, to which many of us owe thanks for their pioneer efforts.

Appendix: Software Listings

This list of over 150 free software programs in molecular biology and related areas is not exhaustive by any means, but includes much of what is available for Macintosh and/or MS Windows computers.

Operating system key: **M** — MacOS, **W** — MS Windows or MS DOS, **O** — Other (UNIX usually)

Software archive abbreviations:

ebi—ftp://ftp.ebi.ac.uk/pub/software/ or http://www.ebi.ac.uk/software/software.html

iubio—ftp://iubio.bio.indiana.edu/molbio/ or http://iubio.bio.indiana.edu/soft/molbio/

Alternate sites for the *IUBio* molecular biology collection:

ftp://ftp.funet.fi/pub/sci/molbio/iubiomolbio

ftp://ftp.sunet.se/pub/molbio

ftp://ftp.nig.ac.jp/pub/mirror/IUBIO/molbio

ftp://ftp.uam.es/pub/mirror/molbio,

ftp://ftp.pasteur.fr/pub/GenSoft/mirrors/IUBio/molbio

http, ftp://mic3.hensa.ac.uk/hosts/iubio.bio.indiana.edu/molbio/

ftp://bioinformatics.weizmann.ac.il/pub/software/mac and software/ibmpc

Some e-mail addresses and home URLs may be out of date. Unless otherwise indicated, all software listed is copyrighted by the author, and is available free for noncommercial use. Specific copyright restrictions should be noted. Some of this software is shareware, the author requesting a fee for use.

ABaCUS — M, W, O

ABaCUS is a no-frills program to investigate the significance of the putative correspondence between exons and units of protein structure.
Author: Arlin Stoltzfus, arlin@is.dal.ca
Archive: iubio/evolve/abacus/

ADE-4 — M

ADE-4 is a multivariate analysis and graphical display software package for Macintosh microcomputers.
Author: Olivier J.M. and others, Jean-Michel.Olivier@biomserv.univ-lyon1.fr
Home: ftp://biom3.univ-lyon1.fr/pub/mac/ADE/ADE4, http://biomserv.univ-lyon1.fr/ADE-4.html

Amplify — M

This Macintosh software is for use in designing, analyzing, and simulating experiments involving the polymerase chain reaction (PCR). *Amplify* will search a target sequence for near matches and display the results of using various primers. It can check oligos for matching sequence and for internal repeats.
Author: Bill Engels, WREngels@macc.wisc.edu
Archive: iubio/mac/amplify*, ebi/mac/

AnalyzeSignalase — M

A Macintosh program for applying the algorithm of von Heijne to the prediction and analysis of mammalian signal sequences. It uses a weight-matrix method to try to predict the site at which signal peptides in secretory peptides are cut off by the signal peptidase.
Author: Ned Mantei
Archive: iubio/mac/analyze-signalase*

ANCESTOR — W

Ancestor is designed to infer ancestral amino acid sequences from a set of homologous amino acid sequences whose phylogenetic relationships are known.
Author: Jianzhi Zhang, `zhang@imeg.bio.psu.edu`
Archive: `iubio/ibmpc/ancestor*`

ANNHYB — W

Annhyb is a little program for Windows 95 that is able to calculate various DNA sequence (degenerated or not) parameters.
Author: O. Friard & G. Stefanuto, `friard@ba.cnr.it`
Home: `http://area.ba.cnr.it/~e105of01/annhyb221.zip+`

ANTHEPROT — W, O

ANalyze THE PROTeins (*ANTHEPROT*) is a package includes study of physico-chemical properties: hydrophobicity, antigenicity, flexibility, solvent accessibility, amphiphilicity; secondary structure prediction: Chou and Fasman, Garnier, Gibrat, Deleage, Levin; prediction of transmembranous regions and of structural domains; multiple alignment; search for biological sites using *PROSITE,* and *PATMAT;* search for homologous protein using *FASTA*; identity level between several sequences. Look and handle protein structures from *PDB*.
Author: G. Deleage, `deleage@ibcp.fr & C. Geourjon, geourjon@ibcp.fr`
Home: `http, ftp://www.ibcp.fr/`
Archive: `iubio/ibmpc/antheprot*`

AUTOMATIC-BLAST — M

This is an AppleScript that automatically sends sequences by e-mail to the *BLAST* server at `ncbi.nlm.nih.gov` at prescribed times, daily or weekly. The script uses the scriptable e-mail program *Eudora.*
Author: Brian Osborne, `bosborne@nature.berkeley.edu`
Home: `http://pgebaker4.pw.usda.-gov/bio/bio.html`
Archive: `iubio/mac/automatic-blast.*`

BCM SEARCH LAUNCHER — M, W, O (PERL)

The *BCM Search Launcher* is an integrated set of Web pages that organize molecular biology-related search and analysis services available on the Web by function, and provide a single point-of-entry for related searches. There is a batch client interface for UNIX and Macintosh computers that allows multiple input sequences to be automatically searched as a background task, with the results returned as individual HTML documents. Requires Perl.
Author: Randall F. Smith et al.
Home: `http://gc.bcm.tmc.edu:8088/search-launcher/launcher.html`

BUFFERSTACK — M

Given the appropriate information, the *BufferStack* will construct a complete recipe for a buffer that is defined in terms of both pH and ionic strength, at the temperature of use.
Author: Rob Beynon
Archive: `iubio/mac/buffer*, ebi/mac/bufstack*`

CABUFFER W

This program allows you to calculate the concentrations of all ionic species present in a mixture of up to four divalent cations and four ligands for these ions. Examples of such buffers are EDTA, EGTA, NTA, HEDTA, citrate, Ca-binding proteins, and so on. Corrections for temperature, ionic strength and pH are provided.
Author: Jochen Kleinschmidt, kleinschmidt@mcclb0.med.nyu.edu
Archive: iubio/ibmpc/cabuf*

CAIC M

CAIC, Comparative Analysis by Independent Contrasts, computes phylogenetically independent contrasts from comparative data, allowing valid statistical testing of adaptational hypotheses. The manual gives guidance on testing hypotheses of correlated evolution among traits, rate variation among traits or taxa, and grade shifts.
Author: Andy Purvis and Andrew Rambaut, Andrew.Rambaut@zoology.ox.ac.uk
Home: http://evolve.zps.ox.ac.uk/CAIC/CAIC.html

CAP M, W, O

Contig Assembly Program (*CAP*) based on sensitive detection of fragment overlaps. C source, command-line program, can be used from *SeqPup,* others.
Author: Xiaoqiu Huang, huang@cs.mtu.edu
Home: ftp://cs.mtu.edu/pub/huang/
Archive: iubio/align/cap*

CGR M

A *HyperCard* stack for presenting nucleotide sequence data using *Chaos Game Representation.*
Author: Heikki Lehva, lehvaslaiho@cc.helsinki.fi
Archive: iubio/mac/cgr*, ebi/mac/cgr*

CLUSTAL W M, W, O

A multiple sequence alignment program, which is widely used and described in the article *Clustal W*: improving the sensitivity of progressive multiple sequence alignment through sequence weighting, position-specific gap penalties and weight matrix choice. *Nucleic Acids Res.* **22**, 4673–4680 (1994).
Author: D. Higgins et al.
Home: ftp://ftp-igbmc.u-strasbg.fr/pub/Clustal*
Archive: ebi/mac/clustalw*, ebi/dos/clustalw*, iubio/align/clustal*

CLUSTAL X M, W, O

CLUSTAL X is a graphic interface for the *CLUSTAL W* alignment program. The sequence alignment is displayed in a window on the screen, with pull-down menus.
Author: Thompson J.D et al., julie@igbmc.u-strasbg.fr
Home: ftp://ftp-igbmc.u-strasbg.fr/pub/ClustalX

Codon Frequency Analyzer **W**

This program helps to identify coding regions of DNA by comparing the codon frequencies in known coding regions with a sequence of DNA, the coding regions of which are unknown. The program will work with any organism.

Author: Ballyclaire Analysis
Archive: `iubio/ibmpc/codon*`

CodonBiasIndex **M**

The *codon bias index* is a statistic created by Bennetzen and Hall to quantify the extent to which more frequently used codons are used in preference to less frequently used codons.

Author: Tom Ritch, `ritch@seas.ucla.edu`
Archive: `iubio/mac/codonbioasindex*`

Comap **W**

A program for helping with the construction of restriction maps of small DNA fragment from digestion data. The program works under a graphical user interface.

Author: Kay Hofmann, `khofmann@cipvax.biolan.uni-koeln.de`
Archive: `iubio/ibmpc/codon*, ebi/dos/`

ConsInspector **M, W, O**

ConsInspector uses a precompiled library of extended weight matrix descriptions (consensus profiles) of transcription factor binding sites to scan nucleic acid sequences for matches to these sites.

Author: K. Frech, et al. `frech@gsf.de`
Home: `ftp://ariane.gsf.de/pub/`

Covariation **M**

A Hypercard stack for phylogenetic comparative analysis of aligned RNA sequences.

Author: James W. Brown, `jwbrown@mbio.ncsu.edu`
Archive: `iubio/mac/covariation*, ebi/mac`

CPrimer **M**

CPrimer evaluates oligonucleotides as possible PCR primers. It shows melting points, interfering structures, and can search for optimum amplification pairs.

Author: Greg Bristol, `gbristol@ucla.edu`
Archive: `iubio/mac/cprimer*`

DCSE **W, O**

Dedicated Comparative Sequence Editor (*DCSE*) is a multiple alignment editor. It can be used to edit protein, DNA or RNA alignments. The structure of the molecules can be incorporated in the alignment. It offers lots of features such as color display of characters and structure, automatic alignment relative to sequences already aligned with others, sequence grouping, sequence or pattern searching, marker system, checking of incorporated RNA structure, on-line hypertext help, macros, and a lot more.

Author: Peter De Rijk, `derijkp@reks.uia.ac.be`
Home: `http://www-rrna.uia.ac.be/~peter/dcse`
Archive: `ebi/dos/dcse`

Digest W

Digest scans DNA sequence files for restriction sites. It prompts the user to specify which enzymes to cut with, and if they are in the enzyme database, it writes out the positions of all the cuts and sorts the fragments by size.
Author: Ramin Nakisa, ramin@ic.ac.uk
Archive: iubio/ibmpc/digest*, ebi/dos/digest*

Digispeak M

A program for reading sequencing gels with the aid of a Graf-Bar or similar sonic digitizer.
Author: Ned Mantei, bcmantei@aeolus.vmsmail.ethz.ch
Archive: iubio/mac/digispeak*, ebi/mac/

Dispan W

DISPAN (*genetic DIStance and Phylogenetic ANalysis*) is designed to compute the following: average heterozygosity and standard error for each population; gene diversity and associated parameters; standard genetic distances and errors; DA distances between populations. It also constructs phylogenetic trees and does bootstrap tests.
Author: Tatsuya Ota, imeg@psuvm.psu.edu
Archive: iubio/ibmpc/dispan*

DNA Runs M

DNA Runs is a program for performing a significance test of the number of runs in DNA sequence polymorphism and divergence data.
Author: John H. McDonald, mcdonald@udel.edu
Home: http://udel.edu/~mcdonald/
Archive: iubio/mac/dna-runs*

DNA Slider M

DNA Slider is a program for performing a significance test of heterogeneity in the ratio of polymorphic sites to fixed differences in DNA sequence data.
Author: John H. McDonald, mcdonald@udel.edu
Home: http://udel.edu/~mcdonald/
Archive: iubio/mac/dna-slider*

DNA Stacks M

DNA Stacks is a software package of HyperCard stacks providing utilities for viewing and manipulating molecular data. *DNA Translator* includes a gene mapping facility, draws and displays two linearized gene maps for comparison. *Aligner* is a stack for editing and display of multiple alignments. *Codon Usage* displays codon and amino acid usage data for a variety of organisms and organelles.
Author: D. J. Eernisse, DEernisse@fullerton.edu
Home: http://biology.fullerton.edu/people/faculty/doug-eernisse
Archive: iubio/mac/dnastacks*, ebi/mac/

DNA WORKBENCH **M, W, O (PERL)**

A program for sequence searching and manipulation. It offers powerful and fast searches on *GenBank* and other databases and client-server access to remote databases and programs. Its many sequence manipulation functions include calculating the reverse complement, displaying reading frames and nucleotide-to-protein translations, editing, searching for restriction enzyme sites, searching for human repeat or vector in a sequence, comparing a sequence against a library or a user file, and searching for a regular expression in a sequence. It requires Perl.
Author: James Tisdall, `tisdall@cbil.humgen.upenn.edu`
Home: `ftp://cbil.humgen.upenn.edu/pub/dnaworkbench`

DNADRAW **M**

DNAdraw is a program designed for preparing DNA and protein sequences for publication. A large selection of highlighting options is available. It has special features for formatting raw data into a style commonly used for publication, and for doing automatic highlighting of aligned sequences.
Author: Marvin Shapiro, `mbs@kias.com`
Archive: `iubio/mac/dnadraw*`

DNAFRAG **W**

This program is used in restriction mapping of DNA or sizing of proteins from gels. It calculates the size of restriction fragments or peptide bands if standards are run on the same gel, and a standard curve of the standard bands using their mobilities.
Author: John Nash, `Nash@biologysx.lan.nrc.ca`
Archive: `iubio/ibmpc/dnafrag*, ebi/dos/dfrag*`

DNASP **W**

DnaSP is a package for Windows that performs extensive population genetics analysis on DNA data, for hundreds of sequences of thousands of bases. It estimates several measures of polymorphism within and between populations, linkage disequilibrium, recombination, gene flow, gene conversion, and neutrality. *DnaSP* can do analyses by a sliding window method, and will make graphic representations.
Author: Julio Rozas & Ricardo Rozas, `julio@porthos.bio.ub.es`
Home: `http://www.bio.ub.es/~julio/DnaSP.html`
Archive: `ebi/dos/dnasp`

DOTPLOT **W**

A *dotplot* program for MS-DOS.
Author: Ramin Nakisa, `ramin@ic.ac.uk`
Archive: `iubio/ibmpc/dotplot*, dprel3*, ebi/dos/dotplot*`

DOTTY PLOTTER **M**

Dotty Plotter is a tool for drawing dot matrix comparisons of sequences in molecular biology. Dot plots are used to view all areas of homology between two nucleic acid or protein sequences.
Author: D. Gilbert, `software@bio.indiana.edu`
Home: `iubio/mac/dottyplot*, ebi/mac/`

Double Digester — M, O

This is a program designed to help researchers in molecular biology assemble restriction maps of DNA using data from double-digest experiments.
Author: L. Wright, wright-lawrence@yale.edu
Archive: iubio/restrict-enz/, ebi/mac/

dPrimer — M

This Macintosh software is for use in calculating T_m values for degenerate primer.
Author: Haoyuan Chen, hchen@bimcore.emory.edu
Archive: iubio/mac/dprimer*

EditView — M

DNA Sequence Viewer for *ABI Sequencer* trace data. *EditView* is a software application that allows you to view and print analyzed sample files containing sequence data from an *ABI PRISM Genetic Analyzer.*
Author: EditView@perkin-elmer.com
Home: ftp://ftp.abd.perkin-elmer.com/pub/public/Sequencing/EditView/EditView1.0.1.sea.hqx
Archive: iubio/mac/editview*

Entrez — M, W, O

Entrez is a molecular sequence retrieval system developed at the NCBI. *Entrez* provides an integrated approach for gaining access to nucleotide and protein sequence information, to the *MEDLINE* citations in which the sequences were published, and to a sequence-associated subset of *MEDLINE.*
Author: various at NCBI
Home: ftp://ncbi.nlm.nih.gov/entrez/

Enzyme Kinetics — M

Enzyme Kinetics is a Hypercard stack for Macintosh computers. It calculates and plots the biochemical values for the kinetics of enzyme-catalyzed reactions.
Author: D. Gilbert, software@bio.indiana.edu
Archive: iubio/mac/enzymekinetic*, ebi/mac/enzymekin*

ESEE — W

Eyeball SEquence Editor (ESEE), for MS DOS.
Author: Eric L. Cabot, cabot@gcg.com
Archive: iubio/ibmpc/esee*

FASTA — M, W, O

The *FASTA* sequence comparison programs, improved versions of the *FASTP* program, originally described in *Science* (Lipman and Pearson, (1985) *Science* **227**, 1435–1441)
Author: Bill Pearson, wrp@virginia.edu
Home: ftp://ftp.virginia.edu/pub/fasta/
Archive: iubio/search/fasta*

FASTDNAML — M, W, O

fastDNAml is a faster version of Joseph Felsenstein's *DNAML* (part of *PHYLIP*). Users should consult the documentation for *DNAML* before using this program.
Author: Gary J. Olsen et al., gary@phylo.life.uiuc.edu
Archive: iubio/evolve/fastdna*

FOLDIT — M

FoldIt (light) is a molecular modeling program to visualize and manipulate proteins. It has an integrated environment in which statistical analysis as well three-dimensional observations can be realized on PDB files. It can analyze proteins up to 1600 residues. It can extract a number of structural features: Ramachandran plots, SS-bond plots, H-bond plots, and statistics on atomic parameters.
Author: Jean-Claude Jesior, jean-claude.jesior@imag.fr
Home: ftp://ftp.imag.fr/pub/TIMC/FoldIt.html
Archive: iubio/mac/foldit*, ebi/mac/

GCUA — M, O

General Codon Usage Analysis (GCUA) is designed to calculate various parameters that might be relevant in accessing the codon usage patterns of a group of genes. The user can look at codon usage (or any other statistic) in the dataset as a whole or for each gene individually. Features of this program include: Multivariate analyses of codon usage (RSCU) and amino acid patterns. Calculation of codon usage frequency, RSCU values, amino acid frequency data, base composition; distances between genes; and ability to analyse complete prokaryotic genomes.
Author: James O. McInerney, J.mcinerney@nhm.ac.uk
Home: ftp://ftp.nhm.ac.uk/pub/gcua/
Archive: iubio/mac/gcua*

GEL — M, W

An application to calculate the size of DNA fragment in an agarose gel.
Author: Jean-Michel Lacroix, lacroix@medac.med.utoronto.ca
Archive: iubio/mac/gel-jml, iubio/ibmpc/gel-jml, ebi/mac/gel-jml, ebi/dos/gel-jml

GEL — W

GEL takes a set of standard DNA fragment sizes and mobilities and predicts the sizes of unknown fragments, using a least squares fit to the relationship of mobility and fragment length.
Author: John R. Thompson
Archive: iubio/ibmpc/gel*, ebi/dos/gel/

GEL FRAG SIZER — M

Gel Frag Sizer is a HyperCard stack which calculates restriction fragment sizes from their mobilities. Two methods for estimating sizes are provided: the local reciprocal method of Elder and Southern or the cubic spline method.
Author: D. Gilbert, software@bio.indiana.edu
Home: iubio/mac/gelfragsizer.*
Archive: ebi/mac/

Gel Manager **W**

Gel Manager is a user-friendly program that runs in MS Windows. It includes techniques of image processing along with options for data analysis. It can deal with different kinds of data such as: RFLP, RAMM, RAPD, microsatellites, and other fingerprinting techniques. It is useful for studies including genetic relationships, taxonomy, and classification, epidemiology, and so on.

Author: Carlos Vaquerizo, Joaquin Dopazo, dopazo@samba.cnb.uam.es
Home: ftp://ftp.cnb.uam.es/software/molbiol/gel_man

GeneDoc **W**

GeneDoc is a full-featured multiple sequence alignment editor and shading utility. It is intended to help you bring your genetics research work to publication with shading, page, and font layout features.

Author: Karl Nicholas, ketchup@cris.com
Home: http://www.cris.com/~ketchup/genedoc.shtml
Archive: iubio/ibmpc/genedoc*

GeneMaster **W**

A small gene analysis package that performs searches for sequences, looks for regions of GC richness, translates using a variety of start codons and genetic codes, and restriction analysis.

Author: Shawn Abigail, ad873@freenet.carleton.ca
Archive: iubio/ibmpc/genemast*

GeneTree **M, W**

GeneTree is a program for comparing gene and species trees using reconciled trees. The program can compute the cost of embedding a gene tree within a species tree, visually display the location and number of gene duplications and losses, and search for optimal species trees.

Author: Roderic D. M. Page, r.page@bio.gla.ac.uk
Home: http://taxonomy.zoology.gla.ac.uk/rod/genetree/

Gepasi **W**

Gepasi is intended for the simulation of the kinetics of systems of chemical and biochemical reactions. *Gepasi* is able to simulate the steady-state and time-course behavior of reactions in several compartments of different volumes. Results can be plotted in two- and three-dimensional graphs directly from the program. Steady states are analyzed with metabolic control analysis and linear-stability analysis.

Author: Pedro Mendes, prm@aber.ac.uk
Home: http://gepasi.dbs.aber.ac.uk/softw/gepasi.html
Archive: ebi/dos/

HDProbe **M**

HDProbe matches a probe sequence against a set of alleles and tabulates stable and unstable mismatches. *HDProbe* accepts probe and allele sequences as input, then displays the sequences with respect to their orientation in a heteroduplex molecule.

Author: Marvin Shapiro, mbs@pa.net
Home: iubio/mac/hdprobe.*

HelixVu **W**

HelixVu illustrates an 80-bp region of DNA as a helix with the sequence listing printed above the helix diagram. This view is useful to see the spatial relationship between DNA modifications.
Author: Richard Seyler
Archive: `iubio/ibmpc/helixvu*`

Hyper **W**

Hyper is a program for the analysis of enzyme kinetic data under MS Windows. Enzyme kinetic data are subjected to nonlinear regression and the results displayed in five standard graphical forms and printed.
Author: J S Eastery, `jse@liverpool.ac.uk`
Archive: `iubio/ibmpc/hyper*`

HyperPCR **M**

A HyperCard stack that calculates the optimal annealing temperature of a PCR reaction according to the algorithm of Rychlik.
Author: Brian Osborne, `bosborne@violet.berkeley.edu`
Archive: `iubio/mac/hyperpcr*, ebi/mac/`

Intron Analyzer **W**

There are basic differences in the base composition of introns from animals and plants, and this program will examine introns to find regularities. From a given list of introns you can explore by aligning them either at the 5′ or at the 3′ end, or study the adjoining exon-parts. The program will build a consensus-sequence that shows the most frequent base in each position, including a graphic plot.
Author: Michael Liss, `LISS@alf1.ngate.uni-regensburg.de`
Archive: `iubio/ibmpc/intron-analyzer*, ebi/dos/intana*`

LalnView **M, W, O**

LalnView is a graphical program for visualizing local alignments between two sequences. Sequences are represented by colored rectangles to give an overall picture of their similarities. It is able to display sequence features (active site, domain, motif, propeptide, and so on) along the alignment. *LalnView* is a useful tool for analysing pairwise alignments and for making the link between sequence homology and what is known about its structure or function.
Author: Laurent Duret, `duret@dim.hcuge.ch`
Home: `ftp://expasy.hcuge.ch/pub/lalnview`

Lines&Kinetics **M**

A graphic way to calculate linear regressions with normal or logarithmic data, the doubling time of a microbial culture, and the kinetic parameters for an enzyme reaction.
Author: Manuel G. Claros, `claros@uma.es`
Home: `http://www.ie.embnet.org/embnet.news/vol5_1/kinetics.html`
Archive: `iubio/mac/lines-kinetics*`

Linkage-1 M

Linkage-1 is designed to aid the geneticist in the detection and estimation of linkage in segregating progenies.
Author: Karl A. Suiter, ksuiter@acpub.duke.edu
Home: iubio/mac/linkage1*

Lintr W, O

These programs are for testing the molecular clock on a given topology of a phylogenetic tree and making linearized trees, using nucleotide or amino acid sequences.
Author: Naoko Takezaki, ntakezak@lab.nig.ac.jp
Archive: iubio/evolve/lintr/

LoopDLoop M, W, O (Java)

loopDloop is a tool for drawing and editing RNA secondary structures in molecular biology. A MacOS-specific and Java version are available. *Mulfold* will generate RNA foldings for display by loopDloop. A related program, *LoopViewer*, lacks the editing features but is simpler to use.
Author: D. Gilbert, software@bio.indiana.edu
Home: iubio/loopdloop/
Archive: ebi/mac/loop*

MACAW M, W

Multiple Alignment Construction & Analysis Workbench (*MACAW*) is a program for locating, analyzing, and editing blocks of localized sequence similarity among multiple sequences and linking them into a multiple alignment. It includes sequence alignment search, editing, and display. It is a very nice program according to many, allowing one to look for blocks of homology in sequences.
Author: Greg Schuler and Stephen Altschul, schuler@ncbi.nlm.nih.gov
Home: ftp://ncbi.nlm.nih.gov/pub/macaw/
Archive: iubio/ncbi/macaw/, ebi/dos/macaw*

MacBoxshade M, O

A program for creating good-looking printouts from multiple aligned protein or DNA sequences. The program does no alignment by itself, it uses files from a multiple alignment program. Output can be PostScript, EPSF, PICT, RTF, or ASCII text. Identical and similar residues in the multiple alignment are represented by different colors or shadings. There are many options of shading, sequence numbering, consensus output, and so on.
Author: Michael D. Baron, michael.baron@bbsrc.ac.uk (macos), Kay Hofmann (original)
Home: ftp://ulrec3.unil.ch/pub/boxshade/macboxshade
Archive: iubio/mac/macboxshade*

MacPattern M

MacPattern is a Macintosh application for protein pattern searches (using *PROSITE*) and block profile searches (using *BLOCKS*). *MacPattern* assists in finding putative functions for new protein sequences by supporting pattern searches using the *PROSITE* database, block searches

using the *BLOCKS* database, and statistical analyses (maximal segment score analysis and Eguchi-Seto method).
Author: Rainer Fuchs, `rainer_fuchs@glaxo.com`
Archive: `iubio/mac/macpattern*, ebi/mac`

MacPlasmap **M**

If your study or research involves preparation of circular plasmid maps, you will find *MacPlasmap* an indispensable tool to have. It draws, stores, and prints high-quality circular plasmid maps with the data you specify.
Author: Jingdong Liu
Archive: `iubio/mac/macplasmap*, ebi/mac/`

MacProt **M**

MacProt is a of a set of programs for analyzing protein sequences for secondary structure, chain flexibility, hydropathy, helical wheels, and so on.
Author: Peter Markiewicz
Archive: `iubio/mac/plota/, ebi/mac/plota_*`

MacStripe **M**

MacStripe is a program for the prediction and analysis of potential coiled-coil regions in protein sequences. *MacStripe* is the ideal tool for anyone who wants to explore potential alpha-helical coiled coils in the sequence of their protein. With a full Macintosh interface, the results of analyses (raw data or publication quality plots) can easily be exported to other software. *MacStripe* uses the algorithm of Andrei Lupas's *COILS2* for detailed and reliable coiled-coil predictions.
Author: Alex Knight, `aek4@york.ac.uk`
Home: `http://www.york.ac.uk/depts/biol/units/coils/coilcoil.html`
Archive: `iubio/mac/macstripe*`

MacT **M**

MacT is a set of programs for the Macintosh to construct and evaluate unrooted trees derived from amino acid sequences using a distance matrix method.
Author: Angela Luettke and Rainer Fuchs
Archive: `ebi/mac/mact_*, iubio/mac/mact.*`

Map Manager **M, W**

Map Manager is a program that helps analyze the results of genetic mapping experiments using intercrosses with codominant markers, backcrosses, or recombinant inbred strains in experimental plants or animals. It is a specialized database program which allows easy storage, retrieval, and display of information from such mapping experiments, and it also has tools for searching and for statistical analysis of the experimental results. These tools assist the user in determining linkage among loci and in determining the order of loci.
Author: Kenneth F. Manly, `kmanly@mcbio.med.buffalo.edu`
Home: `http,ftp://mcbio.med.buffalo.edu/`
Archive: `iubio/mac/map-manager*, ebi/mac/mapmanager*`

MapMaker — M, W, O

MapMaker is a linkage analysis package designed to help construct primary linkage maps of markers segregating in experimental crosses. One version performs full multipoint linkage analysis for dominant, recessive, and codominant (e.g., RFLP-like) markers.
Author: Whitehead Institute for Biomedical Research mapmaker@genome.wi.mit.edu
Home: ftp://genome.wi.mit.edu/distribution/mapmaker3
Archive: iubio/mapmaker/

Materials & Methods — M

A Hypercard stack for the storage and retrieval of laboratory procedures. The stack comes preloaded with many commonly used procedures used in molecular biology.
Author: James W. Brown, jwbrown@mbio.ncsu.edu
Archive: iubio/mac/mandm*, ebi/mac/matmeth*

Matilda — W

Matilda is a specialized DNA database management system that helps scientists extract the high-level information that they need for recombinant DNA experiments from a large sequence and genetic information database. It incorporates functional data and restriction map data. Sequence and functional data are extracted from sequence files and from additional information. As recombinant DNA clones are constructed, their descriptions are added to the database so that they can be used to describe clones constructed later.
Author: Isralewitz, B. and Shalloway, D.
Archive: iubio/ibmpc/matilda*

MatInd and MatInspector — M, W, O

MatInd is a simple but powerful method to derive a matrix description of a consensus from a number of short sequences on which the definition of an IUPAC code would be based. *MatInspector* is a program that uses a large library of predefined matrix descriptions of transcription factor binding sites to locate matches in nucleotide sequences of unlimited length. It assigns a quality rating to matches and thus allows a quality-based filtering and selection of matches.
Author: K.Quandt, et al, quandt@gsf.de
Home: ftp://ariane.gsf.de/pub/
Archive: ebi/mac/matind*, ebi/dos/matind*

Memsat — W, O

MEMbrane protein Structure And Topology.
Author: David T. Jones, jones@bsm.bioc.ucl.ac.uk
Home: ftp://ftp.biochem.ucl.ac.uk/pub/MEMSAT

Metree — W

A package for inferring and testing minimum evolution trees. This package is intended to find the minimum evolution tree that has the smallest value of the sum of branch lengths for a set of sequences, identify a set of trees that are not significantly different from the *ME tree,* and print the trees in a publishable form.
Author: Andrey Rzhetsky and Masatoshi Nei, aur1@psuvm.psu.edu
Archive: iubio/ibmpc/metree*

MitoProt M, O

It supplies a series of parameters that permit theoretical evaluation on mitochondrial targeting sequences and the importability. *MitoProt II* provides the possibility to predict mitochondrial proteins harboring targeting sequences. Chloroplast proteins also can be studied.
Author: Manuel G. Claros, claros@uma.es, Pierre Vincens
Home: ftp://ftp.ens.fr/pub/molbio/, ftp://ftp.rediris.es/software/incoming/science/
Archive: iubio/mac/mitprot*, ebi/mac/

MolWt W

This program calculates molecular weight from an entered chemical formula, and gives concentrations in various units in response to an entered formula and concentration.
Author: John A. Kiernan, jkiernan@julian.uwo.ca
Archive: iubio/ibmpc/molwt*, ebi/dos/molwt*

MFold (Mac) M

A MacOS port of Michael Zucker's *MFold* software for prediction of RNA secondary structure by free energy minimization, including sub optimal folding with temperature dependence. See also *PCFold.*
Author: D. Gilbert (mac port), M. Zuker (MFold)
Home: iubio/mac/mulfold*, ftp://snark.wustl.edu/pub/ (MFold)
Archive: ebi/mac/

NIH Image M

Image can be used to measure the area, average density, center of gravity, and angle of orientation of a user-defined region of interest. It also performs automated particle analysis and can be used to measure path lengths and angles.
Author: Wayne Rasband
Home: ftp://zippy.nimh.nih.gov/pub/nih-image/, http://rsb.info.nih.gov/nih-image/

NJbafd W, O

These programs are for constructing a neighbor joining or UPGMA tree from allele frequencies of microsatellite DNA or other genetic markers, and computing heterozygosities and Gst.
Author: Naoko Takezaki, ntakezak@lab.nig.ac.jp
Archive: iubio/evolve/njbafd/

NJPlot M, W, O

NJPlot is a phylogenetic tree-drawing program that handles files describing trees by the nested parentheses method (e.g. *PHYLIP*-built trees). Features: A graphical interface allows to re-root a tree anywhere and to swap branches. Bootstrap values are displayed next to internal branches. Branch lengths can be displayed optionally. Tree plots can be saved to a PostScript or PICT file.
Author: Manolo Gouy, mgouy@biomserv.univ-lyon1.fr
Home: ftp://biom3.univ-lyon1.fr/pub/mol_phylogeny/njplot
Archive: ebi/mac/

NONCODE W

A program that will read an *ESEE* file containing nucleic acid sequences and produce a distance matrix using the Kimura 2-parameter model.
Author: Eric L. Cabot, cabot@gcg.com
Archive: iubio/ibmpc/noncode*

NUMCLONE W

Estimates the number of clones one has to screen from a genomic library in order to find a desired clone.
Author: John Nash, Nash@biologysx.lan.nrc.ca
Archive: iubio/ibmpc/numclone*

OLIGOBASE W

This is a shareware program for Windows designed to organize and catalog oligonucleotides collection of a biological laboratory. It stores information about oligonucleotides, select subsets according to specified criteria, print out order forms, calculate molecular weight and melting temperature, and manipulate oligonucleotides.
Author: Igor Sidorenkov, sidorenk@rocketmail.com
Home: http://lochfort.com/oligobase
Archive: iubio/ibmpc/obase*

OLIGOCR M

OligoCR is a data management tool for organizing and cataloging oligonucleotide collection. It allows you to store information about the oligos, e.g., the category (PCR, sequencing), type of project the oligo is used for, its application, its description. To search your oligo database, just click the mouse. It has the capability to proofread your sequences.
Author: Yongming Sun, ysun@hdklab.wustl.edu
Home: http://hdklab.wustl.edu/~ysun
Archive: iubio/mac/oligocr*

OLIGOMUTANTMAKER W

OligoMutantMaker simplifies the designing and screening of oligonucleotide-directed single amino acid substitution experiments by searching for nucleotide sequences that introduce a restriction endonuclease recognition sequence into the codon substitution site of the mutant.
Author: Kevin Beadles et al.
Archive: iubio/ibmpc/oligo*, ebi/dos/oligo*

ONIX W

A program for MS Windows, *Onix* allows users to examine proteins with known three-dimensional structure from *PDB*. This program was designed for structure investigation of ligand binding site in proteins. *Onix* is interactive software with a high-performance interface, fast three-dimensional molecular graphics and analysis of water-accessible surface.
Author: A.S.Ivanov et al., ivanov@ibmh.msk.su
Home: ftp://org.chem.msu.su/pub/software/Onix/

P1 Clones — M

This is a simple HyperCard stack for keeping track of clones that were constructed with the bacteriophage P1 cloning system.
Author: Ken Abremski, `sabremske@esvax.dnet.dupont.com`
Archive: `iubio/mac/p1clones*`

PAML — M, W, O

Phylogentic Analysis by Maximum Likelihood (*PAML*) contains three main programs for model fitting and phylogenetic tree reconstruction using nucleotide or amino acid sequence data.
Author: Ziheng Yang, `z.yang@ucl.ac.uk`
Home: `ftp://abacus.gene.ucl.ac.uk/pub/paml`
Archive: `iubio/evolve/paml*`

PcFold — W

A PC version of Michael Zuker's RNA-folding program which uses an energy minimization algorithm to predict stem and loop regions of RNA structures. *See* also *MulFold,* and home site for UNIX versions.
Author: Michael Zuker et al., `zuker@snark.wustl.edu`
Home: `ftp://snark.wustl.edu/pub/`
Archive: `iubio/ibmpc/pcfold*, ebi/dos/pcfold`

PHYLIP — M, W, O

A *PHYLogeny Inference Package* (*PHYLIP*) of many programs for phylogenetic analysis, including parsimony, compatibility, distance matrix invariants ("evolutionary parsimony") and likelihood methods on various kinds of data.
Author: Joseph Felsenstein, `joe@genetics.washington.edu`
Home: `http, ftp://evolution.genetics.washington.edu/`
Archive: `iubio/evolve/phylip*, ftp://ftp.nig.ac.jp/pub/UNIX/phylip, ftp://ftp.bioss.sari.ac.uk/pub/phylogeny/phylip`

Phylodendron — M, W, O (Java)

Phylodendron is an application for drawing phylogenetic trees. It reads data in New Hampshire (Newick) format. Options allow you to adorn and edit the tree.
Author: D. Gilbert
Home: `iubio/java/apps/trees/`

Phyltest — W

A program for testing phylogenetic hypothesis, with comparison of three alternative phylogenetic trees, estimation of average pairwise distances, and others.
Author: Sudhir Kumar, `imeg@psuvm.psu.edu`
Archive: `iubio/ibmpc/phyltest*`

Plasmid Processor — W

Plasmid Processor is a simple tool for plasmid presentation for scientific and educational purposes. It features both circular and linear DNA, user-defined restriction sites, genes, and mul-

tiple cloning site. In addition you can manipulate plasmid by inserting and deleting fragments. Created drawings can be copied to clipboard or saved to disk for later use. Printing from within program is also supported.
Author: T. Kivirauma, P. Oikari, and J.Saarela, Dept. of Biochemistry and Biotechnology, University of Kuopio, `plasmid@uku.fi`
Home: `http://www.uku.fi/~kiviraum/plasmid/plasmid.html`
Archive: `iubio/ibmpc/plasmid-processor*, ebi/dos/plasmid`

PLASMID-MAKER — M

Draws linear and circular plasmid maps, allows borders of various widths and fills of grays, arrows, and other options.
Home: `http://yeamob.pci.chemie.uni-tuebingen.de/Archiv/PlasmidMaker.html`
Archive: `iubio/mac/plasmid-maker*`
Author: Kai-Uwe Froehlich, `kaifr@uni-tuebingen.de`

PRIMERDESIGN — W

PrimerDesign is a DOS program to choose primer for PCR or oligonucleotide probes. It is tailored to check known sequences for repeats and unique sequences and subsequently to create primers according to this data. A lot of constraints are available to meet your conditions. It can handle up to 31,500 base pairs. Additional features: unique sequences, repeats, restriction sites.
Author: Andreas Becker, Joerg Napiwotzki, `becker@ps1515.chemie.uni-marburg.de`
Home: `ftp://ftp.chemie.uni-marburg.de/pub/PrimerDesign`

PRIMER — M, W, O

Primer is a computer program for automatically selecting PCR primers. It tests oligos for annealing temperature, complementarity to genomic repeat sequences, ability to form primer–dimer, and other criteria. Primer annealing temperature calculation is based on thermodynamic parameters.
Author: Steve Lincoln et al., `primer@genome.wi.edu`
Home: `ftp://genome.wi.mit.edu/pub/software/Primer2.2`
Archive: `iubio/primer/primer-wi*`

PRIMER-MASTER — W

Automatically search and selection of optimal primers and primers pairs for various variants of PCR; analysis of oligos supposed to be used as PCR primers or hybridization probes; editor for comfortable typing-in new nucleotide sequences;
Author: Proutski Vitali, Sokur Oleg, `proutski@influenza.spb.su`
Archive: `ebi/dos`

PRIMERS! — M

Primers! is a primer design shareware application, written by the author of Whitehead Institute *Primer2*. It allows users to interactively scroll through lists of forward and reverse primers to pick exactly the primer pair wanted.

Author: Richard Resnick, rjr@applepi.com
Home: http://www.applepi.com/
Archive: iubio/mac/primers*

ProAnal W

ProAnal is for analysis of multiple protein alignments, studying the structure-function and structure-activity relationships in protein/peptide families. The program uses aligned amino acid sequences with data of their activity and searches for correlations between data on activity and various physico-chemical characteristics of different regions in primary structures.
Author: Alexey Eroshkin, eroshkin@vector.nsk.su
Archive: iubio/ibmpc/proanal*, ebi/dos/

ProAnalyst W

ProAnalyst is for investigation of structural differences between proteins divided by functional, evolutionary, or other criteria; structure-activity relationships investigation; searching motifs; protein-engineering experiments; and many other protein analysis functions.
Author: Vladimir Ivanisenko, Alexey Eroshkin, eroshkin@vector.nsk.su
Archive: iubio/ibmpc/panalyst*, ebi/dos/proanalyst

ProAnWin W

Multiple sequence alignment, analysis of protein sequences and structures, structure-activity relationships, design of protein-engineering experiments. Threads multiple alignment onto known three-dimensional structure; searches linear and spatial sites, conservative and variable in changes of specified physico-chemical properties; plots of different physico-chemical profiles for individual or a set of protein sequences; and many other functions.
Author: I. Pika et al., eroshkin@vector.nsk.su
Archive: iubio/ibmpc/paw*, ebi/dos/proanwin

ProfileGraph W

A graphical protein analysis tool.
Author: Kay Oliver Hofmann, khofmann@biomed.biolan.uni-koeln.de
Archive: iubio/ibmpc/prograph*, ebi/dos/pgraph*

PromFind W

PromFind is a DOS program that accepts a DNA sequence, and adds a feature table to annotate the location of putative promoter regions.
Author: Gordon B. Hutchinson, hutch@netshop.bc.ca
Archive: iubio/ibmpc/promfind*

ProMSED W

ProMSED, a Windows application for both automatic and manual DNA and protein sequence alignment, editing, comparison, and analysis. Automatic alignment is based on *Clustal V*; manual alignment and visual analysis are facilitated by group and block operations and amino acid coloring reflecting their similarity.

Author: Anatoly Frolov, Alexey Eroshkin, `eroshkin@vector.nsk.su`
Archive: `iubio/ibmpc/promsed*, ebi/dos/promsed/`

PROPHET **W, O**

Prophet offers advanced, easy-to-use software tools for data management and visualization, and statistical analysis—from simple descriptive statistics to multifactor ANOVA, logistic regression, and nonlinear modeling. It also offers tools for analyzing biological sequences, including multiple sequence alignment, translation, restriction enzyme and proteolytic cleavage analyses, PCR primer design, *BLAST* searches, remote database retrievals, and more.
Author: Prophet software group, BBN, `prophet-info@bbn.com`
Home: `http://www-prophet.bbn.com/`

PROTEIN SEQUENCE ANALYSIS **W**

The program is a sequence editor with the capability to amino acid composition, hydrodynamic calculation, mass for various isotope labeling, isoelectric point, UV spectrum, relative hydrophobicity, secondary structure prediction, and others.
Archive: `iubio/ibmpc/prot-sa*`

PUZZLE **M, W, O**

Puzzle is a maximum likelihood analysis for nucleotide, amino acid, and two-state data. It reconstructs phylogenetic trees from molecular sequence data, and has a fast tree search that allows analysis of large data sets. *Puzzle* is *PHYLIP* compatible.
Author: Korbinian Strimmer, Arndt von Haeseler, `strimmer@zi.biologie.uni-muenchen.de`
Home: `ftp://fx.zi.biologie.uni-muenchen.de/pub/puzzle`
Archive: `iubio/evolve/puzzle/, ebi/mac/puzzle, dos/puzzle, UNIX/puzzle`

RAMHA **W**

Monte Carlo simulation of the random mutagenesis of synthetic cDNAs.
Author: David P. Siderovski, `Siderovski@Galen.OCI.UToronto.CA`
Archive: `iubio/ibmpc/ramha*`

RASMOL **M, W, O**

RasMol is a molecular modeling program for the visualization of proteins and nucleic acids. It reads protein databank (*PDB*) files and interactively renders them in a variety of formats, including wire, stick, stick_and_ball, CPK, and ribbon.
Author: R. Sayle, `ros@dcs.ed.ac.uk`
Home: `ftp://ftp.dcs.ed.ac.uk/pub/rasmol/`
Archive: `ftp://kekule.osc.edu/pub/chemistry/software/X-WINDOWS/rasmol*, ebi/mac/rasmol*, software/dos/raswin*,`

RBINDING W

Calculates the number of binding sites and the affinities of cell surface receptors for ligands (Scatchard analysis).
Author: Nico van Belzen and Joop van Zoelen, `belzen@pa1.fgg.eur.nl`
Archive: `iubio/ibmpc/rbindin*`

READSEQ — M, W, O

A program for converting among several biosequence file formats.
Author: Don Gilbert, `software@bio.indiana.edu`
Home: `iubio/readseq/`
Archive: `ebi/mac/readseq*`

REALIGN — W

A program that realigns a DNA alignment according to a peptide alignment, thereby improving the alignment in areas not too well conserved.
Author: Rasmus Wernersson, `RWer@novo.dk`
Archive: `iubio/ibmpc/realign*`

REPFIND — W

RepFind (promoter find) is a MS DOS program to identify common repetitive elements in DNA sequence. It is also able to identify and mask vector sequence.
Author: Gordon B. Hutchinson, `hutch@netshop.bc.ca`
Archive: `iubio/ibmpc/repfind*, ebi/dos/repfind*`

RESTDATA — W

Restriction data and phylogenetic analysis, computes the numbers of nucleotide substitutions per site for pairs of DNA sequences; constructs phylogenetic trees by using the neighbor-joining method.
Author: Tatsuya Ota, `imeg@psuvm.psu.edu`
Archive: `iubio/ibmpc/restdata*`

RESTSITE — W

Several programs for analyzing restriction site or fragment data for use in molecular systematics studies.
Author: Joyce C. Miller
Archive: `iubio/ibmpc/restsite*`

RNA DOTPLOT — M

RNA Dotplot is a simple utility to print a dot matrix of the potential base pairing interactions in an RNA sequence.
Author: David S. McPheeters, `mcpheeters@biochemistry.cwru.edu`
Archive: `iubio/mac/rna-dotplot*`

RNADRAW — W

Rnadraw offers RNA optimal structure/basepair-probability matrix/heat curve calculation on Intel computers, providing a consistent user interface with many possibilities to view, print, import/export and edit calculation results.
Author: Ole Matzura, `ole@mango.mef.ki.se`
Home: `ftp://broccoli.mfn.ki.se/pub/rnadraw`

RNA_d2 W

RNA_d2 is a user-friendly program developed for interactively generating aesthetic and nonoverlapping drawings of RNA secondary structures. It allows easy untangling and editing of RNA molecules > 1000 nucleotides long.
Author: J. Perochon-Dorisse et al., `rnad2@ibcg.biotoul.fr`
Home: `ftp://hpsrv.biotoul.fr/rna`

Sagittarius Dna W

A package for exon/intron structure revealing on the base of protein k-tuples statistic.
Author: Victor B. Strelets, `strelets@bio.indiana.edu`
Archive: `iubio/ibmpc/sag-exo*, ebi/dos/sag-exo`

Sagittarius Pir W

A highly compact databank variant of original *PIR* database designed to assist individuals in utilization of sequence database information without huge storage space requests. Includes fast homology searches and selection of sequences by fields (name, source, keyword, and so on), or (non)perfect homology with user-defined short sequence.
Author: Victor B. Strelets, `strelets@bio.indiana.edu`
Archive: `iubio/ibmpc/sag-pir*`

Sagittarius Seqanalref W

A dialog shell for storage and manipulation of reference information. This particular variant is oriented on *SEQANALREF* databases compiled by A. Bairoch.
Author: Victor B. Strelets, `strelets@bio.indiana.edu`
Archive: `iubio/ibmpc/seqanalr*, ebi/dos/sag-sar*`

Send W

A program for computing the standard errors of nucleotide diversity and divergence using the algorithm of Nei and Jin.
Author: Li Jin
Archive: `iubio/ibmpc/send*`

Sendbs W, O

A program that computes average nucleotide substitutions within and between populations using the algorithm of Nei and Jin. It computes standard errors with a bootstrap method that differs from Nei and Jin's.
Author: Naoko Takezaki, `ntakezak@lab.nig.ac.jp`
Archive: `iubio/evolve/sendbs*`

Seq-Eudora-Blast M

Macintosh AppleScript applications that automate your *BLAST* searches. Drag-and-dropped sequence files to the *BLAST* server at ncbi.nlm.nih.gov using *Eudora* mail.
Author: Brian Osborne, `bosborne@nature.berkeley.edu`
Home: `http://pgebaker4.pw.usda.gov/bio/bio.html`
Archive:: `iubio/mac/seq-eudora-blast*`

SEQAID II **W**

Seqaid II is a MS-DOS program for DNA and protein sequence analysis. Functions include editing, modified Needleman-Wunsch alignment, dot matrix comparison, fragment sizer, base composition, translations, protein structure, and hydropathicity, restriction site search, and locating potential exons by codon bias.
Author: Donald Roufa and D.D. Rhoads
Archive: `iubio/ibmpc/sequaid*, ebi/dos/sqaid*`

SEQAPP **M**

A Macintosh biosequence editor, analyzer, and network handyman (*see SeqPup*).
Author: D. Gilbert, `seqapp@bio.indiana.edu`
Home: `iubio/seqapp/`

SEQPUP **M, W, O (JAVA)**

The successor to *SeqApp, SeqPup* is a biological sequence editor and analysis program. It includes links to network services and external analysis programs. Features include multiple sequence alignment and editing, support for several file formats, sequence feature editing, manipulation and marking, translate dna/protein, consensus, reverse/complement, and distance methods, pretty print of alignments and sequences with boxed and shaded regions, Internet searches, use of external analysis programs, including *Clustal W* multiple alignment, *CAP* contig assembly, *tacg* restriction map, and a remote client-server interface. The current version runs on any os supporting Java, an older version runs on MacOS, MSWin, and some UNIX.
Author: D. Gilbert, `seqpup@bio.indiana.edu`
Home: `iubio/seqpup/`

SEQSIMPRESENTER **M**

SeqSimPresenter converts a set of aligned sequences to shaded bars of which correspond to the degree of similarity. It presents large alignments in a compact form and allows a fast recognition of the amount, extension, and distribution of conserved regions.
Author: `cbkfr01@mailserv.zdv.uni-tuebingen.de`
Archive: `iubio/mac/seqsimpresent*, ebi/mac/`

SEQUIN **M, W, O**

Sequin is a stand-alone software tool developed by the NCBI for submitting entries to the *GenBank, EMBL,* or *DDBJ* sequence databases. It is capable of handling simple submissions that contain a single short mRNA sequence, and complex submissions containing long sequences, multiple annotations, segmented sets of DNA, or phylogenetic and population studies.
Author: Jonathan Kans, Colombe Chappey, `info@ncbi.nlm.nih.gov`
Home: `ftp://ncbi.nlm.nih.gov/sequin/, http://www.ncbi.nlm.nih.gov/Sequin`

SEQVU **M**

An alignment editor with analysis options that allows you to work quickly and simply with multiple sequences. It is ideal for manually correcting alignments produced using software such as *Clustal V.*
Author: James Gardner, `j.gardner@garvan.unsw.edu.au`
Home: `ftp://gimr.garvan.unsw.edu.au/pub/`
Archive: `iubio/mac/seqvu*`

SHM — W, O

Shm was designed to provide assistance in the analysis of somatic (point) mutations induced in the immunoglobulin genes of B-lymphocytes. It builds clonal trees by parsimony and displays them along with somatic mutations.
Author: Laurentiu Cocea, `cocea@necker.fr`
Archive: `iubio/ibmpc/shm*`

SIGMA — M, O

System for Integrated Genome Map Assembly (*Sigma*) graphical genome map editor. As a viewer, *Sigma* puts full color maps of the genome in the users hands to display, browse, manipulate, and print. It is capable of giving the user a perspective on an entire chromosome map, as well as an arbitrarily detailed view. Features allow the user to find specific parts of a map. *Sigma* allows the user to integrate data from a variety of sources. A convenient user interface makes data entry easy.
Author: Theoretical Biology and Biophysics Group at LANL, `sigma@ncgr.org`
Home: `http://www.ncgr.org/sigma/home.html`
Archive: `ebi/linkage_and_mapping/SIGMA`

SILMUT — W

Silmut helps you to identify regions in a sequence that can be altered to introduce restriction enzyme sites and other sequences by silent mutations.
Author: Raj Shankarappa, `bsh@med.pitt.edu,` K. Vijayananda, `vijay@litsun.epfl.ch`
Archive: `iubio/ibmpc/silmut*`

SIM2 — M, W, O

This program builds local alignments of two sequences, each of which may be hundreds of kilobases long.
Author: Chao K-M et al., `zjing@sunset.nlm.nih.gov`
Home: `ftp://ncbi.nlm.nih.gov/pub/sim2`

SITES — M, W, O

Sites is a program for the analysis of comparative DNA sequences. It is primarily intended for data with multiple closely related sequences.
Author: Jody Hey
Archive: `iubio/evolve/sites/`

SIXCUTTERFREQ — M

The Hypercard stack calculates the frequency of the various six-cutter restriction enzymes in a few genomes, including bacteriophage lambda, *Mus musculus,* wheat, *Escherichia coli, Saccharomyces,* and *Homo sapiens.* The algorithm is based on the frequency of dinucleotide pairs.
Author: Brian Osborne, `bosborne@nature.berkeley.edu`
Archive: `iubio/mac/sixcutterfreq*`

SNEATH ST — W

Statistical programs to screen a matrix of molecular sequences for atypical sequence comparisons, and to simulate the addition of constant sites to randomly placed differences in a molecular sequence comparison.
Author: P.H.A. Sneath, mjs@le.ac.uk
Archive: iubio/ibmpc/sneathst*

SOLUPRED — M, W

This spreadsheet allows one to predict the solubility of recombinant proteins in *E. coli* based on the amino acid content.
Author: Roger Harrison and Dan Diaz, BL275@cleveland.freenet.edu
Archive: iubio/mac/solupred-mac*, ibmpc/solupred*

SORFIND — W

SorFind is a DOS program that adds to DNA sequence files a feature table to annotate the location of putative coding exons.
Author: Gordon B. Hutchinson, hutch@netshop.bc.ca
Archive: iubio/ibmpc/sorfind*, ebi/dos/sorfin*

SPECTRUM — M, W

Spectrum is a Macintosh and MS Windows program to read in phylogenetic data in *Nexus* format, and display the bipartition spectra corresponding to the data. It can also be used to find the tree whose expected spectrum is closest to the observed spectrum. It outputs spectra in Microsoft Excel and other formats.
Author: Michael Charleston, Roderic Page, m.a.charleston@bio.gla.ac.uk
Home: http://taxonomy.zoology.gla.ac.uk/mike/spectrum/

SPOMBE-STRAIN — M

S.pombe strain collection is a Hypercard stack for cataloging the genotypes of *Schizosaccharomyces pombe* yeast strains. It is adapted from Kai-Uwe Frölich's yeast strain.
Author: Doug Drummond, ddrummon@fs2.scg.man.ac.uk
Archive: iubio/mac/spombe-strain*

SSU RRNA — M

This Hypercard stack contains the entire Ribosomal Database Project sequence release 1. The sequences are accessible via a series of linked phylogenetic trees, a list, or by name.
Author: James W. Brown, jwbrown@mbio.ncsu.edu
Archive: iubio/mac/ssu-rrna*

SWISS-PDBVIEWER — M, W

Swiss-PdbViewer is an application that can display *PDB* files. Several proteins can be analyzed and can be piled-up in three-dimensional space. Differences can be calculated on selected amino acids of the aligned proteins, for comparison of active sites. It can also measure distances,

angles, torsions angles between atoms as well as add/remove amino acids from the view. It includes many other features.
Author: Nicolas Guex, Manuel Peitsch, ng45767@ggr.co.uk
Home: ftp://expasy.hcuge.ch/pub/PDBViewers/Prot3Dviewer, http://www.expasy.ch/spdbv/mainpage.html, http://www.pdb.bnl.gov/expasy/spdbv/mainpage.htm

TACG M, W, O

A program for restriction enzyme and other analyses of DNA. This is a command-line program that can be used from MacOS, Wintel through others such as *SeqPup*.
Author: Harry Mangalam, mangalam@uci.edu
Home: http://hornet.bio.uci.edu/~hjm/projects/tacg/
Archive: iubio/restrict-enz/tacg*

TFPGA W

TFPGA (*Tools for population genetic analyses*) is a Windows program for the analysis of allozyme and molecular population genetic data. The program calculates simple descriptive statistics, genetic distances, and F-statistics. It also performs tests for Hardy-Weinberg equilibrium, population differentiation and performs UPGMA clustering and Mantel Tests.
Author: Mark P Miller, mpm2@nauvax.ucc.nau.edu
Home: http://dana.ucc.nau.edu/~mpm2

TopPredII M

Prediction of transmembrane segments in integral membrane proteins, and the putative topologies.
Author: Claros M.G and von Heijne G., claros@cica.es, gvh@cbs.ki.se
Archive: ebi/mac/, iubio/mac/toppred*

TOPS W, O

Program to automatically generate and edit protein topology cartoons. These cartoons are two-dimensional representations of the secondary structure of proteins.
Author: Tom Flores, flores@ebi.ac.uk
Home: ebi/pub/contrib/TOPS

Tree Draw Deck M

A Hypercard stack that draws phylogenetic trees. It is derived from *Drawgram* and *Drawtree* of *PHYLIP* by J. Felsenstein
Author: D. Gilbert, software@bio.indiana.edu
Home: iubio/mac/treedraw*
Archive: ebi/mac/treedraw*

Treecon W

A package for the construction of phylogenetic trees. Its advantages include menu-driven, easy-to-use interface, quick, handles large datasets, large set of distance-measure methods and distance-matrix based tree construction tools, sophisticated options like subset resampling and

test of outgroup influence, additional tools like production of partial alignments and indication of informative positions.
Author: Yves Van de Peer, `yvdp@reks.uia.ac.be`
Home: `ftp://uiam3.uia.ac.be/`
Archive: `iubio/ibmpc/treecon*`

TreeView **M, W**

A program for drawing phylogenies on MacOS and MS Windows. The program reads *NEXUS, PHYLIP, Clustal W* and similar tree formats.
Author: Roderic D M Page, `r.page@bio.gla.ac.uk`
Home: `http://taxonomy.zoology.gla.ac.uk/rod/treeview.html`
Archive: `iubio/mac/treeview*, iubio/ibmpc/treeview*`

Vised **W**

A visual sequence editor/display software for Windows, including effective and easy-to-use interface. Features include edit up to 200 sequences of 18,000 bases; powerful pattern search function; support for many sequence file formats; supports extraction of sequences from library files; figure preparation; boxed output of sequence identities; imports *MACAW* alignments; protein sequence prediction in one or all six frames.
Author: Ken Peters, `kpeters@qb.island.net`
Archive: `iubio/ibmpc/vised*`

Visual BLAST and FASTA **W**

These programs are designed for interactive analysis of full *BLAST* and *FASTA* output files containing protein sequence alignments. They implement analytical tools which automate detailed analysis of *BLAST* and *FASTA* outputs, and include tools for multiple alignment analysis.
Author: Patrick Durand et al., `durand@lmcp.jussieu.fr`
Home: `http://www.lmcp.jussieu.fr/~durand/`

WinDot **W**

A dotplot program for MS Windows.
Author: Ramin Nakisa, `ramin@ic.ac.uk`
Archive: `iubio/ibmpc/windot, ebi/dos/windot`

WinMGM **W**

Visualization and manipulation tools for proteins, nucleic acids, and organic molecules, including manipulation of molecules represented as CPK, stick and ball, ribbons and cylinders, and colorations by atomic type, atoms of a selected area, of an active site, and others. Several other features are included.
Author: Mehdi Rahman, Robert Brasseur, `mehdirah@fsagx.ac.be`
Home: `http://www.fsagx.ac.be/info_faculte/info_dep/info_bp/mehdi/winmgm/winmgmen.htm`

WinSeq W, O

The *ReadSeq* program for converting among biosequence file formats, for MS Windows.
Author: Ramin Nakisa, ramin@ic.ac.uk
Archive: iubio/ibmpc/winseq, ebi/dos/winseq, see iubio/readseq/ for others

WPDB W

WPDB (the Protein Data Bank through MS Windows) is a package with a compressed version of *PDB* and a set of tools to query features of a single structure or perform a comparative analysis on multiple structures with emphasis on sequence alignment and structure superposition.
Author: Ilya N. Shindyalov, Philip E. Bourne, bourne@sdsc.edu
Home: ftp://ftp.sdsc.edu/pub/sdsc/biology/WPDB/

Yeast Strain M

This is a Hypercard stack for cataloging the genotypes of *Saccharomyce cerevisiae* yeast strains.
Author: Kai-Uwe Froehlich, cbkfr01@mailserv.zdv.uni-tuebingen.de
Archive: iubio/mac/yeaststrain*, ebi/mac/

10

Flexible Sequence Similarity Searching with the FASTA3 Program Package

William R. Pearson

1. Introduction

Since the publication of the first rapid method for comparing biological sequences 15 years ago *(1),* DNA and protein sequence comparisons have become routine steps in biochemical characterization, from newly cloned proteins to entire genomes. As the DNA and protein sequence databases become more complete, a sequence similarity search is more likely to reveal a database sequence with statistically significant similarity, and thus inferred homology, to a query sequence. Indeed, even in the archaebacterium *Methanococcus jannaschii,* more than 40% of the open reading frames could be assigned a function based on significant sequence similarity to a protein of known function *(2)*.

This chapter provides a "hands on" overview of the programs in the *FASTA* package (`ftp:/ftp.virginia.edu/pub/fasta`). Rather than discuss in depth the theory and practice of protein and DNA sequence comparison, I focus on more practical questions, such as: "Which *FASTA* program should I use?", "What threshold should I use for statistical significance?", "Which databases should I search?", "When should I use *FASTA* and when should I use *BLAST*?", and "When should I change the scoring matrix and gap penalties?" For an excellent review of similarity searching with *BLAST* and *FASTA* and of local similarity statistics, *see* **ref.** ***3***. For more specific information on how to use the *FASTA* programs to identify distantly related sequences, *see* **refs.** ***4*** and ***5***. A detailed explanation of the statistical estimates in the *FASTA*3 package is provided in **ref.** ***6***.

From: *Methods in Molecular Biology, vol. 132: Bioinformatics Methods and Protocols*
Edited by: S. Misener and S. A. Krawetz © Humana Press Inc., Totowa, NJ

2. Similarity Searching with the *FASTA3* Programs

The *FASTA* program package has evolved significantly since its introduction 10 years ago *(7)*. The original package offered four programs: *fasta, tfasta, lfasta,* and *rdf* (*rdf* was introduced with the first *fastp* program in 1985; **ref. 8**). Today, programs are available for rigorous Smith-Waterman searches (*ssearch3*) and for searches with mixed peptide sequences (*fastf3* and *tfastf3*); the programs for translated DNA:protein sequence comparison have been improved substantially with the introduction of *fastx3, fasty3, tfastx3,* and *tfasty3,* and the program for estimating statistical significance from shuffled-sequence similarity scores (*prss3*) produces accurate statistical estimates. The *FASTA3* programs for database searching are summarized in **Table 1**; the programs for evaluating statistical significance are shown in **Table 2**.

In addition, several programs in the *FASTA2* package are not yet included with the *FASTA3* programs (**Table 3**). As this chapter is written (summer, 1998), *lalign* is the most important program in the *FASTA2* package that is not in the *FASTA3* package. *lalign* (and the related graphical programs *plalign* and *flalign*) can produce multiple local alignments from the same pair of protein sequences, whereas *FASTA3* and *FASTA* produce only one alignment. Multiple local alignments can highlight domains with proteins; i.e., a protein may contain several domains that share strong similarity with a library sequence. When multiple similar domains are present, *FASTA3* shows only the most similar alignment; *lalign* is required to detect the alternative alignments.

In general, programs in the *FASTA3* package are preferred over the older *FASTA2* programs if *FASTA3* has the function you need. Programs in the *FASTA3* package have more robust statistical estimates and error handling, a larger variety of scoring matrices (*FASTA3* has MDM10, MDM20, PAM120, and BLOSUM80 in addition to PAM250, BLOSUM50, and BLOSUM62 in *FASTA2*), and a broader array of comparison functions (*fasty3, fastf3, tfasty3,* and *tfastf3*).

2.1. Which Program Should I Use?

Many investigators who use the *FASTA* program for protein and DNA database searches are unfamiliar with other programs in the package, or are unclear as to when they should be used. **Table 4** suggests some strategies for using the programs in the FASTA3 package.

The suggestions in **Table 4** are based on two rules-of-thumb: use the program that is designed for your problem; and whenever possible, search protein sequence databases before DNA sequence databases. Protein sequence comparison routinely reveals homologous sequences that diverged 2–3 billion years ago; it is difficult for DNA sequence comparison to look back more than

Table 1
Comparison Programs in the *FASTA3* Package

fasta3	Compare a protein sequence to a protein sequence database or a DNA sequence to a DNA sequence database using the FASTA algorithm ***(4,7)***. Search speed and selectivity are controlled with the ktup (word size) parameter. For protein comparisons, *ktup* = 2 by default; *ktup* = 1 is more sensitive but slower. For DNA comparisons, *ktup* = 6 by default; *ktup* = 3 or *ktup* = 4 provides higher sensitivity; *ktup* = 1 should be used for oligonucleotides (DNA query lengths <20).
ssearch3	Compare a protein sequence to a protein sequence database or a DNA sequence to a DNA sequence database using the Smith-Waterman ***(22)*** algorithm. *ssearch3* is about 10-times slower than *FASTA3*, but is more sensitive for full-length protein sequence comparison.
fastx3/ *fasty3*	Compare a DNA sequence to a protein sequence database, by comparing the translated DNA sequence in three frames and allowing gaps and frameshifts. *fastx3* uses a simpler, faster algorithm for alignments that allows frameshifts only between codons; *fasty3* is slower but produces better alignments with poor quality sequences because frameshifts are allowed within codons.
tfastx3/ *tfasty3*	Compare a protein sequence to a DNA sequence database, calculating similarities with frameshifts to the forward and reverse orientations.
tfasta3	Compare a protein sequence to a DNA sequence database, calculating similarities (without frameshifts) to the three forward and three reverse reading frames. *tfastx3* and *tfasty3* are preferred because they calculate similarity over frameshifts.
fastf3	Compare a mixed peptide sequence to a protein sequence database. A mixture of peptides, typically obtained by Edman degradation after cyanogen bromide cleavage without further separation, is compared with protein sequences in a database to identify those sequences that are most likely to produce the peptide mixture.
tfastf3	Compare a mixed peptide sequence to a translated DNA sequence database.

200–500 million years. Thus, protein sequence comparison, or translated DNA sequence comparison, allows one to identify homologs that diverged 5–10 times farther back in evolutionary time (**Table 5**).

In addition, low-complexity regions are relatively easily removed from protein sequence databases and recognized in protein sequence alignments, but they are much more difficult to recognize in DNA sequence alignments. These regions can produce statistically significant similarity scores for nonhomologous sequences because of their unusual amino acid composition. Thus,

Table 2
Statistics Programs in the *FASTA3* Package

prss3	Evaluate the significance of a protein or DNA sequence similarity score by comparing two sequences and calculating optimal similarity scores, and then repeatedly shuffling the second sequence, and calculating optimal similarity scores using the Smith-Waterman algorithm. The characteristic parameters of the extreme value distribution are estimated from the shuffled sequence scores and used to calculate the statistical significance of the unshuffled sequence similarity score.
sc_to_e	Calculate the statistical significance of a similarity score from the raw score, the length of the sequence, the statistical parameters estimated from a search, and the size of the database.
randseq	Produce a random sequence with the same length and amino acid composition as a query sequence. Random sequences are useful in evaluating the accuracy of statistical estimates. In general in a database search, the highest-scoring match to a random query sequence should have an expectation value E of approx 1.

Table 3
Programs Available only with *FASTA2*

lalign/ plalign/ flalign	Find multiple local alignments between two protein or DNA sequences using the *sim* implementation *(23)* of the Waterman-Eggert *(24)* algorithm. *lalign* shows traditional alignments; *plalign* produces graphics, whereas *flalign* produces graphics commands for the GCG figure program. This program performs successive full Smith-Waterman alignments, and is best used for protein alignments. For DNA, try *lfasta* (below).
lfasta/ plfasta/ flfasta	Find multiple local alignments between two protein or DNA sequences using the *fasta* algorithm. *lalign* uses the heuristic *fasta* algorithm with a local band-alignment. *lalign* is preferred for protein alignment, but *lfasta* is much faster for very long DNA sequences. *plfasta* and *flfasta* produce graphical output.
prdf	Like *prss3*, but uses the *fasta* algorithm instead of Smith-Waterman. *prss3* is preferred.
align	Global sequence alignment between two protein or DNA sequences using linear space *(25)*.
aacomp	Reports amino acid composition and molecular weight of a protein sequence.
grease/ tgrease	Calculates the hydropathy plot of a protein sequence using the Kyte-Doolittle method *(26)*. *tgrease* produces tektronix graphics.

Table 4
Which Program When?

Problem	Program	Explanation	Alternative
Identify unknown protein	(1) *fasta3*	General protein comparison. Use *ktup* = 2 (the default) for speed; *ktup* = 1 for a more sensitive search. Search first against the smallest library likely to contain a homolog (i.e. *SwissProt* rather than *Genpept*).	*blastp*
	(2) *ssearch3*	10- to 50-fold slower than *fasta3*, but provides maximum sensitivity. No advantage for DNA comparisons.	*fasta3*/ *blastp*
	(3) *tfastx3*/ *tfasty3*	If a homolog cannot be found in the protein databases, check the DNA databases with *tfastx3* or *tfasty3*. *tfasty3* provides more accurate alignments, but is approx 33% slower.	*tblastn*/ *tfasta*[a]
Identify structural DNA sequence	*fasta3*	If the DNA sequence encodes a protein, use protein sequence comparison first, then try translated protein sequence comparison (*fastx3/fasty3*). For repeated DNA sequences or structural RNAs, search first with *ktup* = 6 (the default), then *ktup* = 3. Search with *ktup* < 3 only for very short sequences (PCR primers).	*blastn*
Identify EST sequence	*fastx3*/ *fasty3*	Protein sequence comparison is far more sensitive than DNA comparison, so check first to see if the EST encodes a product homologous to a known protein.	*fasta3*/ *blastx*/ *tblastx*
Identify new orthologs	*tfastx3*/ *tfasty3*	If possible, search EST sequences from the same species. Use low/close MDM20 scoring matrices to detect close relationships and avoid distant relationships.	*tblastn*/ *tblastx*
Confirm statistical significance	*prss3*	Use 500–2000 shuffles, and remember to normalize the statistical significance to the size of the database originally searched (typically 10,000–100,000 sequences).	
Confirm statistical estimates	*randseq*	Use to generate random sequences; then search using *fasta3* (or *blastp* or *ssearch3*) and look for E approx 1.0.	

[a]No longer recommended.

Table 5
DNA vs Protein Sequence Comparison

The best scores are:		DNA E(188,018)	*tfastx3* E(187,524)	prot. E(331,956)
DMGST	*D. melanogaster* GST1-1	1.3e–164	4.1e–109	1.0e–109
MDGST1	*M. domestica GST-1* gene	2e–77	3.0e–95	1.9e–76
LUCGLTR	*Lucilia cuprina* GST	1.5e–72	5.2e–91	3.3e–73
MDGST2A	*M. domesticus GST-2* mRNA	9.3e–53	1.4e–77	1.6e–62
MDNF1	*M. domestica nf1* gene. 10	4.6e–51	2.8e–77	2.2e–62
MDNF6	*M. domestica nf6* gene. 10	2.8e–51	4.2e–77	3.1e–62
MDNF7	*M. domestica nf7* gene. 10	6.1e–47	9.2e–77	6.7e–62
AGGST15	*A. gambiae GST* mRNA	3.1e–58	4.2e–76	4.3e–61
CVU87958	*Culicoides* GST	1.8e–41	4.0e–73	3.6e–58
AGG3GST11	*A. gambiae GST1-1* mRNA	1.5e–46	2.8e–55	1.1e–43
BMO6502	*Bombyx mori GST* mRNA	1.1e–23	8.8e–50	5.7e–40
AGSUGST12	*A. gambiae GST1-1* gene	2.3e–16	4.5e–46	5.1e–37
MOTGLUSTRA	*Manduca sexta* GST	5.7e–07	2.5e–30	8.0e–25
RLGSTARGN	*R. legominosarum gstA* and *gstR*	0.0029	3.2e–13	1.4e–10
HUMGSTT2A	*H. sapiens* GSTT2	0.32	3.3e–10	2.0e–09
HSGSTT1	*H. sapiens GSTT1* mRNA	7.2	8.4e–13	3.6e–10
ECAE000319	*E. coli* hypothet. prot.	–	4.7e–10	1.1e–09
MYMDCMA	*Methylophilus dichlorometh.* DH	–	1.1e–09	6.9e–07
BCU19883	*Burkholderia maleylacetate* red.	–	1.2e–09	1.1e–08
NFU43126	*Naegleria fowleri* GST	–	3.2e–07	0.0056
SP505GST	*Sphingomonas paucim*	–	1.8e–06	0.0002
EN1838	*H. sapiens* maleylacetoacetate iso.	–	2.1e–06	5.9e–06
HSU86529	Human GSTZ1	–	3.0e–06	8.0e–06
SYCCPNC	Synechocystis GST	–	1.2e–05	9.5e–06
HSEF1GMR	*H. sapiens EF1*g mRNA	–	9.0e–05	0.00065

The primate, other mammal, invertebrate, and bacterial sections of *GenBank* were searched using a *Drosophila* glutathione transferase cDNA (DMGST) and protein (gtt1_drome) sequence using *fasta3* (DNA, *ktup* = 4), *tfastx3*, and *fasta3* (protein, *ktup* = 2). Expectation values for selected high scoring sequences are shown. DNA comparisons with "—" had expectation values E>100 .With this query, DNA sequence comparison detects homologs only in other insects, while protein and translated DNA comparison finds statistically significant similarity with homologs from humans and bacteria.

when seeking to identify a newly sequenced expressed sequence tag (EST) sequence, you should first use *fastx3* or *fasty3* to search a comprehensive protein database like *SwissProt* or *PIR,* then search a larger but more redundant database like the *BLAST*/NCBI (National Center for Biotechnology Information) *nr* or *OWL* ***(9)*** nonredundant protein databases, or *Genpept*, and, only after these searches have failed to turn up statistically significant matches should you look for DNA sequence matches.

Table 6
Comparison of *BLAST2* and *FASTA3* Programs

Program		Function
BLAST	*FASTA*	
blastp	*fasta3*	General protein sequence similarity searches. *blastp* is faster and can show alignments between several domains in the same sequence. *fasta3* displays a Smith-Waterman final alignment and produces more accurate statistical estimates in some cases.
blastn	*fasta3*	DNA sequence comparison. *blastn* is highly optimized for speed; it uses a fixed word size (11 nucleotides) and scoring matrix that are inappropriate for some problems (e.g., searching for PCR primer matches).
blastx	*fastx3/* *fasty3*	Compare a translated DNA to a protein sequence database. Whereas *blastx* does six independent searches (one for each of the six frames), *fastx3* and *fasty3* effectively does a single forward (or backward) search, which allows frameshifts in computing the similarity score and alignments. As a result, *fastx3* and *fasty3* are more sensitive and can produce much better alignments than blastx when the DNA sequence has frameshift errors.
tblastn	*tfastx3/* *tfasty3/* *tfasta*	Compare a protein sequence to a DNA sequence database, translating in the three forward and reverse frames. Again, *tfastx3* and *tfasty3* provide more accurate alignments than *tblastn* or *tfasta* when the DNA sequences have frameshift errors.
	tblastx	Compare a DNA query sequence to a DNA library, translating both sequences in all six frames and scoring using a protein substitution matrix (BLOSUM62). *fasta3* with *ktup* = 6 (the default) provides a similar function, but does not use a protein scoring matrix.

2.2. FASTA *vs* BLAST

The *BLAST* family of sequence comparison programs ***(10,11)*** offers many of the same search capabilities as the *FASTA* programs (**Table 6**). In general, the *BLAST* programs are faster, but the *FASTA* programs can provide more accurate alignments. For most protein sequence database searching, the current *Blastp2.0* (gapped blast, **ref. *11***) will identify an unknown protein as effectively as *fasta3* and even the more rigorous *ssearch3*. *fasta3* and *ssearch3* use different scoring matrices (BLOSUM50) and gap penalties (-12 for the first residue in a gap, -2 for each additional residue) from *blastp2.0* (BLOSUM62, -12 for the first residue in a gap, -1 for each additional residue). The previous *blastp1.4* produced very poor sequence alignments (because of the restriction on gaps); but the current *blastp2.0* version produces protein alignments that are very similar to those obtained with a rigorous Smith-Waterman search.

For translated DNA–protein comparison and DNA database searches, the *FASTA* programs are much better than their *BLAST* counterparts. Although the gapped *blastp2.0* performs very well in protein comparisons, *blastx* performs the three forward-frame searches separately, whereas *fastx3* and *fasty3* calculate a single alignment that allows frameshifts. Treating the all three forward reading frames as a single sequence makes it much easier to produce high-quality alignments that extend across the length of the matched protein sequence and allows similarity from the different reading frames to be combined in a natural way to improve sensitivity. For example, a *blastx* search with a class-mu mouse glutathione transferase cDNA sequence with insertion and deletion errors at 5% of the positions detected only other class-mu glutathione transferases, whereas a search with the same sequence using *fasty3* detected more class-mu protein sequences with $10^{-20} < E() < 10^{-17}$ and an additional eight more distantly related class-pi glutathione transferase sequences ($10^{-5} < E < 0.01$).

The FASTA programs also provide additional flexibility for DNA sequence searches. Searches can be done with any wordsize (*ktup*) from 1–6; small *ktup*'s are particularly appropriate for searches with short sequences, such as PCR primers. In addition the *FASTA* programs can use a variety of scoring matrices, including matrices with very high mismatch penalties that can be used to identify long identities in sequences.

3. Interpreting *FASTA* Statistics

When rapid sequence comparison programs were first introduced in 1983 ***(1)***, it became possible to find similar DNA and protein sequences by searching sequence databases, but there was no formal basis for deciding whether a weak similarity was likely to be biologically significant. A Monte-Carlo shuffling method for evaluating similarity scores (*rdf*) was provided with the *FASTP* program ***(8)***, but the recommended guidelines for significant similarity ($Z > 5$) were not based on the correct statistical model for local similarity scores and did not account for database size. A sequence with a score that is 10 standard deviations ($Z > 10$) above the mean is expected 0.015 times by chance in a search of a 10,000 entry database; the same score would be expected 0.11 times by chance in a search of *SwissProt* (70,000 entries), and thus would not be statistically significant, even at the 0.05 level.

Accurate statistical estimates were introduced into similarity searching with the *blastp* program ***(10)***, based on the recognition that local similarity scores can be described accurately by the extreme value distribution ***(12,13)***. The Monte-Carlo shuffling program introduced with *fastp* now uses the extreme value distribution to calculate the probability of an alignment score, and the library searching programs in the *FASTA2* and *FASTA3* packages provide a value that can be used to infer homology from statistically significant similarity the expectation (E) value ***(6)***.

The E value is the first number that you should look at when deciding whether to analyze further a high-ranking sequence alignment. Investigators often wonder what E value they should use. This is discussed in detail in the next section, but in most cases, and E value between 0.001 and 0.01 can be used to infer homology reliably, but lower (more conservative) values are required when hundreds or thousands of searches are performed (as when characterizing all the genes in a bacterial genome).

The E value calculated by the *FASTA3* programs and *BLAST* programs is a statistical measure of the likelihood that the observed similarity score could have occurred by chance. Like any statistical measure, its usefulness depends on: whether the assumptions of the underlying statistical model are correct, and the kinds of errors that one is willing to accept when using the measure to draw a conclusion. For similarity searching, we infer homology (common ancestry) from statistically significant similarity. However, the threshold for statistical significance will vary, depending on whether we are more concerned about occasionally misidentifying a nonhomolog (labeling a sequence as related when it is not, a false positive or type I error) or missing a likely homolog (labeling a sequence as nonhomologous when a high-scoring homolog has been found, a false-negative or type II error).

3.1. What Threshold Should I Use to Infer Sequence Homology?

For most molecular biologists, the greatest concern in similarity searching is a false-positive error; we do not want to send a letter to *Nature* identifying a yeast homolog of `p53_human` when no evolutionary relationship exists. (The gold-standard test for homology is structural similarity. If the candidate yeast homolog of P53 has a completely different three-dimensional structure, the hypothesis is wrong.) Whereas incorrect assertion of homology was relatively common before accurate similarity statistics became available, it is rare today. (Unfortunately however, once the observation has been published, it is difficult to remove from the literature.) The E value or expectation calculated by *fasta3* is the number of times you would expect to see a score equal or greater by chance in a search of the database. In other words, E < 0.01 says that you expect to see a score that high (or higher) once by chance in 100 searches; E < 0.001 says once in 1000 searches, and so on. E approx 1 says that you expect to see a score that high, simply by chance, every time you do a search.

Older versions of the *blast* programs used a related statistic, the *p* value, to characterize the significance of a similarity score. The E value reported by the *fasta* programs ranges from 0..*D*, where *D* is the number of entries in the database, whereas the blast *p* value ranges from 0..1. The probability [p()-value] of an E value can be found with the Poisson formula: $p(E)=1-e^{-E}$. For values of E < 0.1, $p \sim$ E, thus *p*(E = 0.1) = 0.1; *p*(E = 1.0) = 0.63; *p*(E = 5.0) = 0.99.

Whereas a sensible E value threshold (0.001–0.01) can ensure that researchers avoid false positive errors, little can be done to avoid false negatives, i.e., labeling a sequence as unrelated to anything in the database when in fact a homolog is present. Most diverse protein families contain pairs of related sequences that do not share statistically significant sequence similarity. Fortunately, if those families are large (e.g., globins, serine proteases, glutathione transferases, G-protein-coupled receptors), it is likely that newly discovered family members will share significant similarity with some known members of the family. As the sequence databases grow more complete and protein families expand, the rate of false negatives should decrease.

3.2. Choosing a Database

The expectation value E(S > x) of a similarity score is calculated from the probability of the pairwise similarity score *p*(S > x), which can be calculated using the extreme value distribution ***(12,13)***, and the number of tests (i.e., sequence comparisons) that were performed to find the high-scoring sequence. Thus, E(S > x) = *p*(S > x)*D*, where *D* is the number of sequences in the database. (For DNA sequence comparison, *D* is not the number of sequences in the database but the length of the database in nucleotides divided by the length of the query sequence.)

Because E increases linearly with the number of database entries, a similarity found in a search of a bacterial genome with 1000–5000 entries will be 50- to 250-fold more significant than an alignment with exactly the same score found in the *OWL* nonredundant protein database (**ref.** ***9***; 250,000 entries). Thus, when searching for very distant relationships, one should always use the smallest database that is likely to contain the homolog of interest. If the goal is to find the *Escherichia coli* homolog of the *Bacillus subtilis* DAHP synthase (`arog_bacsu`), one should search the *E. coli* proteome [which finds the

E. coli kdsA homolog with E(4283) < 0.00015] rather than *SwissProt* [kdsa_ecoli E(74,417) < 0.0017] or *OWL* [kdsa_ecoli E(260,784) < 0.0085]. Here, the same alignment, with the same similarity score, is 50-fold less significant against the largest database than with the smallest.

Likewise, a search of *SwissProt* (approx 70,000 entries) will be three- to fivefold more sensitive than either *OWL* (261,000 sequences) or the *BLAST nr* protein database (332,000 sequences), simply because *SwissProt* is smaller. Thus, an efficient strategy for identifying protein homologs should: 1) search smaller databases first; 2) then research a smaller database (like *SwissProt*) with a more sensitive algorithm (*fasta3* with *ktup = 1* or *ssearch3*), and then, if no significant matches are found, 3) search larger databases (*OWL* or *nr*).

Whereas their size reduces search sensitivity, larger databases can be effective when they provide more diverse members of a protein family. For example, the most distant `p53_human` homolog in *SwissProt* is a flounder sequence. *OWL* contains about twice as many novel p53 homologs, including one from squid.

3.3. Thresholds for Large-Scale Sequence Analysis

Genome sequencing centers and other groups that do thousands of similarity searches each day must use more conservative thresholds of statistical significance to avoid false-positive errors. A threshold of E = 0.001, which is conservative for someone who does a few searches a day, should produce 10 scores below the threshold between nonhomologous sequences by chance after 10,000 searches. Indeed, if you do 100 searches with random sequences against the PIR or Swissprot databases, one of those 100 sequences will find a homolog with E < 0.01, 10 will have E < 0.1, and so on ***(6)***. Genome sequencing centers typically use thresholds of $E < 10^{-6}$, or even lower, when characterizing thousands of sequences.

However, using a more conservative threshold of statistical significance ensures that you will make more false-negative (type II) errors when looking at distant relationships. For example, in a comparison of 2608 human proteins from *SwissProt* against the *E. coli* proteome (4289 sequences), 417 obtained E < 0.02, 373 had E < 0.01, 301 had E < 0.001, 256 had E < 0.0001. Of the 72 with 0.001 < E < 0.01, we would expect that about 26 (0.01 2608) shared similarity this high by chance, while the other 45 are truly homologous. (Unfortunately, we cannot identify which 45 sequences are homologs without additional information.) In the human/*E. coli* search, 209 sequences had $E < 10^{-6}$; we would expect all of these matches are genuine homologies. However, using the conservative 10^{-6} threshold would misidentify as unrelated almost 200 probable homologs. Thus, estimates of the number of novel or unidentified proteins in newly sequenced bacterial genomes are generally overestimates, because many of these novel proteins may share significant similarity when searched individually, but not when searched in a group of 2000–4000 sequences.

3.4. Statistical Estimates—What Can You Trust?

If the statistical estimates are accurate, the guidelines in the previous section provide a reliable strategy for identifying related sequences based on sequence similarity. However, with biological sequences (as opposed to fair coins), the assumptions underlying the statistical model may not be met. When the assumptions fail, the highest scoring unrelated sequence may have an expectation value that is much too low [e.g., $E < 10^{-3}$] or much too high [$E > 100$]. If the E value is too low, unrelated sequences will be mistakenly labeled as related (false positives). If the E values are too high, it is likely that the E values of related sequences are too high as well, and related sequences will be missed (false negatives).

In general, inaccurate statistical estimates are caused by either incorrect gap penalties or low complexity regions (runs of simple amino acid composition, e.g., `ggqgppgdaggpg` from a *Caenorhabditis elegans* collagen or `ssggvtfsvss` from a *Drosophila* trypsin) in the query sequence ***(3,14)***. In the first case, the statistical model has failed. The statistical theory behind the estimates for *BLASTP, FASTA* and Smith-Waterman (*ssearch3*) scores assumes that the scores are local, i.e., on average, nonidentical amino acids will have similarity scores $s_{ij} < 0$. If the gap penalties are too low, then the alignment algorithm will choose to insert a gap, rather than to end the alignment, and the alignment will tend to become global, aligning the sequences from end to end. The statistical properties of global alignment scores are different from those of local scores. Local scores follow the extreme-value distribution; the distribution of global alignment scores is not well understood.

The reliability of the sequence statistics can be confirmed quickly by looking at the histogram of observed and expected similarity scores that is displayed after a *FASTA3* search, and by checking the expectation [E] value of the highest-scoring unrelated sequence. These examples show results from running the *FASTA3* and *ssearch3* programs, which are distributed from `ftp://ftp.virginia.edu/pub/fasta/`. These programs available from this site run on most UNIX platforms (Digital UNIX, IBM, AIX, Linus, SGI Irix, and Sun Solaris) as well as *Windows* (*Windows95* and *NT*) and *Macintosh*. The output shown here may differ slightly from the *FASTA* program distributed with the Genetics Computer Group (GCG), but similar information is available from all modern *FASTA* implementations. Although identifying the highest-scoring unrelated sequence seems to presume knoweldge of the protein family, additional searches with candidate unrelated sequences [E ~ 1] can often separate low-scoring related from high-scoring unrelated sequences *(5)*. If there is good agreement between the observed and expected distribution of scores and the E() value of the highest scoring unrelated sequence is approx 1, the statistical estimates should be accurate.

3.4.1. Low Gap Penalties Cause Inaccurate Estimates

For most protein and DNA sequence searches, there is excellent agreement between the observed and expected distribution of scores (**Fig. 1**) and the E value of the highest-scoring unrelated sequence is ~approx 1.0 (**Table 7**; **ref.** ***6***). The *FASTA* programs provide a histogram summarizing the distribution of observed and expected scores after every search (**Figs. 1–3**). **Figure 1** reports that for this search, 788 sequences (**opt** column) in the database obtained scores of 38–39 (left-most column), whereas 692 sequences [**E** column] are expected to have scores in that range for a database of 14,000 sequences. Agreement between observed ("===" graph) and expected ("*" in histogram) is especially important in the shaded area in **Fig. 1**. For many searches, it is also possible to confirm the accuracy of the estimates by looking for the highest-scoring unrelated sequence in the list of high scoring sequences. In **Table 7** the highest-scoring unrelated sequences are S30223 and NOBY2, with expectation values approx 8. [Ideally, these scores would be a bit closer to 1; the highest-scoring unrelated sequence in the same search with *ssearch3* has E < 3.]

Tables 8 and **9**, and **Fig. 2** show two examples of searches where the statistical model has failed. In the first case (**Table 8**), a DNA search was performed with gap penalties of -12 and -2, rather than the default –16, –4. Whereas the histogram (not shown) shows good agreement between the observed and expected distribution of scores, the E value of the highest-scoring unrelated sequence is 0.01. (That the high-scoring unrelated sequence does not contain a homolog was confirmed by scanning it with *tfasty3*). Moreover, the E values for homologous alignments increase by 10^7 (e.g., from 1.2×10^{-12} to 0.0008 for AC002520; **Table 8**) when the gap penalties are reduced from –16/–4 to –12/–2. DNA sequence searches with even lower gap penalties do show sizeable differences between the observed and expected distribution of scores, but the E value of the highest-scoring unrelated sequence is usually the most sensitive measure of the accuracy of the statistical estimates.

3.4.2. Low E() Values from Low-Complexity Regions

Low E() values between nonhomologous sequences are usually caused by low-complexity regions *(3,14)*. The *Drosophila* groucho protein sequence (`grou_drome`) contains only five low-complexity regions (83 of 719 residues as determined by *seg*, **ref.** ***14***), but as comparison of **Figs. 2** and **Fig. 3** shows, matches in these regions significantly distort the distribution of the

```
gtt1_drome.aa: 209 aa
 >gi|121694|sp|P20432|GTT1_DROME GLUTATHIONE S-TRANSFERASE 1-1 (CLASS-THETA)
 vs  NBRF Annotated Protein Database (rel 56) library
searching /seqlib/lib/pir1.seq 5 library

        opt      E()
< 20     13      0:=
  22      0      0:          one = represents 22 library sequences
  24      0      0:
  26      0      0:
  28      1      3:*
  30     11     19:*
  32     46     75:===*
  34    242    204:=========*=
  36    493    419:===================*===
  38    788    692:===============================*====
  40   1055    965:===========================================*====
  42   1275   1180:=====================================================*====
  44   1299   1302:===========================================================*
  46   1251   1326:=========================================================  *
  48   1186   1269:======================================================   *
  50   1077   1158:=================================================   *
  52    907   1018:=========================================    *
  54    849    870:=======================================*
  56    714    727:=================================*
  58    570    596:========================== *
  60    456    483:=====================*
  62    393    387:=================*
  64    313    308:==============*=
  66    268    243:===========*=
  68    219    192:========*=
  70    191    150:======*==
  72    127    117:=====*
  74     93     91:====*
  76     91     71:===*=
  78     44     55:==*
  80     33     43:=*
  82     22     33:=*
  84     32     26:=*
  86     19     20:*
  88     19     16:*          inset = represents 1 library sequences
  90      8     12:*
  92      8      9:*         :========*
  94      5      7:*         :===== *
  96      2      6:*         :==   *
  98      3      4:*         :===*
 100      1      3:*         := *
 102      3      3:*         :==*
 104      0      2:*         :  *
 106      1      2:*         :=*
 108      0      1:*         :*
 110      0      1:*         :*
 112      0      1:*         :*
 114      0      1:*         :*
 116      0      0:          *
 118      1      0:=         *=
>120      7      0:=         *=======
```

Fig. 1. Histogram of *fasta3* similarity scores—results of a search of a *Drosophila* class-theta glutathione transferase (`gtt1_drome`) against the annotated PIR1 protein sequence database. The initial histogram output is shown. The shaded section indicates the region that is most likely to show discrepancies between observed and expected numbers of scores when the statistical model fails.

high-scoring unrelated sequences. In contrast, a search with the five low-complexity regions masked (**Fig. 3**) shows the expected distribution of scores. Examination of the list of high-scoring sequences in the low-complexity search (**Table 9**) shows a large number of significant matches [$0.00013 < E < 0.02$] to unrelated proteins with biased amino-acid compositions, whereas the highest-

Table 7
FASTA Search–High-Scoring Sequences

Name	description	len	initn	opt	z-score	E()
XUFF11	glutathione transferase	209	1399	1399	1626.5	1.2e–84
XUZM32	glutathione transferase	222	133	173	210.9	8.6e–06
XUZM31	glutathione transferase	220	107	164	200.6	3.2e–05
XUZM1	glutathione transferase	213	123	144	177.7	0.00061
RGECSS	string. starv. prot.– *E. coli*	212	106	140	173.1	0.0011
XURTG	glutathione transferase	222	58	139	171.7	0.0013
XURT8C	glutathione transferase	222	39	115	144.0	0.046
XURTG4	glutathione transferase	218	40	93	118.7	1.2
A37378	glutathione transferase	210	40	82	106.2	5.8
S30223	elongation factor eEF-1g	227	34	80	103.5	8.3
NOBY2	*phosphopyruvate hydratase*	437	53	83	103.1	8.8
PWBYD	*H+-transporting ATP synthase*	212	53	79	102.7	9.2

High-scoring sequences from searches of `gtt1_drome` against the annotated PIR1 database *(27)* with *fasta3* (*ktup* = 2). High-scoring unrelated sequences are highlighted in *italics*.

scoring unrelated sequence in the *seg-ed* search has E < 0.047. Perhaps surprisingly, the significance of the related GTP-binding regulatory protein similarity scores improve almost 1000-fold as well (**Table 9**).

For protein–protein database searches, removal of low-complexity sequences is equally effective for either the query sequence or the protein database. However, it is more difficult to remove low-complexity regions from DNA query sequences, such as EST sequences. Unfortunately, high-scoring alignments between low-complexity protein sequences and out-of-frame DNA translations are common ***(15)***. A simple strategy for improving the sensitivity of translated DNA searches (*fastx3, fasty3,* or *blastx*) is to search against a *seg-ed* protein database ***(14)***.

Low-gap penalties and low-complexity regions produce unreliable statistical estimates because the underlying assumptions of the statistical model do not apply. Low gap penalties cause alignments to shift from local to global; extreme-value alignment statistics apply only to local alignments. Low-complexity regions violate implicit assumptions about higher-order structure in the unrelated sequences. With low-complexity sequences the matches are statistically significant but not biologically significant, because the statistical model assumed that each position of a random (unrelated sequence) is independent of all the others.

When the statistical model is valid—local alignments and truly random unrelated sequences—statistically significant similarity scores can be used to infer homology reliably. And one can usually check that the statistical model is correct by looking at the histogram of observed and expected similarity scores,

Table 8
***FASTA* Search–Low Gap Penalties**

The best scores are:		(length)	initn	opt	z-sc	E(−12/−2)	E(−16/−4)
AC002520	Human Chr. 1p13	(11901)	1507	404	173.1	0.0008	1.2e–12
AC000031	Human Chr. 1p13.3	(39043)	1396	394	161.0	0.0011	6.5e–12
HSU47924	*Human chr. 12p13*	(78864)	235	352	138.3	0.01	2.0
AC000032	Human Chr. 1p13	(29867)	1354	345	141.6	0.018	6.6e–09
CACD42	*C.atys CD4 mRNA*	(1189)	69	307	146.1	0.26	—
HUMDXS455A	*Human cosmid*	(38409)	126	274	109.2	0.89	—
HSHS12ENH	*Homo sapiens DNA*	(3735)	151	278	126.1	1.1	0.038
HSV411C11	*Human DNA*	(5637)	165	276	122.5	1.1	—
HUMHSLA	*Human hormone-sens.*	(3255)	63	275	125.7	1.3	—
AF031078	*Human chr. X*	(78864)	188	264	100.2	1.4	0.078
AF035180	*Human chr. 4q35*	(4638)	67	271	121.7	1.5	0.08

High-scoring sequences from a *fasta3* search (*ktup* = 6) of the Primate division of *GenBank* 106 (approx 80,0000 sequences) using the reverse complement of a m*Gstm1* cDNA sequence (MUSGLUTA) using the default substitution matrix (+5/–4) and low (–12/–2) or default (–16/–4) gap penalties. Unrelated sequences are highlighted with *italics*. The low gap penalties improve the E value of the unrelated HSU47924 sequence to E < 0.01 and reduce the significance of the homologous AC002520, AC000031, and AC000032 sequences by 10^7.

and by checking the expectation value of the highest scoring unrelated sequence.

4. *FASTA3* Program Options

The behavior of the programs in the *FASTA* package can be modified with a variety of command line options; options are available to change the scoring matrix and gap penalties, use alternate statistical estimation methods, and change the format of the alignment output. Many of the options apply to all of the programs in the package (**Table 10**); other options are specific to *fasta3* or *tfastx/y3* (**Table 11**). When using the *FASTA* programs distributed from the University of Virginia, command line options must precede other program arguments. The standard invocation of a *FASTA* program is:

```
program -opt1 -opt2 arg2 query_file library ktup-opt
```

specifically:

```
fasta3 -q -f -14 -w 75 -L -m 1 mgstm1.aa /slib/swiss 1
```

In the latter case, the *fasta3* program is run in quiet (`-q`) mode with a penalty for the first residue in a gap of -14 (`-f -14` rather than the default -12),

Table 9
***FASTA* Search–Low- Complexity Regions**

Search with complete grou_drome:		length	initn	init1	opt	z-sc	E(14,212)
RGHUB1	GTP-binding reg. prot.	(340)	161	147	237	197.4	4.9e–05
RGHUB3	GTP-binding reg. prot.	(340)	163	152	233	194.2	7.4e–05
RGBOB2	GTP-binding reg. prot.	(326)	181	149	228	190.5	0.00012
PIHUB6	*salivary proline-rich prot*	(392)	142	142	229	190.1	0.00013
RGKWB	GTP-binding reg. prot.	(340)	159	154	222	185.4	0.00023
RGFFBH	GTP-binding reg. prot.	(340)	169	144	219	183.0	0.00031
PIHUSD	*proline-rich glycoprot.*	(310)	141	141	217	182.0	0.00035
PIRT3	*acidic proline-rich protein*	(206)	138	138	212	180.7	0.00042
WMBEW6	*capsid protein–herpes*	(635)	101	101	206	168.7	0.002
S23447	*annexin XI form B-bovine*	(505)	84	84	202	166.9	0.0024
PIHUPF	*salproline-rich glycoprot.*	(251)	147	147	193	164.3	0.0034
PIHUSC	*proline-rich phosphoprot.*	(166)	88	88	180	156.6	0.0092
CGHU6C	*collagen alpha 1 (II)*	(1487)	104	104	197	156.0	0.0099
RGOOBE	GTP-binding reg. prot.	(341)	156	125	181	152.8	0.015
FOLJSP	*gag polyprotein–foamy vir*	(811)	121	121	187	151.9	0.017
CGBO1S	*collagen alpha 1 (I)-bovine*	(779	88	88	185	150.6	0.02
LUDO7	*annexin VII–slime mold*	(462)	88	88	179	149.2	0.024
CGHU2S	*collagen alpha 2 (I)*	(1366)	88	88	187	148.6	0.026
LUBO11	*annexin XI form A-bovine*	(503)	84	84	177	147.1	0.031
S09257	*Hox A4–chicken*	(309)	116	116	172	146.2	0.035
OZZQMY	*circumsporozoite prot pre.*	(367)	146	146	172	145.1	0.04
Search with seg-ed grou_drome: (low complexity regions removed)							
RGHUB1	GTP-binding reg. prot.	(340)	161	147	237	247.5	8e–08
RGHUB3	GTP-binding reg. prot.	(340)	163	152	233	243.3	1.4e–07
RGHUB2	GTP-binding reg. prot.	(340)	181	149	228	238.1	2.7e–07
RGKWB	GTP-binding reg. prot.	(340)	159	154	222	231.9	5.9e–07
RGFFBH	GTP-binding reg. prot.	(340)	169	144	219	228.7	8.9e–07
RGOOBE	GTP-binding reg. prot.	(341)	156	125	181	189.1	0.00014
BVBYMS	*MSI1 protein–yeast*	(422)	116	74	139	143.9	0.047
ERHUAH	*coatomer complex alpha*	(1224)	109	109	134	131.7	0.23
I37062	*involucrin S–gorilla*	(495)	129	81	115	117.8	1.3

Unrelated sequences are highlighted in *italics*.

alignments are printed at 75 residues per line (`-w 75`), a long description of the library sequence is shown with the alignment (`-L`), and the alignment symbol highlights the differences rather than similarities (`-m 1`). **Figure 4** shows the difference between a conventional alignment (**Fig. 4A**) and one produced with the command line options shown above (**Fig. 4B**).

Command line options can be divided into five general categories: scoring parameter options, statistics options, algorithm-specific options, file specification options, and output options.

```
grou_drome.aa: 719 aa
 >GROU_DROME GROUCHO PROTEIN (ENHANCER OF SPLIT M9/10). - DROSOPHILA MELANOGAS
 vs  NBRF Annotated Protein Database (rel 56) library
searching /seqlib/lib/pir1.seq 5 library

      opt      E()
< 20   13     0:=
  22    0     0:          one = represents 28 library sequences
  24    1     0:=
  26    0     0:
  28    1     3:*
  30   10    20:*
  32   21    76:= *
  34  105   205:====   *
  36  272   422:==========     *
  38  540   697:====================    *
  40  937   972:==================================*
  42 1269  1188:==========================================*===
  44 1645  1311:==============================================*============
  46 1666  1335:===============================================*============
  48 1577  1278:=============================================*===========
  50 1310  1166:=========================================*=====
  52 1056  1025:====================================*=
  54  851   876:===============================*
  56  669   732:========================  *
  58  423   601:================     *
  60  419   487:===============  *
  62  255   390:==========   *
  64  196   310:=======    *
  66  181   245:======= *
  68  154   193:======*
  70   99   151:==== *
  72   74   118:=== *
  74   63    92:===*
  76   60    72:==*
  78   47    56:=*
  80   48    43:=*
  82   36    33:=*
  84   33    26:*=
  86   27    20:*
  88   21    16:*         inset = represents 2 library sequences
  90   18    12:*
  92   20     9:*        :====*=====
  94   20     7:*        :===*======
  96   17     6:*        :==*======
  98    7     4:*        :=*==
 100   10     3:*        :=*===
 102   11     3:*        :=*====
 104   10     2:*        :*====
 106   11     2:*        :*=====
 108    7     1:*        :*===
 110   10     1:*        :*====
 112    6     1:*        :*==
 114    4     1:*        :*=
 116   11     0:=        *======
 118   10     0:=        *=====
>120   70     0:===      *===================================
5446221 residues in 14321 sequences
 Expectation_n fit: rho(ln(x))= 8.0964+/-0.00108; mu= 4.7475+/- 0.061;
 mean_var=157.6967+/-31.622, 0's: 13 Z-trim: 96  B-trim: 33 in 1/62
 Kolmogorov-Smirnov  statistic: 0.0497 (N=29) at  52
```

Fig. 2. Poor statistics: low complexity regions—a *fasta3* search (*ktup* = 2) of the PIR1 database using `grou_drome`. The histogram of sequence similarity scores is shown. In this case, there are clear discrepancies between the observed and expected numbers of sequences with scores in the central part of the distribution and in the tails, and there is an excess of high-scoring sequences. **Table 9** shows that all of these excess high-scoring sequences are unrelated.

4.1. Changing the Scoring Parameters

All the programs in the *FASTA3* package calculate sequence alignments using two types of scoring parameters: a substitution matrix and gap penalties. The default scoring matrix, gap penalties, E value cutoff, and comparison algorithm are shown in **Table 12**. The *fasta3*, *ssearch3, fastx/y3,* and *tfastx/y3*

```
grou_drome.seg: 719 aa
 >GROU_DROME GROUCHO PROTEIN (ENHANCER OF SPLIT M9/10). - DROSOPHILA MELANOGAS
 vs  NBRF Annotated Protein Database (rel 56) library
searching /seqlib/lib/pir1.seq 5 library

      opt      E()
< 20    48      0:==
  22    14      0:=          one = represents 24 library sequences
  24    21      0:=
  26    37      0:==
  28    39      3:*=
  30    65     20:*==
  32    95     76:===*
  34   175    206:========*
  36   348    424:===============  *
  38   591    700:=========================    *
  40   891    977:======================================  *
  42  1141   1194:================================================ *
  44  1328   1317:======================================================*=
  46  1373   1342:=======================================================*==
  48  1395   1285:=====================================================*=====
  50  1227   1172:================================================*===
  52  1107   1031:==========================================*====
  54   888    880:====================================*
  56   723    735:==============================*
  58   602    604:=========================*
  60   490    489:====================*
  62   357    392:=============== *
  64   284    312:============*
  66   246    246:==========*
  68   177    194:========*
  70   131    152:======*
  72   110    119:====*
  74    64     93:===*
  76    76     72:==*=
  78    53     56:==*
  80    41     43:=*
  82    44     33:=*
  84    22     26:=*
  86    26     20:*=
  88    17     16:*          inset = represents 1 library sequences
  90    11     12:*
  92    14      9:*         :========*=====
  94     5      7:*         :===== *
  96     7      6:*         :=====*=
  98    11      4:*         :===*=======
 100     2      3:*         :==*
 102     5      3:*         :==*==
 104     3      2:*         :=*=
 106     1      2:*         :=*
 108     1      1:*         :*
 110     0      1:*         :*
 112     1      1:*         :*
 114     0      1:*         :*
 116     0      0:          *
 118     1      0:=         *=
>120    13      0:=         *=============
5446221 residues in 14321 sequences
 Expectation_n fit: rho(ln(x))= 6.3481+/-0.00105; mu= 10.5411+/- 0.059;
 mean_var=92.0111+/-17.844, 0's: 13 Z-trim: 24  B-trim: 593 in 1/62
                  Kolmogorov-Smirnov  statistic: 0.0129 (N=29) at  42
```

Fig. 3. Accurate statistics with "*seg*-ed" query—the search in **Fig. 3** was performed using the `grou_drome` sequence with low-complexity sequences masked using the *seg* program ***(14)***. With low-complexity sequences removed, the numbers of observed and expected similarity scores agree closely. Identical results are obtained when low-complexity regions are removed from the PIR1 database instead of `grou_drome`.

programs use the BLOSUM50 scoring matrix ***(16)*** for protein sequence (and translated protein sequence) comparisons. Alternate protein scoring matrices can be specified with the `-s` option. Available protein matrices include BLOSUM62 (`-s BL62`) and BLOSUM80 (`-s BL80`), PAM250 (`-s P250`),

Table 10
FASTA3 General Options

-a	show full sequences rather than only overlapping region (*fastx/y3* and *tfastx/y3* do not provide this feature)
-b #	number of best scores to show (must be < -E cutoff)
-d #	number of best alignments to show (must be < -E cutoff)
-E #	expectation value limit for displaying scores and alignments. (By default, 10.0 for protein sequence comparisons; 5.0 for *fastx/y3*, and 2.0 for DNA sequence comparisons.)
-H	turn off histogram display
-i	(DNA only) reverse complement the query sequence (*tfastx/y3*); compare against only the reverse complement of the library sequences.
-L	report long sequence description in alignments
-m 1–6,10	alignment display options (**Table 14**)
-n	force query to nucleotide sequence (default: autodetect)
-N #	read database in chunks of # residues. # should be > 2-times the query sequence length, as the chunks overlap by the length of the query. (default: 80,000-query-length)
-O file	send output to file
-q/-Q	quiet option; do not prompt for input
-R file	save all scores to statistics file
-S #	offset substitution matrix values
-s name	scoring matrix. BLOSUM50 is used by default for proteins, PAM120, PAM250, and BLOSUM62 can be specified by setting -s P120, P250, or BL62. Additional matrices include: BLOSUM80 (BL80), and MDM_10, MDM_20, MDM_40 (M10, M20, M40; **ref. *19***). Alternatively, BLASTP1.4 format scoring matrix files can be specified.
-w #	line width for similarity score and sequence alignment output
-W #	amount of sequence context around the alignment. Default is 30 residues (not used by *fastx/y3, tfastx/y3*).
-x "#,#"	offsets query and library sequence for numbering alignments
-z #	specify statistics calculation. Default is -z 1. **Table 13**.
-Z #	specify the size of the library to be used for statistical significance estimates.

and PAM120 (`-s P120`) ***(17,18)***, and low evolutionary distance matrices MDM10 (`-s M10`) and MDM20 (`-s M20`) ***(19)***. In addition, any scoring matrix can be used by providing a file name for the file containing the substitution values (`-s matrix.file`). Version 3 of the *FASTA* programs uses the same substitution matrix format as the *blastp* programs, and the *pam* program distributed with the *BLAST* package can be used to generate appropriately formatted matrices.

Table 11
Program-Specific Command-Line Options

fasta3, fastx/y3, tfastx/y3, tfasta3 options	
-1	sort by "init1" score
-3	(*tfasta3, tfastx3, tfasty3* only) use only forward frame translations
-A	force Smith-Waterman alignment for output. Smith-Waterman is the default for protein sequences, *fastx/y3*, and *tfastx/y3*, but not for *tfasta3* or DNA comparisons with *fasta3*.
-c #	threshold for band optimization
-f #	penalty for the first residue in a gap
-g #	penalty for additional residues in a gap
-h #	*fastx/y3, tfastx/y3* only–penalty for a frameshift between codons
-j #	*fasty3, tfasty3* only–penalty for a substitution codon
-t #	translation table–*fastx/y3, tfastx/y3*, and *tfasta3* now support the *BLAST* translation tables. See `http://www.ncbi.nlm.nih.gov/htbin-post/Taxonomy/wprintgc?mode=c/`
-y #	Width for band optimization; by default 16 for DNA and protein *ktup* = 2; 32 for protein *ktup* = 1
***ssearch3* command line options**	
-f #	penalty for first residue in a gap
-g #	penalty for additional residues in a gap

For DNA sequence comparisons, the substitution matrix scores +5 for a match and -4 for a mismatch (+2 for match to an ambiguous nucleotide, -1 for a mismatch to an ambiguous residue). Alternate DNA substitution matrices can be specified using the `-s dna-matrix.file` option.

The BLOSUM50 matrix works well for recognizing very distant relationships (and works well for long, closely related sequences as well). Searches with short sequences ***(18)*** or for closely related sequences (e.g., mouse proteins against mouse ESTs) will be more effective with "shallower" scoring matrices like MDM10 and MDM20 that are optimum for small amounts of change in very short sequences.

Gap penalties in the *FASTA* programs can be changed with the **-f** and **-g** options; **-f** specifies the cost of the first residue in a gap and **-g** specifies the cost of each additional residue. An alternate representation of gap penalties takes the form: $q + rk$, where q is the penalty for opening a gap and r is the penalty for each residue in the gap (k is the length of the gap). Thus, `-f -12, -g -2` (the default for protein searches) is equivalent to: $q = 10$, $r = 2$. Protein substitution matrices like BLOSUM50 and PAM250, which are scaled in 1/3-bit units ***(18)***, work well with gap penalties of -12/-2 or -14/-2 ***(20)***, whereas

A

```
>>GTT1_MUSDO GLUTATHIONE S-TRANSFERASE 1 (EC 2.5.1.18) (C (208 aa)
 initn: 1229 init1: 1229 opt: 1230 Z-score: 1472.4 expect() 2.3e-75
Smith-Waterman score: 1230;  85.024% identity in 207 aa overlap

               10        20        30        40        50        60
gi|121 MVDFYYLPGSSPCRSVIMTAKAVGVELNKKLLNLQAGEHLKPEFLKINPQHTIPTLVDNG
        .::::::::.::::.:::.::::.::::::::::::::::::::::::::::::::.
GTT1_M  MDFYYLPGSAPCRSVLMTAKALGIELNKKLLNLQAGEHLKPEFLKINPQHTIPTLVDGD
                10        20        30        40        50
```

B

```
>>GTT1_MUSDO GLUTATHIONE S-TRANSFERASE 1 (EC 2.5.1.18) (CLASS-THETA).    (208 aa)
 initn: 1229 init1: 1229 opt: 1230 Z-score: 1615.1 expect() 2.6e-83
Smith-Waterman score: 1230;  85.024% identity in 207 aa overlap

               10        20        30        40        50        60        70
gi|121 MVDFYYLPGSSPCRSVIMTAKAVGVELNKKLLNLQAGEHLKPEFLKINPQHTIPTLVDNGFALWESRAIQVYLVE
            x       x     x     x x                                xX         x
GTT1_M  MDFYYLPGSAPCRSVLMTAKALGIELNKKLLNLQAGEHLKPEFLKINPQHTIPTLVDGDFALWESRAIMVYLVE
                10        20        30        40        50        60        70
```

Fig. 4. Alternative output formats—Alignments of gtt1_drome with gtt1_musdo are shown using the default **(A)** program parameters and **(B)** the command line options: `-f -14 -w 75 -L -m 1` (see text for details).

scoring matrices like BLOSUM62 and PAM120, which are scaled in 1/2-bit units, work well with a lower initial residue penalty, (`-f -8`).

Just as shallower substitution matrices may be appropriate for comparisons between closely related sequences (e.g., mammals), higher gap penalties may be appropriate as well. Using a MDM20 scoring matrix with gap penalties of –20/–4 will cause the program to recognize, with very high expectation values, sequences that have diverged by approx 20–40%, but the program will probably miss clear homologs that share less than 30% protein sequence identity.

The *fastx3/tfastx3* and *fasty3/tfasty3* programs provide additional gap parameters. *fastx3/tfastx3* uses -h to specify the cost of a frameshift (which must, because of the nature of the *fastx3* algorithm, fall between two codons). *fasty3/tfasty3* uses -h to set the cost of a between-codon frameshift and -j to specify the cost of a frameshift that within a codon. When searching with EST sequences that contain approx 5% errors, the default values -h -20 and -j -20 work well ***(15)***. However, if the DNA sequences are known to be relatively error free, searches with higher frameshift penalties are appropriate, as they will reduce the noise from out-of-frame alignments.

In general, the default gap parameters provided by the *FASTA* programs are at the lower end of the useful range. Reducing the gap penalties more will often cause alignments to shift from local to global, and thus violate the assumptions underlying the statistical estimates. Small increases in the initial residue (-f) penalty will sometimes slightly improve the expectation value of an alignment, but researchers should be suspicious of borderline scores that change dramati-

Table 12
***FASTA* Program Defaults**

program	query	library	scoring(−s) matrix	gap (−f, −g) penalties	frameshift (−h,−j)	-E() cutoff	alignment
fasta3	protein	protein	BLOSUM50	−12/−2		10.0	Smith-Waterman
	DNA	DNA	+5/−4	−16/−4		2.0	band Smith-Waterman[a]
ssearch3	protein	protein	BLOSUM50	−12/−2		10.0	Smith-Waterman
	DNA	DNA	+5/−4	−16/−4		2.0	Smith-Waterman
fastx3	DNA (1 strand)	protein	BLOSUM50	−15/−2	−20	5.0	Smith-Waterman[b]
fasty3	DNA (1 strand)	protein	BLOSUM50	−15/−2	−20/−20	5.0	Smith-Waterman[b]
tfastx3	protein	DNA	BLOSUM50	−15/−2	−20	5.0	Smith-Waterman[b]
tfasty3	protein	DNA	BLOSUM50	−15/−2	−20/−20	5.0	Smith-Waterman[b]
fastf3	mixed peptides	protein	MDM20			5.0	
tfastf3	mixed peptides	DNA	MDM10			5.0	

[a] **Ref.** *28*.
[b]**Ref.** *15*.

cally with different gap penalties. Changes in substitution matrices usually have a greater effect than small changes in gap penalties; the expectation values from searches with the PAM250 matrix are often 10^{-3}–10^{-10} lower than when BLOSUM50 is used. For example, for the scores shown in **Table 7,** the E() values for the alignments of `gtt1_drome` and `xuzm32`, `xuzm31`, and `xuzm1` drop from 8.5×10^{-8}, 2.5×10^{-6}, and 8.8×10^{-5}, to 7.1×10^{-5}, 0.001, and 0.15 when the PAM250 matrix is used. When evaluating the significance of an alignment using the Monte-Carlo *prss3* program, one should be certain to use the same substitution matrix and gap penalties.

4.2. Alternate Statistical Estimates

One of the strengths of the *FASTA3* package is its ability to estimate accurately the statistical significance of a local similarity score, regardless of whether it was calculated from a protein:protein, DNA:DNA, or protein: translated-DNA alignment. The programs in the *FASTA3* package calculate

Table 13
Statistics Options

-z -1	No statistical estimates. Sometimes necessary when there are no unrelated sequences in the database.
-z 0	Unscaled statistical estimates. Estimates are calculated from the mean and and variance of the sequence similarity scores. Typically used when all of the library sequences have about the same length.
-z 1	Regression-scaled estimates. Mean and variance of the similarity scores are calculated after correcting the scores for a log(n) effect.
-z 2	Log-corrected estimates. Provided for historical purposes only; this method is out of date and should not be used.
-z 3	Altschul-Gish estimates (protein only). Instead of estimating the parameters from the data, pre-calculated parameters published by Altschul and Gish *(29)* are used. -z 3 is the only option for estimating the significance of an alignment when unrelated sequences are not the majority of the searched library.
-z 4	An alternative to -z 1 that uses a different method for removing high-scoring, potentially related sequences during the parameter estimating process.
-z 5	An alternative the -z 1 that also uses regression of the score variance with log(n) (library sequence length). Whereas -z 5 is likely to provide somewhat more accurate estimates than -z 1, it is also more sensitive to problems with the data, particularly when relatively small libraries (< 500 entries) are searched.

expectation values based on parameters estimated from the distribution of scores from unrelated sequences. Thus, the statistical estimates are accurate for the typical case of a search against a database containing tens of thousands of unrelated sequences, but they will not be accurate if the database does not contain unrelated sequences. The *FASTA3* programs provide six statistical estimation options (**Table 13**; **ref.** ***6***). The **-z 3** option is of particular interest, as it can be used when searching databases that do not contain unrelated sequences, or even when comparing a pair of sequences.

The dependence of statistical significance on database size can complicate comparisons of searches on different databases. The **-Z number** option can be used to force the program to pretend that a database of size "number" was searched, e.g. `-Z 100000` might be used to reflect the consensus that there are approx 100,000 mammalian genes. ("number" should never be smaller than the actual size of the database searched.) This option is particularly important in combination with -z 3 when searching a small set of preselected sequences.

Table 14
Input Options

@	In addition to using file names, the *FASTA3* programs can accept query sequences from the `stdin` file stream on UNIX and Windows computers. In this case, all information must be given on the command line, e.g.: `fasta3 -q @ /slib/swiss.seq 1 < query.aa` indicates that the input will come from `stdin` (< query.aa) and that the swiss.seq library will be searched with *ktup* = 1. The @ option is most commonly used with perl scripts on Web servers.
:#-#	Specify a subsequence. Query sequence file names can be followed by a ":" and a range of numbers to specify a portion of a sequence. If the first number is not given, 1 is assumed. If the last number is not given, the subsequence extends to the end of the sequence. Thus, `gtt1_drome`.aa:51–150 specifies the 100 residues beginning at residue 51. Subsequence ranges can be given when the query sequence is entered on the command line or when prompted by the program. They can also be entered after an "@" (stdin) symbol. Subsequence ranges can only be used for the first (query sequence).
-i	(DNA queries only) Search with the reverse complement of the query sequence.
-l file	Identify the FASTLIBS file used to locate sequence databases.
-n	Force the input (query) sequence to be read as DNA (*fasta3* and *ssearch3* only).
-N #	Read long library sequences (such as bacterial genomes) in chunks of "#" residues; e.g. -N 5000 would read long sequences in 5000 residue portions.
-q/Q	Quiet. Do not prompt for input.

4.3. Input Options

The *FASTA* programs provide a number of options that change how the query sequence is used and how the database is selected (**Table 14**). The most commonly used input option is **-i,** which causes a DNA search to use the reverse complement of the query sequence. (Unlike *BLASTN* and the GCG version of *FASTA,* the University of Virginia *FASTA* programs did not before Version *FASTA* 32, December 1998, search automatically with both the forward and reverse DNA strands when a DNA query is used.)

The *FASTA* programs make it easy to specify a search with only part of the query sequence with the ":" modifier to the query sequence file name. The command:

```
fasta3 gtt1_drome.aa:1-100 s
```

searches the database specified by the "s" abbreviation with the first 100 residues of the query sequence `gtt1_drome`.

fasta3 and *ssearch3* use a simple algorithm to decide if a query sequence is likely to be protein or DNA. If the sequence is more than 85% A+C+G+T, it is assumed to be DNA; otherwise it is treated as a protein sequence. The **-n** option forces a query sequence to be treated as DNA; the **-n** option is required for DNA sequences provided through the **stdin** (@) option (**Table 14**). Unlike the *BLAST* programs, the *FASTA* programs currently report only the best alignment between the query sequence and the library sequence, even when the library sequence is very long and may contain hundreds of genes. By default, *FASTA* breaks up long DNA sequences into approx 80,000 nucleotide pieces, but this size is too large for gene-dense bacterial, yeast, and *C. elegans* genomes. The **-N 5000** option tells *fasta3* and *tfastx/y3* to read long DNA sequences in chunks of 5000 nucleotides. This is essential when scanning large, gene-dense DNA sequences.

4.4. Changing the Output Appearance

Many of the *FASTA* command line options change the appearance of the alignment output (**Tables 15 and 16**). Options are available to change the number of residues displayed on an alignment line, to change the numbering of the residues, and to change the format of the alignment. Two options are of particular interest (**Table 16**): **-m 5** provides both the sequence alignment and a crude graphical mapping of the aligned region against the query sequence. This graph makes it much easier to see quickly the parts of the query that align with the different library sequences, and thus can highlight query sequences with separable domains. The **-m 6** option is identical to **-m 5,** but provides HTML mark-up commands and links to *Entrez* and other sites for researching to confirm relationships with the library sequence.

5. Beyond Sequence Homlogy—Identifying New Paralogs

The use of the *FASTA* and *BLAST* programs for identifying distantly related sequences has been reviewed extensively *(3–5)*, so in this last section we will consider a slightly different problem that exploits the flexibility of the *FASTA* programs and the high quality of their alignments.

Here, we seek to identify new paralogs of known human or mouse families from EST databases. For example, two human prostaglandin synthase enzymes are known, COX1 (`pgh1_human`) and COX2 (`pgh2_human`), in humans, mice, rats, and other mammals. Prostaglandin synthases are targets of nonsteroidal anti-inflammatory drugs, including aspirin and ibuprofen. Thus, there is

Table 15
Output Options

-a	(*fasta3* and *ssearch3* only) show the query and library sequences in their entirety, not just the portion that aligns.
-A	(*fasta3* DNA only) *fasta3* does a full Smith-Waterman ***(22)*** alignment for protein sequences (and translated *fastx/y3* and *tfastx/y3* alignments) but only a band-limited alignment for DNA:DNA alignments. The -A option forces *fasta3* to do a full Smith-Waterman alignment for DNA sequences. This can slow the program down substantially if one of the sequences is quite long.
-b #	The number of high-scoring library sequences scores to be shown.
-d #	The number of high-scoring alignments to be shown.
-E #	The expectation [E] value cutoff for showing scores and alignments. By default, -E 10 for protein:protein comparisons, -E 5 for translated DNA:protein comparisons, and -E 2 for DNA:DNA comparisons. The -E cutoff overrides the -b and -d options; to ensure that at least 20 scores and 5 alignments are shown, the options: -E 1000.0 -b 20 -d 10 would be used.
-F #	A lower-bound expectation value cutoff that prevents very closely related sequences from being shown. -F 1e-4 will prevent the programs from showing library sequences with E $<10^{-4}$. This option is useful for focussing on distant homologues in large protein families with many close homologs.
-H	Do not show the histogram.
-L	Provide long sequence descriptions with the alignment. Some sequence library formats (particularly reformatted GCG libraries) include a lot of uninformative text before the actual sequence description. With the -L option, all the sequence description available is displayed with the alignment.
-m #	See **Table 16**.
-O file	Send results to **file**. UNIX and Windows users should use the "**> file**" method for output redirection.
-R file	Send intermediate results for all sequences to **file**.
-w #	Width of alignment output. The FASTA programs display alignments with 60 residues per line by default; this width can be increased to 200 residues with the -w option.
-W #	Amount of sequence context. *fasta3* and *ssearch3* provide neighboring sequence context in the alignment (translated *fastx/y3* and *tfastx/y3* do not). The amount of context is typically one half of an output line, but this amount can be increased or reduced with the -W option.
-x "# #"	Sequence coordinates. Normally, the *FASTA* programs assume that each sequence begins at residue 1. On occasion, it is useful to use a different initial coordinate, such as when comparing a cDNA to the encoding gene or when working with only a portion of a sequence. -x "1, –751" would tell *fasta3* to begin the numbering of the library sequence at "–751" rather than "1". On UNIX, DOS, and Macintosh systems, the two numbers must be surrounded by double quotation (". . .") marks.

great interest in finding additional members of this family and it is certainly possible that additional prostaglandin synthases have been sequenced, either by large-scale EST sequencing or by genomic sequencing.

Table 16
Alignment Options

-m 0	Highlight identical aligned residues with ":", conservative replacements with "."
-m 1	Identities are not highlighted. Highlight conservative replacements with "x", nonconservative replacements with "X".
-m 2	Highlight identities with ".", non-identical residues with the residue.
-m 3	The alignments are printed as two fasta format sequence entries with "-" indicating gaps. These files are sometimes useful as input to other programs.
-m 4	Do not show an alignment; show a graph (-----) of where the aligned region maps onto the query sequence. Useful for highlighting different domains in proteins.
-m 5	A combination of -m 0 and -m 4 that shows both the mapping and the alignment.
-m 6	Similar to -m 5, but includes HTML commands for a Web browser like *Netscape* or *Internet Explorer* and links to simplify looking up the library sequence and researching the database.
-m 10	Parseable output designed to be read by other computer programs. Each alignment is a series of labeled tags that specify the beginning, end, score, search parameters, and other information.

5.1. Overall Strategy

Paralogs are members of a gene family (and are thus related or homologous) that differ from other sequences in the family because of gene duplication events. (Orthologous genes differ because they are found in different species.) A search of the *SwissProt* database (**Table 17**) shows the two prostaglandin synthase (PGH) subfamilies, but also shows distantly related peroxidases. The human PGH1 and PGH2 isoenzymes share approx 65% sequence identity [$E < 10^{-165}$]. (In contrast, orthologous human and mouse PGH1 sequences share 89.3% identity.) We expect a new human PGH synthase to share very strong similarity to PGH1 and PGH2 [$E < 10^{-20}$] but to share less than 80% identity to either PGH1 or PGH2. Because we will be scanning EST databases to find the new paralogs, we expect that sequences with > 90–95% identity are probably from mRNAs for known proteins that have sequencing errors, but that sequences that are 50–90% identical are candidate paralogs.

To identify new `pgh1_human` paralogs, we will search the human EST database (obtained from `ftp://ncbi.nlm.nih.gov/blast/db/`) with the `pgh1_human` and `pgh2_human` protein sequences using the *tfasty3* program. *tfasty3* is used because: 1) we wish to compare a protein query to a DNA (EST) database; and 2) we will use both the expectation value E and the percent identity to characterize matches, so a high-quality protein:DNA alignment is

Table 17
Prostaglandin Synthase Search Results

The best scores are:		len	E(74357)
PGH1_HUMAN	prostaglandin G/H synthase 1	599	3.9e–264
PGH1_SHEEP	prostaglandin G/H synthase 1	600	2.3e–244
PGH1_MOUSE	prostaglandin G/H synthase 1	602	9.5e–237
PGH2_CHICK	prostaglandin G/H synthase 2	603	1.2e–168
PGH2_HUMAN	prostaglandin G/H synthase 2	604	1.9e–165
PGH2_MOUSE	prostaglandin G/H synthase 2	604	2.4e–164
PGH2_CAVPO	prostaglandin G/H synthase 2	604	1.7e–163
PGH2_RAT	prostaglandin G/H synthase 2	604	1.4e–162
PERM_MOUSE	myeloperoxidase prec.	718	0.0001
PERO_DROME	peroxidase prec.	690	0.00024
PERT_HUMAN	thyroid peroxidase prec.	933	0.0003
PERM_HUMAN	myeloperoxidase prec.	745	0.00034
PERT_PIG	thyroid peroxidase prec.	926	0.0029
PERL_BOVIN	lactoperoxidase prec.	712	0.016
PERT_MOUSE	thyroid peroxidase prec.	914	0.02
PERL_HUMAN	lactoperoxidase LPO	324	0.027
PERT_RAT	thyroid peroxidase prec.	914	0.089
FBP1_STRPU	fibropellin I prec.	1064	0.16
PGCN_RAT	neurocan core prot. prec.	1257	0.21
FBP3_STRPU	fibropellin C prec.	570	0.31
PGCN_MOUSE	neurocan core prot. prec.	1268	0.33
PERE_MOUSE	eosinophil peroxidase prec.	716	0.51
NOTC_DROME	neurogenic locus notch prot.	2703	0.74
DLK_MOUSE	delta-like prot. prec.	385	0.86
PERE_HUMAN	eosinophil peroxidase prec.	715	0.92
NTC1_MOUSE	neurogenic locus notch homolog	2531	0.94

Results of a *fasta3* (*ktup* = 2) search with pgh1_human against the *SwissProt* protein sequence database.

required (*tfastx3* is faster but produces a lower quality alignment, **ref.** ***15***). We will then examine the EST sequences that share significant similarity and categorize them as orthologous to pgh1_human, pgh2_human, or a new paralog.

5.2. Statistical Significance and Percent Identity

Whereas our goal is to identify sequences that are similar to, but not identical with known prostaglandin synthases, conventional similarity criteria [E value and percent identity] do not fully capture the information we seek. As the results of the pgh1_human and pgh2_human *tfasty3* searches demonstrate (**Table 18**), EST sequences that share higher sequence identity do not necessarily have better E values.

The discrepancy between E value and percent identity reflects the dependence of E value on alignment length. EST sequences tend to be partial, so that an orthologous 100% match to the carboxy-terminal 30 amino acids in `gb|N79146` can have a worse expectation value (2.9×10^{-6}) than a 59% identity to a paralogous gene [$E < 6.7 \times 10^{-19}$]. However, percent identity is a poor criterion for similarity, because unrelated sequences (e.g., `gb|AA485017`) can share high identity (66.1% over 62 codons) that does not produce a statistically significant similarity score. Nevertheless, for sequences that share significant similarity, percent identity is a useful measure of sequence difference. Thus, among the statistically significant matches in **Table 18,** orthologous matches always had percent identities > 90%, with one possible exception (`gb|AA223896`, *see* **Subheading 5.3.**).

5.3. Shifting Evolutionary Horizons with Scoring Matrices

Examination of the high-scoring ESTs found with `pgh1_human` and `pgh2_human` in **Table 18** suggests that all but one of the ESTs share > 90% identity with either `pgh1_human` or `pgh2_human`. The exception, `gb|AA223896`, shares only 80% identity with `pgh1_human` and 50% identity with `pgh2_human`, and thus is a candidate novel paralog prostaglandin synthase.

However, the `gb|AA223896` EST sequence is very short (97 nucleotides), and there are only six mismatches, half of which are within 20 nucleotides of one end of the sequence. Thus, we must consider whether this is truly a novel paralog, or simply a short, poor-quality sequence of a `pgh1_human` mRNA that has several errors at one end (as is expected with high-throughput EST sequencing). Whereas the end-sequence error problem could be reduced by ad hoc changes to the alignment code that down-weighted end-mismatches, a simpler approach is to use shallower scoring matrices.

Additional searches with very shallow scoring matrices (MDM20 and MDM10, **ref.** ***19***; **Table 19**) show slightly different, potentially more interesting perspectives. When shallower scoring matrices are used, both orthologous and paralogous alignments become more statistically significant, and, as expected, the percent identities increase (shallower scoring matrices give more positive scores to identities and more negative scores to nonconservative replacements). Of greater interest are two sequences `gb|AA223896` and `gb|AA485017`, which show significant similarity with `pgh1_human` with MDM20 and MDM10. Both sequences are tantalizing candidates for new paralogs (as orthologs consistently have percent identities higher than 90% with MDM20. However, the alignments of both sequences show a large number of frameshifts (that do not affect the percent identity calculation), suggesting that

Table 18
Prostaglandin Synthase ESTs

pgh1_human:		len	[f/r]	opt	E(10^6)	%ident.	I/II
gb|R96180	Pineal_gland_N3HPG	355	[f]	654	3e–38	98.0	I
gb|AA296431	Umbilical vein endothelial	279	[f]	380	6.7e–19	59.1	II
gb|T29235	Human Bone	257	[f]	358	2.2e–17	63.3	II
gb|AA037294	Senescent_fibroblasts_NbHSF	471	[f]	304	3.1e–13	98.0	I
gb|AI022012	Senescent_fibroblasts_NbHSF	537	[r]	248	3.5e–09	64.5	II
gb|N79146	Multiple_sclerosis_2NbHMSP	544	[f]	207	2.9e–06	100.0	I
gb|AA223896	NT2 neuronal precursor	97	[f]	185	1.3e–05	80.0	??
gb|AA485017	NCI_CGAP_GCB1	208	[f]	124	0.72	66.1	
pgh2_human:		**len**	**[f/r]**	**opt**	**E(10^6)**	**%ident.**	**I/II**
gb|AA296431	Umbilical vein endothelial	279	[f]	574	1.4e–35	96.8	II
gb|T29235	Human Bone	257	[f]	536	1e–32	92.9	II
gb|AI022012	Senescent_fibroblasts_NbHSF	537	[r]	541	1.1e–32	95.8	II
gb|R96180	Pineal_gland_N3HPG	355	[f]	410	6.3e–23	65.8	I
gb|AA223896	NT2 neuronal precursor	97	[f]	136	0.01	50.0	??
gb|AA885610	NCI_CGAP_Lu5	320	[f]	141	0.018	46.3	
gb|AA911293	NCI_CGAP_Lu5	172	[f]	131	0.049	43.6	

Results from searches with pgh1_human and pgh2_human against the *BLAST* est_human database using *tfasty3* and with the default BLOSUM50 scoring matrix. pgh1 (COXI) or *pgh2* (COXII) orthologs are labeled in the right column.

these sequences may have percent identities < 90% because of a poor quality sequence, rather than a novel gene.

The last two entries (gb|AA885610 and gb|AA911293) in the pgh2_human search shows that shallow scoring matrices can also be used to quickly rule out high-scoring unrelated sequences. The expectation values for those two sequences, which were marginally significant (0.018 and 0.049) scores with BLOSUM50 and were not significantly similar to pgh1_human, became very high [E > 5] when MDM20 and MDM10 were used. Thus, shallower scoring matrices can be used to provide a more stringent test for sequence similarity when near-identity is expected for at least one of the query sequences. Whereas MDM20 and MDM10 can serve to provide more stringent alignments, they are not the best matrices, because they were built assuming an evolutionary model. More accurate matrices could be derived from looking at large numbers of EST sequencing errors, and building a matrix that was based on a sequencing error model, rather than evolutionary divergence.

6. Summary

The *FASTA3* and *FASTA2* packages provide a flexible set of sequence-comparison programs that are particularly valuable because of their accurate

Table 19
Searching with Shallow Scoring Matrices

pgh1_human:	len	E(BL50)	%	E(M20)	%	E(M10)	%	I/II
gb|R96180	355	3e–38	98.0	2.3e–72	99.0	6.5e–75	100.0	I
gb|AA296431	279	6.7e–19	59.1	6.8e–25	61.3	1.3e–22	62.4	II
gb|T29235	257	2.2e–17	63.3	5.3e–22	64.8	2.6e–18	66.2	II
gb|AA037294	471	3.1e–13	98.0	3e–30	98.0	3.3e–31	97.8	I
gb|AI022012	537	3.5e–09	64.5	1.2e–15	58.8	3.4e–13	60.8	II
gb|N79146	544	2.9e–06	100.0	2.6e–16	100.0	3.0e–17	100.0	I
gb|AA223896	97	1.3e–05	80.0	8.4e–13	87.1	2.8e–12	87.1	??
gb|AA485017	208	0.72	66.1	4.8e–14	84.7	4.1e–14	88.9	??
pgh2_human:								
gb|AA296431	279	1.4e–35	96.8	2.2e–69	96.8	8.0e–72	98.9	II
gb|T29235	257	1e–32	92.9	2.9e–61	94.1	9.1e–63	95.2	II
gb|AI022012	537	1.1e–32	95.8	1.6e–68	96.0	1.1e–70	97.0	II
gb|R96180	355	6.3e–23	65.8	1.0e–30	56.9	9.1e–27	60.3	I
gb|AA485017	208	—[a]	–	2.4e–05	75.6	3.3e–4	79.1	??
gb|AA223896	97	0.01	50.0	0.01	69.0	0.2	79.2	??
gb|AA885610	320	0.018	46.3	—	—	—	—	
gb|AA911293	172	0.049	43.6	—	—	—	—	

[a]E() values indicated as — were >5.0.

statistical estimates and high-quality alignments. Traditionally, sequence similarity searches have sought to ask one question: "Is my query sequence homologous to anything in the database?" Both *FASTA* and *BLAST* can provide reliable answers to this question with their statistical estimates; if the expectation value E is < 0.001–0.01 and you are not doing hundreds of searches a day, the answer is probably yes.

In general, the most effective search strategies follow these rules:

1. Whenever possible, compare at the amino acid level, rather than the nucleotide level. Search first with protein sequences (*blastp, fasta3*, and *ssearch3*), then with translated DNA sequences (*fastx, blastx*), and only at the DNA level as a last resort (**Table 5**).
2. Search the smallest database that is likely to contain the sequence of interest (but it must contain many unrelated sequences for accurate statistical estimates).
3. Use sequence statistics, rather than percent identity or percent similarity, as your primary criterion for sequence homology.
4. Check that the statistics are likely to be accurate by looking for the highest-scoring unrelated sequence, using *prss3* to confirm the expectation, and searching with shuffled copies of the query sequence [*randseq*, searches with shuffled sequences should have E approx 1.0].

5. Consider searches with different gap penalties and other scoring matrices. Searches with long query sequences against full-length sequence libraries will not change dramatically when BLOSUM62 is used instead of BLOSUM50 ***(20)***, or a gap penalty of -14/-2 is used in place of -12/-2. However, shallower or more stringent scoring matrices are more effective at uncovering relationships in partial sequences ***(3,18)***, and they can be used to sharpen dramatically the scope of the similarity search.

However, as illustrated in the last section, the E value is only the first step in characterizing a sequence relationship. Once one has confidence that the sequences are homologous, one should look at the sequence alignments and percent identities, particularly when searching with lower quality sequences. When sequence alignments are very short, the alignment should become more significant when a shallower scoring matrix is used, e.g., BLOSUM62 rather than BLOSUM50 (remember to change the gap penalties).

Homology can be reliably inferred from statistically significant similarity. Whereas homology implies common three-dimensional structure, homology need not imply common function. Orthologous sequences usually have similar functions, but paralogous sequences often acquire very different functional roles. Motif databases, such as PROSITE ***(21)***, can provide evidence for the conservation of critical functional residues. However, motif identity in the absence of overall sequence similarity is not a reliable indicator of homology.

Acknowledgments

W.R.P. is supported by a grant from the National Library of Medicine (LM04961).

References

1. Wilbur, W. J. and Lipman, D. J. (1983) Rapid similarity searches of nucleic acid and protein data banks. *Proc. Natl. Acad. Sci. USA* **80,** 726–730.
2. Bult, C. J., White, O., Olsen, G. J., Zhou, L., Fleischmann, R. D., Sulton, G. G., Blake J. A., Fitzgerald, L. M., Clayton, R. A., Gocayne, J. D., Kerlavage, A. R., Dougherty, B. A., Tomb, J.-F., Adams, M. D., Reisch, C. I., Overbeek, R., Kirkness, E. F., Weinstock, K. G., Merrick, J. M., Glodek, A., Scott, J. L., Geoghagen, N. S. M., Weidman, J. F., Fuhrmann, J. L., Nguyen, D., Utterback, T. R., Kelley, J. M., Peterson, J. D., Sadow, P. W., Hanna, M. C., Cotton, M. D., Roberts, K. M., Hurst, M. A., Kaine, B. P., Borodovsky, M., Klenk, H.-P., Fraser, C. M., Smith, H. O., Woese, C. R., and Venter, J. C. (1996) Complete genome sequence of the methanogenic archaeon, methanococcus jannaschii. *Science* **273,** 1058–1073.
3. Altschul, S. F., Boguski, M. S., Gish, W., and Wootton, J. C. (1994) Issues in searching molecular sequence databases. *Nat. Genet.* **6,** 119–129.
4. Pearson, W. R. (1996) Effective protein sequence comparison. *Meth. Enzymol.* **266,** 227–258.

5. Pearson, W. R. (1997) Identifying distantly related protein sequences. *Comput. Appl. Biosci.* (now *Bioinformatics*) **13,** 325–332.
6. Pearson, W. R. (1998) Empirical statistical estimates for sequence similarity searches. *J. Mol. Biol.* **276,** 71–84.
7. Pearson, W. R. and Lipman, D. J. (1988) Improved tools for biological sequence comparison *Proc. Natl. Acad. Sci. USA* **85,** 2444–2448.
8. Lipman, D. J. and Pearson, W. R. (1985) Rapid and sensitive protein similarity searches. *Science* **227,** 1435–1441.
9. Bleasby, A. J., Akrigg, D., and Attwood, T. K. (1994) Owl-a non-redundant composite protein sequence database. *Nucleic Acids Res.* **22,** 3574–3577.
10. Altschul, S. F., Gish, W., Miller, W., Myers, E. W., and Lipman, D. J. (1990) A basic local alignment search tool. *J. Mol. Biol.* **215,** 403–410.
11. Altschul, S. F., Madden, T. L., Schaffer, A. A., Zhang, J., Zhang, Z., Miller, W., and Lipman, D. J. (1997) Gapped blast and psi-blast: a new generation of protein database search programs. *Nucleic Acids Res.* **25,** 3389–3402.
12. Arratia, R., Gordon, L., and Waterman, M. S. (1986) An extreme value theory for sequence matching. *Ann. Stat.* **14,** 971–993.
13. Karlin, S. and Altschul, S. F. (1990) Methods for assessing the statistical significance of molecular sequence features by using general scoring schemes. *Proc. Natl. Acad. Sci. USA* **87,** 2264–2268.
14. Wootton, J. C. and Federhen, S. (1993) Statistics of local complexity in amino acid sequences and sequence databases. *Comput. Chem.* **17,** 149–163.
15. Pearson, W. R., Wood, T., Zhang, Z., and Miller, W. (1997) Comparison of DNA sequences with protein sequences. *Genomics* **46,** 24–36.
16. Henikoff, S. and Henikoff, J. G. (1992) Amino acid substitutions matrices from protein blocks. *Proc. Natl. Acad. Sci. USA* **89,** 10,915–10,919.
17. Schwartz, R. M. and Dayhoff, M. (1978) Matrices for detecting distant relationships, in *Atlas of Protein Sequence and Structure,* vol. 5, suppl. 3 (Dayhoff, M., ed.) National Biomedical Research Foundation, Silver Spring, MD, pp. 353–358.
18. Altschul, S. F. (1991) Amino acid substitution matrices from an information theoretic perspective. *J. Mol. Biol.* **219,** 555–565.
19. Jones, D. T., Taylor, W. R., and Thornton, J. M. (1992) The rapid generation of mutation data matrices from protein sequences. *Comp. Appl. Biosci.* (now *Bioinformatics*) **8,** 275–282.
20. Pearson, W. R. (1995) Comparison of methods for searching protein sequence databases. *Protein Sci.* **4,** 1145–1160.
21. Bairoch, A. (1991) PROSITE: a dictionary of sites and patterns in proteins. *Nucleic Acids Res.* **19** (suppl) 2241–2245.
22. Smith, T. F. and Waterman, M. S. (1981) Identification of common molecular subsequences. *J. Mol. Biol.* **147,** 195–197.
23. Huang, X. and Miller, W. (1991) A time-efficient, linear-space local similarity algorithm. *Adv. Appl. Math.* **12,** 337–357.
24. Waterman, M. S. and Eggert, M. (1987) A new algorithm for best subsequences alignment with application to tRNA-rRNA comparisons. *J. Mol. Biol.* **197,** 723–728.

25. Myers, E. W. and Miller, W. (1988) Optimal alignments in linear space. *Comp. Appl. Biosci.* **4,** 11–17.
26. Kyte, J. and Doolittle, R. F. (1982) A simple method for displaying the hydropathic character of a protein. *J. Mol. Biol.* **157,** 105–132.
27. Barker, W. C., Garavelli, J. S., Haft, D. H., Hunt, L. T., Marzec, C. R., Orcutt, B. C., Srinivasarao, G. Y., Yeh, L. S. L., Ledley, R. S., Mewes, H. W., Pfeiffer, F., and Tsugita, A. (1998) The PIR-International protein sequence database. *Nucleic Acids Res.* **26,** 27–32.
28. Chao, K.-M., Pearson, W. R., and Miller, W. (1992) Aligning two sequences within a specified diagonal band. *Comp. Appl. Biosci.* (now *Bioinformatics*) **8,** 481–487.
29. Altschul, S. F. and Gish, W. (1996) Local alignment statistics. *Meth. Enzymol.* **266,** 460–480.

11

The Use of CLUSTAL W and CLUSTAL X for Multiple Sequence Alignment

Ashok Aiyar

1. Introduction

Multiple protein and nucleic acid sequences are aligned for two principal purposes: to identify common motifs in sequences with a conserved biological function and to identify motifs in a newly characterized sequence that may provide insight into its biological functions. This is typically performed by scanning the newly identified sequence against a database.

CLUSTAL W and *CLUSTAL X* are two related programs used to align multiple protein and nucleic acid sequences rapidly and reliably. In this chapter I will describe using these programs to identify common sequence patterns and motifs in protein and nucleic acid sequences through multiple alignment. I shall also illustrate the use of these programs to align a new sequence to a previously aligned set of sequences, or profile. Both programs can also be used to construct phylogenetic trees from the aligned set of sequences. As there are several other programs and packages written expressly for phylogenetic analysis, I shall not discuss the use of CLUSTAL W and *CLUSTAL X* for this purpose in detail.

CLUSTAL was written originally for IBM PC-compatible computers running *MSDOS* *(1)*. Later, an extensively rewritten version of *CLUSTAL*, *CLUSTAL V*, was made freely available as binaries and source code for a wide variety of computers and operating systems *(2)*. The two most recent releases of *CLUSTAL*, *CLUSTAL W* and *CLUSTAL X*, contain several improvements over *CLUSTAL V* that increase the reliability and sensitivity of multiple alignments, without sacrifices in speed *(3,4)*. The primary difference between *CLUSTAL W* and *CLUSTAL X*, is that the former has a simple text-mode

From: *Methods in Molecular Biology, vol. 132: Bioinformatics Methods and Protocols*
Edited by: S. Misener and S. A. Krawetz © Humana Press Inc., Totowa, NJ

interface and the latter has an elegant graphical user interface built using the National Center for Biotechnology Information (NCBI) *VIBRANT* toolkit. *CLUSTAL X* also offers the user additional options during multiple and profile alignments. At the time of writing, the most recent releases of these programs are *CLUSTAL W* version 1.75 and *CLUSTAL X* version 1.65b. Despite the difference in version number, both programs share the same pairwise and multiple alignment algorithms. The *CLUSTAL* programs are available as pre-built binaries for a wide variety of operating systems, including DOS, Linux, MacOS, various versions of UNIX, VMS, and *Windows 95/NT*. The full source code to both *CLUSTAL W* and *CLUSTAL X* is made freely available, and the programs can thus be recompiled easily for other operating systems for which an ANSI C compiler, such as the GNU C compiler (gcc), is available. All the examples in this chapter were obtained using *CLUSTAL W* and *CLUSTAL X* under Linux. I note, however, that the interfaces presented by these programs are very similar under other operating systems. When I discuss features that are common to both programs I shall refer to the programs as *CLUSTAL* (*W/X*). I will refer to the programs by their individual names when describing features specific to each.

CLUSTAL (*W/X*) aligns sets of sequences through a variation of the progressive multiple alignment method of Feng and Doolittle ***(5,6)***. This is done in four stages within the program as illustrated in **Fig. 1**. Initially, pairwise alignments of those sequences designated to be multiply aligned are performed. Second, these pairwise alignments are used to calculate similarity scores (percent identity), from which an unrooted tree is created using the Neighbor Joining (NJ) method ***(7)***. These unrooted trees have branch lengths proportional to estimated divergence from each branch node. Third, this unrooted tree is converted to a rooted tree using the mid-point method ***(8)***. At this step, branch lengths in the rooted tree are used to calculate a weight for each sequence, as explained in detail by Thompson et al ***(3)***. Finally, to produce multiple alignments, the guide rooted tree is used to align increasingly larger groups of sequences proceeding from the tips of the tree toward the tree root. At each step, a dynamic programming algorithm is used together with a residue-specific weight matrix, and gap-opening/gap-extension penalties, to align the alignments that were created in the previous iteration of the algorithm ***(3,4)***.

Multiple different residue-weight matrices can be used for protein alignments including a simple identity matrix, or the matrices from the BLOSUM ***(9)***, PAM ***(10)***, and GONNET ***(11)*** series. For nucleic acid alignments, the user can choose between the IUB matrix or the *CLUSTAL W* (1.6) identity matrix ***(3)***.

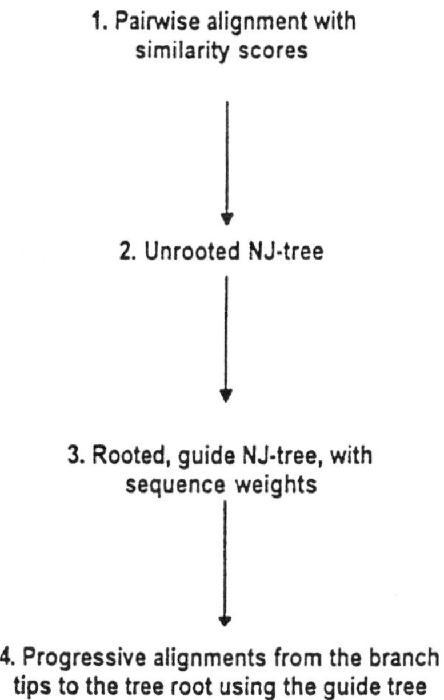

Fig. 1. Outline of the progressive alignment method used by *CLUSTAL* (*W/X*).

2. Materials

2.1. *Obtaining* CLUSTAL *(W/X)*

CLUSTAL (*W/X*) binaries and source code are freely available along with installation instructions. Binaries and source code can be obtained from the following URLs:

CLUSTAL W: `ftp://ftp-igbmc.u-strasbg.fr/pub/ClustalW`
CLUSTAL X: `ftp://ftp-igbmc.u-strasbg.fr/pub/ClustalX`

In addition, *CLUSTAL* (*W/X*) for various operating systems are available from:

UNIX/Linux:
`ftp://ftp.ebi.ac.uk/pub/software/:>unix/clustalw`
`ftp://ftp.ebi.ac.uk/pub/software/:>unix/clustalw/clustalx`

DOS, Windows 95/NT:
`ftp://ftp.ebi.ac.uk/pub/software/dos/clustalw`
`ftp://ftp.ebi.ac.uk/pub/software/dos/clustalw/clustalx`
`http://www.csc.fi/molbio/progs/clustalw`

MacOS:
`ftp://ftp.ebi.ac.uk/pub/software/mac/clustalw`
`ftp://ftp.ebi.ac.uk/pub/software/mac/clustalw/clustalx`
`http://www.csc.fi/molbio/progs/clustalw`

VMS:
`ftp://ftp.ebi.ac.uk/pub/software/vax/clustalw`

2.2. *Installing* CLUSTAL *(W/X)*

The provided *CLUSTAL* (*W/X*) binaries should be adequate for most purposes. To install *CLUSTAL W*, copy the provided binary to a directory that can be seen by all users, such as a directory that is on the PATH environment variable. The on-line help file "clustalw_help" should be placed in the same directory as the executable. For *CLUSTAL X*, the file "clustalx_help" should be placed in the same directory as the binary executable. In addition, the parameter files (*.par) should also be copied to this directory.

2.3. *Compiling* CLUSTAL *(W/X)*

If you need to alter program parameters that cannot be set when the program is run, or the alignment algorithm for a specific purpose, then you will need to compile the *CLUSTAL* (*W/X*) source. This procedure is described below.

2.3.1. *Compiling* CLUSTAL W

You will need an ANSI C compiler, such as gcc or egcs, to compile *CLUSTAL W*.

1. Create a directory called "clustalw" and copy the distribution archive to that directory. On UNIX/Linux systems this will be the file called "clustal 1.75. UNIX.tar.Z".
2. Uncompress and untar the distribution in the CLUSTAL W directory as follows:

```
cat clustalw1.75. UNIX.tar.Z | uncompress | tar xvf -
```

3. Compile the program using the provided Makefile, which is compatible with GNU make.
4. Install the compiled binaries as described in **Subheading 2.2.**

2.3.2. *Compiling* CLUSTAL X

As with *CLUSTAL W*, you will need an ANSI C compiler to compile *CLUSTAL X*. In addition, you will need to compile and install the NCBI *VIBRANT* toolkit.

1. Source code for the Windows 95/NT, Macintosh, and UNIX/Linux versions of the NCBI toolbox is available from: `ftp://ncbi.nlm.nih.gov/toolbox/ncbi_tools`. Download and install the toolbox that is appropriate for your operating system. On UNIX/Linux systems, compiling the *VIBRANT* libraries from the NCBI toolkit requires *OSF/Motif* version 2.0 or higher. *CLUSTAL X* is linked against the *VIBRANT* libraries, and thus it is essential that a suitable version of *Motif* is installed. If *Motif 2.0* is not already installed on your computer, it is usually available from the vendor of your operating system, or from the computing services office of your institution. Compile the NCBI libraries using the instructions that accompany the NCBI toolbox distribution.
2. Uncompress and untar the *CLUSTAL X* distribution in the directory "clustalx" as described above for *CLUSTAL W*. Edit the provided Makefile to note the location of the directory that the NCBI toolbox is installed in.
3. Compile *CLUSTAL X* using the edited Makefile and GNU make.
4. Install the compiled binaries as described in **Subheading 2.2.**

3. Method

I will begin by describing the use of *CLUSTAL W* and *CLUSTAL X* for multiple alignments in **Subheadings 3.1.** and **3.2.**, respectively. I will then describe the use of *CLUSTAL* (*W/X*) for profile alignments in **Subheading 3.3.**, and finally describe the construction of phylogenetic trees using *CLUSTAL* (*W/X*) in **Subheading 3.4.**

When describing the use of *CLUSTAL* (*W/X*) to create multiple alignments, I shall use as examples multiple alignments of the nucleocapsid proteins (NC) of bovine immunodeficiency virus (BIV), bovine leukemia virus (BLV), equine infectious anemia virus (EIAV), human immunodeficiency virus type 1 (HIV1), mouse mammary tumor virus (MMTV), Mason-Pfizer monkey virus (MPMV), ovine lentivirus (OLV), and Rous sarcoma virus (RSV) ***(12)***. NC is a small basic RNA-binding protein, with either one or two copies of the zinc-binding motif $CX_2CX_4HX_4C$, referred to as a cys–his box. Except for the cys–his box(es), there are no other regions of conserved sequence identity between retroviral NC proteins ***(13,14)***.

3.1. Multiple Alignments of Protein Sequences using CLUSTAL W

1. The sequences to be aligned should be part of a single file. This file must be in one of the following formats: *EMBL/SwissProt **(15)***, *NBRF/PIR **(16)***, *FASTA **(17)***, *GCG/MSF **(18)***, *GDE **(19)***, *GCG/RSF **(18)***, or *CLUSTAL*.
2. To execute the program, type `clustalw` and hit **ENTER**. If the program does not execute, ensure that you are in the directory that contains the clustalw binary, or that the executable is on your path. Once the program executes, you will see the menu shown in **Fig. 2**. Online help can be obtained from this menu and many

```
***************************************************************
******** CLUSTAL W (1.74) Multiple Sequence Alignments  ********
***************************************************************


     1. Sequence Input From Disc
     2. Multiple Alignments
     3. Profile / Structure Alignments
     4. Phylogenetic trees

     S. Execute a system command
     H. HELP
     X. EXIT (leave program)


Your choice: []
```

Fig. 2. The *CLUSTAL W* main menu.

of the submenus by picking option **H**. Load the file containing the sequences to be aligned using option 1.

3. Once the sequences have been loaded, you can enter the multiple alignment menu by picking option **2**. Upon picking option **2**, the user is presented the menu shown in **Fig. 3**. From this menu, pick option **1** to begin a multiple alignment with the default parameters set in the program. These parameters can be altered using other options from this menu. Of particular interest for this purpose are options **5** and **6**, which can be used to define parameters for pairwise and multiple alignments respectively, as described in **items 5** and **6**, below.
4. Once option **1** is picked, an interactive menu allows the user to pick the name for the output dendrogram and alignment files. By default, the program picks the name "input_file.dnd" and "input_file.aln", where "input_file" is the name of the input file without any extension. The multiple alignment output files are described in **item 7**.
5. The pairwise alignment parameter menu (**Fig. 4**), allows the user to set parameters for slow and fast pairwise alignments. For slow and accurate alignments, penalties for gap opening and extension, and the residue weight matrix can be set by the user. By default, the BLOSUM matrix is used for proteins and the IUB matrix for nucleic acid sequences. If fast pairwise alignments are selected (option **4** in the multiple alignment menu shown in **Fig. 3**), then the user may alter the k-tuple size, gap penalty, window size, and number of top diagonals. For maximum sensitivity, use a small k-tuple size (word size) along with a large window size. For maximum speed, large k-tuple and small window sizes should be used.
6. In the multiple alignment parameter menu, the user can adjust properties such gap opening and extension penalties, and the residue-weight matrix to be used, as illustrated in **Fig. 5**. The "delay divergent sequences" option is of particular interest when distantly related sequences are among those being aligned. Sequences that are less identical than this level will be aligned later to make optimal use of consensus alignments and gap insertions that are created earlier during the alignment process using less divergent sequences from the input sequence set.

```
****** MULTIPLE ALIGNMENT MENU ******


     1.  Do complete multiple alignment now (Slow/Accurate)
     2.  Produce guide tree file only
     3.  Do alignment using old guide tree file

     4.  Toggle Slow/Fast pairwise alignments = SLOW

     5.  Pairwise alignment parameters
     6.  Multiple alignment parameters

     7.  Reset gaps before alignment? = OFF
     8.  Toggle screen display          = ON
     9.  Output format options

     S.  Execute a system command
     H.  HELP
     or press [RETURN] to go back to main menu


Your choice: []
```

Fig. 3. *CLUSTAL W* multiple alignment menu.

```
********* PAIRWISE ALIGNMENT PARAMETERS *********


    Slow/Accurate alignments:

    1. Gap Open Penalty       :10.00
    2. Gap Extension Penalty  :0.10
    3. Protein weight matrix  :BLOSUM series
    4. DNA weight matrix      :IUB

    Fast/Approximate alignments:

    5. Gap penalty            :3
    6. K-tuple (word) size    :1
    7. No. of top diagonals   :5
    8. Window size            :5

    9. Toggle Slow/Fast pairwise alignments = SLOW

    H. HELP


Enter number (or [RETURN] to exit): []
```

Fig. 4. Pairwise alignment parameter menu.

```
********* MULTIPLE ALIGNMENT PARAMETERS *********

     1. Gap Opening Penalty              :10.00
     2. Gap Extension Penalty            :0.05
     3. Delay divergent sequences        :40 %

     4. DNA Transitions Weight           :0.50

     5. Protein weight matrix            :BLOSUM series
     6. DNA weight matrix                :IUB
     7. Use negative matrix              :OFF

     8. Protein Gap Parameters

     H. HELP


Enter number (or [RETURN] to exit):
```

Fig. 5. Multiple alignment parameter menu.

```
CLUSTAL W (1.74) multiple sequence alignment


mmtv-nc         AAAMRGQKYSTFVKQTYGGGKGGQGAEGPVCFSCGKTGHIRKDCKDEKGSKRAP--PGLC
mpmv-nc         AAAFSGQTVKDFLNNKNKE-KGG------CCFKCGKKGHFAKNCHEHAHNNAEPKVPGLC
blv-nc          -VHTPGPKMPGPRQPAPKRPPPG------PCYRCLKEGHWARDCP--TKTTGPP--PGPC
hiv1-nc         ------MQRGNFRNQRKIV----------KCFNCGKEGHIARNCR-------APR-KKGC
rsv-nc          AVVNRERDGQTGSGGRARG----------LCYTCGSPGHYQAQCPKKRKSGNSR---ERC
                                              *: * . **   :*               *

mmtv-nc         PRCKKGYHWKSECKSKFDKDGNPLPPLETNAENSKNL-------
mpmv-nc         PRCKRGKHWANECKSKTDNQGNPIPPHQGNGWRGQPQAPKQAYG
blv-nc          PICKDPSHWKRDCPTLKSKN------------------------
hiv1-nc         WKCGKEGHQMKDCTERQAN-------------------------
rsv-nc          QLCNGMGHNAKQCRKRDGNQGQRPGKGLSSGPWPGPEPPAVS--
                  *    *   :*     :
```

Fig. 6. Multiple alignment of five retroviral nucleocapsid proteins.

7. Different output formats can be chosen by picking option **9** from the multiple alignment menu shown in **Fig. 3**. By default, aligned output is in the *CLUSTAL* format. An example of an output alignment in this format is shown in **Fig. 6**. In this example, the NC proteins of BLV, HIV1, MMTV, MPMV, and RSV were aligned using the default alignment parameters in *CLUSTAL W*. The output alignment file (.aln file) shown in **Fig. 6** is an ASCII text file that may be imported into an alignment editor or printed. Positions that are conserved in all five NC proteins (the cys–his boxes) are marked by asterisks. The output alignment can be stored in other formats such as *GCG/MSF*, *NBRF/PIR*, *GDE*, and *PHYLIP* ***(20)*** formats. *GCG/MSF*-formatted output can be used as input by GCG programs such as *PRETTY* and *PROFILEMAKE* ***(18)***. Output in the *PHYLIP* format can be used as aligned sequence input by the *PHYLIP* package for phylogenetic analyses ***(20,21)***.

3.2. Multiple Alignment of Protein Sequences Using CLUSTAL X

1. *CLUSTAL X* has all the functions of *CLUSTAL W* with a facile graphical user interface. *CLUSTAL X* also has additional features not present in *CLUSTAL W* that are illustrated in this section. To execute *CLUSTAL X* under UNIX or Linux, you must have access to an X-Terminal. This is not necessary for the Windows 95/NT and MacOS versions.
2. The main display window in *CLUSTAL X* loads in multiple alignment mode by default, but can be set to profile alignment mode. In profile alignment mode the user can align new sequences against previously aligned sequences, or align two sets of previously aligned sequences. Profile alignments are described in **Subheading 3.3.**
3. After starting *CLUSTAL X*, the file containing the sequences to be aligned can be loaded using the **Load sequences** dialog box under the **File** menu. The restrictions on the format of the input file are identical to those described in **Subheading 3.1., item 1** for *CLUSTAL W*. Upon loading the sequences to be aligned, a window similar to that shown in **Fig. 7**, will be displayed to the user. As with *CLUSTAL W*, there is an online help function, accessible under the **Help** menu in *CLUSTAL X*. The order of sequences in the main window can be altered by selecting the sequence name and then using the **cut** and **paste** options under the **Edit** menu. The main display window also has a ruler indicating residue position, and a graph at the bottom that depicts the quality of the alignment at each position. This alignment graph is better illustrated in **Figs. 9**, **10**, and **11**, below.
4. Under the **Alignment** menu, the user can either perform a multiple alignment using the default parameters within the program, or alter the various alignment parameters using interactive dialog boxes equivalent to the *CLUSTAL W* menus shown in **Figs. 3**, **4**, and **5.** As an example, the dialog box used to define protein gap parameters for multiple alignments is shown in **Fig. 8**. The choices in this dialog box allow the user to adjust the penalties that place a particular amino acid residue in a gap opening position. Other options in this menu define the minimal distance between gaps, and whether end gaps should be treated like internal gaps while conforming to the gap distance separation chosen. A menu equivalent to this dialog box is available in *CLUSTAL W* by picking option **8** from the multiple alignment parameters menu shown in **Fig. 5**.
5. After multiple alignment, conserved and aligned residues are colored within the *CLUSTAL X* main display window as shown in **Fig. 9**. Colors are designated by two types of rules. In the first type of rule, a residue is assigned a residue-specific color that is independent of its position in the alignment. In the second type of rule, residues are colored based on the alignment consensus at each position. This coloring system permits highlighting conserved positions in the alignment. Residue color parameter files are provided with the *CLUSTAL X* distribution, but user-specified files can also be used. The format of the color parameter files is described in the *CLUSTAL X* online help.

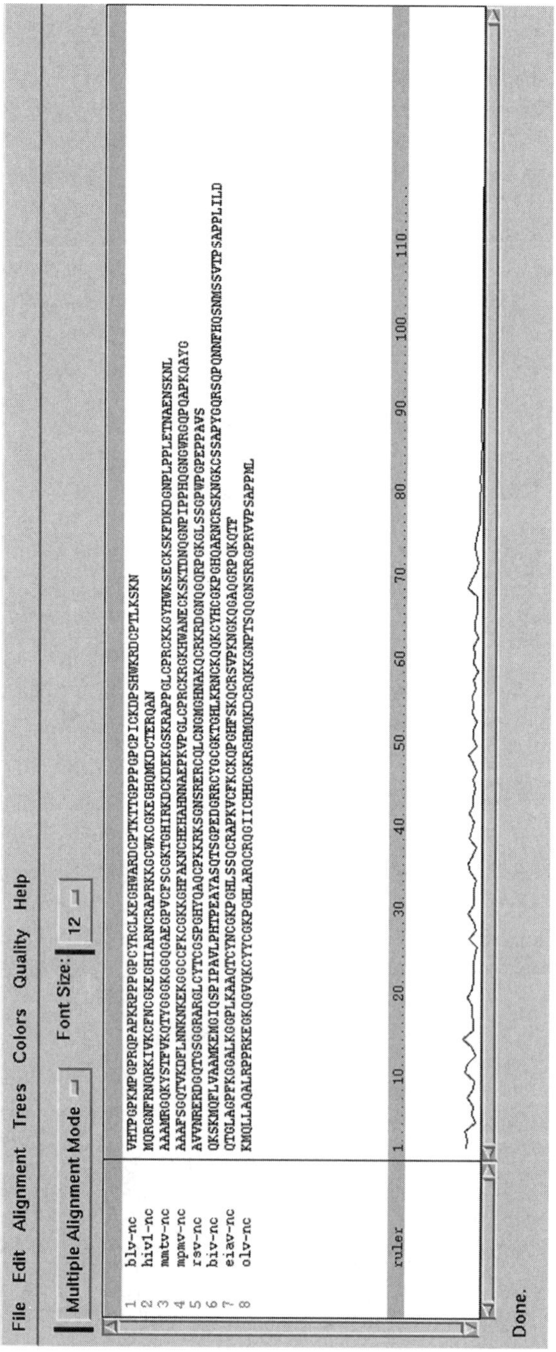

```
File  Edit  Alignment  Trees  Colors  Quality  Help

Multiple Alignment Mode     Font Size: 12

1  blv-nc    VHTPGPKMPGPRQPAPKRPPPGPCYRCLKEGHWARDCPTKTTGPPPGPCPICKDPSHWKRDCPTLKSKN
2  hiv1-nc   MQRGNFRNQRKIVKCFNCGKEGHIARNCRAPRKKGCWKCGKEGHQMKDCTERQAN
3  mmtv-nc   AAAMRGQKYSTFVKQTYGGGKGGQGAEGPVCFSCGKTGHIRKDCKDEKGSKRAPPGLCPRCKKGYHWKSECKSKFDKDGNPLPPLETNAENSKNL
4  mpmv-nc   AAAFSGQTVKDFLNNKNKEKGGCCFKCGKKGHFAKNCHEHAHNNAEPKVPGLCPRCKRGKHWANECKSKTDNQGNPIPPHQGNGWRGQPQAPKQAYG
5  rsv-nc    AVVNRERDGQTGSGGRARGLCYTCGSPGHYQAQCPKKRKSGNSRERCQLCNGMGHNAKQCRKRDGNQGQRPGKGLSSGPWPGPEPPAVS
6  biv-nc    QKSKMQFLVAAMKEMGIQSPIPAVLPHTPEAYASQTSGPEDGRRCYGCGKTGHLKRNCKQQKCYHCGKPGHQARNCRSKNGKCSSAPYGQRSQPQNNFHQSNMSSVTPSAPPLILD
7  eiav-nc   QTGLAGPFKGGALKGGPLKAAQTCYNCGKPGHLSSQCRAPKVCFKCKQPGHFSKQCRSVPKNGKQGAQGRPQKQTF
8  olv-nc    KMQLLAQALRPPRKEGKQGVQKCYYCGKPGHLARQCRQGIICHHCGKRGHMQKDCRQKKGNPTSQQGNSRRGPRVVPSAPPML

   ruler     1.......10........20........30........40........50........60........70........80........90.......100.......110......

Done.
```

Fig. 7. Sequence loaded in *CLUSTAL X* prior to multiple alignment.

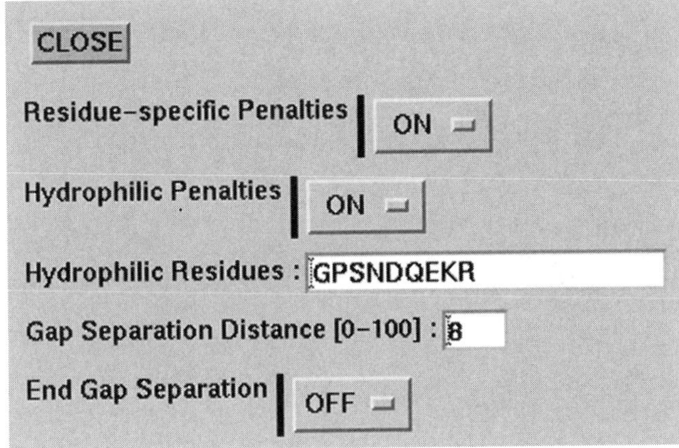

Fig. 8. Protein gap parameters in *CLUSTAL X*.

6. *CLUSTAL X* has the capacity to realign divergent regions within a multiple alignment. This permits correction of misalignments that are inadvertently introduced during alignment of highly divergent sequences. This is accomplished by the following two options. Specific sequences within the multiply aligned sequences can be selected by clicking on their names with the mouse as depicted in **Fig. 10**. These selected sequences, which appear as white text on a black background, are removed from the multiple alignment set, and then realigned to the sequences remaining in the alignment set. The other option is for the user to specify a residue range from the alignment to be realigned for all the sequences in the alignment set. This is accomplished by selecting the desired residues using the mouse, as also shown in **Fig. 10**. The selected range, which is highlighted in gray, is removed from the multiple alignment, realigned using the progressive alignment method diagrammed in **Fig. 1**, and then appropriately inserted back into the full alignment. These two realignment options allow the original alignment to be improved and refined. This is illustrated by comparing the alignment of the cys–his boxes in **Fig. 9** and **Fig. 11**. Initial alignment of all eight NC proteins resulted in alignment of the first cys–his box, but not the second. Realigning the selected residues and sequences highlighted in **Fig. 10**, resulted in aligning the second cys–his box for all eight NC proteins, as shown in **Fig. 11**.
7. Unlike *CLUSTAL W, CLUSTAL X* can create a color PostScript output file of the multiple alignment, which may be suitable for publication or presentation. To use this function, choose the option **Write Alignment as PostScript** under the **File** menu. Various output parameters such as the paper size and orientation, residue color, and alignment output layout can be specified in the postscript output dialog box. An example of the postscript output is shown in **Fig. 12**. Other output formats available in *CLUSTAL W* (**Subheading 3.1., item 7**) are also available in *CLUSTAL X*.

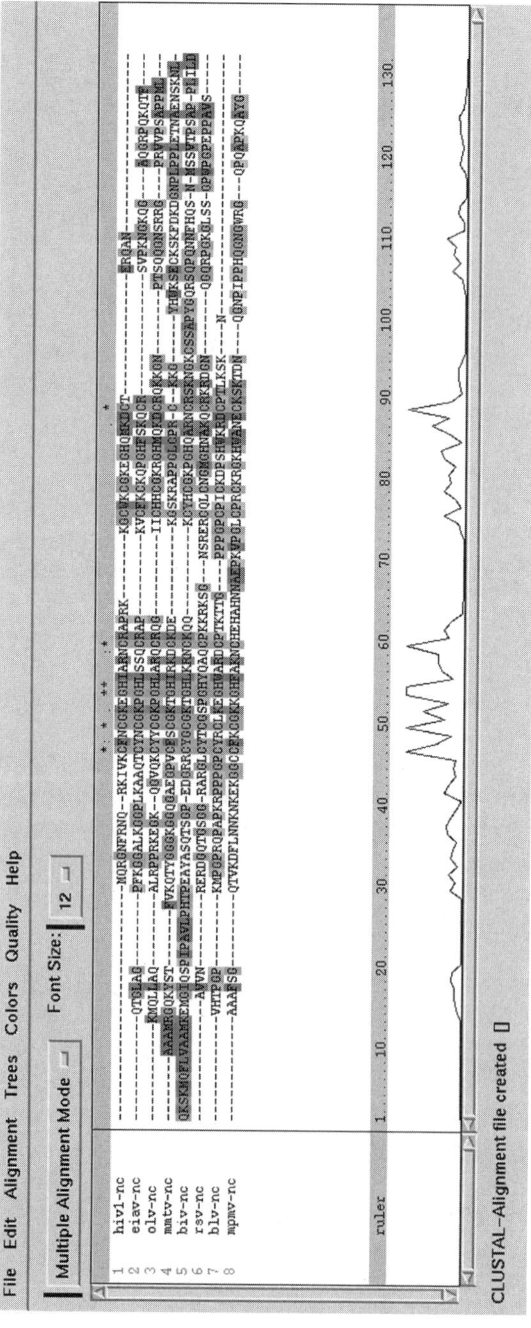

Fig. 9. Multiply aligned sequences displayed in the *CLUSTAL X* alignment window.

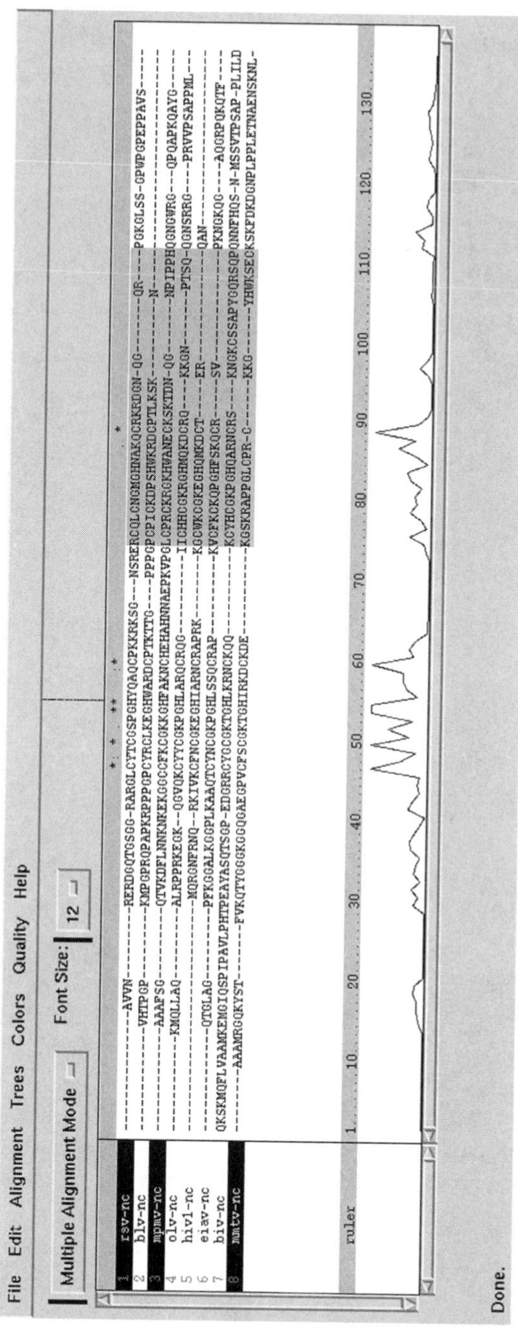

Fig. 10. Selecting a subset of sequences for realignment in *CLUSTAL X*.

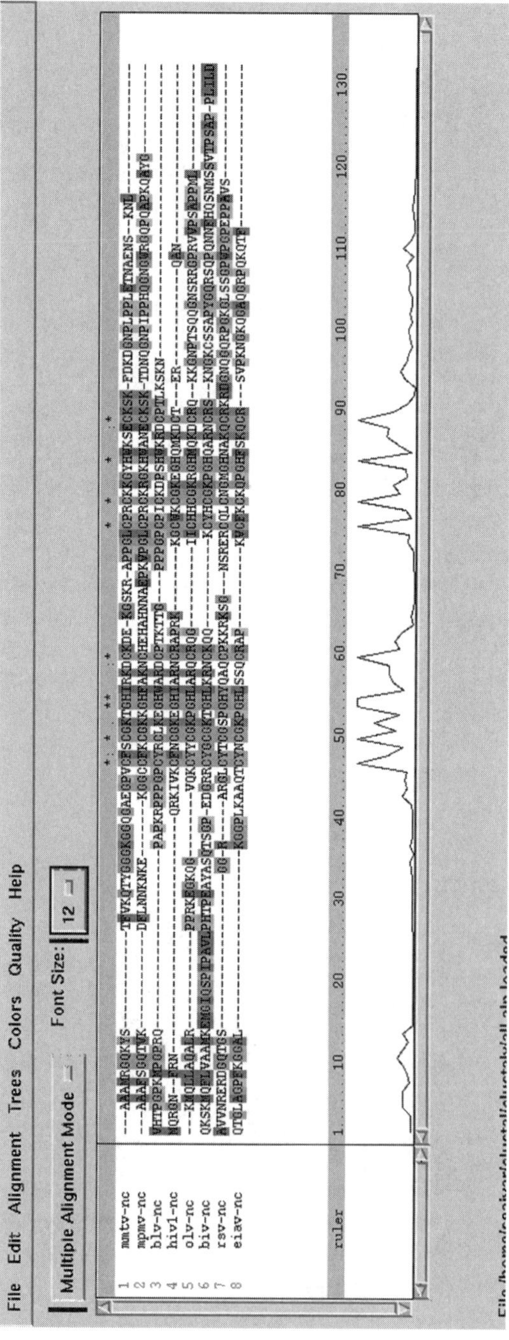

Fig. 11. Multiple alignments can be reiteratively refined in *CLUSTAL X*.

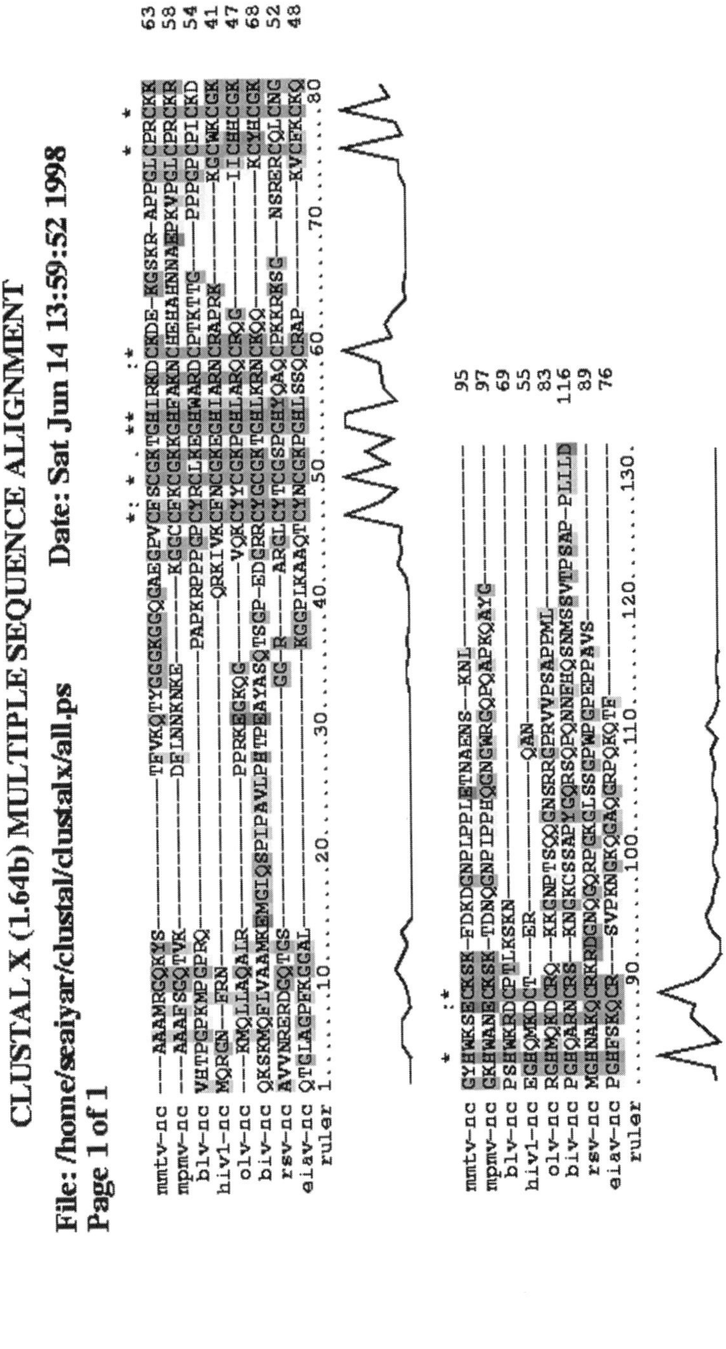

CLUSTAL X (1.64b) MULTIPLE SEQUENCE ALIGNMENT

File: /home/seaiyar/clustal/clustalx/all.ps **Date: Sat Jun 14 13:59:52 1998**

Page 1 of 1

```
                                                      *: *  . **   :*               *  *
mmtv-nc ---AAAMRGQKYS------------TFVKQTYGGGKGGQGAEGPVCFSCGKTGHIRKDCKDE-KGSKR-APPGLCPRCKK   63
mpmv-nc ---AAAFSGQTVK------------DFLNNKNKE-------KGGCCFKCGKKGHFAKNCHEHAHNNAEPKVPGLCPRCKR   58
 blv-nc VHTPGPKMPGPRQ----------------------PAPKRPPPGPCYRCLKEGHWARDCPTKTTG----PPPGPCPICKD   54
hiv1-nc MQRGN--FRN-----------------------------QRKIVKCFNCGKEGHIARNCRAPRK--------KGCWKCGK   41
 olv-nc ----KMQLLAQALR-----------PPRKEGKQG--------VQKCYYCGKPGHLARQCRQG----------IICHHCGK   47
 biv-nc QKSKMQFLVAAMKEMGIQSPIPAVLPHTPEAYASQTSGP-EDGRRCYGCGKTGHLKRNCKQQ-----------KCYHCGK   68
 rsv-nc AVVNRERDGQTGS-------------------GG-R-----ARGLCYTCGSPGHYQAQCPKKRKSG---NSRERCQLCNG   52
eiav-nc QTGLAGPFKGGAL----------------------KGGPLKAAQTCYNCGKPGHLSSQCRAP----------KVCFKCKQ   48
  ruler 1.......10.......20.......30.......40.......50.......60.......70.......80

          *   :*
mmtv-nc GYHWKSECKSK-FDKDGNPLPPLETNAENS--KNL----------------    95
mpmv-nc GKHWANECKSK-TDNQGNPIPPHQGNGWRGQPQAPKQAYG-----------    97
 blv-nc PSHWKRDCPTLKSKN------------------------------------    69
hiv1-nc EGHQMKDCT---ER------------QAN----------------------    55
 olv-nc RGHMQKDCRQ--KKGNPTSQQGNSRRGPRVVPSAPPML-------------    83
 biv-nc PGHQARNCRS--KNGKCSSAPYGQRSQPQNNFHQSNMSSVTPSAP-PLILD   116
 rsv-nc MGHNAKQCRKRDGNQGQRPGKGLSSGPWPGPEPPAVS--------------    89
eiav-nc PGHFSKQCR---SVPKNGKQGAQGRPQKQTF--------------------    76
  ruler ........90.......100.......110.......120.......130.
```

Fig. 12. Postscript output of multiply aligned sequences from *CLUSTAL X*.

3.3. Profile Alignments Using CLUSTAL (W/X)

CLUSTAL (*W/X*) can be used to align two existing alignments (profiles) to each other, or add new sequences to an existing alignment.

1. To perform profile alignments, switch to profile alignment mode in *CLUSTAL X*, or pick option **3** from the main menu in *CLUSTAL* W (*see* **Fig. 2**).
2. The user can begin a profile alignment by loading the first alignment using the **Load Profile 1** option under the **File** menu in *CLUSTAL X*, or by picking option **1** from the profile alignment menu in *CLUSTAL W*, shown in **Fig. 13**. The file that is loaded should be a multiple alignment output file (*.aln) created previously by *CLUSTAL* (*W/X*). The second file loaded can either be another aligned profile or a set of unaligned sequences. This file is loaded using the **Load Profile 2** option under the **File** menu in *CLUSTAL X*, or option **2** from the profile alignment menu in *CLUSTAL W*.
3. If the second file contains one or more unaligned sequences to be aligned to profile 1, the user can pick the option **Align Sequences to Profile 1** under the **Alignment** menu in *CLUSTAL X*, or option **4** in the profile alignment menu in *CLUSTAL W*.
4. If the second file is also a profile, then the two profiles can be aligned by picking the option **Align Profile 2 to Profile 1** under the **Alignment** menu in *CLUSTAL X*, or by picking option **3** from the profile alignment menu in *CLUSTAL W*.
5. An added feature of *CLUSTAL X* is that sequences can be cut from one profile and pasted in the other using the **Edit** menu. This manipulation allows the user to pick specific sequences from profile **2** to be added to profile 1 for alignment.
6. If a solved structure is available, it can be used to guide the alignment. This is done by raising gap penalties within secondary structural elements, such that gaps are preferentially inserted into surface loops and other less structured regions.
7. The output from profile alignments is similar to the multiple alignment output described in **Subheading 3.1., item 7**.

3.4. Phylogenetic Trees

CLUSTAL (*W/X*) can draw a phylogenetic tree using a previously calculated multiple alignment. Trees are created using the **Trees** menu in *CLUSTAL X*, or by selecting option **4** from the main menu in *CLUSTAL W* (*see* **Fig. 2**). Sequences *must* be aligned before a tree can be drawn.

1. The user can load an alignment into memory using the **File** menu in *CLUSTAL X*, or option **1** from the *CLUSTAL W* phylogenetic tree menu shown in **Fig. 14.**
2. To calculate a tree using default options, choose **Draw tree now** under the **Trees** menu in *CLUSTAL X*, or option **4** from the phylogenetic tree menu shown in **Fig. 14**. This will create an output file with the extension ".ph".
3. The user may choose to exclude positions where any of the alignment sequences has a gap during tree calculation. This option removes the more ambiguous (gapped) portions of the alignment during tree calculation, and is particularly

```
****** PROFILE AND STRUCTURE ALIGNMENT MENU ******

     1. Input 1st. profile
     2. Input 2nd. profile/sequences

     3. Align 2nd. profile to 1st. profile
     4. Align sequences to 1st. profile (Slow/Accurate)

     5. Toggle Slow/Fast pairwise alignments = SLOW

     6. Pairwise alignment parameters
     7. Multiple alignment parameters

     8. Toggle screen display              = ON
     9. Output format options
     0. Secondary structure options

     S. Execute a system command
     H. HELP
     or press [RETURN] to go back to main menu

Your choice:
```

Fig. 13. Profile alignment menu in *CLUSTAL W*.

```
****** PHYLOGENETIC TREE MENU ******

     1. Input an alignment
     2. Exclude positions with gaps?        = OFF
     3. Correct for multiple substitutions? = OFF
     4. Draw tree now
     5. Bootstrap tree
     6. Output format options

     S. Execute a system command
     H. HELP
     or press [RETURN] to go back to main menu

Your choice:
```

Fig. 14. Phylogenetic tree menu.

useful when the aligned sequences are highly divergent. This option is available under the **Trees** menu in *CLUSTAL X*, and is option **2** in the *CLUSTAL W* phylogenetic tree menu shown in **Fig. 14**.

4. The **correct for multiple substitutions** option allows the user to correct the distance calculations for the effects of multiple substitutions. As sequences diverge, multiple substitutions occur in each position. However only one of these many possible substitutions is represented in each position of a given sequence in the alignment set.
5. Trees may be bootstrapped to give a measure of the reliability of the groupings within the tree. The user is prompted for a seed number for the random number generator, and for the number of bootstrap samples (iterations) to be used.

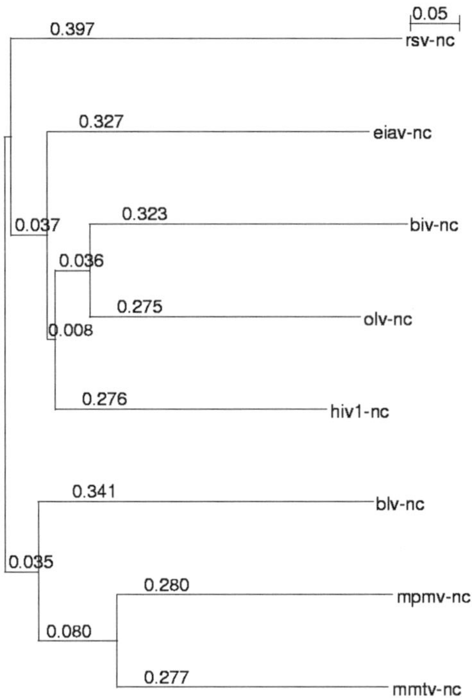

Fig. 15. Phylogenetic tree generated from a *CLUSTAL X* multiple alignment.

Bootstrapped trees are written to output files with the ".phb" extension when the *PHYLIP* output format is chosen, and to output files with the extension ".njb" when the *CLUSTAL* ouput formats is chosen.

6. Trees can be displayed using Manolo Gouy's *NJPLOT* application that is distributed along with *CLUSTAL X*. *NJPLOT* uses as input the "*.ph", or "*.phb" Newick-format file output by *CLUSTAL* (*W/X*). Trees are displayed on the screen, and can be written to PostScript output files. An example of tree output created using the alignment shown in **Fig. 12** is shown in **Fig. 15**. Note that the NC proteins from lentiviruses (HIV1, OLV, EIAV, BIV) appear marginally more closely related to each other than they are related to onco-retroviral NC proteins.

4. Notes

1. The reader is referred to an excellent set of articles by Higgins, Thompson, and coworkers for a detailed description of the pairwise and multiple alignment algorithms used by *CLUSTAL* (*W/X*) ***(1–4)***.
2. The seven sequence formats that can be used for sequence input are described in the online *CLUSTAL* (*W/X*) help files. All nonalphanumeric characters are ignored, except for "-" which is used to designate a gap in *CLUSTAL* format ".aln" file, and "." that is used to designate a gap in GCG format MSF/RSF files.

The reader is referred to the "*readseq*" utility, written by Don Gilbert, that can convert sequences from a variety of formats to input format used by *CLUSTAL* (*W/X*). *Readseq* is available from the URL: `ftp://ftp.bio.indiana.edu/molbio/readseq`.

3. *CLUSTAL* (*W/X*) automatically determines whether the input files contain protein or nucleic acid sequences. A nucleic acid sequence is assumed if more than 85% of the characters in the input file are A, G, C, T, and U. If a protein sequence is very short, with a highly biased amino acid composition, the user should be aware that it may be misidentified as a nucleic acid sequence.
4. Five different output formats can be selected. If necessary, all five output formats can be chosen at the same time. *CLUSTAL* format output can be read in again at a later time for profile alignments or to add another sequence to the alignment. This output file is in plain ASCII text, and can be imported into word-processors as a text-format file. This format is also compatible with some sequence alignment editors such as *SeaView* ***(22)*** (`ftp://biom3.univ-lyon1.fr/pub/mol_phylogeny/seaview`). To import the multiple alignment into other alignment editors such as *GeneDoc* (`http://www.cris.com/~ketchup/genedoc.shtml`), the output alignment should be saved in the *MSF* format.
5. Pairwise alignments can be performed using a slow/accurate dynamic programming method, or a fast/approximate method ***(23)***. The slow/accurate method works well for short sequences, but will likely be very slow when a very large number (>30) of long sequences (>1000 residues) have to aligned. For such alignments the fast/approximate method of pairwise alignments is recommended. When this method is chosen, the user should adjust the k-tuple and top-diagonal sizes. Increase the k-tuple size and decrease the top-diagonals value for increased speed.
6. For the final multiple alignment, the user can adjust the gap opening and extension penalties, and choose the residue weight matrix. Increasing the gap opening and extension penalties will make gaps less frequent and shorter. These values have no effect on terminal gaps. For proteins, four weight matrices are offered: identity, BLOSUM, PAM and GONNET. The GONNET matrices are essentially an up-to-date version of the PAM (Dayhoff) matrices, that are based on a much larger data set. The BLOSUM matrices are used by default. For nucleic acids, either the IUB matrix or the *CLUSTAL W* (1.6) matrix can be used. When the IUB matrix is used, unknown nucleotides are treated as matches to ambiguous IUB symbols (i.e., RYMWSKDHVBN). When the *CLUSTAL* W (1.6) matrix is used, matches to ambiguous IUB symbols are treated as mismatches.
7. If a solved structure is available, it can be used to guide the alignment by raising gap penalties within structured elements. Thus gaps are preferentially inserted into unstructured regions such as surface loops. A user-supplied gap penalty mask may also be used for this purpose. A gap penalty mask consists of a number between 1 and 9 for each position in the alignment. The basic gap opening penalty set is multiplied by this number to obtain the gap opening penalty for a given position. Gap penalty masks can be read from *CLUSTAL*, *GDE,* and *SwissProt* input files.

8. Phylogenetic trees are calculated by the neighbor-joining method of Saitou and Nei *(7)*. *CLUSTAL (W/X)* can produce trees in three output formats, none of which visually display the tree. Instead, trees are described in ASCII text files that can be read by other programs to draw trees. The *CLUSTAL* format output is a descriptive format that lists all of the pairwise distances between the multiply aligned sequences, and the number of alignment positions used for each. This format also lists the sequences that are joined at each alignment step and the branch lengths. *PHYLIP* format output is the New Hampshire format, wherein trees are listed as a series of nested parentheses, with descriptions of the branching order, branch lengths and sequence names. These output files can be read by the *NJPLOT* program distributed with *CLUSTAL X*. It can also be used by programs from the *PHYLIP* package to visually display the trees. Files of this format can also be displayed by other phylogenetic tree display programs such as *TREEVIEW AND PHYLO_WIN (22)*. When multiple alignments are saved in the *PHYLIP* format, they can be used as input files by programs in the *PHYLIP* package to generate phylogenetic trees using methods other than neighbor joining.

References

1. Higgins, D. G. and Sharp, P. M. (1988) CLUSTAL: a package for performing multiple sequence alignments on a microcomputer. *Gene* **73,** 237–244.
2. Higgins, D. G., Bleasby, A. J., and Fuchs, R. (1992) CLUSTAL V: improved software for multiple sequence alignment. *Comput. Applic. Biosci.* (now *Bioinformatics*) **5,** 151–153.
3. Thompson, J. D., Higgins, D. G., and Gibson, T .J. (1994) CLUSTAL W: improving the sensitivity of progressive multiple sequence alignment through sequence weighting, position-specific gap penalties and weight-matrix choice. *Nucleic Acids Res.* **22,** 4673–4680.
4. Thompson, J. D., Gibson, T. J., Plewniak, F., Jeanmougin, F., and Higgins, D. G. (1997) The CLUSTAL X windows interface: flexible strategies for multiple sequence alignment aided by quality analysis tools. *Nucleic Acids Res.* **25,** 4876–4882.
5. Feng, D.-F. and Doolittle, R. F. (1987) Progressive alignment of amino acid sequences and construction of phylogenetic trees from them. *Meth. Enzymol.* **266,** 368–382.
6. Feng, D.-F. and Doolittle, R. F. (1996) Progressive alignment of amino acid sequences and construction of phylogenetic trees from them. *Meth. Enzymol.* **266,** 368–382.
7. Saitou, N. and Nei, M. (1987) The neighbor-joining method: a new method for reconstructing phylogenetic trees. *Mol. Biol. Evol.* **4,** 406–425.
8. Thompson, J. D., Higgins, D. G., and Gibson, T. J. (1994) Improved sensitivity of profile searches through the use of sequence weights and gap excision. *Comput. Applic. Biosci.* (now *Bioinformatics*) **10,** 19–29.

9. Henikoff, S. and Henikoff, J. G. (1992) Amino acid substitution matrices from protein blocks. *Proc. Natl. Acad. Sci. USA* **89,** 10,915–10,919.
10. Dayhoff, M. O., Schwartz, R. M., and Orcutt, B. C. (1978) in *Atlas of Protein Sequence and Structure*, vol. 5, supplement 3 (Dayhoff, M. O., ed.), NBRF, Washington, DC, pp. 345–352.
11. Benner, S. A., Cohen, M. A., and Gonnet G. H. (1994) Amino acid substitution during functionally constrained divergent evolution of protein sequences. *Protein Eng.* **7,** 1323–1332.
12. Sequences were obtained from *GenBank* with the following accession numbers: 120810 (BIV), 120812 (BLV), 120814 (EIAV), 3023824 (HIV1), 120873 (MMTV), 120876 (MPMV), 120879 (OLV), and 120880 (RSV).
13. Katz, R. A. and Jentoft J. E. (1989) What is the role of the cys–his motif in retroviral nucleocapsid (NC) proteins? *BioEssays* **11,** 176–181.
14. Darlix, J. L,, Lapadat-Tapolsky, M., de Rocquigny, H., and Roques, B. P. (1995) First glimpses at structure-function relationships of the nucleocapsid protein of retroviruses. *J. Mol. Biol.* **254,** 523–537.
15. Bairoch, A. and Boeckmann, B. (1991) The SWISS-PROT protein sequence data bank. *Nucleic Acids Res.* **19,** 2247–2248.
16. Barker, W. C., George, D. G., Hunt, L. T., and Garavelli, J. S. (1991) The PIR protein sequence database. *Nucleic Acids Res.* **16,** 1869–1871.
17. Pearson, W. R. and Lipman, D .J. (1988) Improved tools for biological sequence comparison. *Proc. Natl. Acad. Sci. USA* **85**, 2444–2448.
18. Devereux, J., Haeberli, P., and Smithies, O. (1984) A comprehensive set of sequence analysis programs for the VAX. *Nucleic Acids Res.* **12,** 387–395.
19. Smith, S. Harvard University Genome Center.
20. Felsenstein, J. (1985) Confidence limits on phylogenies: an approach using the bootstrap. *Evolution* **39,** 783–791.
21. Felsenstein, J. (1996) Inferring phylogenies from protein sequences by parsimony, distance, and likelihood methods. *Meth. Enzymol.* **266,** 418–427.
22. Galtier, N., Gouy, M., and Gautier, C. (1996) SEAVIEW and PHYLO_WIN: two graphic tools for sequence alignment and molecular phylogeny. *Comput. Appl. Biosci.* (now *Bioinformatics*) **12,** 543–548.
23. Wilbur, W. J. and Lipman, D. J. (1983) Rapid similarity searches of nucleic acid and protein data banks. *Proc. Natl. Acad. Sci. USA* **80,** 726–730.

12

Phylogenetic Analysis Using PHYLIP

Jacques D. Retief

1. Introduction

Phylogenetic analysis is a powerful tool to study the relationships among sequences. From such relationships the origins, evolution, and possible changes in structural and functional properties of genes can be inferred.

PHYLIP (*Phylogeny Inference Package*) is an extensive collection of tools for phylogenetic analysis covering almost every method used in phylogenetic analysis today. The program suite is distributed by Joseph Felsenstein at the Department of Genetics, University of Washington, Seattle. It runs on a wide variety of platforms including UNIX workstations as well as IBM an Mac personal computers and, best of all, the program suite is provided at no cost by its author.

The variety of phylogenetic analysis programs available today is bewildering and there is no consensus as to which method is clearly superior. Even though the veracity of various methods may be hotly debated, in practice every set of sequences or gene family may have a particular protocol that will produce the optimum results. This is not intended as a course in phylogenetics, there are a number of excellent handbooks on the subject *(1–3)*. The documentation files provided with the programs also offer detailed descriptions of the algorithms and program options. This chapter, although by no means exhaustive, is intended as an introduction and a guide to find a suitable method to analyze your particular set of genes.

2. Materials

1. Installation: Precompiled versions of the *PHYLIP* package are available for UNIX, PC (DOS, Windows, and Windows NT), and Mac platforms. The instructions in this chapter are written for UNIX, but can be readily translated to other platforms.

From: *Methods in Molecular Biology, vol. 132: Bioinformatics Methods and Protocols*
Edited by: S. Misener and S. A. Krawetz © Humana Press Inc., Totowa, NJ

Detailed information on the supported hardware and procedures to download the program package are available from `http://www.ibb.waw.pl/docs/PHYLIPdoc/main.html`. To run *DRAWGRAM* and *DRAWTREE* you need to have a font file, called fontfile, in your local directory. A series of font files, called font1, font2 and so on is included with the package. Copy one of the files to your local directory where you intend to run the programs and rename it to fontfile.

2. Sequence alignment: Most programs in the *PHYLIP* package require a set of aligned sequences as input. There are several sequence alignment packages available on the web such as *CLUSTAL W* (`http://www2.ebi.ac.uk/clustalw/`) *(4)*. The *CLUSTAL W* user interface is intuitive and simple to use. The Wisconsin Package© (Genetics Computer Group, [GCG] Inc. Madison WI) includes *PILEUP* and *SEQLAB*, a powerful multisequence alignment program and editor.
3. Sequence reformatting: The *READSEQ* program will reformat most sequence formats to *PHYLIP* format. *READSEQ* was developed by Don Gilbert at the University of Indiana and is available via anonymous ftp (`ftp.bio.indiana.edu/molbio/readseq`). There are several Web sites available that provide a Web-based interface to *READSEQ*. The following reformatting instructions are for GCG's *MSF* sequence format, because the GCG program is one of the most commonly used sequence analysis packages. The procedures are universal and should be easily adaptable to most sequence formats.
4. Figure preparation programs: A large number of commercial programs, such as *Macromedia FreeHand*® and *Corel Draw*® allow the importation and editing of graphics produced by *PHYLIP*.

3. Methods

3.1. Selecting the Appropriate Sequences

1. A basic assumption of all phylogenetic analysis is that orthologous genes are being compared. This may seem obvious, but genes subject to horizontal transfer or orphan genes, for example, will produce spurious results because they are subject to different evolutionary constraints from the ancestral genes. Make sure that the sequences that are being compared belong together.
2. Outgroups provide a reference used to measure distances and help determine the root of a tree when an actual ancestral sequence is not available. An outgroup is the closest relative that does not belong to the group under study. For example, to build a tree of mammalian sequences, a bird sequence may provide a suitable outgroup. In this case a plant sequence would be a poor choice, because plants are very distant relatives and would degrade the alignments and the distance estimations.
3. DNA or protein? When too many mutations accumulate sequences become saturated with mutations. Consider a position in state A mutating to state B. As more mutations occur the chances increase that it may mutate to a third state C or back to state A, making us underestimate the number of mutations. Apart from the

redundancy in the genetic code, the protein is usually the functional product of the gene and the preservation of the protein function is a driving force for sequence conservation. Protein sequences therefore change much more slowly than DNA sequences and are our first choice for studying distant relationships or genes that change very rapidly. In some cases, when a gene changes very slowly, or when very close relationships are being examined, or when the genes are very small, the peptide sequences may not contain enough information to resolve trees and DNA sequences may be a better choice.

3.2. Sequence Alignment

The alignment of the input sequences forms the foundation of the analysis. A poor alignment can render even the most sophisticated protocol useless. Unfortunately sequence alignment programs are not perfect and it may take more effort to get a satisfactory alignment than any other part of the analysis. This is particularly true when there are many gaps in an alignment. The time spent on getting a good alignment is time well spent. Look out for the following common sequence alignment problems:

1. Sequence features such as start and stop codons as well as intron junctions should be carefully considered. *PILEUP* and *CLUSTAL* do not consider the relative importance of an ATG sequence of a start codon, or an ATG coding for a methionine, or simply an out of frame ATG sequence. Consequently sequence features, such as start codons may not align without intervention. The evolutionary implication is profound when a start codon shifts, because it could be caused by a frame shift or a deletion that will change the size of the gene product. These events are much less likely than the possibility that an adjacent gap was simply not well aligned. Inspect regions around misaligned start and stop codons and intron junctions to confirm that their alignment cannot be improved if you consider their importance. The same applies to peptide sequences in which known functional domains may be misaligned.
2. Dealing with gaps is a significant problem. Gaps, when coded with a "-" character, are considered as additional characters by some programs. For example, *DNAPARS* considers five characters A, T, G, C and gaps. When two sequences share a gap of, say, 50 nucleotides the program considers it a perfect match of 50 nucleotides, yet in reality the gap may have been produced by a single event. Sequences that share large gaps will group together if no compensation is made. The "?" character is used to code for missing information, such as at the ends of sequences when sequences of unequal length are compared. If all but one of the "-" characters in a gap is replaced by "?" characters the gaps will only count as a single event which may reduce the influence of gaps on the phylogenetic trees. If sequences contain many gaps it is good practice to compare results where all gaps are coded with "-" characters compared to "?" characters to determine the influence of gaps on the trees produced. Gaps alone should not determine the final tree.

A

```
T  Y  R  R  S  R
T  Y  R  R  S  R
T  Y  R  -  S  R
T  Y  R  -  S  R
T  Y  R  R  S  R
```

B

```
ACA TAC AGG CGA AGC CGG
 T   Y   R   R   S   R
ACA TAC AGG CGA AGC CGG
 T   Y   R   R   S   R
ACA TAC AGG --- AGC CGG
 T   Y   R   -   S   R
ACA TAC --- CGA AGC CGG
 T   Y   -   R   S   R
ACA TAC AGG CGA AGC CGG
 T   Y   R   R   S   R
```

C

```
gatttgggggtggg
gattattggggaag
aattat---ggtgg
gatctagtttatgg
gatt-tgggggtgg
g--tattggggtgg
gatttgggggaggg
```

Fig. 1. Comparing the alignment of coding regions with the alignment of their protein sequences may resolve ambiguous gaps. **(A)** The gap placement in column 4 is clearly arbitrary and could just as well be in column 3. **(B)** Comparison of the protein sequence with the nucleotide sequence resolves the alignment and shows, unexpectedly, that the gap is split between the protein columns 3 and 4. This results in the loss of an informative site that may influence the structure of a parsimony tree. **(C)** A sequence domain in which a large number of equally valid alignments are possible. Such an alignment produces what amounts to random noise in the alignment. It may improve the analysis to remove such domains.

3. The alignment of coding regions should always be compared with the alignment of their protein sequences. The alignment of the protein sequences and the alignment of the nucleotide coding regions should then be reconciled. There is no program currently available that will do this automatically, so the sequences may need to be edited manually in a sequence or text editor. In the DNA alignment this will ensure that the gaps are in the form of triplets. In the protein sequences the codon usage may help to resolve ambiguous alignments (**Fig. 1A** and **B**).
4. Low-information regions such as introns may contain long runs of dinucleotides or mononucleotides. Such runs may be inserted or extended very rapidly and should therefore be considered very carefully. Similarly, areas in which a large number of equally valid sequence alignments are possible may affect results (**Fig. 1C**). Low information regions may fruitfully be deleted from sequence alignments to remove the random bias they may create. Deleting parts of an alignment will affect branch length estimates.

3.3. Formatting Sequences

The *PHYLIP* package uses a proprietary sequence format (**Fig. 2**). The format is relatively simple and it is possible to reformat a set of aligned sequences to *PHYLIP* format with nothing more than determination and a text editor. However, it is much simpler and less error prone to use a sequence reformatting program such as *READSEQ*. *READSEQ* will accept most common sequence formats, but *READSEQ* is not perfect and the sequences require a certain amount of preparation before conversion. The following protocol will reformat a multisequence alignment of DNA or protein sequences in GCG's

```
6 50
HSP1_PANTR   ARYRCCRSQS RSRCYRQRQR SRRRKRQSCQ
HSP1_GORGO   ARYRCCRSQS RSRCYRQRQT SRRRRRRSCQ
HSP1_HYLLA   ARYRCCRSQS RSRCYRRGQR SRRRRRRSCQ
HSP1_HUMAN   ARYRCCRSQS RSRYYRQRQR SRRRRRRSCQ
HSP1_RABIT   ?????CRSQS RSRCRRRRRR CRRRRRRCCQ
HSP1_SAGIM   ARYRCCRSQS RSRCYRQRRR GRRRRRRTCR

TQRRAMRCCR RRSRMRRRRH
TQRRAMRCCR RRNRLR????
TRRRAMRCCR PRYRLRR???
TRRRAMRCCR PRYRPRCRRH
-RRRVRKCCR RTYTLRCRR?
-RRRASRCCR RRYKLTCRR?
```

Fig. 2. Sequence format used by *PHYLIP*. A typical interleaved input sequence. The numbers at the top of the alignment indicate the number of sequences and the number of characters in the sequence. The characters are all in upper case. The gaps at the ends of the sequences are padded with missing data "?" characters, while gaps inside the sequences are "-" characters. Spaces are ignored and may be included to improve the readability of the sequences.

MSF format to *PHYLIP* format. Most of these considerations will also apply to other sequence formats.

1. Load the *MSF* file into any text editor, such as *JOVE*, *EMACS* or *VI*, that can automatically find and replace characters. For the following steps it is important not to modify the header. The sequence is separated from the header by a text line consisting of "//". Modify only the characters below that line and save the file when the modifications are complete.
   a. Replace all lowercase sequence characters with uppercase characters. If you use GCG it is much more effective to remember to change the case of the sequences with the **ONECASE** command (UNIX: `onecase -men=u filename`), before aligning them.
   b. Replace all "~" characters with either "`?`" or "`-`" characters. See **Subheading 3.2., item 2** dealing with gaps.
   c. Replace all "`.`" characters with "`-`" characters. It is very important not to replace the "`. .`" in the GCG comment line.
   d. Make sure there are no ambiguous or unknown characters in the sequences. For example DNA sequences should only consist of A, T, G, C and gap characters. Spaces and numbers will be ignored.
2. To reformat an *msf* file called filename.msf do the following:
   a. Type: `readseq filename.msf`.
   b. Supply the name of the output filename, for example filename.inf when prompted.
   c. Choose output file format option **12. PHYLIP**.
   d. Make sure all the sequence names are listed and type "`all`" when prompted.

3. If a file is improperly formatted you will usually get a enigmatic "Memory allocation error" message in *PHYLIP*. Not all the *PHYLIP* modules are equally stringent about the file format. It is possible to successfully run *SEQBOOT* and get a file format error when running *PROTPARS*. Load the file in a text editor, or in UNIX type: `more filename.inf`. Check the input file format for the following:
   a. If you used a word processor, such as *Microsoft Word*, make sure the file was saved as a text file.
   b. If the alignment was destroyed during reformatting, you probably forgot to replace the "~" characters.
   c. Make sure you are using the right type of file. Not all programs require sequence files. *FITCH*, for example uses a file containing distance matrices produced by *DNADIST* or *PROTDIST*. The flow diagrams indicate the input file requirements.
   d. All sequence characters must be upper case.
   e. Gaps in sequences must be indicated by "-" or "?" characters ("." characters are not allowed.)
   f. Sequences must only contain legitimate characters. Failure to do so will produce an "Illegal character" error.
   g. The first line contains the number of sequences followed by the number of characters in a sequence.
   h. All sequences must be the same length. They may be padded with "-" or "?" characters.
   i. The sequence names must be exactly 10 characters long. They may be padded with spaces.
   j. If you still get a sequence format error, run the sequences through *READSEQ* again, it will sometimes fix the error the second time around.

3.4. Statistical Methods

Without statistical methods it is impossible to judge the validity of branch points in a particular tree (**Fig. 3A**). Felsenstein introduced bootstrapping to phylogenetic analysis *(5)*. Bootstrapping is a method of resampling the original dataset to produce a series of datasets, each with some of its data randomly changed. Changing, as opposed to deleting, data ensures that the dataset remains the same size. Setting option **b** (in *PHYLIP 3.6*), to **3** ensures that every codon position is sampled with the same frequency. In its simplest form, we can imagine that a tenuous grouping which is produced by a few sites will be easily disrupted, whereas robust associations will persist when some sites are corrupted. Typically 100 datasets will be created with the *SEQBOOT* program. The analysis is run with the **m** option set to **100** to produce 100 trees. From these 100 trees a consensus tree is then calculated with the *CONSENSE* program. The bootstrap values in the outfile indicate the number of occurrences of a particular branchpoint out of 100 trees. If we insert the bootstrap values in **Fig. 3B** it becomes clear which branch points are valid. Bootstrap values higher

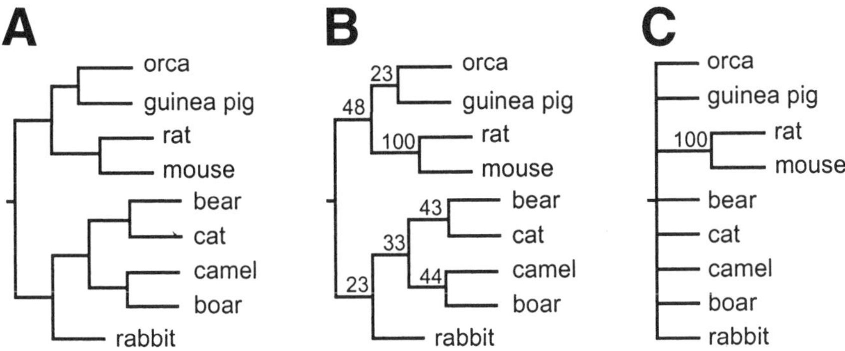

Fig. 3. Bootstrap values indicate the number of times a particular branch appears in a randomly resampled dataset and hence the validity of nodes in a tree. **(A)** This protein parsimony tree shows a number of highly unusual associations, for example grouping orca (killer whale) with the rodents. Without statistical methods it is impossible to judge the validity of the tree. **(B)** The same tree with bootstrap values from 100 datasets added to the nodes. (The bootstrap values were transcribed from the outfile produced by *CONSENSE*.) It is now obvious that all the associations, except for rat and mouse, were produced by chance. **(C)** The same tree with all nodes with bootstrap values less than 50 collapsed into a polytomy. This is a faithful representation of the real tree.

that 90 out of a 100 datasets are considered statistically significant, whereas values below 50 are essentially random chance. In **Fig. 3**, the rodents represent the only significant grouping. A more faithful representation of the tree can be created by collapsing the branches into a polytomy (**Fig. 3C**). This can be done by removing the parentheses in the tree file. *RETREE* (particularly in *PHYLIP 3.6*) allows you to easily remove insignificant branch points.

3.5. Running Programs

To execute a program in UNIX type the program name followed by return. In a graphical user interface, such as Windows, simply double click on the program icon. The program will take its input from a file and write its output to a file. Usually the files are called infile, treefile, or intree and outfile, treefile or outtree. When a program is executed it presents a menu of options. To change an option, type the option character followed by return. The program will then prompt for input if it is required. When all the settings are correct type "y" followed by a return. The number of programs and the large number of options presented can be daunting. Follow the flow diagrams in **Figs. 4–6** as a guide to the input and output file requirements. Usually the programs will present sensible defaults that may be used as a starting point.

All the instructions that follow are for *PHYLIP 3.5c*. At time of writing *PHYLIP 3.6* was in a "pre alpha" release. *PHYLIP 3.6* is functionally the same

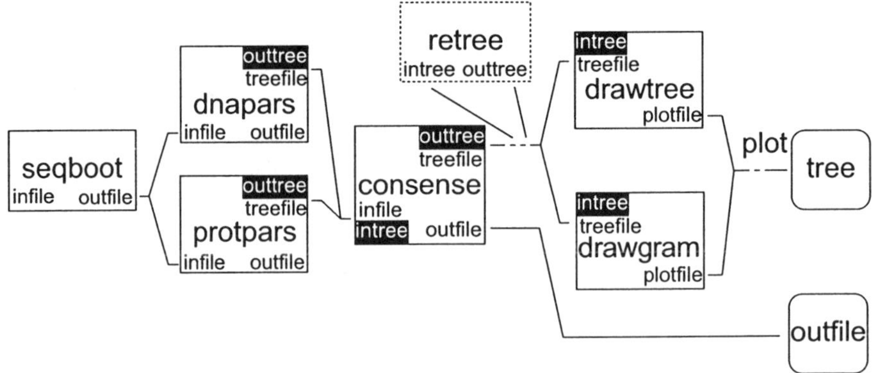

Fig. 4. A flow diagram indicating the combination of programs used to produce a parsimony tree. The diagram reads as follows: The *SEQBOOT* program takes its input from a file called infile and writes its output to a file called outfile. The next module is chosen to match the type of sequence file you started with. To proceed to the next module, say *DNAPARS*, the outfile produced by seqboot needs to be copied to infile, because *DNAPARS* takes its input from a file called infile. In turn, *DNAPARS* will produce an outfile and a treefile. *SEQBOOT* and *CONSENSE* are two optional programs used for bootstrapping. *RETREE* is also an optional step to manipulate the tree. *DNAPARS* and *PROTPARS* are selected for DNA or protein sequences and *DRAWTREE* and *DRAWGRAM* are selected depending on the style of tree desired. The final outfile contains the bootstrap values. The white text on black background are the names of the files produced and required by *PHYLIP 3.6*

as the previous version with the addition of some new features and options. The important changes are indicated in the text. A useful change is the option to rename output files. Importantly, tree files are now consistently called intree and outtree. These changes are indicated in the flow diagrams where the version 3.6 file names are indicated in white text. To find out which version you are running, look for the release number at the top of the menu.

3.6. Parsimony Methods

Parsimony is a character-based analysis. Each character is considered independent of its neighbor. Only informative sites are considered. For a site to be informative the same mutation must appear in at least two sequences. For a detailed description of maximum parsimony analyses, *see* ***(1,3,6)***. The parsimony programs calculate the order of the branches of the tree and do not give branch-length estimates. The advantage of parsimony methods is that they use a logical model and the calculations are rapid. The disadvantage is that a large amount of data is discarded because only informative sites are considered. This

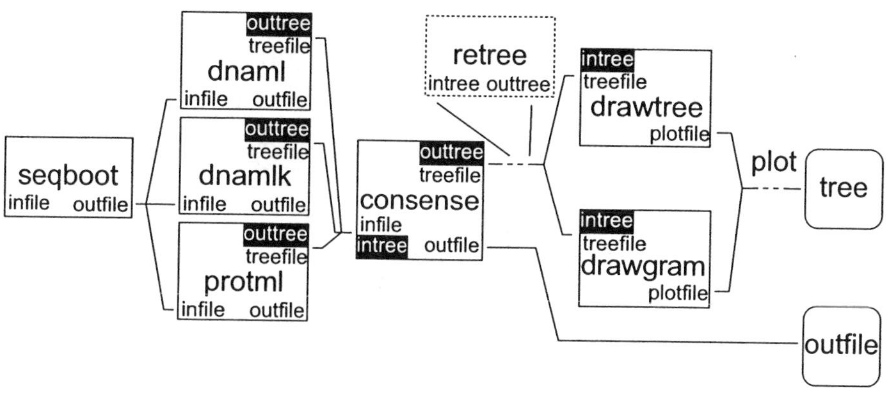

Fig. 5. A flow diagram to produce a maximum likelihood tree. The method is very similar to the parsimony tree in **Fig. 4**.

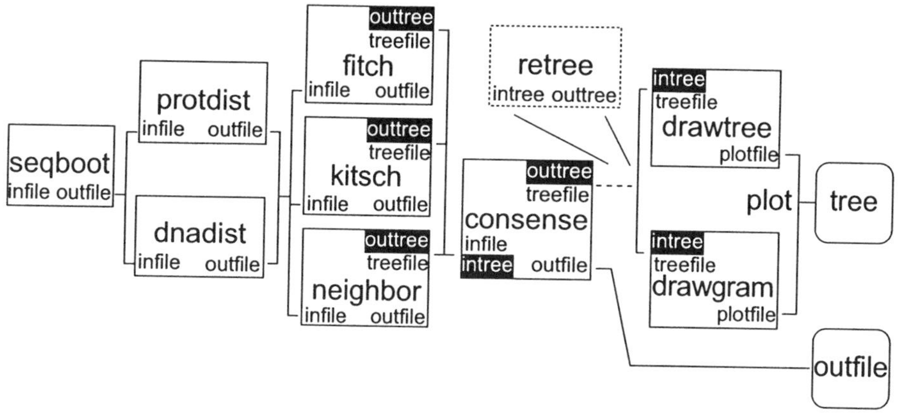

Fig. 6. A flow diagram used to produce a distance tree. The diagram is used as in **Fig. 4**. It contains one extra step to produce a matrix before the tree can be built with *FITCH*, *KITSCH*, or *NEIGHBOR*.

may be a problem if you are using short sequences, or sequences without a many informative sites. A flow diagram for parsimony analysis is shown in **Fig. 4**. The following steps are for a typical parsimony calculation.

1. Copy the sequence alignment in *PHYLIP* format to a file called infile (UNIX: `cp yourfile infile`). *See* **Subheading 3.3.** for the file format.
2. Optional, if you want to use a bootstrapped dataset run *SEQBOOT* (UNIX: `seqboot`). Provide a random number and accept the defaults to produce 100 datasets (*see* **Subheading 3.4.** for discussion). Copy the outfile produced by *SEQBOOT* to infile. (UNIX: `cp outfile infile`).

3. Do the parsimony calculation for a DNA or protein dataset by running *DNAPARS* or *PROTPARS*. (UNIX for DNA: `dnapars`, UNIX for protein: `protpars`). A window will appear with several options:
   a. Option **U** allows you to specify your own tree. See the *PHYLIP* documentation files for details on the file format. It is possible to calculate a parsimony tree and then use the branch lengths of a distance tree. Accept the default to let the program find the best tree.
   b. Option **J** allows the order of the sequences to be randomized. A tree is usually constructed from the top sequence down. Closely grouped sequences should be placed at the top of the alignment where they will be grouped first. More distant sequences are placed at the bottom of the alignment where they will be added last to the tree, with less chance of disturbing the closer groupings. Sequences aligned by GCG's *PILEUP* are arranged in this way by default. Randomizing the order of the sequences will eliminate any bias the sequence order may have on the construction of the tree.
   c. Option **O** allows you to specify an outgroup. The selection of an outgroup is discussed in **Subheading 3.1., item 2**.
   d. Option **T** will specify threshold parsimony. This limits the number of steps used to calculate distant branches.
   e. Option **M** specifies the number of bootstrapped sequence sets. Remember to set this option to 100 if you use bootstrapped data.
   f. Option **I** specifies the default interleaved sequence format. This is the sequence format in **Fig. 1**.
   g. Options **0–6** specify system parameters that do not affect the tree data. The defaults should suit most environments.
4. If you elected to use a bootstrapped dataset, copy the treefile to infile. (UNIX: `cp treefile infile`) (*PHYLIP 3.6*: copy the outtree to intree, UNIX: `cp outtree intree`). Now run *CONSENSE* (UNIX: `consense`). The outfile produced by *CONSENSE* contains the bootstrap values.
   a. Options **O** and **R** allows you to specify an outgroup root or treat the tree as rooted.
   b. Options **0–4** are system parameters and usually do not need to be changed.
5. Draw the tree by using *DRAWGRAM* or *DRAWTREE*. These programs require the single tree file produced by *CONSENSE*. (*PHYLIP 3.6:* copy the outtree to intree, UNIX: `cp outtree to intree`). This is the example for *DRAWGRAM* (UNIX: `drawgram`). The results are written to plotfile. (*PHYLIP 3.6:* the following three menus are combined into one, but the options remain basically the same.)
   a. The first menu allows you to select the device where you want to plot the tree. Set the option to select PostScript printer.
   b. The second menu selects the device where the tree will be previewed. If your terminal or computer can only display text, select **N**.
   c. The third menu selects the plotting options for the tree. To plot a phenogram, select the option for tree style and then select **P** for phenogram.
   d. The remaining options may be changed according to your preferences.

6. The method used to plot the file would vary depending on the operating system you are using and the options you selected in *DRAWGRAM* or *DRAWTREE*. If you have a PostScript printer and selected LaserWriter, all you need to do is to copy the plotfile to the printer. (UNIX: `lpr -Pprintername plotfile`.)

3.7. Maximum Likelihood Methods

The basic procedure for calculating maximum likelihood is the same as for parsimony. Every site is considered and the likelihood of the replacement of a particular nucleotide from pools of nucleotides is calculated *(7,8)*. The algorithm has also been adapted for protein sequences. The advantage of the method is that it considers every site. Even unchanged sites have a chance to have changed and then changed back to its original state. *DNAMLK* is the same as *DNAML* but assumes a molecular clock. There are advanced options that allow the determination of the likelihood of a particular tree. It is important to include all sites in the analysis, even those that did not change, to give an accurate estimate of branch lengths. The disadvantage of this method is that it is very slow to calculate. The following is the procedure for a typical DNA maximum likelihood analysis. (*see* also **Fig. 5**).

1. Copy the sequence alignment in *PHYLIP* format to a file called infile. In this example the sequence alignment is called yourfile (UNIX: `cp yourfile infile`). *See* **Subheading 3.3.** for information on the file format.
2. Optional, if you want to use a bootstrapped dataset, run *SEQBOOT* (UNIX: `seqboot`). The bootstrap options are discussed in **Subheading 3.4.** Copy the outfile produced by *SEQBOOT* to infile. (UNIX: `cp outfile infile`).
3. Run *DNAML*. (UNIX for DNA: `dnaml`). There are several options in the menu:
   a. Option **U** allows you to specify your own tree. For the standard analysis accept the default to let the program search for the best tree.
   b. Option **T** sets the transition/transversion ratio. The default **2** is well established.
   c. Option **F** allows you to set your own frequencies for the bases. The frequencies must add up to 1 and be typed in one line separated by blanks. The default, empirical frequencies, is calculated from the input sequences, and although it is not a true maximum likelihood, it is usually very close.
   d. Option **C** (*PHYLIP 3.6*: also **R** and **W**) allows the user to set the number of categories and the rates. These options are for advanced users and it is imperative that you understand the implications before you use them. See *PHYLIP* documentation files for details.
   e. Option **S** (*PHYLIP 3.6*), provides a rapid, but less rigorous estimate of the best tree. In practice this may be suitable for most analyses.
   f. Option **G**, removes and then adds every grouping in the tree. This makes sure that every branch gets reconsidered and its position optimized.
   g. Option **M**—set option **M** to 100 if you are using a bootstrapped dataset.
   h. Options **J**, **O**, **I**, and **1–4** are the usual system options that are discussed elsewhere.

4. If you elected to use a bootstrapped dataset, copy the treefile to infile and run *CONSENSE* (UNIX: `cp treefile infile`) (UNIX: `consense`). (*PHYLIP 3.6*: copy the outtree to intree, UNIX: `cp outtree to intree`) Use options **O** or **R** to root the tree or use an outgroup as a root.
5. The outfile produced by *CONSENSE* contains the bootstrap values. The tree file is written to treefile. In *PHYLIP 3.6* the treefile will be written to outtree.
6. Draw the tree by using *DRAWGRAM* or *DRAWTREE*. These programs require the single tree file produced by *DNAML* or *PROTML*, or in the case of bootstrapped data, *CONSENSE*. *See* **Subheading 3.6., item 5** for an example of *DRAWGRAM* (UNIX: `drawgram`, DOS `drawgram`). Note in *PHYLIP 3.6 DRAWGRAM* and *DRAWTREE* will take their input from a file called intree, so it will be necessary to copy the outtree file to intree (UNIX: `cp outtree intree`).
7. The results from *DRAWGRAM* and *DRAWTREE* are written to plotfile. The method used to plot the file would vary depending on your operating system and the options you selected in *DRAWGRAM* or *DRAWTREE*. If you have a postscript printer and selected LaserWriter, all you need to do is to copy the plotfile to the printer. (UNIX: `lpr -Pprintername plotfile`).

3.8. Distance Methods

Distance methods calculate the total number of changes, scored according to the type of change, between every pair of sequences in the alignment. The results are written to a distance matrix which is then used to construct the tree. Distance methods calculate branch lengths that visually represent the amount of change between sequences. In distance calculations, gaps are scored as unknown characters and effectively ignored. Removing ambiguous alignments or unchanged characters will influence the length estimates of the branches. The flow diagram in **Fig. 6** is a guide to the procedures and file requirements. Use the following steps to build a distance tree:

1. Copy the sequence alignment in *PHYLIP* format to a file called infile. (UNIX: `cp yourfile infile`) *See* **Subheading 3.3.** for information on the file format.
2. Optional if you want to use a bootstrapped dataset run *SEQBOOT* (UNIX: `seqboot`). For a basic analysis accept the defaults. Copy the outfile produced by the *SEQBOOT* to infile (UNIX: `cp outfile infile`). Note that if you use bootstrapping, you will lose the branch lengths. You can run a single tree without bootstrapping to find the branch lengths. With most trees it is a simple matter to combine the bootstrap values with the scaled distance tree. For more complex trees it is possible to specify the bootstrapped tree as a user tree when you run *FITCH*. See the program documentation for details on how to specify user trees.
3. Create a distance matrix for a DNA dataset or protein dataset by running *DNADIST* or *PROTDIST*. (UNIX for DNA: `dnadist`, UNIX for protein: `protdist`). There are several options available:

a. Option **P**: In *PROTDIST* it allows you to choose different models for changes. PAM, the default, is well established, empirical model ***(9)***. The Kimura method is much faster to calculate, but contains some compromises ***(10)***. The categories model groups amino acid residues in functional categories. Although it is slower, the default Dayhoff PAM matrix is a more conservative choice. All three methods should give similar results. In the *DNADIST* program there are similar options. The Kimura two-parameter model is very simple and assumes a two to one ratio between transitions and transversions. This model usually works well, but there are other models available, Jin/Nei ML (maximum likelihood) and J-C (Jukes and Cantor). The maximum likelihood model is similar to the model used in *DNAML* and is very slow.
b. Option **T**: Transition/transversion ratio option is only in *DNADIST* and allows you to specify your own ratio. The default, **2**, is well established.
c. Options **C** and **W** (*PHYLIP 3.6*) is also only in *DNADIST* and allows you to specify areas that may be changing at different rates.
d. Option **M**: Set this option to the number of datasets if you are using a bootstrapped data.
e. Options **L**, **I**, **0**, **1**, and **2** are the usual system parameters and do not affect the analysis. Use the defaults.

4. Copy the distance matrix produced in the outfile to infile (UNIX: `cp outfile infile`).
5. Build a tree with either *FITCH*, *KITSCH*, or *NEIGHBOR*. *KITSCH* is the same as *FITCH,* but it assumes a molecular clock and only calculates rooted trees. *NEIGHBOR* is a very fast, basic program for building trees. The commands are similar for all three programs. The following example is for *FITCH* **(11)** *(UNIX: `fitch`)*.

a. Option **D**: Minimum evolution (*PHYLIP 3.6*)—This method may be subject to artifact when the tree contains negative branch lengths. Leave the option set to Fitch-Margoliash.
b. Option **U** allows you to specify your own tree. See the *PHYLIP* documentation files for details on the file format. It is possible to calculate a parsimony tree and then use the branch lengths of a distance tree.
c. Option **P** sets the power of the equation. Leave it at two to calculate the standard deviation.
d. Option **-**: Some trees may produce negative branch lengths. Leaving this default will set all negative branches to zero.
e. Option **O** allows you to specify an outgroup. The selection of an outgroup is discussed in **Subheading 3.1., item 2**.
f. Option **G** allows global rearrangements. Each grouping is removed and then added back to the tree to ensure that all arrangements are considered.
g. Option **J** allows the sequences to be randomized. This is the same as for **Subheading 3.6., item 3b**.
h. Option **M** sets the number of bootstrapped sequence sets. Remember to set this option if you use bootstrapped data.

i. Option **I** specifies the default interleaved sequence format. This is the sequence format in **Fig. 2.**

j. Options **L**, **R**, and **0–4** specifies system parameters that do not affect the tree data. The defaults should suit most environments.

6. If you elected to use a bootstrapped dataset, copy the treefile to infile (UNIX: `cp treefile infile`) and run *CONSENSE* (UNIX: `consense`). (*PHYLIP 3.6*: `copy the outtree to intree`, UNIX: `cp outtree intree`) The outfile produced by *CONSENSE* contains the bootstrap values.
7. Draw the tree by using *DRAWGRAM* or *DRAWTREE*. These programs require the single tree file produced by *FITCH*, *KITSCH*, or *NEIGHBOR*, or in the case of bootstrapped data, *CONSENSE*. (*PHYLIP 3.6*: `copy the outtree to intree`, UNIX: `cp outtree intree`) *See* **Subheading 3.6, item 5** for an example of *DRAWGRAM* (UNIX: `drawgram`). The results are written to plotfile.
8. The method used to plot the file would vary depending on the operating system you are using and the options you selected in *DRAWGRAM* or *DRAWTREE*. If you have a PostScript printer and selected LaserWriter, all you need to do is to copy the plotfile to the printer. (UNIX: `lpr -Pprintername plotfile`).

4. Interpreting Results

1. A tree is read, starting at the base and following the progression of branch points or nodes (**Fig. 7A, B** and **C**). The program *RETREE* is provided to reorder trees or to change the root. *RETREE* reads the input from a file called intree and writes the results to a file called outtree. Tree files (**Fig. 7E**), can also easily be edited by with a text editor.
2. The tree construction is usually dependent on the order of the input sequences. The effect of the input order can be tested by running the program with the **J** (jumble) option set to at least 10 and then comparing the new tree with original.
3. Use a plot style that is appropriate to the data. Although unrooted trees may be plotted as phenograms by selecting option **2p** in *DRAWGRAM*, they are more easily interpreted when plotted by *DRAWTREE* (**Fig. 7D**).
4. If a tree produces branches where the bootstrap values range from 90 to 100, the results are statistically significant and virtually every method used to analyze the tree will give similar results, although the actual bootstrap values will differ. When the bootstrap values are low, it helps to try several different protocols. It is usually soon evident which tree is the most stable and occurs in most of the analyses.
5. For all trees, it is prudent to build a distance tree and at least one of the character based trees such as parsimony or maximum likelihood.
6. It is important to remember that producing the same tree by a number of different methods does not infer a statistical confidence level. For example, a tree that is favored by 51 out of 100 characters, will be produced by every method. This would create the impression of a high level of confidence, whereas the actual confidence level is approx 50.

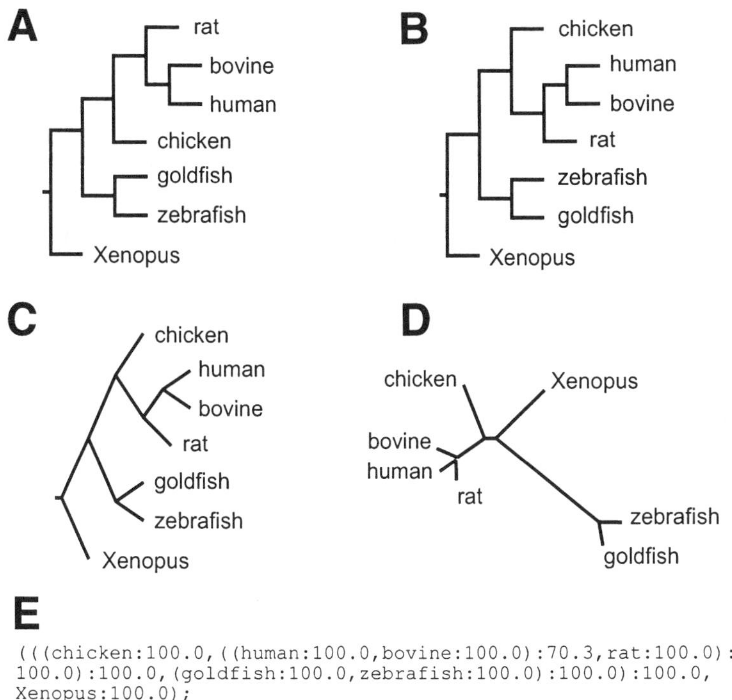

Fig. 7. Various trees: (**A** and **B**) phenograms; (**C**) cladogram; **A**, **B** and **C** are different displays of the same, identical tree and produced by *DRAWGRAM*. (**D**) An unrooted distance tree with scaled branch lengths produced by *DRAWTREE*. The long branch that contains the fishes, represents the relatively large number of changes in the sequences between fishes and the other animals. (**E**). The tree file used to produce **C**.

5. Transferring Results to Other Programs

Because the results are first written to a file, it is very simple to transfer plotfiles to any of a large number of graphic programs. Select PostScript as an option to import the plotfile into programs such as *Adobe Photoshop*®, *Microsoft Word*®, *Macromedia FreeHand*®. It is usually not possible to edit PostScript files. If you select HPGL as a printer format, the plotfile is produced in Hewlett Packard's plotter description language. Most Windows-based programs, such as *Macromedia FreeHand* can import these files as vector files that can be edited. It is usually only necessary to replace the text. In these programs the line thickness and color can be edited at will. *PHYLIP 3.6* will also generate images in the widely used Windows bitmapped and PICT file formats.

References

1. Li, W. -H. and Grauer, D. (1991) *Fundamentals of Molecular Evolution.* Sinauer Associates, Sunderland, MA.
2. Ridley, M. (1993) *Evolution.* Blackwell Scientific Publications, MA.
3. Li, W. -H. (1997) *Molecular Evolution.* Sinauer Associates, Sunderland, MA.
4. Higgins, D., Thompson, J., Gibson, T., Thompson, J. D., Higgins, D. G., and Gibson, T. J. (1994) CLUSTAL W: improving the sensitivity of progressive multiple sequence alignment through sequence weighting, position-specific gap penalties and weight matrix choice. *Nucleic Acids Res.* **22,** 4673–4680.
5. Felsenstein, J. (1985) Confidence limits on phylogenics: an approach using the bootstrap. *Evolution* **39,** 783–791.
6. Fitch, W. M. (1971) Toward defining the course of evolution: minimum change for a specified tree topology. *Sys. Zool.* **20,** 406–416.
7. Felsenstein, J. (1981) Evolutionary trees from DNA sequences: a maximum likelihood approach. *J. Molec. Evol.* **17,** 368–376.
8. Kishino, H. and Hasegawa, M. (1989) Evaluation of the maximum likelihood estimate of the evolutionary tree topologies from sequence data, and the branching order in Hominoidae. *J. Molec. Evol.* **29,** 170–179.
9. Dayhoff, M. O. (1979) *Atlas of Protein Sequence and Structure,* volume 5, suppl. 3, National Biomedical Research Foundation, Washington, DC.
10. Kimura, M. (1980) A simple model for estimating evolutionary rates of base substitutions through comparative studies of nucleotide sequences. *J. Molec. Evol.* **16,** 111–120.
11. Fitch, M. and Margoliash, E. (1967) Construction of phylogenetic trees. *Science,* **155,** 279–284.

13

Annotating Sequence Data Using Genotator

Nomi L. Harris

1. Introduction: What is *Genotator*?

As the amount of sequence data continues to grow exponentially, it is increasingly clear that automated methods are needed to help biologists with the task of sequence annotation. Many researchers have developed tools for analyzing DNA sequences, but running multiple tools and interpreting the results can be tedious and confusing.

Genotator (1), a workbench for automated sequence annotation, provides a flexible, transparent system for automatically running a series of sequence analysis programs on genetic sequences. It also has a graphical display that allows users to view all of the automatically generated annotations and add their own. *Genotator*'s display allows annotated sequences to be examined at multiple levels of detail, from an overview of the entire sequence down to individual bases. By displaying the aligned output of multiple types of sequence analysis, *Genotator* provides an intuitive way to identify the significant regions (for example, probable exons) in a sequence.

Genotator consists of two main portions, a back end and a browser. The back end runs a series of sequence analysis tools on a DNA sequence, handling all of the input and output formats. The analysis tools run by *Genotator* include five different gene finding programs, three homology searches, and searches for promoters, splice sites, and open reading frames (ORFs). The results of the analyses run by *Genotator*'s back end can be viewed with the interactive graphical browser. The browser displays color-coded sequence annotations on a canvas that can be scrolled and zoomed, allowing the annotated sequence to be explored at multiple levels of detail. The user can view the actual DNA sequence in a separate window; when a region is selected in the map display, it

From: *Methods in Molecular Biology, vol. 132: Bioinformatics Methods and Protocols*
Edited by: S. Misener and S. A. Krawetz © Humana Press Inc., Totowa, NJ

is automatically highlighted in the sequence display, and vice versa. Users ca interactively add personal annotations to label regions of interest. Addition capabilities of *Genotator* include primer design and pattern searching *Genotator* can also retrieve the *GenBank* ***(2)*** records for sequences that hav significant homologies to the sequence being annotated.

Genotator runs on UNIX workstations. The back end is written in *perl* an *Tkperl* and calls various sequence analysis programs. The front end, which i also written in *perl* and *Tkperl,* uses Gregg Helt's *bioTkperl* widgets ***(3)*** *Genotator* was developed on SUNs, and has also been installed on SGIs an DEC Alphas.

Genotator is not the only sequence-annotation tool available (others are dis cussed in ***ref. 1***), but it is easy to use, freely available, includes source (so it ca be modified to more closely suit your needs), and does not require a relationa database.

2. Obtaining *Genotator*

Dozens of researchers around the world are currently using *Genotator* *Genotator* is available for free to academic sites; please contact Nomi Harri (`nlharris@lbl.gov`) for information about obtaining it. `http://www-hgc.lbl.gov/inf/genotator/need.html` lists the program *Genotator* needs and describes how to obtain them. Some are included with th *Genotator* distribution; for others, you will have to contact the relevant authors

2.1. Installing Genotator

After you register to obtain *Genotator*, you will be sent instructions fo downloading and installing it. Installation of *Genotator* itself is done by sim ply running the **install-genotator** script included with the distribution. Install ing the programs that *Genotator* needs can be more complicated. The browse requires *perl5* and *Tkperl*. The back end can function with any subset of th sequence analysis programs that it knows about; you will have to obtain an set up any of the programs and sequence databases you want it to use.

The remainder of this chapter describes how to use *Genotator* once it i installed.

3. Running the *Genotator* Back End

The *Genotator* back end runs a series of analyses on a sequence file an saves the results for later browsing. Out of the many available sequence analy sis tools, a reasonable subset was integrated into *Genotator*. The analysis pro grams called by *Genotator* fall into three main categories: gene finders (*Geni* *[4]*, *GRAIL* *[5]*, *GeneFinder* *[6]*, *xpound* *[7]*, and *GENSCAN* *[8]*); databas homology searches (*BLASTN* *[9]* against a database of human or *Drosophil*

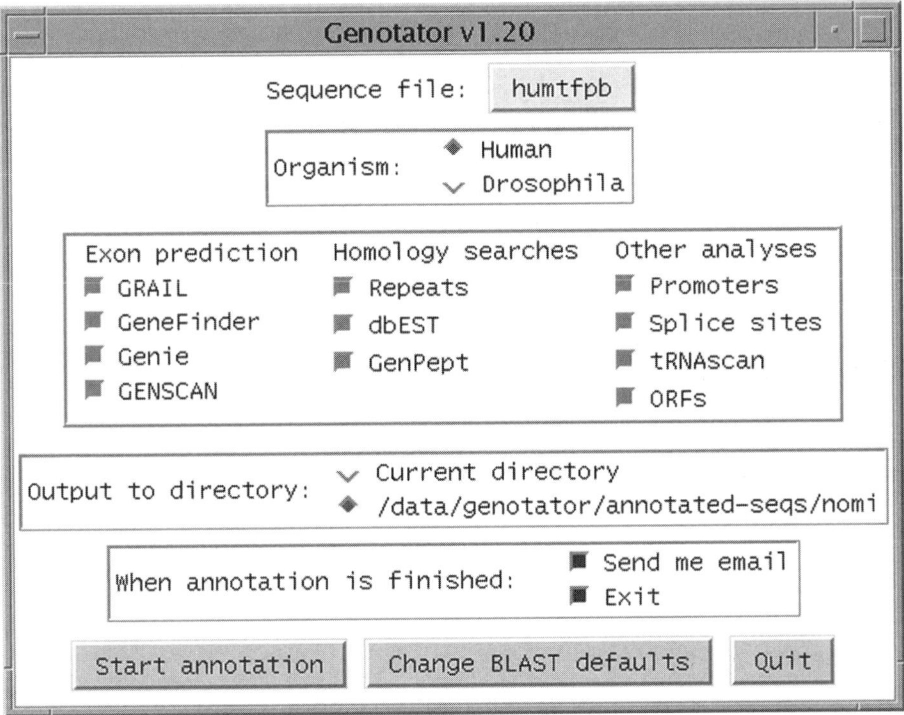

Fig. 1. The graphical user interface for *Genotator*'s back end. The user can select the sequence to be annotated and which analyses are to be performed.

repeat sequences, which are then masked out using *xblast [10]*; *BLASTN* on *dbEST [11]*; and *BLASTX [9]* on *GenPept [2]*); and sequence feature predictors (start/stop codons, ORFs, promoters *[12]*, and splice sites *[13]*). *Genotator* supplies each analysis program with its desired input format, and parses the output into a simple human-readable text format similar to that used by *ACeDB [14]*. The output files are organized hierarchically by sequence and by user.

The *Genotator* back end can be invoked via the graphical user interface (GUI) or with command-line options. The GUI is shown in **Fig. 1**. To invoke *Genotator* via the GUI (assuming you've installed *Genotator* in /home/genotator) type:

```
/home/genotator/genotator
```

If you know the name of the sequence file that you want to annotate, you can put it on the command line, e.g.:

```
/home/genotator/genotator humtfpb
```

3.1. Sequence File Formats

To select the sequence file you wish to annotate, click the file selection box (here labeled "humtfpb" because the user has already selected humtfpb as the sequence to be annotated) to bring up a file selection menu. Acceptable formats for the sequence file are:

Plain (just the sequence, no line numbers or anything).
FASTA (one header line starting with >, followed by lines of sequence).
FASTA-like but with a ; instead of a > at the beginning of the header line.
GenBank format.

Here is an example of a sequence in *FASTA* format:

```
>gb|J02846|HUMTFPB Human tissue factor gene,
complete cds.
GAATTCTCCCAGAGGCAAACTGCCAGATGTGAGGCTGCTCTTCCTCAGTCACTATCTCTG
GTCGTACCGGGCGATGCCTGAGCCAACTGACCCTCAGACCTGTGAGCCGAGCCGGTCACA
[etc.]
```

3.2. Genotator *Options*

Options that can be configured via the *Genotator* GUI include:

- Organism (human or *Drosophila*; default is human).
- Which analyses to perform (default is all of them; uncheck the boxes of analyses you do *not* wish to have performed).
- Where the output will go (default is the directory where all the Genotated sequences go (subdivided by user name), but you may wish to create your own *Genotator* directory hierarchy and save the annotations in your current directory).
- Whether you wish to be notified by e-mail when the analysis process has been completed (default is yes).
- Default blast cutoffs (click the **Change BLAST defaults** button to change them). You can also control whether repeats (such as *Alu*) are screened out before the other *BLAST* searches are performed.

3.3. *Running* Genotator *in Batch Mode*

If you wish to run *Genotator* on a group of sequence files, it may be easier to do it in batch mode with command-line options. Usually, you will simply type something like:

```
/home/genotator/genotator -batch seq1 seq2 seq3 . . .
```

where seq1, etc. are the names of plain or *FASTA*-format sequence files, and `-batch` tells *Genotator* not to bring up the GUI.
Invoking the *Genotator* back end with the **-h** (help) option causes it to print out a list of legal command-line options:

```
Usage: genotator [seqfile1 [seqfile2 . . .]]
[-human or -drosophila] [-none] [-nomail] [-exit]
    [-d(ebug)] [-noblast] [-nomask] [-dir
    output_dir] [-exon] [-all] [-batch] [-grail]
    [-genefinder] [-genie] [-genscan]
    [-xpound] [-genemark] [-genpept] [-est]
    [-repeats] [-promoters] [-splice] [-trnascan]
    [-orf]
```

`[-h(elp)]:` Print this help message.

`[seqfile1 . . .]:` Name of sequence file (in plain, *FASTA,* or *GenBank* format). You may specify multiple sequence files.

`[-human or -drosophila]:` Which organism your sequence is from (human is the default).

`[-none]:` Start with no analysis boxes checked.

`[-nomail]:` Don't send e-mail upon completion (default is to send e-mail).

`[-exit]:` Exit upon completion (default is not to exit).

`[-d(ebug)]:` Debug mode (for developers)—print what *Genotator* would do, but don't really do it.

`[-noblast]:` Try to reuse old *BLAST* output, but redo *BLAST* postprocessing.

`[-nomask]:` Don't mask out repeats before *BLAST*ing *dbEST* and *GenPept* (default is to mask).

`[-dir output_dir]:` Store results in (subdirectory of) output_dir.

`[-exon]:` Run gene finders only (in batch mode).

`[-all]:` Run all analyses in batch mode.

`[-batch]:` Run some analyses in batch mode; analyses to run will be specified by other arguments.

The remaining arguments are the names of sequence analyses that can b specified in conjunction with the `-batch` option.

4. The *Genotator* Browser

4.1. Invoking the Browser

After a sequence has been run through *Genotator*, the *Genotator* browse provides an interactive graphical view of the annotations. The *Genotato* browser can be invoked with the name of an annotated sequence file as a argument, e.g.:

```
/home/genotator/genotator-browser humtfpb
```

If it is invoked with no arguments, a list of annotated sequences is displayed with the sequences annotated by the invoking user listed first. The othe sequence directories are collapsed and indicated by . . ., as shown in **Fig. 2**. T see the sequences in a collapsed directory, double-click on the directory nam (e.g. "liepe . . .") and the sequences (or subdirectories) in that directory wil appear in the list. If there are numerous sequence names in the list, you ca use the **Find** button to help you search for the one you want. When you fin the sequence name of intent, double-click it (or single-click and hit **Select**) The selection list will disappear, and the browser will load the annotation for the selected sequence.

4.2. Map Display

Genotator's main display is called the map display. In the center of the ma display is a horizontal axis representing the sequence, with forward-stran annotations displayed above the axis and reverse-strand annotations below the axis. The numbers along the axis indicate kilobases of sequence. Each type o annotation (for example, *GRAIL* exons) is displayed on its own row, in its own color. The display can be zoomed and scrolled to examine interesting regions in more detail. To zoom, you can drag the zoom bar with your mouse, or position the cursor next to the zoom bar and click to zoom in gradually. To scroll, drag the scroll bar that is under the map display.

In **Fig. 3**, the *Genotator* browser is shown displaying the annotations on *HUMTFPB* ***(15)***, a human tissue factor gene sequence obtained from *GenBank*. (The figures in this chapter may be seen in color at `http://www-hgc.lbl.gov/homes/nomi/chapter.html`.)

Each colored rectangle on the map represents a sequence region that has been annotated. The type of each annotation is identified by the color of the rectangle and also by the row in which it appears. The row labels on the left (e.g. "GenPept hits") can be clicked for more information about the type of

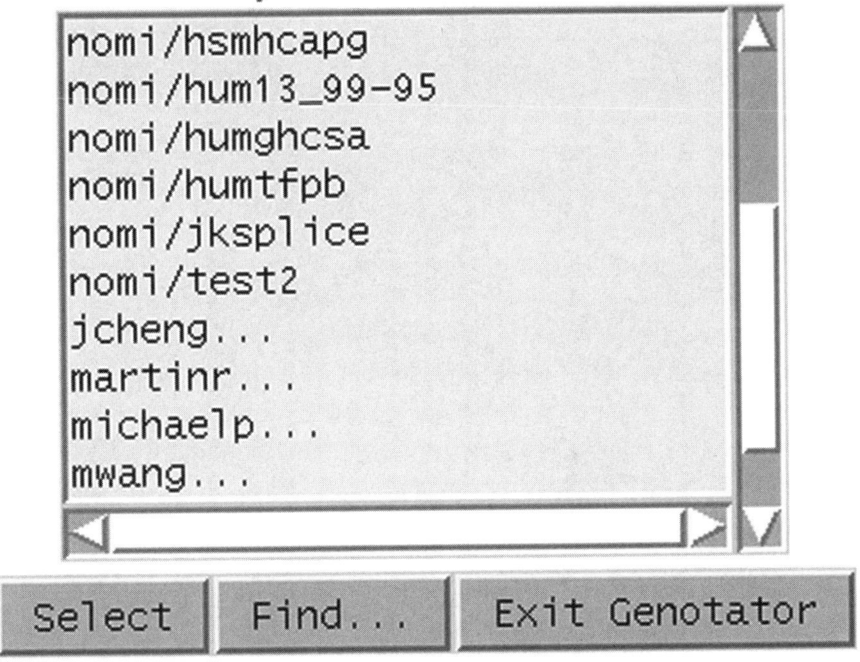

Fig. 2. If the *Genotator* browser is invoked with no arguments, it brings up a list of annotated sequences. Directories that are not owned by the current user are collapsed and indicated with "...".

annotation in the row. Clicking on an annotation rectangle puts a black box around the selected rectangle and displays additional information about that particular annotation in the text window at the top of the browser. This includes the start and end positions of the annotation, possibly a score, and other relevant information. For example, if a *BLAST* hit is clicked, the text window might say, "BLASTX GenPept hit from 864 to 1112 with sequence gp|K01228|HUMCG1PA1_1 (33% identity)". This concise description identifies the program that was used (*BLASTX*), the database that was searched (*GenPept*), the database sequence that was hit (gp|K01228|HUMCG1PA1_1 is its *GenPept* ID), the region that was found to be similar to this database sequence (bases 864 to 1112), and the percentage sequence identity for the hit (33%).

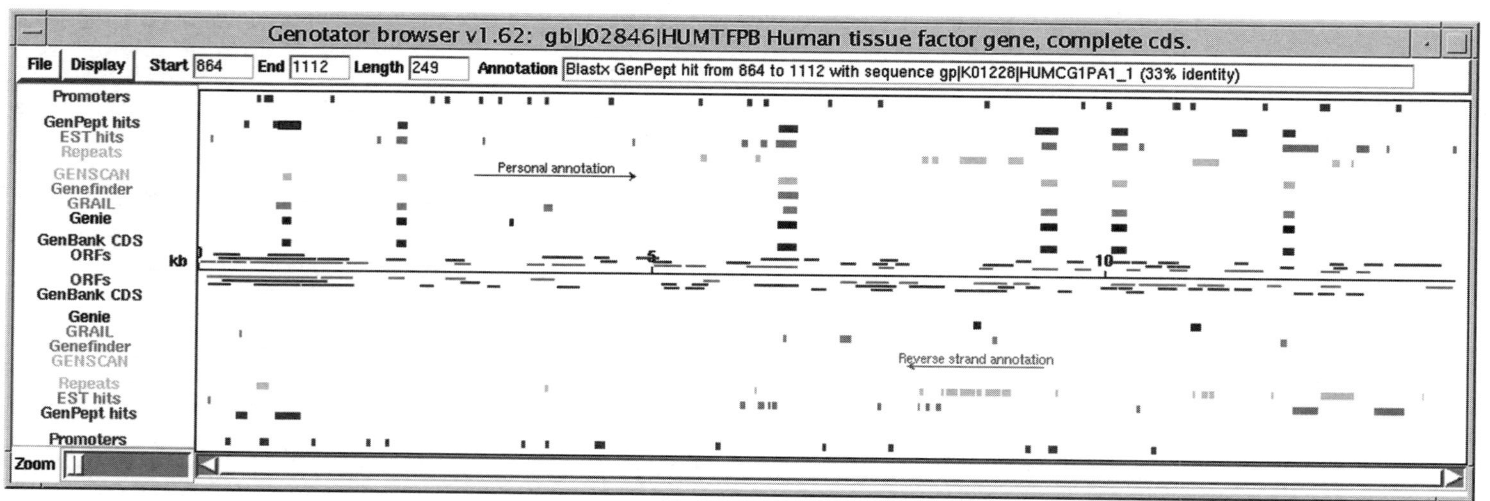

Fig. 3. The *Genotator* map browser provides an overview of the annotations on a sequence. Annotations on the forward and reverse strand are indicated by colored rectangles above and below the center axis.

4.2.1. Viewing BLAST *Hits in More Detail*

BLAST hits can be double-clicked to view them in more detail. (Note *Tkperl* requires that you double-click quickly and carefully. It may prove easier to get the cursor centered in an annotation rectangle if you zoom in first.) For *BLASTN* hits (against nucleotide sequences), the complete alignment appears in a separate window (**Fig. 4**), which can be saved or printed.

When *BLASTX* hits against *GenPept* are double-clicked in the *Genotator* display, *Blixem* ***(16)***, a *BLAST* hit viewer from the Sanger Centre, is invoked (**Fig. 5**).

Blixem shows horizontal black lines to represent hits in the region near where you clicked. The vertical position of the lines represents their percent identity. A blue box shows the region that is expanded below to show the actual hit alignments. You can move the blue box with your middle mouse button. Because *BLASTX* compares your DNA sequence with an amino acid database, the hits are shown in all three frames. The exact and similar matches are highlighted in color. To close the *Blixem* window, click your right mouse button on any empty gray area, which will pop up a menu in which one of the choices is **Quit**.

4.2.2. Genotator *Browser Functions:* ***File*** *Menu*

Open: Choose a different annotated sequence to be displayed.

Reload: Reload the current sequence (often useful if you have mistakenly added or deleted personal annotations).

Print: This captures the map display as a PostScript file and (optionally) sends it to your default printer. (Please note that this functionality may not be available on all systems—you must have *xwd* and *xwd2ps* in order to print *Genotator* displays.) To change your default printer, set your **PRINTER** environment variable *before* invoking *Genotator* by typing:

```
setenv PRINTER myfavoriteprinter
```

Summary report: Generates a long text report describing all annotations, which can then be saved in a file and/or sent to your default printer.

Submit comment: Lets you submit a comment directly to the *Genotator* developer.

Output selected region: This takes any region you have selected (e.g., by sweeping it out with the mouse, or by clicking on an annotation rectangle) and saves it to a file (in *FASTA* format) so you can do further analysis on it.

Quit: Exit the browser. If you have added or deleted personal annotations, you will be asked whether you want to save the changes.

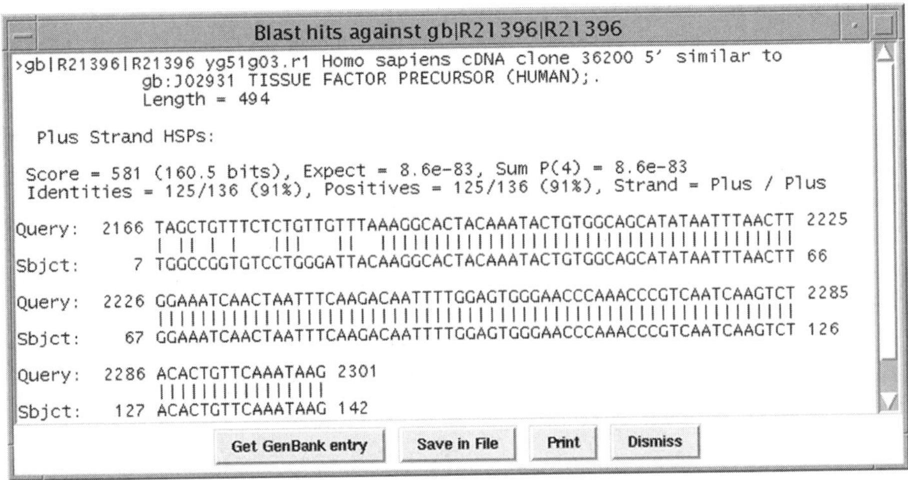

Fig. 4. When a *BLASTN* hit is double-clicked in the *Genotator* window, a window like this pops up, showing the actual alignment.

4.2.3. Genotator *Browser Functions:* ***Display*** *Menu*

Hide/Show complement: Hide (or show) the complementary strand (below the central axis) in the map display. This is mostly useful if you want to save an image showing just the forward strand.

Display sequence: Display the forward strand bases in a separate window. This is discussed in more detail in the next section.

Display sequence complement: Display the complementary strand bases.

Show/hide splice sites: When the browser comes up, the splice sites are not displayed. You can turn on their display with this function.

Show/hide start/stop codons: When the browser comes up, the start/stop codons are not displayed. You can turn on their display with this function.

Delete selection: Delete from the screen all annotations within your selection box. It is important to note that these annotations are *not* permanently gone—only personal annotations can be permanently deleted. If you reload the sequence, all of the automatically generated annotations will reappear. The delete function is mostly useful for making slides.

Get *GenBank* record for hit: This command is only enabled if you have clicked on an *EST* or *GenPept* hit. It queries *GenBank* to try to find the record for the subject sequence (the sequence that was hit). If it can find *Netscape*, it brings up the *GenBank* record in *Netscape*; otherwise, it opens a new text window. (Note that it can take a little while for *Netscape* to come up.) Sometimes the *GenBank* record cannot be displayed because *Genotator* is unable to find a unique record for the subject sequence.

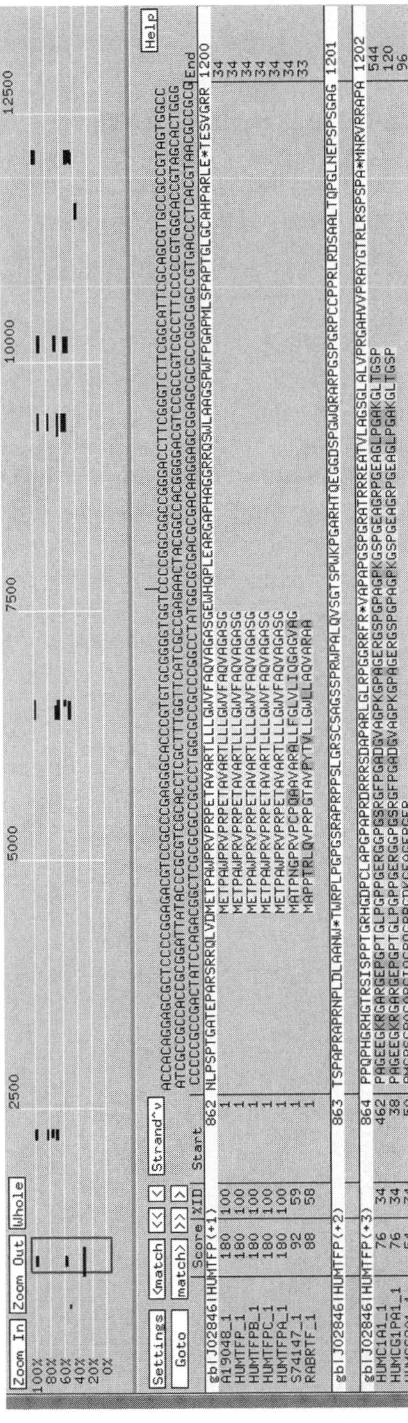

Fig. 5. When a *BLASTX* hit (against *GenPept*) is double-clicked in the *Genotator* window, *Blixem* is used to display the hit (as well as neighboring hits).

Personal annotations...: Brings up a control window for adding and deleting personal annotations, which are described in more detail in **Subheading 4.4.**

Design primers: Lets you design primers for the selected region (*see* **Subheading 4.5.**).

4.3. Sequence Display

The map display shows an overview of the entire sequence. The *Genotator* browser can also display the actual DNA sequence (or its complement) in a separate window; this is shown in **Fig. 6**. When a user selects an annotation in the map display, the corresponding region is highlighted in the appropriate color in the sequence display. In **Fig. 6**, for example, the *GenPept* hit that was selected in the map display is highlighted in the sequence display.

Interaction between the map and sequence displays is bidirectional: When a region is selected in the sequence display, it is boxed in the map display. (If you happen to release the mouse when the cursor is midway between two rows when mousing out a region in the sequence display the boxed region in the map display will erroneously start at zero. This is a known *Tkperl* bug and cannot be fixed within *Genotator*.) If the forward strand is displayed in the sequence window, then clicking on annotations in the forward strand will highlight them in the sequence window, but clicking on annotations in the reverse strand will not.

4.3.1. Functions on the Sequence Display

Show complement/forward strand: Toggle between display of the forward and complementary strand (this will close the sequence display window and open a new one). Please note that for compatibility with the map display, when the complement is displayed, it is not the reverse complement. As a visual reminder of which strand is currently displayed, when the forward strand is being displayed, the bars in the sequence display window are light blue; if the complement is displayed, the bars are pink.

Sequence highlights: This controls whether new sequence highlights (e.g., those that appear when you click on an annotation in the map display) replace the highlights that were already there, or are added to them (the default behavior is “replace”). This does not apply to the yellow highlight that appears when you mouse out a region of the sequence display, which always replaces any previous highlights. If you want to clear all highlights, select the “**replace**” radio button and then click the mouse (without dragging) somewhere in the sequence display.

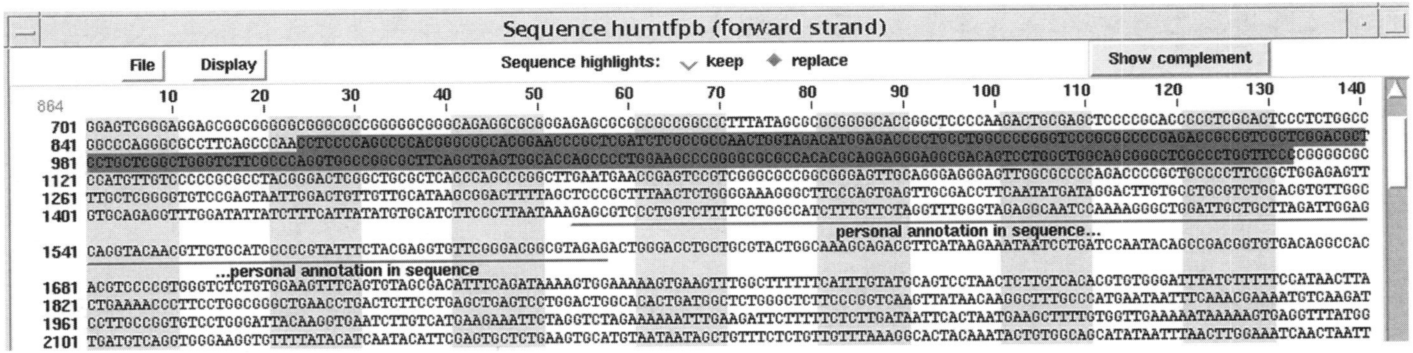

Fig. 6. The sequence display. The user has clicked on an annotation in the map display, the corresponding region is highlighted in the sequence display.

4.3.1.1. File Menu

Output selected region: As with this function on the map display, thi allows you to output the selected region to a file. If you select a region in th complement, it will be saved as the *reverse* complement (even though th sequence displayed in the sequence window is not the reverse).
Print this window: Saves the sequence window as PostScript and (option ally) sends it to your default printer (if you have *xwd* and *xwd2ps*).
Close: Close the sequence window.

4.3.1.2. Display Menu

Find pattern...: This pops up a window that lets you type in a string o regular expression for which to search in the sequence. A string is an sequence of A, C, T, and G, e.g., `CCGCGTTG`. It might represent a restric tion site or a motif. You can also search for UNIX-style regular expressions For example, suppose you wanted to find all instances of an A followed b either a C or a G followed by one or more Ts followed by an A. The UNIX style regular expression for that pattern is `A[CG]T+A`. In **Fig. 7**, *Genotato* has found and highlighted all the subsequences that match that pattern Information about UNIX-style regular expressions can be obtained by typin `man regexp` or at the URL: `http://www.wiley.com/compbooks/unixshell/appendix-i.html`)
Personal annotations...: Enables a control panel for adding or deletin personal annotations as described in the next section.
Highlight stop codons: This highlights stop codons in one or all three frames, color-coded by frame.

4.4. Adding Personal Annotations

The *Genotator* browser allows users to add new annotations to either the map or the sequence display. These personal annotations are saved along with the precomputed annotations. **Figure 8** shows the interface for adding or deleting personal annotations. To add a personal annotation to the map or sequence display, the user selects some region of the sequence, types the annotation text in the text box, and then clicks **Add Annotation to Map** or **Add Annotation to Sequence**. The color of each personal annotation can be specified when it is created. Clicking on the button that says **forestgreen** brings up a menu of color choices. (Changing the annotation color only affects annotations you are about to add; it does not change those you have already added.) You may wish to use different colors to represent different types of personal annotations.

Annotations that refer to a sizable portion of the sequence are generally added to the map; those referring to a small region (such as a primer) are more

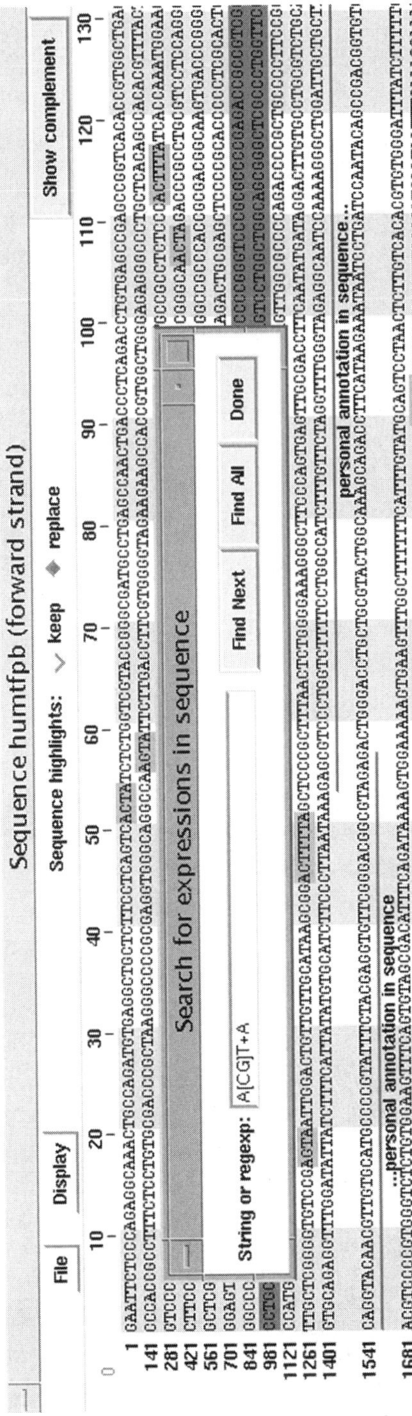

Fig. 7. Finding patterns: All of the subsequences matching the pattern `A[CG]T+A` are highlighted.

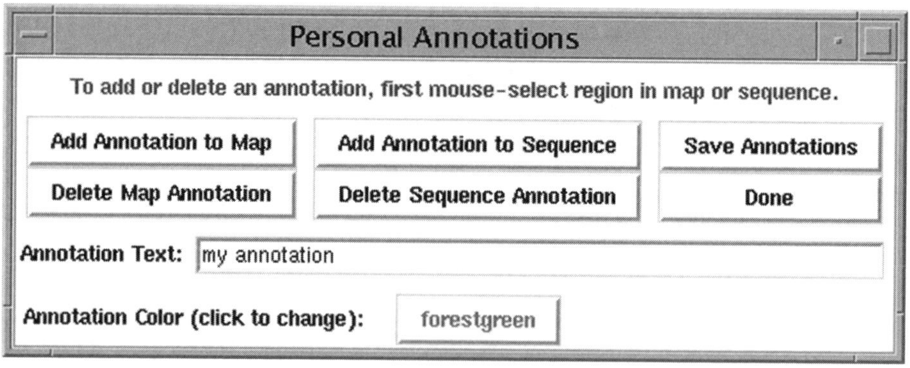

Personal Annotations

To add or delete an annotation, first mouse-select region in map or sequence.

Add Annotation to Map | Add Annotation to Sequence | Save Annotations

Delete Map Annotation | Delete Sequence Annotation | Done

Annotation Text: my annotation

Annotation Color (click to change): forestgreen

Fig. 8. The interface for adding annotations to the map or sequence display.

appropriately added to the sequence. All personal annotations are saved in the database along with the automatically generated annotations. Examples of personal annotations can be seen on the map display in **Fig. 3** ("Personal annotation" and "Reverse strand annotation") and the sequence display in **Fig. 6** ("personal annotation in sequence").

4.4.1. Deleting Personal Annotations

To delete personal annotations in the map display, mouse-select a box around the annotation you wish to delete (this may be easier to do if you zoom in first). The annotation will disappear from the display; however, it is not permanently gone until you hit **Save Annotations**. (If you quit the browser without saving, you will be asked if you want to save your personal annotations.) There is no **undo** function for deletion, but if you mistakenly delete an annotation, you can use **reload** to reload all the personal annotations that have been saved in the database (of course, any new ones you have not yet saved will be lost).

Because personal annotations in the sequence display may overlap, the procedure for deleting them is slightly different. Selecting **Delete Sequence Annotation** will invoke a window showing the positions and labels on the sequence annotations. If you single-click on one of the annotations, that annotation will be highlighted in cyan on the sequence display. Double-clicking, or pressing the **Delete** button, will delete that annotation. The sequence display window will vanish and reappear without the deleted annotation.

4.5. Primer Design

To help the user design primers for a region of interest, *Genotator* can call *Primer3* ***(17)***. First select a sequence region (this can be done by mousing out a region in the map or sequence display, or by clicking on an annotation in the map display.) Then select **Design Primers** from the menu, and change any of the default *Primer3* options if desired. The best forward and reverse primers are printed to the terminal (so that they can be cut and pasted into a primer order form) and are also indicated in the sequence display. For more information about *Primer3*, please consult Chapter 20.

5. Customizing *Genotator*

The previous sections described the options that help to make *Genotator* user-configurable. *Genotator* is also programmer-configurable. A competent *Perl* programmer should be able to modify or add to *Genotator*'s functionality. One of the easiest aspects to configure is the choice of gene finders—which ones are run, and in which order they are displayed. Adding a new gene finder to *Genotator*'s suite would involve copying and modifying the functions for dealing with one of the known gene finders, such as *GRAIL* and writing a parser to convert the output format of the new gene finder into a format *Genotator* can parse. It would also be fairly straightforward to make *Genotator* run *BLAST* on another database—for example, all of *GenBank* (which would entail choosing a color and offset position for the results).

Acknowledgments

I am grateful for the assistance of Gregg Helt, who wrote the *bioTkperl* widgets as well as a graphical browser (*AnnotP1*) that inspired many of the features in the *Genotator* browser; Martin Reese, who wrote several of the sequence analysis programs called by *Genotator*, helped me debug earlier versions of *Genotator*, and offered insightful suggestions on a previous *Genotator* paper; Colin Collins, who enthusiastically supported the use of *Genotator* in his group; Suzanna Lewis, director of the Berkeley *Drosophila* Genome Project informatics group, for which I now work; and Judith R. Harris, who suggested the name *Genotator*.

References

1. Harris, N. L. (1997) Genotator: a workbench for sequence annotation. *Genome Res.* **7,** 754–762. To obtain *Genotator*, email the author, `nlharris@lbl.gov`.
2. Benson, D. A., Boguski, M. S., Lipman, D. J., Ostell, J., and Ouellette, B. F. (1998) GenBank. *Nucleic Acids Res.* **26,** 1–7.
3. Helt, G. (1997) Data visualization and gene discovery in *Drosophila melanogaster*, PhD thesis, University of California at Berkeley.
4. Kulp, D., Haussler, D., Reese, M. G., and Eeckman, F. H. (1996) A generalized hidden Markov model for the recognition of human genes in DNA, in *Proceedings of the Conference on Intelligent Systems in Molecular Biology '96,* AAAI/MIT Press, St. Louis, MO., pp. 134–142.

5. Xu, Y., Mural, R. J., Shah, M. B. and Uberbacher, E. C. (1994) Recognizing Exons in Genomic Sequence Using GRAIL II, in *Genetic Engineering: Principles and Methods*, vol. 15 (Setlow, J., ed.), Plenum, New York, NY, pp. 241–253.
6. Green, P. (1994) Ancient conserved regions in gene sequences. *Curr. Opin. Struct. Biol.* **4,** 404–412.
7. Thomas, A. and Skolnick, M. H. (1994) A probabilistic model for detecting coding regions in DNA sequences. *IMA J. Math. Appl. Med. Biol.* **11,** 149–160.
8. Burge, C. and Karlin, S. (1997) Prediction of complete gene structures in human genomic DNA. *J. Mol. Biol.* **268,** 78–94.
9. Altschul, S. F., Gish, W., Miller, W., Myers, E. W., and Lipman, D. J. (1990) Basic local alignment search tool. *J. Mol. Biol.* **215,** 403–10.
10. Claverie, J. M. and States, D. J. (1993) Information enhancement methods for large scale sequence analysis. *Comp. Chem.* 17:191–201.
11. Boguski, M. S. (1995) The turning point in genome research. *Trends Biochem. Sci.* **20,** 295–296.
12. Reese, M. G. and Eeckman, F. H. (1994) New neural network algorithms for improved eukaryotic promoter site recognition, in *The Seventh International Genome Sequencing and Analysis Conference*, Hilton Head Island, South Carolina, September 16–20, 1995.
13. Reese, M. G., Eeckman, F. H., Kulp, D., and Haussler, D. (1997) Improved splice site detection in Genie, in *First Annual International Conference on Computational Molecular Biology* (*RECOMB*), 1997, Santa Fe, Waterman, M., ed., ACM Press, New York, NY.
14. Durbin, R. and Thierry-Mieg, J. (1991) A *C. elegans* Database. Documentation, code and data available from anonymous FTP servers at `lirmm.lirmm.fr`, `cele.mrc-lmb.cam.ac.uk` and `ncbi.nlm.nih.gov`.
15. Mackman, N., Morrissey, J. H., Fowler, B., and Edgington, T. S. (1989) Complete sequence of the human tissue factor gene, a highly regulated cellular receptor that initiates the coagulation protease cascade. *Biochemistry* **28,** 1755–1762.
16. Sonnhammer, E. L. L. and Durbin, R. (1994) A workbench for large scale sequence homology analysis. *Comput. Applic. Biosci.* **10,** 301–307.
17. Rozen, S. and Skaletsky, H. J. (1996) Primer3. Code available at `http://www-genome.wi.mit.edu/genome software/other/primer3.html`

14

Low Cost Gel Analysis

Jeffry A. Reidler

1. Introduction

Gel documentation is a broad term, ranging from a $300 film camera to a $15,000 complete gel-analysis station. We will focus here on the essential parts needed to assemble an entry level low-light gel image acquisition, computer archive, and analysis system. This consists of an illuminator, a camera, a means to generate a digital signal, analysis software, a computer archive, and finally a printer. Earlier systems consisted of a Polaroid camera with a resulting photograph, but without the digital signal required for the computer. These images can be scanned or the system adapted to digital signal output for the cost of a camera/lens with digital signal output, plus possibly the cost of a new filter for some gel applications. A new system can be assembled for under $2500, consisting of the camera with digital signal, lens, filter, and analysis software. Illuminators and computers are often present in the lab, and hoods or copy stands used for mounting the camera system are sometimes available or easily adapted from existing lab materials. Video printers cost $1300, but recent low-cost ink jet printers can be used for under $200. Digital consumer cameras will soon allow further cost reductions as digital camera performance and ease-of-use approach current camera systems.

Contact information may be found in **Subheading 6.** for manufacturers of products described here.

2. Materials

Complete systems can be purchased from a variety of sources, such as Bio-Rad, UVP, Ultra-Lum, Fotodyne, Stratagene, Alpha Innotech, and many more. Systems generally consist of an illuminator, hood, camera system, computer,

From: *Methods in Molecular Biology, vol. 132: Bioinformatics Methods and Protocols*
Edited by: S. Misener and S. A. Krawetz © Humana Press Inc., Totowa, NJ

software, and printer, and range in price from $6000–15,000. A Web directory is available in **Subheading 6.** Many regional vendors are also available for system integration, custom applications, and service. Science-Intl offers a consulting service for system transitions.

Illuminators can be purchased from many of these same vendors for approx $1000. See the Web directory in **Subheading 6.** In many cases, the illuminator in the lab can be adapted for digital use. Current illuminators now often consist of just the UV lamps, and use white light converters to transform UV to white. We used the UVP illuminator for this test.

Copy Stands are widely available that allow the mounting of the camera for use in a dark room. This can consist of a tripod or post mounted in the lab bench with a movable mount for the camera. Many cameras have a standard 1/4 × 20-in mounting thread common on 35-mm SLR cameras. Hoods provide the same function, with the added ability to work in low light, and to shield the gels from interference from room light. Prices start at $500.

Cameras for normal light gels and bright fluorescence gels can consist of a black and white videocamera that might be present in the lab. Videocamera output is translated to digital signals with a video frame grabber, available from many suppliers in **Subheading 6.**, including vendors of frame grabbers, software, and systems. We used the Scion LG-3 frame grabber for this application, and these are compatible with *NIH-Image* and *Scion-Image* on PowerMac and Pentium platforms.

Low-light cameras allow use also with fluorescence gels, such as ethidium bromide (EtBr). Digital consumer camera capabilities and resolution are advancing, and prices are falling; the DC120, DC210, and DC260 from Kodak are good examples. Low-light videocameras are commonly used because of their speed and performance, and prices below $1500, including the frame grabber. An example package is the GMS300 available from Scion for $1695 that can be used with both visible and UV gels.

Resolution can be enhanced in many ways. European B/W CCIR (PAL color) format for video is 768 × 576 pixels, and this format provides approx 30% more resolution than the American B/W RS170 (NTSC color) format of 640 × 480. Oversampling techniques (e.g., Scion Series-7) can provide an additional 30% resolution. The combination of a CCIR format camera with oversampling can provide 1000-pixel resolution in the long axis of the image, and this meets the needs for most gel analysis. Additional resolution can be obtained in the DC260 at 2000 × 1600 pixels for under $1000, or with high resolution digital cameras starting at $6000.

Filters vary with application, but three-cavity interference filters for EtBr gels can be obtained from Chroma Technology and Omega Optical in **Subheading 6.** Many are available with filter rings attached, which allow the filter to screw directly onto the zoom lens. These interference filters provide UV and IR blocking, as well as 80% transmission of the wavelength of interest. New filters may be required when converting from Polaroid camera gel-imaging systems because of the different sensitivities of film-type Polaroid cameras compared to silicon-based CCDs in videocameras.

Lenses can consist of an 8- to 48-mm F1.0 C-mount zoom, with a minimum field of view of 2 in (35 mm) on the long axis at the minimum working distance (14 in, 35 cm) at the maximum zoom. An alternative lens is the 11- to 69-mm zoom at F1.4. This type of lens is commonly used with hoods in gel-imaging systems. Rainbow and Canon both offer this type of lens with a 46-mm screw thread for mounting filters; Toyo and Cosmicar are also widely used. Prices range from $250 to $500.

A less expensive option is a 12-mm, 16-mm, or 25-mm C-mount lens, and this is suitable if a copy stand, tripod, or mounting post is used, so that the height can be adjusted for varying gel sizes. These lenses can be transformed to macro lenses by placing a thin ring between the lens and the camera body. A 1-mm thick ring will reduce the field of view to about one-half of its previous. Prices for these lenses range from $80 to $150.

If your application is low-light fluorescence, select the lens with the largest light throughput, or lowest F number at a given focal length. We have found that a 25-mm F-1.4 25-mm lens provides more light throughput than a 8- to 48-mm zoom lens F-1.0 at 25-mm focal length, and macro rings can be used for minigels. Prices, light throughput, and minimum working distance vary greatly. Compare lens F-numbers at comparable focal lengths.

Computers are now prevalent in the laboratory, so most can be adapted, shared, or purchased locally. Computer choice is more determined by the software to be used. Power Macintosh or Pentium platforms with 32 mb computer RAM will be sufficient for *NIH Image* and *Scion Image*. Performa computers will require a shorter frame grabber (VG-5); PowerMac models 7100, 8100, and 9150 are NuBus slots, so they require frame grabbers which are out of production.

Software for analysis of gels is widely available, starting with freeware like *NIH-Image* for Macintosh and the PC port in *Scion-Image* for Windows. These programs are free, as well as the gel analysis macros and extended versions available through the *NIH-Image* website. Tutorials on the use of *Image* in gel analysis are also available at this site (e.g. `http://scrc.dcrt.nih.gov/imaging/tutorials/gel_density/short/index.html`). Links to gel analysis freeware, as well as many vendors of software and systems can be made through this site. Scion is developing additional gel analysis software to be used in conjunction with the Scion frame grabbers. A newsgroup is available. To subscribe to a digested version of the list, send e-mail to `nih-image-d-request@biomed.drexel.edu`. The subject of the message should contain `subscribe`.

Additional software analysis capability can be purchased through specific software programs from Media Cybernetics and Scanalytics. Both programs offer ease-of-use and additional analysis not available with *NIH-Image*. RFLP and two-dimensional gel analysis can also be found at these sites. Links to both companies can be found in **Subheading 6.**

Archive capability can be as simple as creating a new folder, but might also require a protocol for naming gels as well as a means to display and store large numbers of gels. Several image archive programs are now available, and *Thumbs-Plus* by Cerious is attractive because it offers versions for both Macintosh and PC for a nominal fee with a 30-day free trial. Media Cybernetics *Gel-Pro* provides an image archive database within the program, and connection to each experiment automatically.

Output to printers has undergone dramatic cost reductions recently with the introduction of 1400-dpi ink jet printers ($200) and of small format dye sublimation printers ($500). In many cases, adequate output can be obtained from current printers in the laboratory. Examples include Epson 600 and Hewlett-Packard 722. Dye sublimation printers are best suited for lower volume gel documentation due to the cost per print ($0.50). Ink jet printers are best suited for labs on a limited budget. Labs with higher volumes of gels might prefer a video thermal printer ($1500) because of the rapid output, low cost per print ($0.10), and size format designed for the lab notebook.

3. Methods

3.1. EtBr Gel Example Configurations

NIH Image and *Scion Image* are equivalent programs, with Macintosh versions available from both NIH and Scion, and Windows 95, 98, and NT versions written by Scion. Some gel documentation macros are written into *NIH Image*, and many more are available free. Data and procedures for these two programs are largely interchangeable. This setup will work for either brightfield or fluorescence gels. For this test, we used the UVP Transilluminator, Cohu 4912-5010 Camera, Scion CG-7 frame grabber, *NIH-Image* software, Macintosh 9500 computer. For Pentium, most equipment was unchanged, except we used a Gateway 2000 with *Scion-Image*.

1. Install frame grabber into the computer and connect the frame grabber to the camera. This essentially consists of finding an open PCI slot and inserting the board into the slot. Optional cables are needed for low-light use, consisting of video going out from the camera, and integration control coming in from the frame grabber.

2. Mount the camera lens 14 in (30 cm) to 30 in (75 cm) from the gel samples on the illuminator. A UVP #97-0063-01 and Bogan TC-2 (Item 1882) copystand were used in this test. A 46- to 49-mm filter will cover the zoom lens, depending on the lens. Some interference filters have adequate UV- and IR-blocking capability out to several seconds of integration. Three cavity filters are adequate for normal EtBr gels, but very dim gels may require additional blocking to eliminate UV and IR from the illumination source. The zoom lens will accept a screw-in filter (e.g. ChromaTech, Corion, and Omega Optical). The Rainbow H6X8 has a 46 × 0.75-mm thread, and many vendors offer filters with attached threaded rings in various sizes. We also used the filter from AAB with a 49-mm thread and a 46- to 49-mm adapter ring purchased at a local photo shore (*see* **Fig. 1**).
3. Download, install, and run *NIH-Image* or *Scion Image*.
4. Select **Specials**, and **Start Capturing** to see the live video image on the computer screen. We have found that many gels are visible in live video and require no additional integration capability. In this case, many videocamera options are available.
5. For gels not visible in **Start Capturing**, on-chip integration enables up to 1,000× sensitivity increase. Select **Specials**, **Load Macros**, and select the macro **Video** in the Macros folder. If using a Scion Series-7 board, then select the macro **Video Series-7**.
6. Acquire the image using the integration on-chip feature. If maximum resolution is needed, orient the gel so that the lane is horizontal. Select **Specials**, **Continuous Integration On-Chip** (*see* **Fig. 2**). This will provide an image in integration mode with an update after each four frames of integration. Place the mouse over the lower part of the image, click and hold the mouse in this region to increase the integration time. Place the mouse over the upper part of the image, click and hold the mouse in this region to decrease the integration time. If using a Scion Series-7 frame grabber, select **Continuous Integration On-Chip Hi-Res** to obtain the oversampled image with 1536 × 1152 CCIR resolution. Alternatively, the commands **Special**, **Multi-Frame Operations** can be used. The last integration time is copied from the macro into the **Multi-Frame Operations** for ease-of-use.
7. Snap the integrated image by selecting **Apple, period (.)** on the Mac, or **Esc** on the PC. Save the image if desired using **File**, **Save**. Adopt a uniform numbering scheme, such as UserID-YrMoDay-Gel#, or YrMoDay-Expt# when saving the images. Archive the gels with a program such as *Thumbs Plus* by Cerious or another image database program that provides thumbnail images. Some image database programs allow layering, so that annotations are layered onto, but not a part of, the original image.
8. Calibrate your optical density. Each pixel has a brightness value with a range of 2^8 steps or 256. A conversion is needed between OD and brightness using a set of optical density standards. Select **Analyze** and **Calibrate** in *Scion Image* or *NIH Image*. Sample values are measured directly from the standards on the image, using the selection tool, in the image window. Then, select **Calibrate**, and enter the known OD values. You should generally obtain a logarithmic curve of the

Fig. 1. Photo of hardware building blocks.

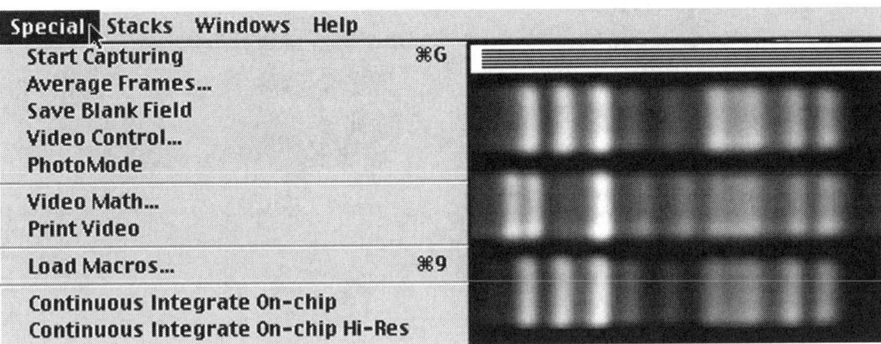

Fig. 2. Integration macro menus with gel image.

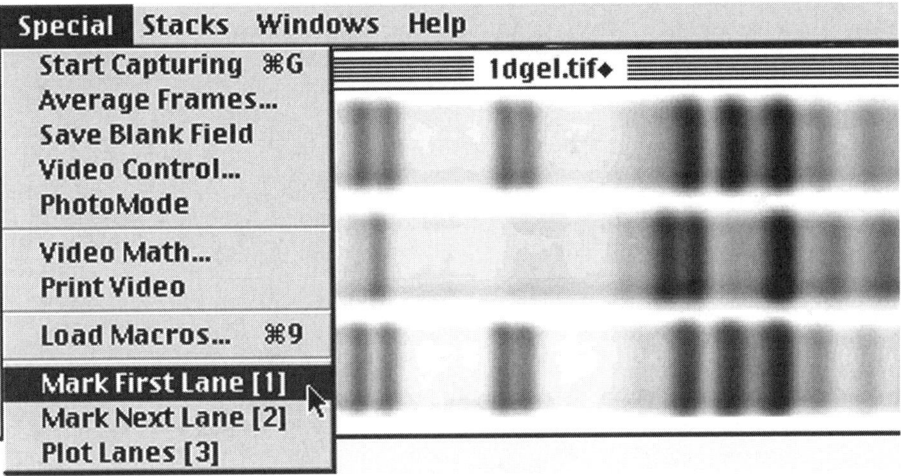

Fig. 3. Gel analysis macro with inverted gel image.

data points. Negative images, where the background is black and the bands are white, should be inverted before calibration. Go to the **Edit** menu to find the **Invert** item.

9. Load the gel analysis macro 2 (*see* **Fig. 3**).
10. Highlight each lane by drawing a box around the lane of interest, starting at the top, and working down. If your gel is longer than the screen width, try increasing the monitor resolution to the maximum. For gels that have curved lanes, try to use a narrow band that falls within the rectangular area. Use the macro command **Plot Lanes** to see the plotted curves. The NIH program can only highlight the length of the lane shown on the screen, so that oversampled images will require a screen resolution to match (*see* **Fig. 4**).

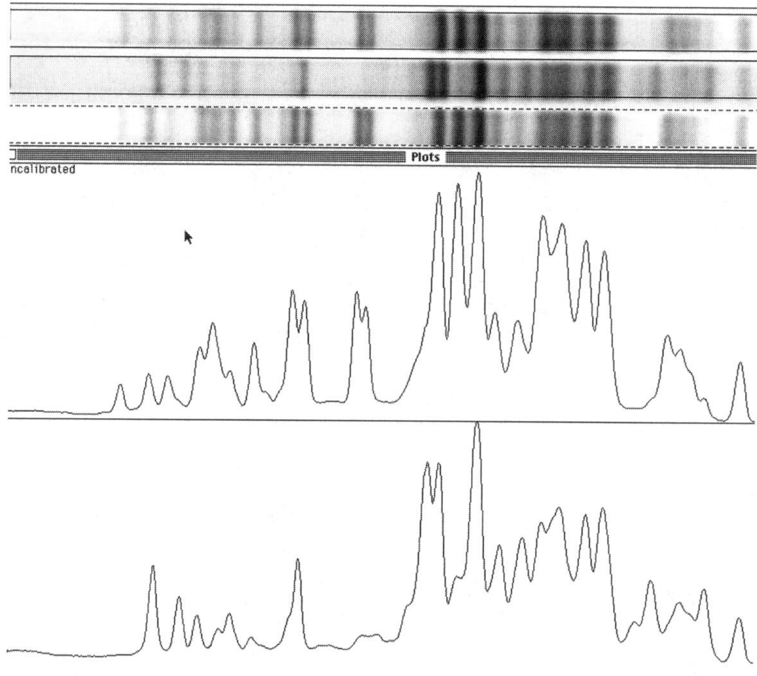

Fig. 4. Plotted gel curves.

11. Use the **line** drawing tool to draw base lines and drop lines so that each peak defines a closed area as shown above. Note that you can hold the **shift** key down to constrain lines to be vertical. You can establish a baseline of background by taking a region-of-interest in a blank lane and adding it to the plot window.
12. Measure the areas of the peaks by clicking inside each one in succession with the 'wand' tool.
13. Option-click with the 'text' tool to automatically label the peaks, in *reverse* order, with the area measurements (*see* **Fig. 5**). (Use **Scroll Lock**-Click on the PC.) The area measurements are also recorded in tabular form, and can be displayed (**Show Results**, *see* **Fig. 6**), printed (**Print**), or exported (**Export**) to a spreadsheet. The combined windows are shown in **Fig. 7**. *Scion Image* for PC is still in beta form, and limitations exist for printing to PostScript printers. Check with Scion at their website for the latest list of printers that work as the system is assembled, or print to a video printer using the VG-5.

3.2. Marburg Macro Options

Marburg Macros are available (`http://www.chemie.uni-marburg.de/~becker/image.html`) and have some alterations to the existing *NIH Image* macros. These macros work on vertical gels and find the lanes automati-

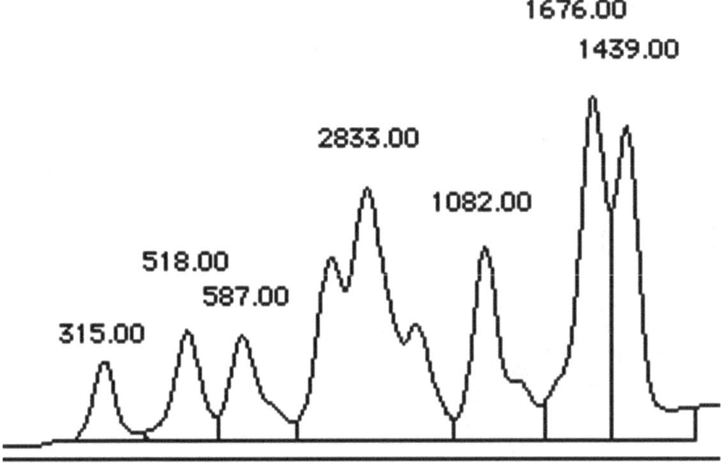

Fig. 5. Integrated intensity measurements.

Results

	Area
1.	315.00
2.	518.00
3.	587.00
4.	2833.00
5.	1082.00
6.	1676.00
7.	1439.00

Fig. 6. Results in text format.

cally. Variations that do not compress the plot windows are also available from *NIH Image* website links to gel analysis.

3.3. Advanced Features

More advanced packages are available from most vendors of gel analysis systems or software. Complete systems are widely available. Some attractive

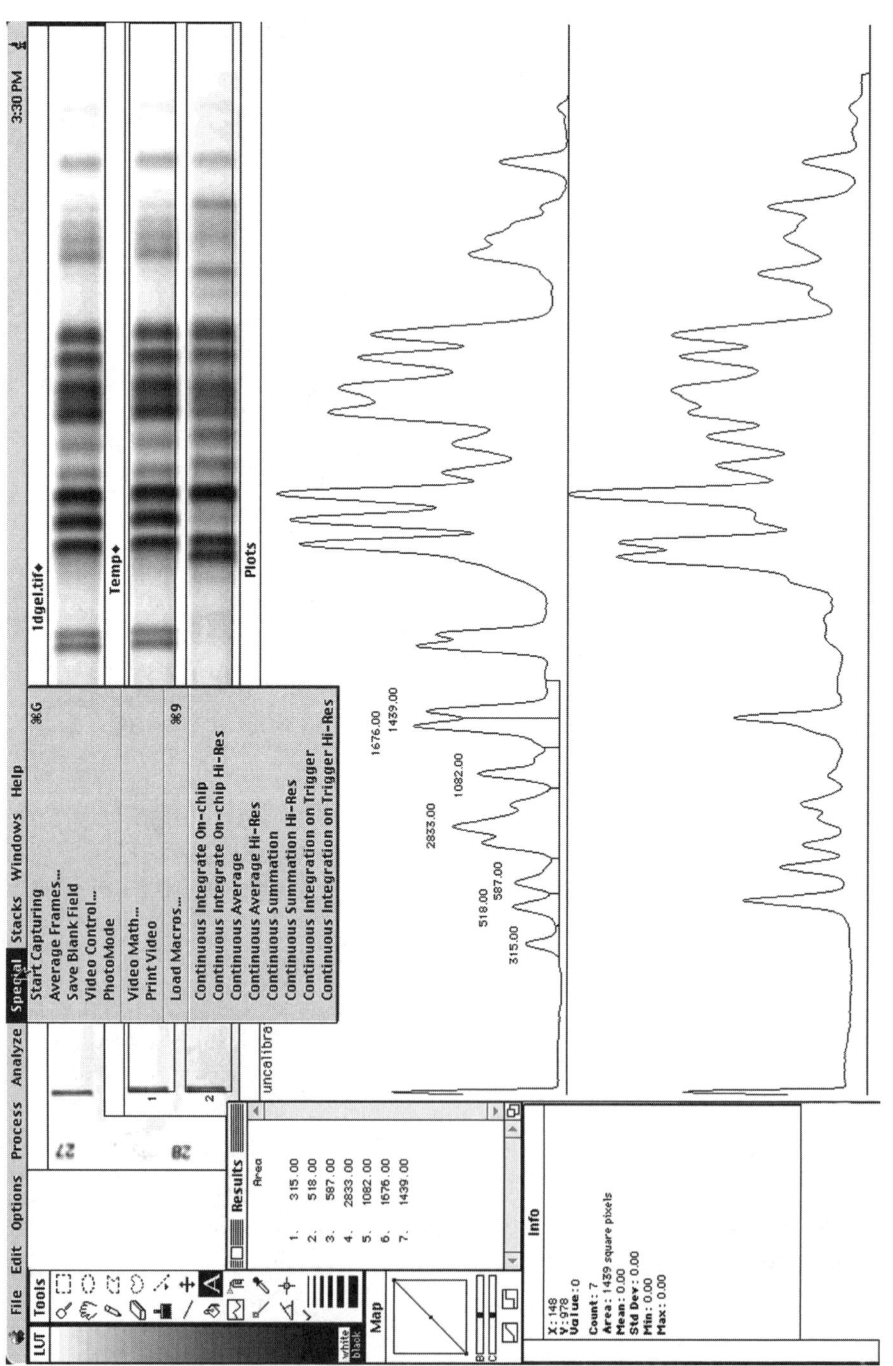
File Edit Options Process Analyze Special Stacks Windows Help
3:30 PM
LUT
Tools
Map
white
black
1dgel.tif
Temp
Plots
uncalibra
Start Capturing ⌘G
Average Frames...
Save Blank Field
Video Control...
PhotoMode
Video Math...
Print Video
Load Macros... ⌘9
Continuous Integrate On-chip
Continuous Integrate On-chip Hi-Res
Continuous Average
Continuous Average Hi-Res
Continuous Summation
Continuous Summation Hi-Res
Continuous Integration on Trigger
Continuous Integration on Trigger Hi-Res
315.00
518.00
587.00
2833.00
1082.00
1676.00
1439.00
Results
Area
1. 315.00
2. 518.00
3. 587.00
4. 2833.00
5. 1082.00
6. 1676.00
7. 1439.00
Info
X: 148
Y: 978
Value: 0
Count: 7
Area: 1439 square pixels
Mean: 0.00
Std Dev: 0.00
Min: 0.00
Max: 0.00

features found in more advanced packages include automatic lane finding, ability to work with curved and slanted lanes, and high resolution digital cameras with >8 bit dynamic range See **Subheading 6.** for alternatives to the *NIH Image* package (and the *Scion Image* port).

3.4. Sample Equipment List (does not include illuminator or computer)

Scion GMS300	1700
Rainbow 8–48 mm zoom	330
Omega EtBr filter	250
Bogan TC-2 (#1882) copy stand	400
Scion Image analysis software	free
Cerious *Archive* software	65
Epson 600 printer (example only)	200
TOTAL	$2945

4. Summary

The progression from gel films to digital image storage and analysis is straightforward and can be cost effective. This method allows a simple progression to digital acquisition and analysis with a minimum of capital expenditure. Data in digital format are easily stored, recalled for reanalysis, printed, or sent by e-mail to collaborators. Additional resolution can be obtained with higher resolution cameras, and digital cameras are available with wider dynamic ranges are available. Additional sensitivity can be obtained with more sensitive dyes such as SYBR series from Molecular Probes. The current description provides a roadmap for an entry-level digital gel documentation system with capability to upgrade to more advanced features as funds allow.

5. Conclusions

A gel documentation system can be assembled for under $3000 comprised of existing parts from the laboratory combined with a few commercial parts and software for $100 or less. More advanced packages offer additional ease-of-use and analysis with prices increasing to $15,000. Several vendors offer kits, and other vendors offer software packages which can be added as the volume and sophistication of the laboratory advances.

6. Equipment and Software Directory

AAB, `http://www.aabi.com`, 714-870-0290

Alpha Video, `http://www.alphavideo.com`, 800-388-0008, 612-896-9898

Bio-Rad, `http:// bio-rad.com/27355.html`, 510-741-1000

Cerious, http://www.cerious.com, 704-529-0200
Chroma Tech, http://www.chroma.com, 800-824-7662, 802-257-1800
Cohu 4910, http://www.cohu.com, 619-277-6700
Fotodyne, http://www.fotodyne.com, 800-362-3686, 414-369-7000
Kodak Scientific, http://www.kodak.com/go/scientific
Media Cybernetics, http://www.mediacy.com, 800-992-4256, 301-495-3305
Molecular Probes, http://www.probes.com, 541-465-8300
NIH Image, http://rsb.info.nih.gov/nih-image
Omega Optical, http://www.omegafilters.com, 802-254-2690
Scanalytics, http://www.iplab.com, 703-208-2230
Science-Intl, http://www.science-intl.com, 301-631-0157
Scion Corporation, http://www.scioncorp.com, 301-695-7870
Ultra-Lum, http://www.ultralum.com, 562-529-5959
UVP, http://www.uvp.com, 800-452-6788 or 909-946-3197

Acknowledgement

Many thanks to Tod Weinberg and Tom Morton at Scion Corporation for helpful discussion.

3

Web-Based Resources

15

Computer Resources for the Clinical and Molecular Geneticist

Yuval Yaron and Avi Orr-Urtreger

1. Introduction

Over the course of the last decade, genetics has become one of the most rapidly expanding sciences. The overwhelming and evergrowing pool of knowledge makes it virtually impossible to follow new and recent discoveries using conventional methods such as reviewing journals or scientific textbooks. Computer resources have therefore become critical for both the clinical and molecular geneticist. The purpose of this chapter is to provide the reader with examples of such resources.

2. Molecular Resources on the Internet (World Wide Web)

The Human Genome Project (HGP) is an international effort initiated in 1990 whose goal is to discover all the 50,000–100,000 genes in the human genome, and make them accessible for further biological study. Therefore, one of the goals of the HGP is the development of analysis algorithms and integration of genetic databases (informatics) for managing and interpreting genome data. These resources have been made available to health professionals and researchers through the internet on the World Wide Web (WWW). Universities, reasearch centers, and health organizations have also generated numerous databases that have been made readily available for searching at no charge. Some have made registration a requirement to allow access only to health professionals, and some require a fee for registration. Searching the internet is rewarding because many databases are linked to each other and offer access to various other related sites.

From: *Methods in Molecular Biology, vol. 132: Bioinformatics Methods and Protocols*
Edited by: S. Misener and S. A. Krawetz © Humana Press Inc., Totowa, NJ

Fig. 1. *Online Mendelian Inheritance in Man* (*OMIM*)—Search OMIM page.

2.1. Online Mendelian Inheritance in Man (OMIM™)

This is a catalog of human genes and genetic disorders, compiled and indexed by Victor A. McKusick and others at Johns Hopkins University and elsewhere. The effort was initiated in the early 1960s with the *Catalog of X-Linked Traits in Man*. After numerous printed editions, it has been developed by the National Center for Biotechnology Information (NCBI) for the World Wide Web. The database contains abstracted information from numerous published sources, references, and links to other internet resources such as *Entrez*: the NCBI's *MEDLINE* and *GenBank* retrieval systems, *Online Mendelian Inheritance in Animals* (*OMIA*), *The Cardiff Human Gene Mutation Database* (*HGMD*), *MitoMap*: the Emory University mitochondrial genome database, and others. This valuable resource is now available on the WWW at: `http://www.ncbi.nlm.nih.gov/Omim`.

In the Search *OMIM* page, key words are entered in the appropriate window. Key words may include syndrome name, gene name or designation, or certain clinical features. In the example presented in **Fig. 1**, we search for "bone dysplasia." The Results page provides a list of all the *OMIM* database entries that satisfy the criteria entered in the Search *OMIM* page. In the example presented, 201 entries were found, of which 50 are presented (**Fig. 2**). Clicking on one of the options opens up the specific entry, which includes an abstract with the clinical features and molecular information, as well as links to other databases such as *Genome Database* (*GDB*) (**Fig. 3**).

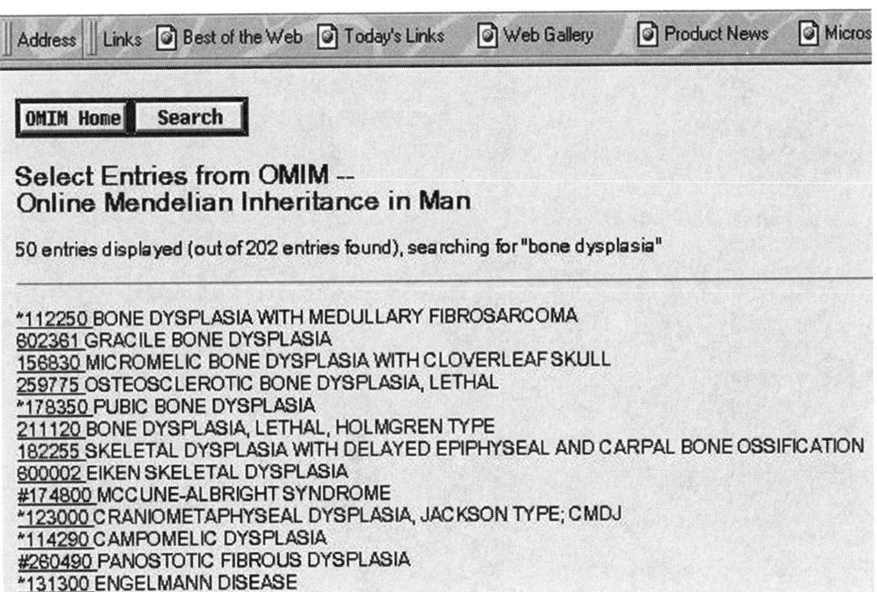

Fig. 2. *Online Mendelian Inheritance in Man* (*OMIM*)—Results page including all matching results.

2.2. Entrez

This resource provides molecular biology data and bibliographic citations from the NCBI's integrated databases. These include: DNA sequences from *GenBank*, *EMBL*, and *DDBJ*; Protein sequences from *SwissProt*, *PIR*, *PRF*, *PDB*; and translated protein sequences from the DNA sequence databases; genome and chromosome mapping data; Three-dimensional protein structures derived from *PDB*, and incorporated into NCBI's *Molecular Modeling Database* (*MMDB*). In addition, a bibliographic database (*PubMed*) containing citations for nearly 9 million biomedical articles is available via the National Library of Medicine's *MEDLINE* and *Pre MEDLINE* databases. The internet address is `http://www.ncbi.nlm.nih.gov/Entrez`.

2.3. National Center for Biotechnology Information

This website provides numerous links to gene sequence databases, such as *Entrez, Gene Map of the Human Genome, dbEST* (database of Expressed Sequence Tags). It also provides links to other NCBI resources such as *PubMed* (free MEDLINE) provided by the National Library of Medicine (NLM), The Internet address is `http://www.ncbi.nlm.nih.gov/`.

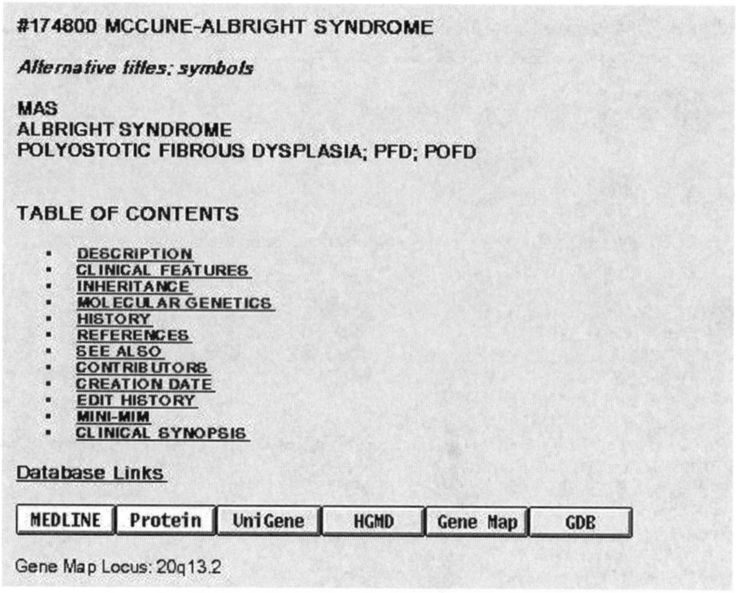

#174800 MCCUNE-ALBRIGHT SYNDROME

Alternative titles; symbols

MAS
ALBRIGHT SYNDROME
POLYOSTOTIC FIBROUS DYSPLASIA; PFD; POFD

TABLE OF CONTENTS

- DESCRIPTION
- CLINICAL FEATURES
- INHERITANCE
- MOLECULAR GENETICS
- HISTORY
- REFERENCES
- SEE ALSO
- CONTRIBUTORS
- CREATION DATE
- EDIT HISTORY
- MINI-MIM
- CLINICAL SYNOPSIS

Database Links

MEDLINE | Protein | UniGene | HGMD | Gene Map | GDB

Gene Map Locus: 20q13.2

Fig. 3. *Online Mendelian Inheritance in Man* (*OMIM*)—Specific entry, title, alternative names, table of contents, database links, test, and references.

2.4. The Cardiff Human Gene Mutation Database

This database includes known gene mutations responsible for a variety of human inherited diseases. It comprises various types of mutations within the coding regions of human nuclear genes that are known to cause inherited disease. The database does not include polymorphisms that do not have obvious phenotypic consequences. The database is maintained at the Institute of Medical Genetics, University of Wales College of Medicine, in Cardiff, by D. N. Cooper, E. V. Ball, P. D. Stenson, M. Krawczak and other. The internet address is `http://www.uwcm.ac.uk/uwcm/mg/hgmd0.html`.

2.5. GeneCards*: Encyclopedia of human genes, proteins, and diseases*

The *GeneCards Encyclopedia* integrates information about human genes and their products, stored in major databases. The data are extracted from the following databases: *GDB*, which contains information about genes and other genomic features; *MGD* (*Mouse Genome Database*) which contains information on the experimental genetics of the laboratory mouse, including markers, mammalian homologies, probes, and clones; *OMIM*: a catalog of human genes and genetic disorders; *SwissProt* which contains information about proteins,

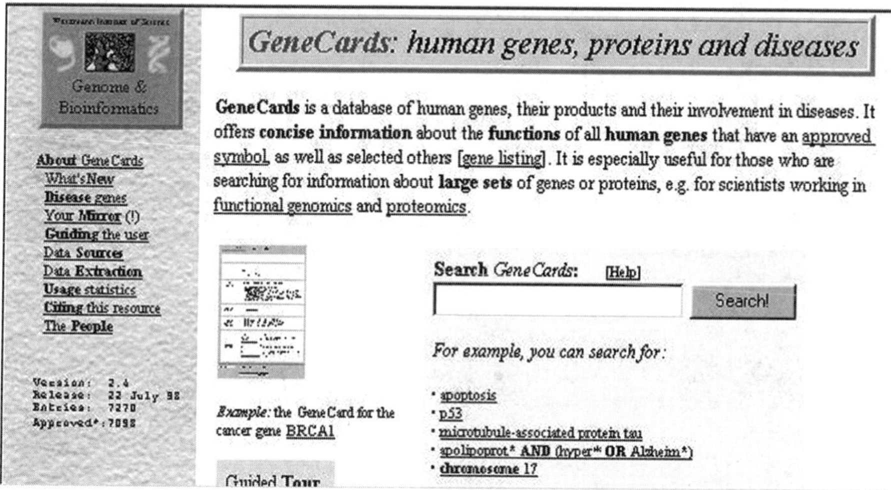

Fig. 4. *GeneCards*: Encyclopedia of human genes, proteins, and diseases.

their sequence, and cellular functions; *HGMD*: a source of information about disease-causing mutations in genes; *Doctor's Guide to the Internet*: A Web service presenting news about biomedical research and its applications. And *Genatlas*: a catalog of genes, markers, and phenotypes with many links to major data sources. Most of the information is automatically added by scripts developed to search other databases for information about those genes. After data is acquired, it is analyzed for relevance, extracted, and transformed into active entries of the database. *GeneCards* is a project of the Weizmann Institute Genome Center and the Weizmann Institute Bioinformatics Unit. The *GeneCards* concept, scripts, and Web interfaces have been developed by Michael Rebhan and Jaime Prilusky, in collaboration with Vered Chalifa-Caspi, Marilyn Safran , Liora Yaar, and Doron Lancet. The internet addres is `http://bioinformatics.weizmann.ac.il/cards/` (**Fig. 4**).

3. Clinical Resources on the Internet

3.1. Information for Genetic Professionals—*University of Kansas Medical Center*

This resource provides regularly updated information for genetic professionals, with links to clinical and research resources, such as: genetic societies, support groups for various genetics conditions, clinical genetic databases (*OMIM*, *GeneTests*), and genetic computer resources. The internet address is `http://www.kumc.edu/gec/geneinfo.html` (**Fig. 5**).

Fig. 5. *Information for Genetic Professionals*—University of Kansas Medical Center.

3.2. GeneTests™

GeneTests™ is a directory of laboratories performing disease-specific diagnostic and/or research testing for genetic disorders. It contains listings for over 300 laboratories, testing for over 550 genetic disorders. The resource is funded by the National Library of Medicine and is maintained by the Children's Hospital, Regional Center and University of Washington School of Medicine. The service is free of charge but is restricted to healthcare professionals, who must register to gain access by password. The resource is available at the internet address `http://www.genetests.org`. It is also accessable by e-mail, phone, or fax.

3.3. Reprotox®

This database provides up-to-date information regarding the reproductive effects of prescription, over-the-counter, and recreational drugs as well as industrial and environmental chemicals. The abstracts provided for each entry offer available data on human, animal, and *in vitro* studies pertaining to the agent queried. The information covers all aspects of human reproduction including fertility, male exposure, and lactation, with a particular focus on embryonic and fetal consequences of exposure. The **Search** button opens up the search screen that allows a search on keywords such as brand names, generic names, street names, chemicals, and other environmental exposures. The Search Results page provides a list of all possible entries that may have relevance to the keyword searched, sorted by confidence level. Choosing any one of the options provides the information summary for the entry.

The service is provided for an annual subscription fee by the Reproductive Toxicology Center (RTC), which is located at Columbia Hospital for Women Medical Center Columbia Hospital for Women Medical Center, 2440 M Street, NW, Suite 217, Washington, DC 20037-1404. The database is also available for a yearly fee at `http://reprotox.com`. It can also be obtained on disk or CD-ROM, in a DOS or Windows version.

4. Computer Resources on Disk or CD-ROM

4.1. POSSUM *TM*

POSSUM is an acronym for "Pictures of Standard Syndromes and Undiagnosed Malformations" (*see* **Note 1**). It is a software tool for the diagnosis of clinical syndromes that is now available on CD-ROM (version 5.0) for use with a standard PC without the need for a special laser disk as with previous versions. The new version also merges information from OSSUM (a system about skeletal dysplasias linked with thousands of X-ray images). The user provides a list of traits exhibited by the patient, and the software offers a list of possible syndromes. It contains 3000 syndromes, and 2000 patients manifesting patterns of birth defects. The database includes a large number of pictures including radiological and clinical features, and has automated links to OMIM. The development was led by David Danks and Agnes Bankier at the Murdoch Institute for Research into Birth Defects and the Victorian Clinical Geneticists services at the Royal Children's Hospital in Melbourne, Australia.

4.2. London Dysmorphology and Neurogenetics Databases (LDDB, LNDB *TM*)

These databases are produced by the Oxford Medical Databases, and authored by Robin Winter and Michael Baraitser, of the Mothercare Unit of Clinical Genetics and Fetal Medicine, Institute of Child Health, 30 Guilford Street, London, WC1N IEH (*see* **Note 2**). The databases offer the clinical geneticist a tool for the clinical diagnosis of congenital anomalies and neurogenetic syndromes. They are compiled from over 1000 journals with many references to the syndromes. The *Dysmorphology Database* includes over 2750 single-gene disorders, sporadic conditions, and those caused by environmental agents. Chromosomal anomalies are not included. The *Neurogenetic Database* contains information on over 2500 syndromes invoving the central and peripheral nervous systems. An additional disk with a Photo Library containing images is available. A Windows version has been available since 1996. The opening screen allows a choice between the *Dysmorphology* and the *Neurogenetics* databases. In each database, the main screen contains a list of all the syndromes in the database. A specific syndrome may be retrieved directly if the name is known. In the clinical setting however, the correct diagnosis is often unknown. The databases therefore are queried by a **Keyword Search** or a **Search on Features** option. The **Keyword Search** option allows a search by keywords appearing in the title, abstract or references, by chromosomal location, by *OMIM* number, or by mode of inheritance.

The "**Search on Features**" option is the most robust feature of this software. Different criteria are specified in appropriate boxes, as chosen from a hierarchical list of clinical features. The feature list begins with general system categories, followed by a second level of generalized clinical features, finally, the lower level consists of specific clinical features. The "**Find Feature**" option allows homing in on a specific criterion, provided it is specified within the category list. A search can be performed on all criteria specified, or on just a few of them. Moreover, some criteria may be marked as mandatory, assuring that all syndromes found will have the criterion specified.

Once a syndrome is located, the syndrome details can be entered. These include a textual abstract, a list of features, and references. Other syndrome details include the cytogenetic location, the *OMIM* number, a list of synonyms, and details of the syndromes inheritance.

On the screen a set of "thumbnails" is shown that represent the matching images in the Photo Library.

4.3. Human Cytogenetic Database

The *Human Cytogenetic Database* has been compiled by Albert Schinzel of the Institute of Medical Genetics in Zurich. It is published on disk by the Oxford University Press, as a part of the Oxford Medical Database Series, under the general editorship of Michael Baraitser and Professor Robin Winter. This database enables retrieval of clinical and cytogenetic data relating to over 1000 chromosomal aberrations. The chromosomal aberrations may be selected by searching a list of clinical features. An important feature of this database is its ability to be extended by adding ones own data.

Notes

1. *POSSUM*: Anne Cronin The Murdoch Institute Royal Children's Hospital, Flemington Road, Parkville, Victoria Australia, 3052, Tel: +61 3 9345 5045, E-mail: `cronin@cryptic.rch.unimelb.edu.au`

2. *Human Cytogenetic Database*, *LDDB*, and *LNDB*:, Janet Caldwell or Rachel Rains, Electronic Publishing, Oxford University Press, Great Clarendon Street, Oxford OX2 6DP, UK., Tel: (01865) 267979, E-mail: `ep.info@oup.co.uk`.

WWW URLs of Cited Resources

1. *On-Line Mendelian Inheritance in Man (OMIM)*, `http://www.ncbi.nlm.nih.gov/Omim`
2. Entrez, `http://www.ncbi.nlm.nih.gov/Entrez`
3. National Center for Biotechnology Information (NCBI), `http://www.ncbi.nlm.nih.gov/`
4. *The Cardiff Human Gene Mutation Database*, `http://www.uwcm.ac.uk/uwcm/mg/hgmd0.html`
5. *GeneCards*: Encyclopedia of human genes, proteins and diseases, `http://bioinformatics.weizmann.ac.il/cards/`
6. *Information for Genetic Professionals*, University of Kansas Medical Center, `http://www.kumc.edu/gec/geneinfo.html`
7. *GeneTests*, `http://www.genetests.org/`
8. *REPROTOX*, `http://reprotox.com`

16

The NCBI

Publicly Available Tools and Resources on the Web

Jack P. Jenuth

1. Introduction

The National Center for Biotechnology Information (NCBI) was established in November 1988, at the National Library of Medicine (NLM) in the United States. The NLM was chosen because it had experience in creating and maintaining biomedical databases and as part of the National Institutes of Health (NIH), it could establish an intramural research program in computational molecular biology. The mission of the NCBI is to develop new information technologies to aid in the understanding of fundamental molecular and genetic processes that control health and disease. It was set up to perform these four major tasks as quoted from the NCBI web site (`http://www.ncbi.nlm.nih.gov/`):

1. Create automated systems for storing and analyzing knowledge about molecular biology, biochemistry, and genetics.
2. Perform research into advanced methods of computer-based information processing for analyzing the structure and function of biologically important molecules.
3. Facilitate the use of databases and software by biotechnology researchers and medical personnel.
4. Coordinate efforts to gather biotechnology information worldwide.

The results of these tasks have been made available to the broad scientific community in a variety of ways over the last decade. These include distribution of databases and software via magnetic media or CD-ROMs and via the Internet using protocols such as ftp, gopher, e-mail, and the World Wide Web

From: *Methods in Molecular Biology, vol. 132: Bioinformatics Methods and Protocols*
Edited by: S. Misener and S. A. Krawetz © Humana Press Inc., Totowa, NJ

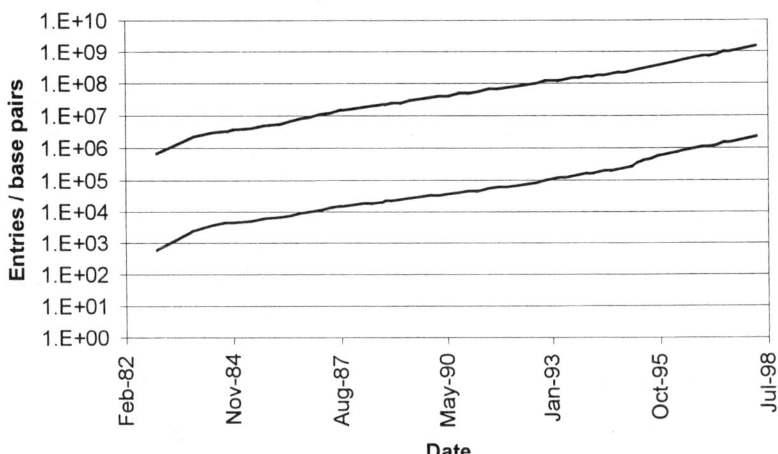

Fig. 1. Increase in the size of *GenBank*. The top line represents the number of nucleotides and the bottom one the number of entries in *GenBank*.

(WWW). This chapter will describe tools and resources that have been made available to users via the WWW.

2. *GenBank*

2.1. Introduction to GenBank

GenBank **(1)** is a database of DNA sequences that was established in the early 1980s. Intelligenetics maintained the database from the later half of the 1980s until October 1992 after which the responsibility for maintaining *GenBank* was assumed by the NCBI. The growth of *GenBank* in terms of both the number of nucleotides and entries has been logarithmic over this time as shown in **Fig. 1**. There are approx 1,500,000,000 bases in 2,209,000 sequence records as of April 1998.

Trained indexers with graduate-level biology experience enter data records into *GenBank* from the scientific literature. This information is augmented by data submissions directly from authors. Collaborations and exchange of data exist with the members of the International Nucleotide Sequence Database Collaboration which include the European Molecular Biology Laboratory (*EMBL*) and the *DNA Database of Japan* (*DDBJ*). These organizations exchange data on a daily basis. Arrangements with the National Agricultural Library and the U.S. Patent and Trademark Office enable the incorporation of plant and patent sequence data.

The *GenBank* database can be accessed in several different ways, via e-mail using list servers, ftp, and through the WWW using most available Web browsers. The Web interface allows one to submit genetic information, and perform a variety of queries on both the nucleotide sequences and the annotations found within the entries.

2.2 Data Submission and Revision

Data submitted to *GenBank* can be done via a local program called *Sequin* or an online-based Web service called *BankIt*. The PC-based program, *Sequin*, allows one to annotate a sequence, perform some analysis, and submit it to *GenBank*. This program is available for a variety of platforms including Mac, Windows, and Unix. When using *Sequin*, the output files for direct submission can be sent to *GenBank* by e-mail to: `gb-sb@ncbi.nlm.nih.gov` or by mailing the submission file copied to floppy disk and mailed to *GenBank* submissions. The Web-based form is called *Bankit*, which can be found by clicking on "BankIt" from the NCBI home page (`http://www.ncbi.nlm.nih.gov/BankIt/index.html`). From here one has the choice to enter a new sequence or revise a sequence. Step-by-step instructions for entering new or updating existing sequences can be found at the *Bankit* home page.

2.3. Searching GenBank

2.3.1. BLAST

Researchers at the NCBI have developed a program called the *Basic Local Alignment Search Tool* (*BLAST*) to rapidly compare an amino acid or nucleotide query sequence to databases of sequences. *BLAST* is not guaranteed to find the best hit between a query sequence and a database; it may miss matches. This is because it uses a strategy (an approximation method) that is expected to find most matches, but sacrifices complete sensitivity in order to gain speed. The most recent version of *BLAST* uses a new algorithm that speeds searches over the initial *BLAST* release, produces gapped alignments, (which was not available in the first version), and the sensitivity is increased in the *psi-BLAST* version.

BLAST is available on the WWW by going to the NCBI home page and clicking on the *BLAST* icon (`http://www.ncbi.nlm.nih.gov/BLAST/`). From here you will have a choice to use a few different versions of *BLAST* to perform a search against a database:

1. *BLAST*: Searches from these pages use *BLAST* version 1.x *(2)*. These searches are slower than *BLAST* 2.0 (below) and will not introduce gaps into the alignment of hits reported.
2. *BLAST 2.0*: This is a new version of *BLAST* that significantly enhances the search speed *(3)*. As well, *BLAST 2.0* is able to introduce gaps into the alignments of reported hits.

Table 1
Valid Queries Using the *BLAST* Family of Programs

Program	Query Sequence	Database type	Program for comparison
BLAST/BLAST2	nucleotide	nucleotide	*blastn*
	nucleotide (translated)	protein	*blastx*
	protein	nucleotide (translated)	*tblastn*
	nucleotide (translated)	nucleotide (translated)	*tblastx*
	protein	protein	*blastp*
Psi-BLAST	protein	protein	*blastp*

3. *Psi-BLAST*: This is the acronym for *Position Specific Iterative BLAST* that uses information from any significant alignments from a *BLAST* search to construct a position-specific score matrix, which replaces the query sequence for a new round of database searching. Using this method one can increase the sensitivity of searches to find distantly related sequences as long as the alignments from the initial *BLAST* search are valid.

The types of searches that can be performed are summarized in **Table 1**. The databases available are derived from a number of sources and do not necessarily correspond directly to what is presently available in *GenBank*. The name of each database and a short description is shown in **Table 2**.

To perform a search, select the *BLAST* search page, cut and paste your sequence of interest, choose the program and database and submit it. The only options available for a basic *BLAST* search are to perform a gapped alignment for *BLAST2* and to filter the sequence for low-complexity regions. The filtering option simply replaces regions such as poly-A tails, poly-glutamine tracts, repeats, and so on (i.e., regions of low complexity) with Ns for nucleotide or Xs for peptide sequences. Low-complexity regions commonly give spuriously high scores that reflect compositional bias rather than significant position-by-position alignment. Simply turn off this option if this not suitable for your query. For most searches the default parameters need not be modified. Users do, however, have the option to modify certain parameters using the advanced Blast pages. These options are described in **Table 3**.

A three-part result is returned for each search performed. Part one shows a graphical overview of the database sequences aligned to the query sequence. The score of each alignment is divided into five groups indicated by one of five different colors. A striped line connects multiple alignments on the same database sequence. Mousing over a hit causes the definition and score to be shown in the window at the top. Clicking on a hit takes the user to the associated alignments. Part two is a short description of the hits, the accession number, the expect value and/or the gi number and the score. Clicking on the score

Table 2
***BLAST* databases at the NCBI**

Database	Description
PEPTIDE SEQUENCES	
nr	All nonredundant *GenBank* CDS translations+*PDB*+*Swiss Prot*+*PIR*
month	All new or revised *GenBank* CDS translation+*PDB*+*SwissProt*+*PIR* released in the last 30 days.
swissprot	The last major release of the *SwissProt* protein sequence database
yeast	Yeast (*Saccharomyces cerevisiae*) protein sequences.
E. coli	*Escherichia coli* genomic CDS translations
pdb	Sequences derived from the three-dimensional structure *Brookhaven Protein Data Bank*
kabat [kabatpro]	Kabat's database of sequences of immunological interest
alu	Translations of select Alu repeats from *REPBASE*, suitable for masking Alu repeats from query sequences.
NUCLEOTIDE SEQUENCES	
nr	All nonredundant *GenBank*+*EMBL*+*DDBJ*+*PDB* sequences (but no *EST*, *STS*, *GSS*, or *HTGS* sequences)
month	All new or revised *GenBank*+*EMBL*+*DDBJ*+*PDB* sequences released in the last 30 days.
dbest	Nonredundant database of *GenBank*+*EMBL*+*DDBJ* EST Divisions
dbsts	Nonredundant database of *GenBank*+*EMBL*+*DDBJ* STS Divisions
htgs	High throughput genomic sequences
yeast	Yeast (*Saccharomyces cerevisiae*) genomic nucleotide sequences
E. coli	*Escherichia coli* genomic nucleotide sequences
pdb	Sequences derived from the three-dimensional structure
kabat [kabatnuc]	Kabat's database of sequences of immunological interest
vector	Vectors
mito	Mitochondrial sequences
alu	Select Alu repeats from *REPBASE*, suitable for masking Alu repeats from query sequences.
epd	*Eukaryotic Promotor Database*
gss	*Genome Survey Sequence*, includes single-pass genomic data, exon-trapped sequences, and Alu PCR sequences.

takes one to the alignment of the query sequence and the hit (part three). One can also click on the gi or accession number, which launches a query to the *Entrez* system and returns the complete *GenBank* entry for the selected hit. Each *GenBank* entry contains links to other databases that cross-reference this entry. These links allow users to quickly gather additional information for each

Table 3
Options for the *BLAST* Program

Option	Description
Expect	Statistical significance threshold for reporting matches. Matches will only be shown with a value less than the number entered.
NCBI-gi	Show the NCBI-gi number for hits in the output in addition to the accession and/or locus name. The gi changes when a sequence is updated but the accession number never changes.
Descriptions	The maximum number of hits returned
Alignments	The maximum number of alignments returned (is always less than the number of descriptions returned)
Graphical overview	An overview of the database sequences aligned to the query sequence is shown. The score of each alignment is divided into five groups indicated by one of five different colors. A striped line connects multiple alignments on the same database sequence. Mousing over a hit sequence causes the definition and score to be shown in the window at the top, clicking on a hit sequence takes the user to the associated alignments.
Other advanced options	
Cost to open gap, cost to extend gap	These values only apply to the generation of alignments. It is beyond the scope of this discussion to fully discuss these options
Reward for nucleotide match and penalty for nucleotide mismatch	The ratio of the match:mismatch determines the sensitivity for which nucleotide hits are found. Increase this ratio for species that are more divergent. Defaults are reward=1, penalty=-3. This parameter does not apply to peptide database searches.
Word size (W)	The word size restricts the program to finding only sequences that share a sequence stretch of 'W' nucleotides or amino acids of 100% identity with the query. The default is 11 for *blastn* and 3 for the other programs. For nucleotide sequences, increased sensitivity can be gained by reducing this number.

hit. Useful links that may be included are: *MEDLINE*, *OMIM*, sequence features databases, corresponding peptide, or nucleic acid entries, and so on.

BLAST searches can also be performed on the incompletely sequenced genomes of a number of species. To access these pages click on *Entrez* from the NCBI home page followed by genomes. A multi-page frame appears from which you can select **BLAST WITH UNFINISHED MICROBIAL GENOMES**. This page is similar to the other BLAST pages with the excep tion that one has the choice of specifying the microbial genome(s) one would like to query.

2.3.2. Entrez

Entrez is a system that allows users to perform text searches on a number of databases that the NCBI maintains. These databases include the following:

1. DNA sequences from *GenBank*, *EMBL*, and *DDBJ*.
2. Protein sequences from *Swiss-Prot*, *PIR*, *PRF*, *PDB*, and translated protein sequences from the DNA sequence databases.
3. Genome and chromosome mapping data.
4. Three-dimensional protein structures derived from *PDB*, and incorporated into NCBI's *Molecular Modeling Database* (*MMDB*).
5. *PubMed* bibliographic databases from the National Library of Medicine's *MEDLINE* and *pre-MEDLINE* databases.

To access *Entrez* simply click on the **Entrez** button from the NCBI home page. The ensuing page gives one the option to search one of the above databases. The form presented is similar for the different databases. Two modes are available to perform the searches: automatic mode, which will take the specified terms and automatically map them to terms that have been indexed from the database and list-term mode, which will give one the option of selecting the indexed terms for each search. Once a search term is selected and submitted, a page is returned that displays the number of hits and a dialog to further refine one's search. Placing an asterisk at the end of a term will cause *Entrez* to search for all terms that begin with that word. Phrases that have a space in the word that occurs after the asterisk will *not* be included. *Entrez* can be forced to search for a phrase by enclosing the words in quotes and will do its best to find logical groupings in your input. To refine the search simply select the field(s) you would like to search, type in the new terms(s), and submit the query. A similar form is returned that shows the new number of hits. One can continue to refine the search until the number of articles is small. At this time just click on the retrieve documents button and the hits are returned.

When documents or sequences are retrieved from the query page, one is presented with a summary of each hit. Each hit can be viewed in a number of different formats such as *GenBank*, *FASTA* or *ASN.1* for nucleic acid sequences and *GenPept*, *FASTA*, and *ASN.1* for peptides. A graphical view can also be presented which shows the position of features that are known for the peptide or nucleotide sequence. One can click on any feature to display additional details. Sequences with significant homologies have been previously calculated for each entry in *GenBank*, and these can be accessed by clicking on the **nucleotide** or **protein neighbors** button.

Additionally, each *GenBank* entry is annotated with cross-references to other databases. Many of these cross-references can be browsed directly by clicking on the blue-underlined text. This feature allows one to quickly gather additional information for any sequence.

2.4. Genome Resources

With the ever-increasing amount of genetic data being accumulated from the many prokaryotic and eukaryotic organisms, a number of useful tools have been made available to elucidate the function of proteins by homology searches; to establish the evolutionary relationship of proteins; and to determine the interrelationships of genes and proteins between different species. These resources include Clusters of orthologous groups and *UniGene*.

2.4.1. Clusters of Orthologous Groups (COGs)

Orthologous groups of genes (genes in different species that have evolved from a common ancestral gene by speciation) have been calculated by comparing protein sequences encoded in seven complete genomes, representing five major phylogenetic lineages *(4)*. These orthologous genes normally retain the same function in the course of evolution. This may be useful for delineating the function of related proteins in other species and, for example, to define classes of proteins that are exclusively found in bacteria and hence might serve as useful targets for new antibiotics.

One can browse the results of the current analysis of the seven complete genomes. These include *Escherichia coli*, *Haemophilus influenzae*, *Mycoplasma genitalium*, *Mycoplasma pneumoniae*, *Cyanobacteria–Synechocystis*, *Methanococcus jannaschii*, and *Saccharomyces cerevisiae*. The proteins encoded by the sequenced DNAs have been placed into four broad categories: information storage and processing, cellular processes, metabolism, and poorly characterized. Each one of these categories has been further subdivided into groups in which the COGs are contained. One can browse the COGs by clicking on **List of all COG**s or **Table** under functional and phylogenetic analysis. The pattern, size, and number of members for each COG is displayed. If one drills down further by clicking on the **ID**, pairwise alignments of peptides from each COG are shown along with a graphical display of the regions of overlap, the scheme of best hits and a cluster dendogram.

To compare a sequence of your own to the available COGs, select the **Cognior** button. A page is displayed on which you can cut and paste your sequence into a box and submit the query. A *BLAST* search is performed against the COG database and you are presented with both the textual *BLAST* result and a graphical display of the alignments of individual hits with the query sequence. As well, you are presented with the scheme of best hits, displaying the relationship of your query to the COG that best matches your sequence. The black solid lines show symmetrical best hits (BeTs) and broken colored lines show asymmetrical BeTs. For example, a green line coming from a yeast protein to an *E. coli* one indicates that the given *E. coli* protein is the best hit for the given yeast protein, but that the reverse is not true.

2.4.2. UniGene

UniGene is a database of unique human and mouse sequences that have been derived by clustering overlapping ESTs. *UniGene* also contains nonredundant cDNAs and CDSs found in *GenBank*. As of *GenBank* release 106 there are 47,000 *UniGene* clusters, 14,000 unique cDNAs, and CDSs. Of the cDNAs only 500 do not have an associated EST. *UniGene* is updated every two months, approx one week after a new *GenBank* release. Files can be downloaded from the NCBI's FTP site in the **repository/UniGene** directory.

To query *UniGene* select **UniGene: Unique Gene Sequence Collection for Human and Mouse**. The next screen contains a description of *UniGene* and two buttons in the upper left-hand corner, one of which takes you to the mouse *UniGene* set and the other to the human clusters. The query screens for both the human and mouse *UniGene* sets are identical. You can search the annotations of *UniGene* by typing in a search term in the box provided. Valid queries include search terms such as phosphatase, and *GenBank* accession numbers. In addition, a number of @functions are provided for specialized purposes which are described by clicking on **query tips**. By clicking on a chromosome number you can see a list of *UniGene* entries for a specific chromosome, or use the Library Browser to see a list of cDNA libraries that have been used in EST projects.

3. *Online Mendelian Inheritance in Man (*OMIM*) (5)*

This resource is a comprehensive and frequently updated database of information about genes that are known to cause human disease. The database is searchable using a simple Web-based interface and keywords. Each search returns a summary of the hits found. Clicking on the any one of the items takes one to the actual entry. Each entry is divided into a number of sections that can be browsed quickly by clicking on the text. These sections may contain descriptions of the disorder, summaries of the results of research performed, clinical symptoms, genetic information, animal models, inheritance, and so on. Each entry is highly referenced with links to many databases, shown near the top of the entry. Mousing over the database buttons shows the number of links to that particular database. Each cited reference has a link to *PubMed* for easy cross-reference. Each entry in *OMIM* also contains a section that shows who created the entry and when, a list of contributors, and a list of the dates and individuals who have updated the record.

4. Molecular Modeling Resources

The NCBI's protein structure database is called the *Molecular Modeling Database (MMDB)*. It is a compilation of three-dimensional structures obtained from the *Brookhaven Protein DataBank (PDB)* ***(6)*** that have been converted to ASN.1-formatted records. The *MMDB* was designed to be capable of archiving conventional structural data as well as future descriptions of biomolecules. By using the *Entrez* system, the NCBI has made this information easily accessible to users interested in structural biology. To access the *Entrez* system, just click on *Entrez* at the NCBI home page (`http://www.ncbi.nlm.nih.gov/`) and select the three-dimensional structure search. From here one is able to perform an annotation search of all known three-dimensional structures. One can also perform an *Entrez* search from the NCBI structure home page (`http://www.ncbi.nlm.nih.gov/Structure/`). A list of hits is presented when an *Entrez* search is performed. If one clicks on the structure summary, hyperlinks to the *GenBank* entry, to *PubMed*, and to the *Taxonomy* databases are presented. As well one can browse the nucleotide neighbors calculated using BLAST, and the structural neighbors calculated using the *Vast* ***(7,8)*** algorithm (*see* `http://www.ncbi.nlm.nih.gov/Structure/iucrabs.html`). The three-dimensional representation for any solved structure can be viewed by using one of three programs supported by the NCBI. These applications are *Rasmol*, *Mage*, and the NCBI's own *Cn3D* ***(9)***. *Cn3D* can be downloaded from: `http://www.ncbi.nlm.nih.gov/Structure/cn3ddown.html`. *Rasmol* from the University of Massachusetts *Rasmol* home page: `http://klaatu.oit.umass.edu/microbio/rasmol/index2.htm`.

5. Useful Tools for Molecular Biologists

As well as database searching and sequence retrieval utilities, the NCBI has made a number of other tools available for molecular biologists.

5.1. E-PCR

Electronic PCR or *e-PCR* is a tool that allows one to determine if any sequence tagged sites (StSs) are located within a query sequence. PCR-based STS's are short DNA sequences that have been mapped to the genomes of a number of organisms. The STSs are useful for quickly mapping new DNA segments of interest. Currently this method of analysis is most useful for the human genome, which has over 45,000 STSs. *Drosophila melanogaster* (fruitfly) has 3203; *Bos taurus* (cattle) 1015; *Gallus gallus* (chicken) 552; *Mus musculus* (house mouse) 343; *Plasmodium falciparum* (malaria parasite) 339. Many other organisms have STSs, but far fewer are available for each.

5.2. ORF Finder

For analysis of sequencing data, the *Open Reading Frame* (ORF) *Finder* may be useful. This tool enables one to determine where potential ORFs are within their query sequence. After submitting a sequence, a graphical display of all of the ORFs are displayed in all six reading frames. The size of each ORF is shown to the left of the graphical display with the largest ORF first and smallest last. The user has the option of changing a few parameters such as the minimum size of the ORF displayed and displaying either the ORFs or stop codons in each reading frame. Clicking on any of the ORFs can launch a standard or advanced *BLAST* search.

5.3. Human–Mouse Homology Maps

This resource allows one to view the regions of synteny between the mouse and human genomes. Each chromosome can be viewed by clicking on the chromosome number and the user is able to see the chromosomal locations of the human or mouse genes. The data for each chromosome is displayed in a table that shows the gene name and the associated mapping information in rows. The information associated with each gene includes:

1. The Genethon map location.
2. The method by which the gene was mapped.
3. The cytogenetic map location of disease genes and other expressed genes described in *OMIM* (*in situ* column).
4. Hyperlinks to the genome informatics site at the Jackson Laboratories for the mouse gene (click on mouse gene) and *OMIM* for human genes (by clicking on human gene).
5. The corresponding mouse or human chromosome.
6. The radiation hybrid mapping field which connects the user to the *Gene Map* of the human genome. Subsequent hyperlinks on the marker will provide marker details and hyperlinks to the actual human gene map region (by selecting the interval defined by the Genethon map loci intervals). By selecting an interval one is presented with all the markers available in that region.
7. The map position of the syntenic mouse or human gene.
8. The cross column, which indicates which laboratory mapped a given cross.

5.4. Taxonomy

This database allows one to view the names of all organisms for which at least one nucleotide or protein sequence is represented in the genetic databases at the NCBI. This database is searchable by selecting the taxonomy browser. By entering any part of the taxonomic description or even the common name a hierarchical view of that taxa will be shown. For example, by typing in '*Mus*' all the subspecies with proteins or genes in *GenBank* are shown. Clicking on any sub species will give the user further information about the species and will show the number of DNA or protein sequences or three-dimensional structures that are available in *GenBank* based on an annotation search using *Entrez*. The sequences for each group can be viewed using the *Entrez* system.

6.0 Summary

As computing technology advances and new sources of valuable biological information are collected the numbers of databases and the tools that are used to analyze the data they contain will change. From the time this article was originally written to the time (about 6 months) this summary was put together another 4 releases of Genbank have been made availalbe, two new BLAST search engines, PHI-BLAST and organism-specific BLAST, a database of Human Genetic Variation (dbSNP), and a new version of Cn-3D. The NCBI will, for the forseeable future, provide the research community with a wealth of information and computing tools. made a number of other tools available for molecular biologists.

Reference

1. Benson, D. A., Boguski, M. S., Lipman, D. J., Ostell, J., and Ouellette, B. F. (1998) GenBank. *Nucleic Acids Res.* **26,** 1–7.
2. Altschul, S. F., Gish, W., Miller, W., Myers, E. W., and Lipman, D. J. (1990) Basic local alignment search tool. *J. Mol. Biol.* **215,** 403–410.
3. Altschul, S. F., Madden, T. L., Schäffer, A. A., Zhang, J., Zhang, Z., Miller, W., and Lipman, D. J. (1997) Gapped BLAST and PSI-BLAST: a new generation of protein database search programs. *Nucleic Acids Res.* **25,** 3389–3402.
4. Tatusov, R. L., Koonin, E. V., and Lipman, D. J. (1997) A genomic perspective on protein families. *Science* **24,** 631–637.
5. Online Mendelian Inheritance in Man, OMIM (TM). (1997) Center for Medical Genetics, Johns Hopkins University (Baltimore, MD) and National Center for Biotechnology Information, National Library of Medicine (Bethesda, MD).
6. Abola, E. E., Bernstein, F. C., Bryant, S. H., Koetzle, T. F., and Weng, J. C. (1987) Protein data bank, in *Crystallographic databases: information content, software systems, scientific applications.* (Allen, F. H., Bergerhoff, G, Sievers R., eds.) International Union of Crystallography, Chester, Cambridge, UK, pp. 107–132.
7. Madej, T., Gibrat, J-F., and Bryant, S. H. (1995) Threading a database of protein cores. *Protein Struct. Funct. Genet.* **23,** 356–369.
8. Gibrat, J-F., Madej, T., and Bryant, S. H. (1996) Surprising similarities in structure comparison. *Curr. Opin. Struct. Biol.* **6,** 377–385.
9. Hogue, C. W. V. (1997) Cn3D: a new generation of three-dimensional molecular structure viewer. *Trends Biochem. Sci.* **22,** 314–316.

17

Resources at EBI

Patricia Rodriguez-Tomé

1. Introduction

The European Bioinformatics Institute (EBI) is an EMBL Outstation located at the Wellcome Trust Genome Campus in Hinxton (UK). The EBI maintains and distributes the *EMBL Nucleotide Sequence* database, Europe's primary nucleotide sequence data resource, the *SwissProt Protein Sequence* database, in collaboration with Amos Bairoch of the Swiss Institute for Bioinformatics (SIB), TrEMBL—a SwissProt complement consisting of translations from *EMBL* database coding sequences, *MSD*, the *Molecular Structure Database* in collaboration with the *Protein Data Bank* (*PDB*) in Brookhaven National Laboratory (Brookhaven, NY), the *Radiation Hybrid database* (*RHdb*), and other additional molecular biology databases produced in collaboration with other groups. The EBI also provides network services that allow access to the most up-to-date data collections via the Internet through its World Wide Web interfaces and ftp services, also providing database and sequence similarity searches facilities.

2. The Databases

2.1. EMBL—*the* Nucleotide Sequence Database

The *EMBL Nucleotide Sequence Database (1)* is a central activity of the EBI. It was first established in 1980 to collect, organize, and distribute a database of nucleotide sequence data and related information. Since 1982 this work has been done in collaboration with *GenBank (2)* (NCBI, Bethesda, MD) and later *DDBJ (3)* (*DNA Database of Japan*, Mishima, Japan) joined the collaboration. Data are collected and updated at each center, and exchanged between the three groups on a daily basis.

From: *Methods in Molecular Biology, vol. 132: Bioinformatics Methods and Protocols*
Edited by: S. Misener and S. A. Krawetz © Humana Press Inc., Totowa, NJ

Because of the high throughput sequencing technology used in many centers, sequences are being deposited currently in the *EMBL* database at a rate of one sequence per minute. There is an ongoing collaboration between the EMBL and the major sequencing projects producing large quantities of data (**Fig. 1**).

2.1.1. Database Divisions

The *EMBL* database is divided into sections based mainly on taxonomy with a few exceptions like the *HTG* (*High Throughput Genome Sequences*) or *GSS* (*Genome Survey Sequences*). The divisions are defined using the three letter codes shown below :

Division	**Code**
Bacteriophage	PHG
Constructed Sequences	CON
ESTs	EST
Fungi	FUN
High throughput genome	HTG
Genome survey sequences	GSS
Human	HUM
Invertebrates	INV
Organelles	ORG
Other mammals	MAM
Other vertebrates	VRT
Plants	PLN
Prokaryotes	PRO
Rodents	ROD
STSs	STS
Synthetic	SYN
Unclassified	UNC
Viruses	VRL

2.1.2. The Special Sections

- CON: This section holds information on very long sequences, such as complete genomes and parts of chromosomes that are built from sequences already in the database. Entries in the CON section do not contain feature, table, or sequence data, but they include information on how the full sequence is built from its components.
- EST: This section contains sequences denoted as "expressed sequences tags" or "transcribed sequences fragments" or "partial cDNA" independent of their taxonomy classification. It is the largest and fastest growing division of the database.
- HTG: This section is used for genomic sequences that are produced by high-throughput sequencing projects. The records consist of long sequences. It is important to note that these data are unfinished and do not necessarily represent the correct sequence. The release of these data is based on the understanding that the sequence may change as work continues.

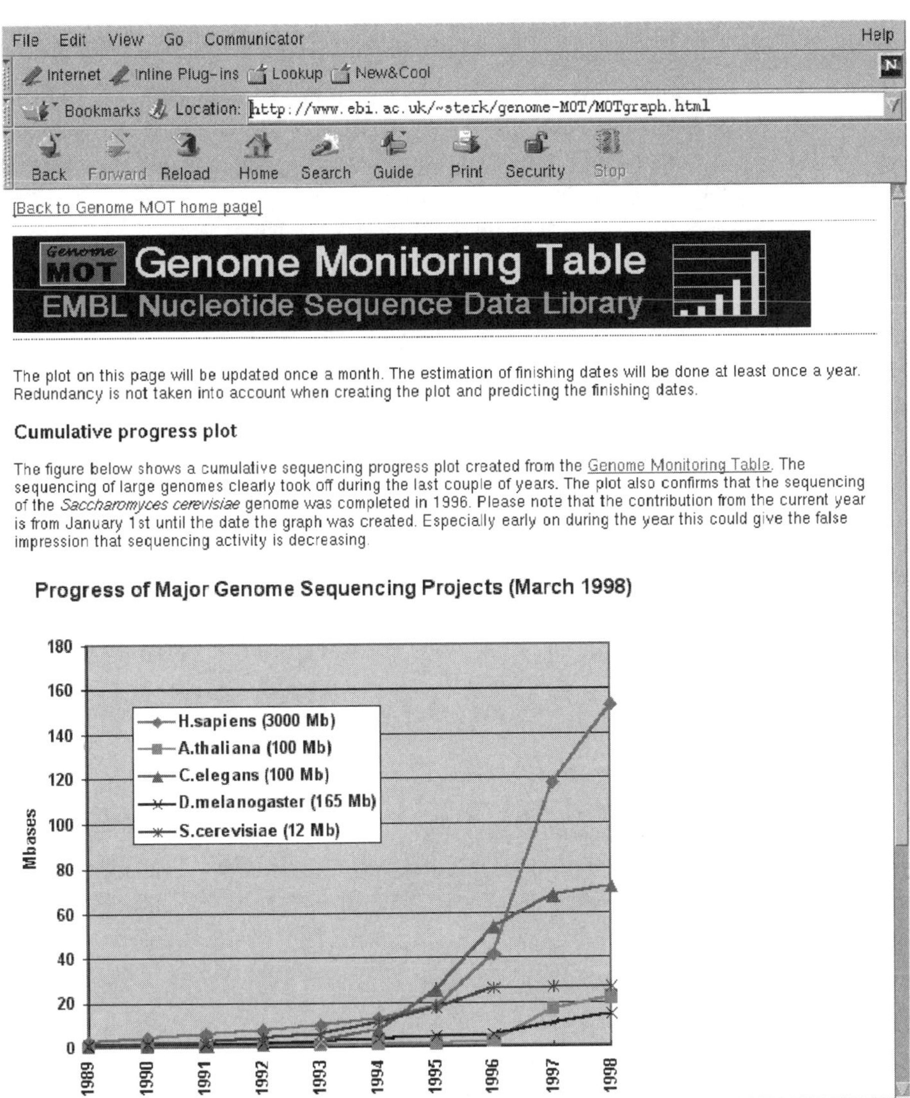

Fig. 1. The *Genome Monitoring Table*, which reports sequencing progress from the genome centers.

- GSS: This section is similar to the EST one, except that its sequences are genomic rather than cDNA. Entries include sequence data generated by 'single pass reads' from random genome survey sequences, exon trapped products, Alu PCR sequences, BAC, or YAC end clones.

- Patents: This section, not cited in the list, incorporates sequence data from the patents offices. The European Patent Office releases data to the public 18 months after the patent application date. These data are immediately incorporated in the EMBL database and made public.
- NEW: This is not a specific section of the databse, but contains all data incorporated since the last full release. This section is updated daily. It disappears at release freezing time (i.e., when a release is created), and is recreated when new data is submitted.

2.1.3. Database Entry Structure

Databases entries are distributed in EMBL flat-file format (**Fig. 2**), a format that is supported by most sequence-analysis software packages. This format provides a structure usable by human readers and is composed of different line types which are used to record the different types of data that compose an entry. A typical database entry contains the following line types:

1. AC: contains a unique identifier (accession number) for that entry.
2. DE: a brief description line.
3. OS: the taxonomic description of the source organism.
4. Reference information (RA line for authors, RT line for the title, RL line for the journal information).
5. NI: an identifier for the nucleic acid sequence, which changes every time the sequence itself is changed, whereas the accession number remains the same. This identifier will be replaced by a proper version number in 1999.
6. FT: the feature table describing locations of coding regions and other biologically significant sites. The feature table follows the unified *DDBJ/EMBL/GenBank* feature table definition and is described in a document available at URL `http://www.ebi.ac.uk/ebi_docs/embl_db/ft/feature_table.html`.
7. SQ: the sequence itself.

2.1.3.1. The Accession Number

Each entry in the database is given a unique identifier—its accession number. This identifier (contained in the AC line) is of the form one prefix letter followed by five digits (the 1+ 5 format) for the older entries, and 2 (prefix letters) + 6 (digits) format for the newer entries. Because of the large amount of data being produced by the genome projects, the accession number space had to be extended. This accession number is unique accross the *EMBL*, *DDBJ*, and *GenBank* databases, and will point to the same entry in all three databases.

2.1.3.2. The PID

The protein identifier (PID) refers to the translated product of a coding sequence of a specific entry. This PID is referenced in the CDS feature of the

```
ID   HSIGHAF    standard; RNA; HUM; 1089 BP.
XX
AC   J00231;
XX
NI   g185041
XX
DT   13-JUN-1985 (Rel. 06, Created)
DT   17-DEC-1994 (Rel. 42, Last updated, Version 6)
XX
DE   Human Ig gamma3 heavy chain disease OMM protein mRNA.
XX
KW   C-region; gamma heavy chain disease protein;
KW   gamma3 heavy chain disease protein; heavy chain disease; hinge exon;
KW   immunoglobulin gamma-chain; immunoglobulin heavy chain;
KW   secreted immunoglobulin; V-region.
XX
OS   Homo sapiens (human)
OC   Eukaryota; Metazoa; Chordata; Vertebrata; Mammalia; Eutheria; Primates;
OC   Catarrhini; Hominidae; Homo.
XX
RN   [1]
RP   1-1089
RX   MEDLINE; 82247835.
RA   Alexander A., Steinmetz M., Barritault D., Frangione B., Franklin E.C.,
RA   Hood L., Buxbaum J.N.;
RT   "gamma Heavy chain disease in man: cDNA sequence supports partial gene
RT   deletion model";
RL   Proc. Natl. Acad. Sci. U.S.A. 79:3260-3264(1982).
XX
DR   GDB; 119339; IGHG3.
DR   GDB; G00-119-339.
DR   IMGT/LIGM; J00231; Release 98.03.
DR   SWISS-PROT; P01860; GC3_HUMAN.
XX
CC   The protein isolated from patient OMM is a gamma heavy chain
CC   disease (HCD) protein. It has a large 5' internal deletion
CC   consisting of most of the variable region and the entire ch1
CC   domain. [1] suggests that the protein abnormality is from a partial
CC   gene deletion rather than from defective splicing. NCBI gi: 185041
XX
FH   Key             Location/Qualifiers
FH
FT   source          1. .1089
FT                   /organism="Homo sapiens"
FT   mRNA            <1. .1089
FT                   /note="gamma3 mRNA"
FT   CDS             23. .964
FT                   /codon_start=1
FT                   /db_xref="PID:g567112"
FT                   /db_xref="SWISS-PROT:P01860"
FT                   /note="OMM protein (Ig gamma3) heavy chain; NCBI gi:
FT                   567112"
FT                   /gene="IGHG3"
FT                   /map="14q32.33"
FT                   /translation="MKXLWFFLLLVAAPRWVLSQVHLQESGPGLGKPPELKTPLGDTTH
FT                   TCPRCPEPKSCDTPPPCPRCPEPKSCDTPPPCPRCPEPKSCDTPPPCPXCPAPELLGGP
FT                   SVFLFPPKPKDTLMISRTPEVTCVVVDVSHEDPXVQFKWYVDGVEVHNAKTKLREEQYN
FT                   STFRVVSVLTVLHQDWLNGKEYKCKVSNKALPAPIEKTISKAKGQPXXXXXXXXXXXXXE
FT                   EMTKNQVSLTCLVKGFYPSDIAVEWESNGQPENNYNTTPPMLDSDGSFFLYSKLTVDKS
FT                   RWQQGNIFSCSVMHEALHNRYTQKSLSLSPGK"
FT   sig_peptide     26. .79
FT                   /codon_start=1
FT                   /note="OMM protein signal peptide"
FT   mat_peptide     80. .961
FT                   /codon_start=1
FT                   /note="OMM protein mature peptide"
XX
SQ   Sequence 1089 BP; 240 A; 358 C; 271 G; 176 T; 44 other;
     CCTGGACCTC CTGTGCAAGA ACATGAAACA NCTGTGGTTC TTCCTTCTCC TGGTGGCAGC        60
     TCCCAGATGG GTCCTGTCCC AGGTGCACCT GCAGGAGTCG GGCCCAGGAC TGGGGAAGCC       120
     TCCAGAGCTC AAAACCCCAC TTGGTGACAC AACTCACACA TGCCCACGGT GCCCAGAGCC       180
     CAAATCTTGT GACACACCTC CCCCGTGCCC ACGGTGCCCA GAGCCCAAAT CTTGTGACAC       240
     ACCTCCCCCA TGCCCACGGT GCCCAGAGCC CAAATCTTGT GACACACCTC CCCCGTGCCC       300
     NNNGTGCCCA GCACCTGAAC TCTTGGGAGG ACCGTCAGTC TTCCTCTTCC CCCCAAAACC       360
     CAAGGATACC CTTATGATTT CCCGGACCCC TGAGGTCACG TGCGTGGTGG TGGACGTGAG       420
     CCACGAAGAC CCNNNNGTCC AGTTCAAGTG GTACGTGGAC GGCGTGGAGG TGCATAATGC       480
     CAAGACAAAG CTGCGGGAGG AGCAGTACAA CAGCACGTTC CGTGTGGTCA GCGTCCTCAC       540
     CGTCCTGCAC CAGGACTGGC TGAACGGCAA GGAGTACAAG TGCAAGGTCT CCAACAAAGC       600
     CCTCCCAGCC CCCATCGAGA AAACCATCTC CAAAGCCAAA GGACAGCCCN NNNNNNNNNN       660
     NNNNNNNNNN NNNNNNNNNN NNNNNGAGGA GATGACCAAG AACCAAGTCA GCCTGACCTG       720
     CCTGGTCAAA GGCTTCTACC CCAGCGACAT CGCCGTGGAG TGGGAGAGCA ATGGGCAGCC       780
     GGAGAACAAC TACAACACCA CGCCTCCCAT GCTGGACTCC GACGGCTCCT TCTTCCTCTA       840
     CAGCAAGCTC ACCGTGGACA AGAGCAGGTG GCAGCAGGGG AACATCTTCT CATGCTCCGT       900
     GATGCATGAG GCTCTGCACA ACCGCTACAC GCAGAAGAGC CTCTCCCTGT CTCCGGGTAA       960
     ATGAGTGCCA TGGCCGGCAA GCCCCCGCTC CCCGGGCTCT CGGGGTCGCG CGAGGATGCT      1020
     TGGCACGTAC CCCGTGTACA TACTTCCCAG GCACCCAGCA TGGAAATAAA GCACCCAGCG      1080
     CTGCCCTGG                                                              1089
//
```

Fig. 2. An *EMBL* entry.

entry. This identifier remains the same as long as the translation of the coding sequence remains the same for that entry. It can thus be used by external databases as an identifier onto which cross-references can be built.

2.1.4. Data Submission

The EBI provides a number of ways to submit new sequence data:

1. *WEBin* (**Fig. 3**) is the WWW sequence submission tool. It is available at URL `http://www.ebi.ac.uk/submission/webin.html`. This tool takes the submitter through all the steps allowing the submission of sequence data and descriptive information in an interactive and easy manner.
2. *Sequin* is a tool developed by the NCBI for submitting entries to the *EMBL*, *GenBank* or *DDBJ* databases. This tool is available from the EBI anonymous ftp server at `ftp://ftp.ebi.ac.uk/pub/software/sequin/`.
3. A data submission form is also available for users whose only access to the Internet is by e-mail. This form solicits all the information needed to create a database entry. This form is available by electronic mail via the EBI file server. The user should send an e-mail to `netserv@ebi.ac.uk` with the words

   `GET DOC:DATASUB.TXT`

   in the body of the message.
4. If a user wishes to submit more than 25 entries, this can become very cumbersome using an interactive tool. The EBI then encourages the submittors to contact the database before submitting the data. Database staff will assist in making the submission of the data as convenient as possible.

When a sequence has been submitted to *EMBL* and accepted, database staff will provide the submittor with its unique identifier, the accession number to identify the entry. The entry will automatically appear in *GenBank* and *DDBJ*, with the same accession number, because submissions to any of the three databases are forwarded daily to one another.

2.1.5. Data Confidentiality

Sequences submitted to the database can be released either immediately after processing or upon publication, depending on the specifications given by the submitter. An entry that is not immediately public is not forwarded to the other databases until either the date of release asked by the submittor or its publication (whichever comes first).

2.1.6. Data Updates

The only way to keep entries correct and up to date is if the authors communicate their new finding or corrections to the database staff. This information can be communicated either via:

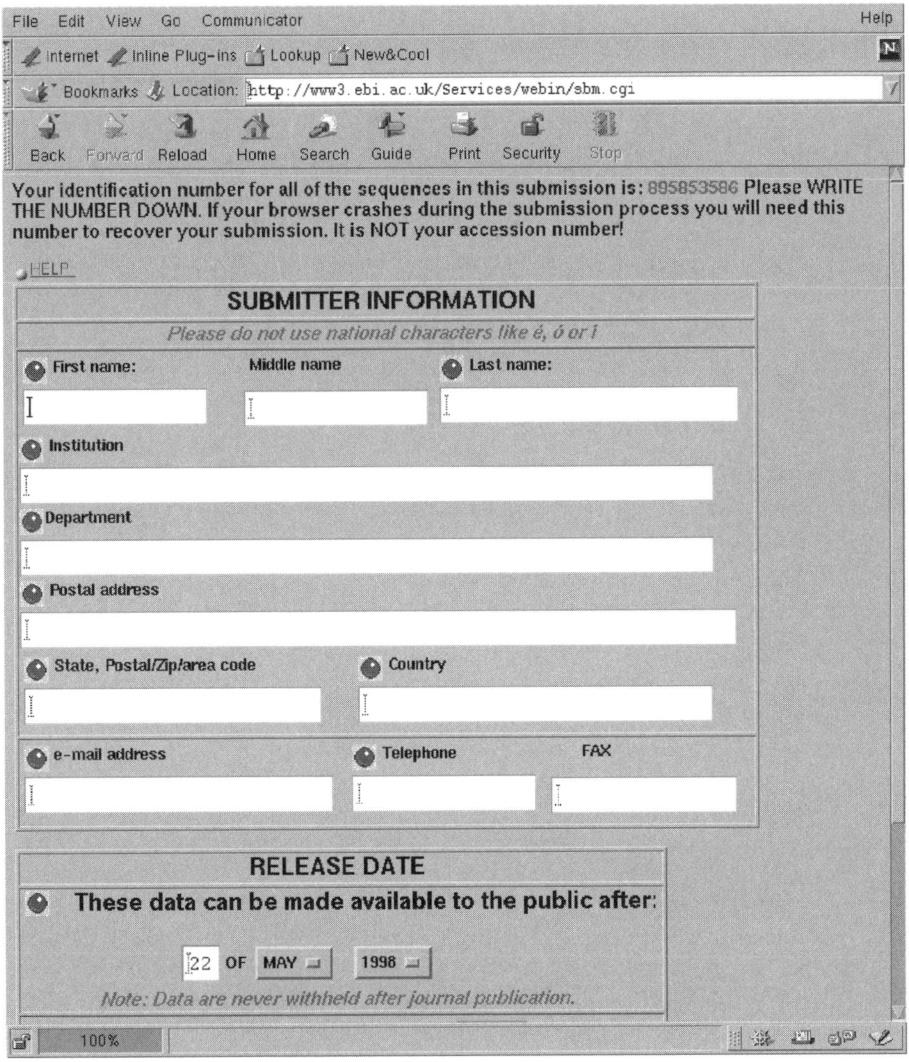

Fig. 3. *WEBin* the *EMBL* submission tool on the WWW.

1. WWW at URL: `http://www.ebi.ac.uk/ebi_docs/update.html`.
2. E-mail to: `update@ebi.ac.uk`.
3. FTP: a document is available at: `ftp://ftp.ebi.ac.uk/pub/databases/embl/release/update.doc`. Updates can be also be faxed to the database.

Users are welcome to report any errors they find in the database, but should be aware that only the original submittor can authorize sequence updates and major annotation changes. Updates are also forwarded daily to the other two databases.

2.1.7. Database Distribution

The EMBL database is released and distributed quarterly on CD-ROM. Full releases are also available on the FTP server at `ftp://ftp.ebi.ac.uk/pub/databases/embl/`. New and updated entries from the database are added daily to this server, making it possible for users to keep an up-to-date copy on their sites. Specific services available at the EBI for querying the *EMBL* nucleotide sequence database are described in the second part of this chapter.

2.2. *The* SwissProt *Database*

SwissProt *(4)* is an annotated protein sequence database, produced in collaboration between the EBI and the Department of Medical Biochemistry of the University of Geneva (Switzerland). It contains high-quality annotated data, is nonredundant, and is cross-referenced to many other databases. For standardization purposes, the format of a *SwissProt* entry follows as closely as possible the format of an *EMBL* entry. A sample of a *SwissProt* entry is shown in **Fig. 4**.

A typical entry consists of the sequence data, the reference, the taxonomy information and the annotation itself which includes the following items:

- Function(s) of the protein.
- Posttranslational modification(s).
- Domains and sites.
- Secondary structure.
- Quaternary structure.
- Similarities to other proteins.
- Disease(s) associated with protein defects.
- Sequence conflicts, variants, etc.

SwissProt entries are produced from translations of sequences in *EMBL*, extracted from the literature or submitted directly by researchers. To build the annotation, the *SwissProt* curators review not only the publications referenced by the author, but also related articles to periodically update the annotations of the families or groups of proteins. External collaborators supply their own expertise to specific domains. In *SwissProt*, the annotation is mainly found in the comment lines (CC lines), in the feature table (FT lines), and in the key word lines (KW lines).

Minimal redundancy is achieved by merging different sequences corresponding to the same protein but corresponding to different literature reports.

```
ID   GC3_HUMAN      STANDARD;      PRT;   290 AA.
AC   P01860;
DT   21-JUL-1986 (REL. 01, CREATED)
DT   21-JUL-1986 (REL. 01, LAST SEQUENCE UPDATE)
DT   01-FEB-1991 (REL. 17, LAST ANNOTATION UPDATE)
DE   IG GAMMA-3 CHAIN C REGION (HEAVY CHAIN DISEASE PROTEIN) (HDC).
GN   IGHG3.
OS   HOMO SAPIENS (HUMAN).
OC   EUKARYOTA; METAZOA; CHORDATA; VERTEBRATA; TETRAPODA; MAMMALIA;
OC   EUTHERIA; PRIMATES.
RN   [1]
RP   SEQUENCE (DISEASE PROTEIN WIS).
RX   MEDLINE; 81021548.
RA   FRANGIONE B., ROSENWASSER E., PRELLI F., FRANKLIN E.C.;
RL   BIOCHEMISTRY 19:4304-4308(1980).
RN   [2]
RP   NORMAL GAMMA-3 CHAINS, REVISIONS TO 12-97 OF PROTEIN WIS.
RX   MEDLINE; 77118561.
RA   MICHAELSEN T.E., FRANGIONE B., FRANKLIN E.C.;
RL   J. BIOL. CHEM. 252:883-889(1977).
RN   [3]
RP   DISEASE PROTEIN ZUC, REVISIONS TO 59-289 OF PROTEIN WIS.
RX   MEDLINE; 77021516.
RA   WOLFENSTEIN-TODEL C., FRANGIONE B., PRELLI F., FRANKLIN E.C.;
RL   BIOCHEM. BIOPHYS. RES. COMMUN. 71:907-914(1976).
RN   [4]
RP   SEQUENCE FROM N.A. (DISEASE PROTEIN OMM).
RX   MEDLINE; 82247835.
RA   ALEXANDER A., STEINMETZ M., BARRITAULT D., FRANGIONE B.,
RA   FRANKLIN E.C., HOOD L., BUXBAUM J.N.;
RL   PROC. NATL. ACAD. SCI. U.S.A. 79:3260-3264(1982).
CC   -!- SUBUNIT: DIMER LINKED BY 12 DISULFIDE BONDS; IT HAS AN EXTRA
CC       INTERCHAIN DISULFIDE BOND AT POSITION 7 IN ADDITION TO THE 11
CC       NORMALLY PRESENT IN THE HINGE REGION.
CC   -!- THE HEAVY CHAIN DISEASE PROTEIN WIS IS SHOWN.
CC   -!- THE SEQUENCE OF RESIDUES 42-76 WAS TAKEN FROM THE REF. 2.
CC   -!- DISEASE PROTEIN WIS IS LACKING MOST OF THE V REGION AND ALL OF THE
CC       CH1 REGION.
CC   -!- DISEASE PROTEIN ZUC LACK MOST OF THE V REGION, ALL OF THE CH1
CC       REGION, AND PART OF THE HINGE COMPARED WITH NORMAL GAMMA-3 HEAVY
CC       CHAINS.
CC   -!- DISEASE PROTEIN OMM MAY REPRESENT AN ALLELIC FORM OR ANOTHER GAMMA
CC       CHAIN SUBCLASS.
CC   -!- THE HINGE REGION IN GAMMA-3 CHAINS IS ABOUT FOUR TIMES AS LONG
CC       AS IN OTHER GAMMA CHAINS AND CONTAINS THREE IDENTICAL 15-RESIDUE
CC       SEGMENTS PRECEDED BY A SIMILAR 17-RESIDUE SEGMENT (12-28).
DR   EMBL; J00231; G567112; ALT_SEQ.
DR   PIR; A02149; G3HUWI.
DR   HSSP; P01857; 1FC1.
DR   MIM; 147120; -.
DR   PROSITE; PS00290; IG_MHC; 1.
KW   IMMUNOGLOBULIN C REGION; GLYCOPROTEIN.
FT   DOMAIN       12     73       HINGE.
FT   DOMAIN       74    183       CH2.
FT   DOMAIN      184    289       CH3.
FT   REPEAT       29     43
FT   REPEAT       44     58
FT   REPEAT       59     73
FT   MOD_RES       1      1       PYRROLIDONE CARBOXYLIC ACID.
FT   CARBOHYD      6      6
FT   DISULFID      7      7       INTERCHAIN (WITH HEAVY CHAIN DIMER).
FT   DISULFID     24     24       INTERCHAIN (WITH HEAVY CHAIN DIMER).
FT   DISULFID     27     27       INTERCHAIN (WITH HEAVY CHAIN DIMER).
FT   DISULFID     33     33       INTERCHAIN (WITH HEAVY CHAIN DIMER).
FT   DISULFID     39     39       INTERCHAIN (WITH HEAVY CHAIN DIMER).
FT   DISULFID     42     42       INTERCHAIN (WITH HEAVY CHAIN DIMER).
FT   DISULFID     48     48       INTERCHAIN (WITH HEAVY CHAIN DIMER).
FT   DISULFID     54     54       INTERCHAIN (WITH HEAVY CHAIN DIMER).
FT   DISULFID     57     57       INTERCHAIN (WITH HEAVY CHAIN DIMER).
FT   DISULFID     63     63       INTERCHAIN (WITH HEAVY CHAIN DIMER).
FT   DISULFID     69     69       INTERCHAIN (WITH HEAVY CHAIN DIMER).
FT   DISULFID     72     72       INTERCHAIN (WITH HEAVY CHAIN DIMER).
FT   CARBOHYD    140    140
FT   MOD_RES     290    290       REMOVED POST-TRANSLATIONALLY.
FT   VARIANT     126    127       QV -> EB (IN ZUC).
FT   VARIANT     134    134       P -> L (IN OMM).
FT   VARIANT     139    139       F -> Y (IN OMM).
FT   VARIANT     182    182       T -> A (IN OMM).
FT   VARIANT     227    227       S -> N (IN OMM).
FT   VARIANT     227    227       MISSING (IN ZUC).
FT   VARIANT     279    279       F -> Y (IN OMM).
SQ   SEQUENCE   290 AA;  32331 MW;  C5E7BE05 CRC32;
     QMQGVNCTVS SELKTPLGDT THTCPRCPEP KSCDTPPPCP RCPEPKSCDT PPPCPRCPEP
     KSCDTPPPCP RCPAPELLGG PSVFLFPPKP KDTLMISRTP EVTCVVVDVS HEDPEVQFKW
     YVDGVQVHNA KTKPREQQFN STFRVVSVLT VLHQNWLDGK EYKCKVSNKA LPAPIEKTIS
     KTKGQPREPQ VYTLPPSREE MTKNQVSLTC LVKGFYPSDI AVEWESSGQP ENNYNTTPPM
     LDSDGSFFLY SKLTVDKSRW QQGNIFSCSV MHEALHNRFT QKSLSLSPGK
//
```

Fig. 4. A *SwissProt* entry.

Conflicts that may exist between these reports are indicated in the feature table of the entry.

2.2.1. Integration with Other Databases

SwissProt is currently cross-referenced with 30 different databases. Cross-references appear in the DR section, in the form:

```
Database_name; primary_identifier; secondary_
identifier.
```

A *SwissProt* entry can point to various entries in the same database.

2.2.2. Data Submission

To submit new sequence data or update entries in *SwissProt* users should contact the following e-mail address at the EBI: `datasubs@ebi.ac.uk`.

2.2.3. TrEMBL

The current growth of the nucleotide sequence databases is too fast for the manual annotation process of *SwissProt* to follow, yet keeping the same quality standards. In March 1999, *EMBL* contains 3.2 million entries. This number is doubling every year. *SwissProt* contains 77,977 entries.

TrEMBL (translation of EMBL nucleotide sequence database) was introduced in early 1997 as a database of computer annotated entries in *SwissProt* format derived from the translation of all coding sequences (CDS) in *EMBL*, excluding CDS already included in *SwissProt*. Entries in *TrEMBL* are progressively merged with *SwissProt* entries.

The DR lines of a *SwissProt* and *TrEMBL* entries pointing to an *EMBL* entry cite the *EMBL* accession number as primary identifier and the PID as a secondary identifier.

2.2.4. Database Distribution

SwissProt and *TrEMBL* are both distributed quarterly on CD-ROM. The full release of each database is available from the EBI anonymous ftp server at URL `ftp://ftp.ebi.ac.uk/pub/databases/swissprot` and `ftp://ftp.ebi.ac.uk/pub/databases/trembl`.

2.3. The MSD Database

The *Molecular Structure Database* (*MSD*) is a project involving developments around the *Protein DataBank* (*PDB*, Brookhaven, NY). *PDB (5)* is an archive of experimentally determined three-dimensional structures of biological macromolecules, serving a global community of researchers, educators, and students. The archives contain atomic coordinates, bibliographic citations, primary and secondary structure information as well as crystallographic structures, and NMR experimental data. The database is available at `http://www2.ebi.ac.uk/pdb/`, which is a mirror of the North American WWW site.

2.3.1. Data Submission

Users can now submit data either to the EBI or to the *PDB* at Brookhaven. The European site at the EBI facilitates the submission procedure for European users. The collaboration between the EBI and the Brookhaven Laboratory includes the development of a new submission tool, *AutoDep* (**Fig. 5**), available at URL `http:/autodep.ebi.ac.uk/autodep-basepage.shtml`.

2.3.2. Database Distribution

The *PDB* is released quarterly. The full database release is available from the EBI anonymous ftp server at `ftp://ftp.ebi.ac.uk/pub/databases/pdb`. Weekly updates are also made available in the same directory.

2.4. *The* Radiation Hybrid Database

The *Radiation Hybrid database* (*RHdb*) *(6)* is a repository of raw data relevant to radiation-hybrid mapping. *RHdb* stores data on panels, experimental conditions, STSs, and experimental results of assays, mapping information, and maps. An important aspect of *RHdb* is extensive cross-referencing to other databases. Each entry in *RHdb* is given a unique identifier of the form RHn, where *n* is a digit.

2.4.1. Data Submission

A simple Web-based form is available at URL: `http://www.ebi.ac.uk/RHdb/submission_form.html`. Because experimental results often come in large batches, a tagged-field format has been developed to handle large quantities of data. This format is explained at URL `http://www.ebi.ac.uk/RHdb/data_input.html`.

2.4.2. Database Distribution

RHdb home page is at URL `http://www.ebi.ac.uk/RHdb` and contains all the current information about the database. Direct database access is gained through various Java-based clients . The full release (and updates) is available for the EBI anonymous ftp server at URL `ftp://ftp.ebi.ac.uk/pub/databases/RHdb`.

2.5. *The* BioCatalog

The *BioCatalog (7)* is a database of information on software of interest in molecular biology and genetics. The different programs are grouped by domain

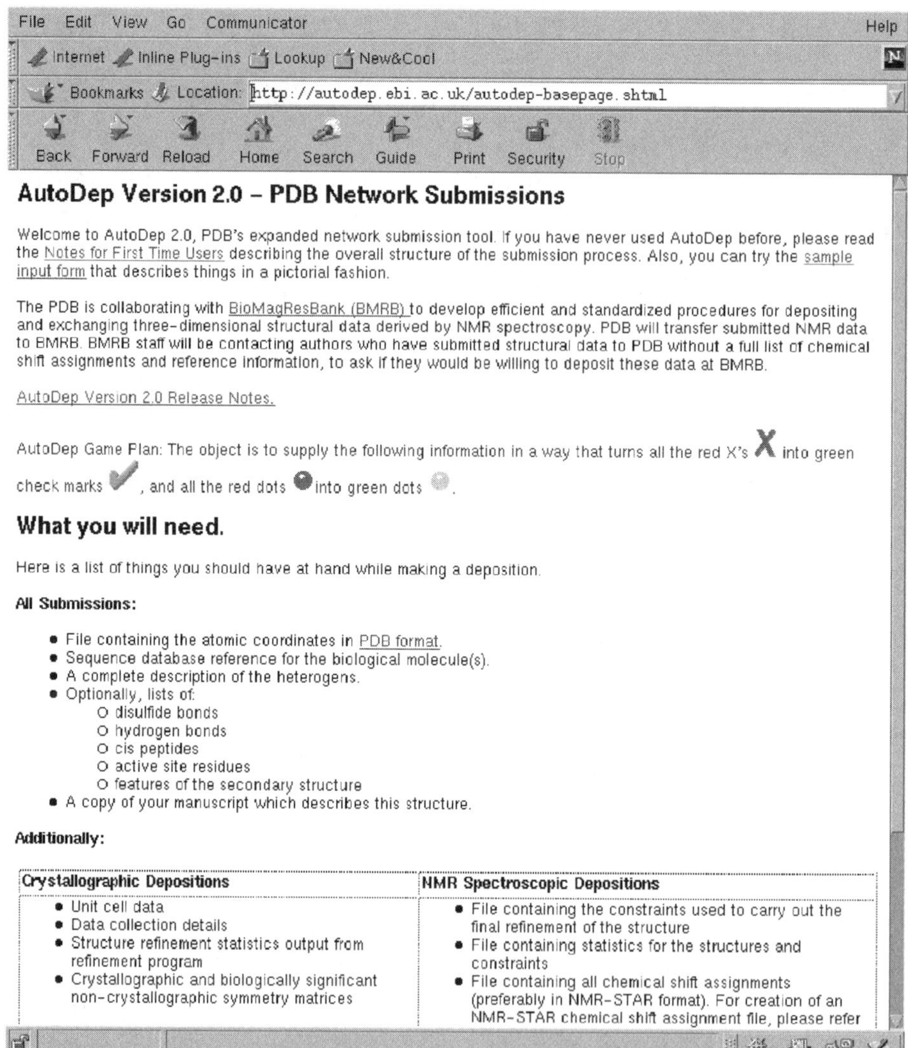

Fig. 5. *AutoDep*, the *PDB/MSD* submission tool on the WWW.

of interest. Each entry is given a unique identifier of the form BCn, where *n* is a digit.

2.5.1. Data Submission

New information can be added to the *BioCatalog* using a Web-based form available at URL `http://www.ebi.ac.uk/biocat/biocat_form.html`.

2.5.2. Database Distribution

The *BioCatalog* is freely distributed as ASCII files, available from the EBI anonymous ftp server at URL `ftp://ftp.ebi.ac.uk/pub/databases/biocat`. The catalog is available from the EBI WWW server at `http://www.ebi.ac.uk/biocat`. Lists are generated by domains/subclasses. The user can follow the links to the original site to retrieve the software. It is also possible to search the *BioCatalog* using keywords.

3. EBI Network Services

In addition to maintaining various databases, the EBI provides an ever-expanding number of free network services to external users.

3.1. Anonymous ftp Server

This is the main route for retrieving full releases or updates of the EBI databases, as well as software from the software repository. The main entry point for the ftp server is at URL `ftp://ftp.ebi.ac.uk/pub/`. From there the user can navigate through the different directories:

- **databases**: all databases distribution.
- **doc**: EBI documentation.
- **help**: help files for databases and services.
- **software**: the software archive.

3.2. Network File Server

The file server enables access via electronic mail (e-mail) to the full collection of databases, public domain software and documentation maintained by the EBI. Items are retrieved from the server by sending a command in the body of the e-mail message that is sent to the file server address (`netserv@ebi.ac.uk`). Sending the word **help** in the message will return the most current help file on how to use the server.

3.3. World Wide Web Server

EBI home page at URL `http://www.ebi.ac.uk/` provides access to all EBI services and databases. This page is updated constantly to follow the progress in technology and the changes in the services. But the philosophy is always the same, to list information about the different areas of the EBI: basic research, services, industry program, and general information. All databases and services described in this chapter can be found under the services heading.

3.3.1. Retrieving Sequences and Related Information: The SRS *System*

SRS (8) is the *Sequence Retrieval System*, a package that was first developed to provide query facilities on the annotation parts of a sequence entry. It has evolved into providing a complete workbench for sequence analysis, integrating the common analysis software. The system is built on top of indices that take into account the links (i.e., cross-references) between the databases.

SRSWWW is a user friendly Web interface to *SRS*. It can be found at URL `http://srs.ebi.ac.uk/`.

The buttons on this page link to various documents : the *SRS Manual* which provides full information on using the system, *SRS WorldWide* which lists the SRS servers on the Web, *SRS newsgroup* giving access to the *SRS*-specific newsgroup (`bionet.software.srs`), *SRS Developers* citing the *SRS* team.

We will illustrate the basic use of *SRS*. However the *SRS* user manual gives full and up-to-date information on all *SRS* features.

To use *SRS*, the user begins a session by clicking on the **Start** button. A session maintains the users selections and configuration throughout the current connection.

3.3.1.1. The Top Page

The page that appears next is called the "SRS top page." This page is used to select the database(s) to be searched. The user can select databases by clicking the check boxes appearing next to the database name. The name itself is linked to a page that provides information about the indices used to search the respective database. Every time a user wishes to change the database search set-up, the user returns to this top page. In the example (**Fig. 6**) both the *EMBL* and *EMnew* (the cumulative updates since last release) databases have been selected.

Clicking on the **reset** button will deselect all selections on that page. All **reset** buttons on *SRS* pages have the same behavior. Clicking on the **continue** button brings up the "query form page."

3.3.1.2. The Query Form Page

The **choice** button, left of the text entry field (**Fig. 7**) lists all fields available for querying the selected databases. In this example the Description, Organism, Description and All Text fields are available. This list depends on which fields have been indexed by the *SRS* server. Different sites may index different fields, so this list may vary for the same databases for different *SRS* servers. If the "**show only fields that selected databases have in common**" check box

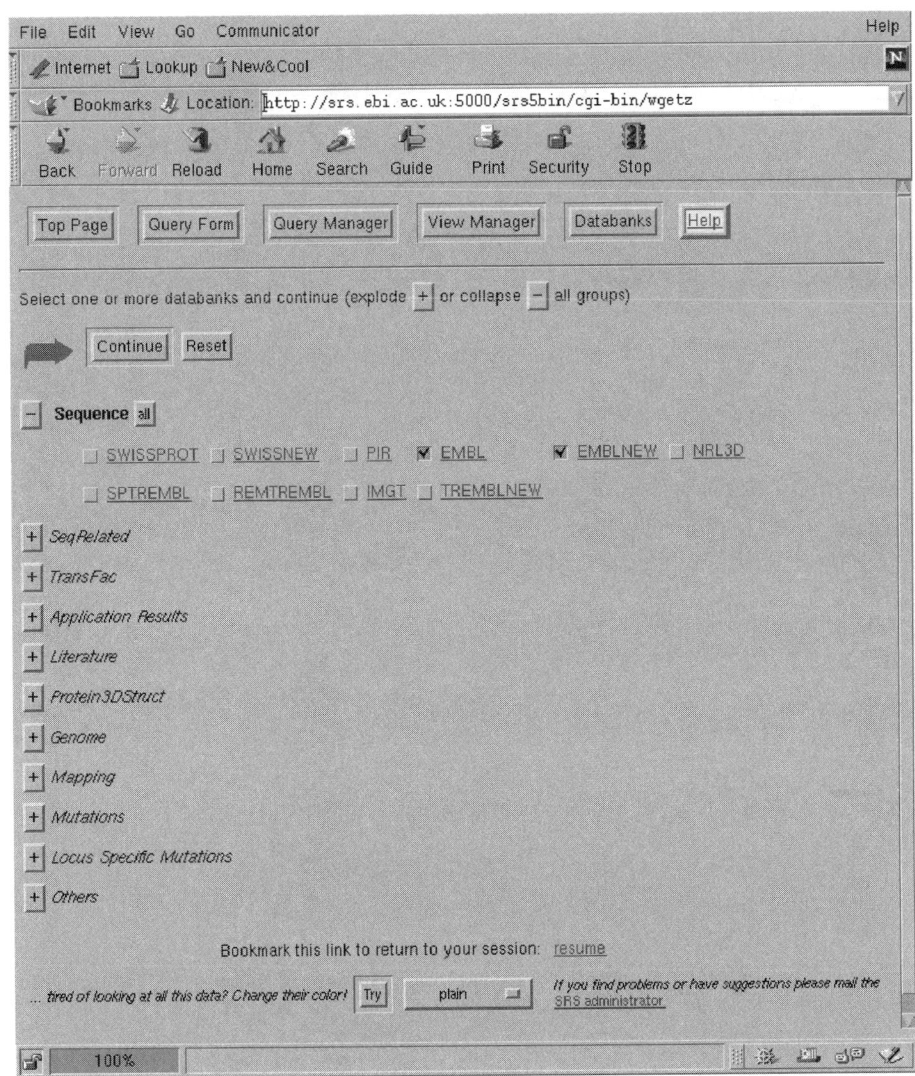

Fig. 6. The SRS top page. Here both *EMBL* and *EMBLnew* databases have been selected for searching.

from the top page has been selected, then only the common subset of indexed fields will be listed. If not all indexed fields from both databases will appear. The user will enter the relevant keywords on the search fields. The "**include fields in output**" list allows the user to select which field will show up in the "query result page." Clicking on the **do query** button will execute the query. The next page appears: the "query result page."

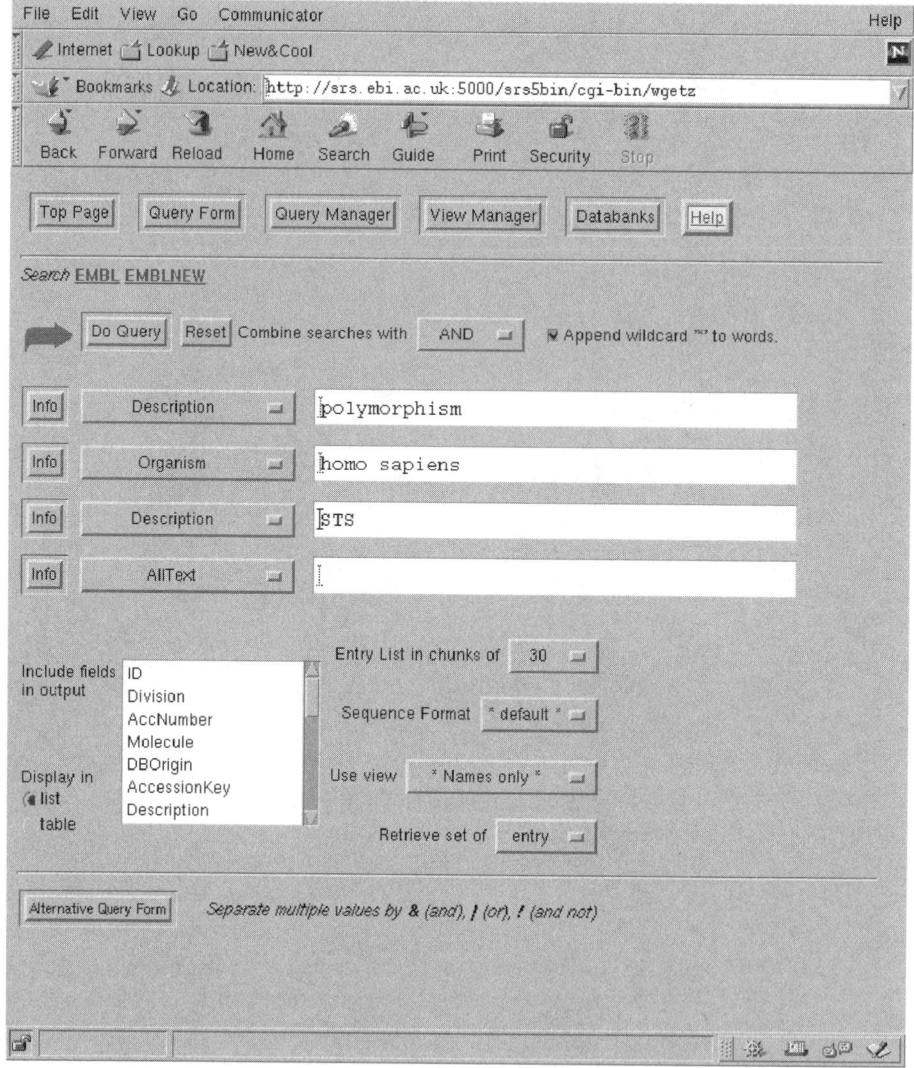

Fig. 7. A query is built in the query page.

3.3.1.3. The Query Result Page

This page (**Fig. 8**) shows the result of the submitted query. The query string, the number of "hits," and found entries with check boxes and included fields (as selected in the "query form page" are indicated. The found entries are listed by searched database and entry ID. Each entry listed has a link to the complete entry in the corresponding database. Clicking on the entry opens the "single

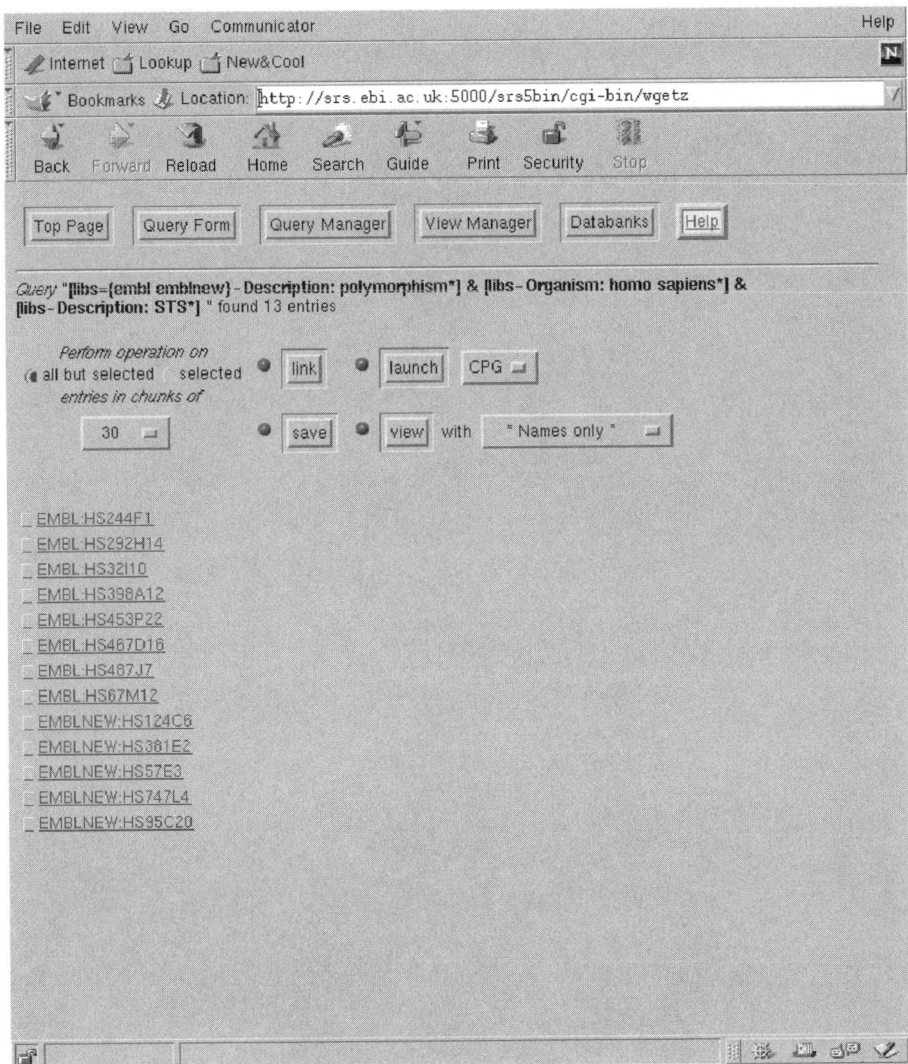

Fig. 8. The results from the query are displayed in the query result page. Clicking on an entry will bring the single entry page that shows the full entry.

entry page" which shows the full entry. Cross-references to other databases are represented as hypertext links to the relevant entries. Entries can be selected with the checkboxes to be viewed together. The other choices on this page are for advanced users. A full description is available in the *SRS* manual.

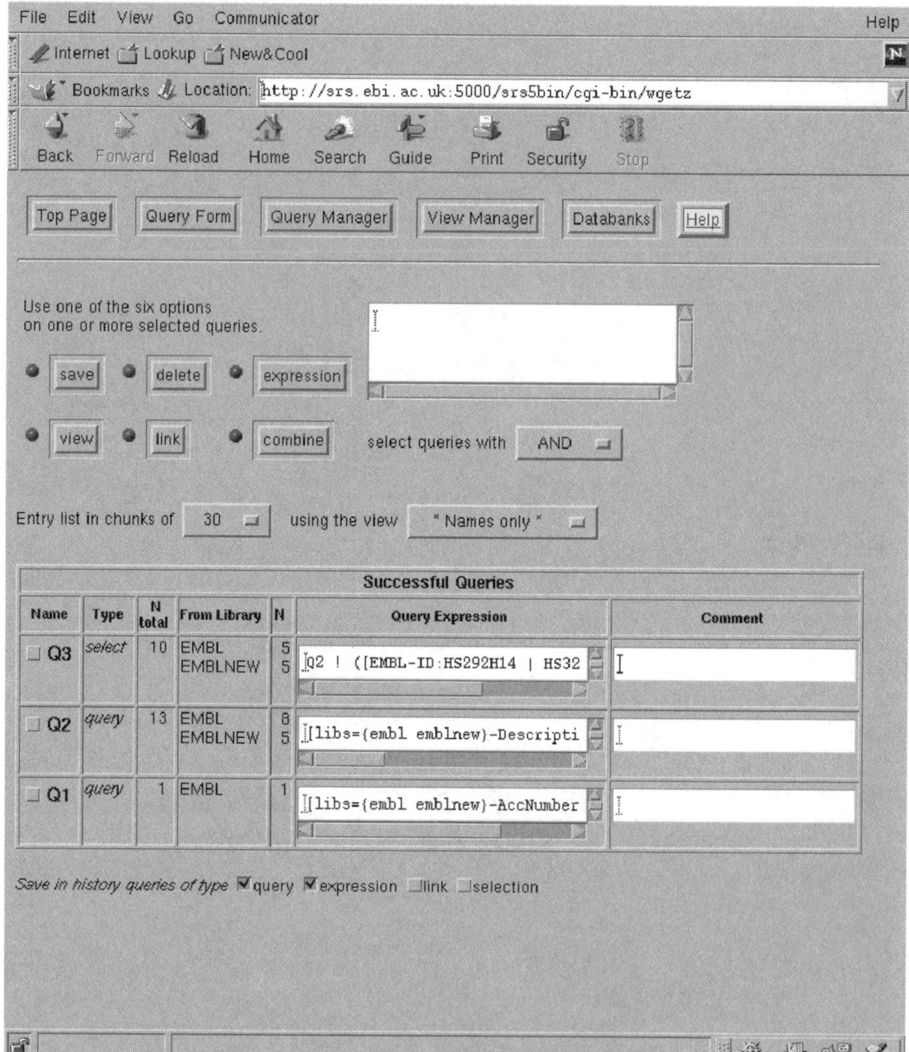

Fig. 9. The *Query Manager* shows all the queries that have been used in the current session. These queries can be reused, combined, or modified.

3.3.1.4. The Query Manager Page

Clicking on the **query manager** button brings the "query manager page." This page (**Fig. 9**) has two functions. First, it stores the completed queries as in a table. Every query is listed with a check-box, the query name in the form 'Qn' (i.e., Q1, Q2, ...), a type (i.e., 'select,''query,''link,'...), the total number of entries found, the database name(s), the number of entries for each, the query expression in *SRS* query syntax and a comment. The second function of the "query manager" is to make further queries and links. This is explained in the *SRS* manual.

3.3.2. Sequence Search Facilities

The EBI provides also a number of services that allow users to compare their own sequences against the most currently available data in the EBI databases. The main access page for these search facilities is at URL `http://www2.ebi.ac.uk/services.html`.

This page is always updated to display the current services available at EBI. All services pages have the same general appearance (**Fig. 10**):

- An E-mail address field: Depending on the load of the servers or the specific query parameters, the search will be run interactively or in batch. In the latter case, the result will be sent to the user by e-mail.
- Various fields, check-boxes, menu choices to enter the specific parameters for the query.
- A large text area to cut and paste or type the sequence to be analyzed or a file upload field and corresponding button to browse the directory structure.
- A help button that gives access to a detailed description on how to use the service.
- A submit button to execute the service.

3.3.2.1. *FASTA3*

This is both an interactive and e-mail service that provides sequences searches using W. Pearson Fasta ***(9,10)*** similarity searches algorithm. The *FASTA* tools are widely used for sequence database similarity searches, for identifying distantly related DNA and protein sequences, for identifying similar structures in sequences. It is available at URL: `http://www2.ebi.ac.uk/fasta3/`.

3.3.2.2. *BIC_SW*

This service provides an implementation of the full Smith and Waterman algorithm that runs on Compugen's Bioccelerator at URL `http://www2.ebi.ac.uk/bic_sw/`.

3.3.2.3. *SCANPS*

Scanps is a protein sequence database search tool developed by G. Barton ***(11)***. It compares protein or nucleic acids sequences to database sequences. It uses various dynamic programming algorithms, such as the Smith and Waterman local alignment method. It can be accessed at URL `http://www2.ebi.ac.uk/scanps/`.

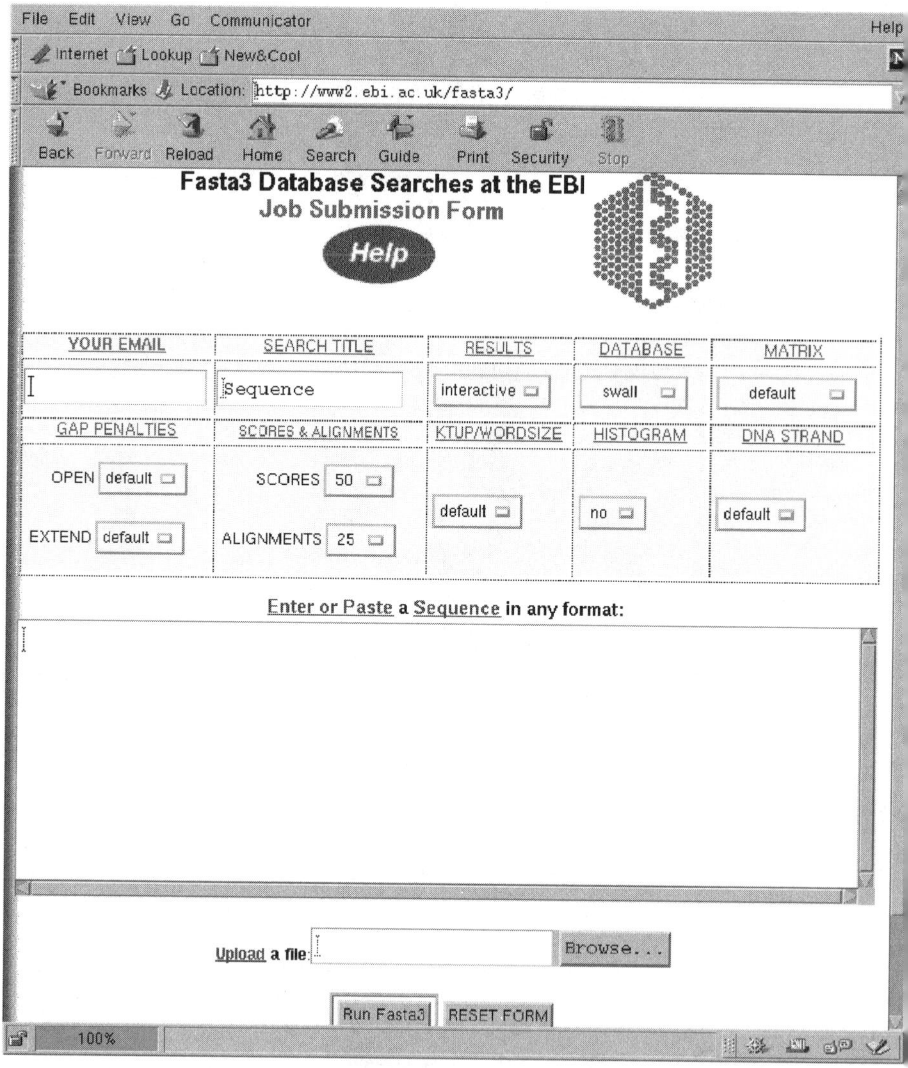

Fig. 10. The *FASTA* service page, as an example of the analysis pages: The top part of the form is used to enter the analysis parameters, the text area serves to enter the sequence to analyze. The sequence can also be uploaded from a file by typing its file name in the **Upload a file** text field. At the bottom of the page are two buttons: to run the analysis tool or to reset the form to its initial values.

3.3.2.4. *BLAST*

This is both an interactive and e-mail service offering searches on the protein sequence databases. Both NCBI ***(12)*** and Washington University ***(13)*** versions of the *BLAST* algorithm are available at URL `http://www2.ebi.ac.uk/blast2`. Blast offers rapid sequence comparison by an algorithm that approximates alignments by optimizing a measure of local similarity. It can be applied in a variety of contexts including DNA and protein sequence searches, gene identification searches, and in the analysis of multiple regions of similarity in a long DNA sequence.

3.3.2.5. *PPSEARCH*

Prosite database patterns searches at URL `http://www2.ebi.ac.uk/ppsearch`.

3.3.3. Analysis Tools

Some of the analysis tools available at the EBI, are described below.

3.3.3.1. CLUSTAL W

CLUSTAL W ***(14)*** is a multiple sequence alignment tool for sequences. Giving the tool a set of sequence, you will get back an alignment at URL `http://www2.ebi.ac.uk/clustalw`.

3.3.3.2. *PRATT*

PRATT ***(15)*** is a tool that allows the user to search for patterns conserved in a set of protein sequences. The user can specify what kind of patterns should be searched for, and how many sequences should match a pattern to be reported. This tool is available at URL `http://www.ebi.ac.uk/pratt`.

3.3.3.3. *GeneMark*

GeneMark ***(16)*** is a gene-prediction service for predicting protein-coding regions in prokaryotic and eukaryotic organisms. It is based on a special type of Markov chain model of coding and noncoding nucleotide sequences. It is accessible at URL `http://www2.ebi.ac.uk/genemark`.

3.3.3.4. Dali server

The Dali ***(17)*** server (`http://www2.ebi.ac.uk/dali`) is a network service for comparing three-dimensional structures. The user submits the coordinates of a query protein structure and Dali compares them against those in the *Protein Data Bank* (*PDB*). A multiple alignment of structural neighbours is mailed back. In favourable cases, comparing three-dimensional structures may reveal biologically interesting similarities that are not detectable by comparing sequences. Structural neighbors of a protein already in *PDB* are maintained in the *FSSP* database at URL `http://www2.ebi.ac.uk/dali/fssp/fssp.html`.

4. Future Developments and Services

Molecular biological data and genome data are inherently complex and rapidly evolving. New algorithms are constantly being developed, old ones updated, and EBI services are constantly upgraded as new technologies are explored. One such technology is *CORBA*, the *Common Object Broker Architecture*, which is becoming standard that should enable the development of new methods facilitating database and service access.

5. How to Contact the European Bioinformatics Institute

The postal address for the EBI is : EMBL Outstation, the EBI, Wellcome Trust Genome Campus, Hinxton, Cambridge CB10 1SD, UK.

For the Network services :

General enquiries: `support@ebi.ac.uk`
Databases enquiries: `datalib@ebi.ac.uk`
RHdb submissions and enquiries: `rhdb@ebi.ac.uk`
BioCatalog enquiries: `biocat@ebi.ac.uk`
Data submission by e-mail (*EMBL* and *SwissProt*): `datasubs@ebi.ac.uk`
Data submissions by WWW (*EMBL*): `http://www.ebi.ac.uk/submission/webin.html`
Corrections to *EMBL* entries by e-mail: `update@ebi.ac.uk`
Corrections to *EMBL* entries by WWW: `http://www.ebi.ac.uk/ebi_docs/update.html`
EBI network file server: `netserv@ebi.ac.uk`
EBI ftp server: `ftp://ftp.ebi.ac.uk`
EBI WWW server: `http://www.ebi.ac.uk`
EBI WWW search and analysis services: `http://www2.ebi.ac.uk/Services`

References

1. Stoesser, G., Moseley, M. A., Sleep, J., McGowran, M., Garcia-Pastor, M., and Sterk, P. (1998) The EMBL nucleotide sequence database. *Nucleic Acids Res.* **26,** 8–15.
2. Benson, D. A., Boguski, M. S., Lipman, D. J., Ostell, J., and Ouellette, F. (1998) GenBank. *Nucleic Acids Res.* **26,** 1–7.
3. Tateno, Y., Fukami-Kobayashi, K., Miyazaki, S., Sugawara, H., and Gojobori, T. (1998) DNA data bank of Japan at work on genome sequence data. *Nucleic Acids Res.* **26,** 16–20.
4. Bairoch, A. and Apweiler, R. (1998) The SWISS-PROT protein sequence databank and its supplement TrEMBL in 1998. *Nucleic Acids Res.* **26,** 38–42.

5. Benrnstein, F. C., Koetzle, T. F., Williams, G. J. B., Meyer, E. F. Jr., Brice, M. D., Rodgers, J. R., Kennard, O., Shimanouchi, T., and Tasumi, M. (1977) The Protein Data Bank: a computer based archival system file for macromolecular structures. *J. Mol. Biol.* **112,** 535–542.
6. Lijnzaad, P., Helgesen, C., and Rodriguez-Tomé, P. (1998) The Radiation Hybrid Database. *Nucleic Acids Res.* **26,** 102–105.
7. Rodriguez-Tomé, P. (1998) The BioCatalog. *BioInformatics* **14,** 469–470.
8. Etzold, T. and Argos, P. (1993) SRS—an indexing and retrieval tool for flat file data libraries. *Comput. Appl. Biosci.* **9,** 49–57.
9. Pearson, W. R. and Lipman, D. J. (1988) Improved tools for biological sequence analysis. *Proc. Natl. Acad. Sci. USA* **85,** 2444–2448.
10. Pearson, W. R. (1990) Rapid and sensitive sequence comparison with FASTP and FASTA. *Meth. Enzymol.* **183,** 63–98.
11. Barton, G. J. (1997) *SCANPS version 2.3.11 User Guide*. University of Oxford, UK.
12. Altschul, S. F., Gish, W., Miller, W., Myers, E. W., and Lipman, D. J. (1990) Basic alignment search tool. *J. Mol. Biol.* **215,** 403–410.
13. Altschul, S. F. and Gish, W. (1996) Local alignment statistics. *Meth. Enzymol.* **266,** 460–480.
14. Higgins, D., Thomson, J., and Gibson, T. J. (1994) ClustalW: improving the sensitivity of progressive multiple sequence alignment through sequence weighting, position-specific gap penalties and weight matrix choice. *Nucleic Acids Res.* **22,** 4673–4680.
15. Jonassen I. (1997) Efficient discovery of conserved patterns using a pattern graph. *Comput. Appl. Biosci.* **13,** 509–522.
16. Borodvski, M. and McIninch, J. D. (1993) GeneMark: parallel gene recognition for both DNA strands. *Comp. Chem.* **17,** 123–133.
17. Holm, L. and Sander, C. (1993) Protein structure comparison by alignment of distance matrices. *J. Mol. Biol.* **233,** 123–138.

18

Computer-Assisted Analysis of Transcription Control Regions

MatInspector and Other Programs

Thomas Werner

1. Introduction

Computer-assisted analysis of DNA sequences used to be a task for the "local expert." However, various genome sequencing projects as well as the availability of sequencing services are providing a constantly increasing number of scientists with DNA sequence data. These data need to be interpreted to include the information into experimental schemes.

Fortunately, a number of software tools have been developed to facilitate this task. Several of these programs are available on the World Wide Web (WWW). This relieved researchers from the necessity to have all programs locally installed and maintained (no easy thing for a wet-lab-based laboratory). Browser-oriented programs also provide user-friendly interfaces. Typically, use of WWW-based programs requires no more than a so-called Web browser (programs like *Netscape* or the *Internet Explorer*) which are standard on most high-level PCs and workstations. This is the major reason that WWW-based programs are the most popular in molecular biology and I will place emphasis on these tools in this chapter.

Analysis of genomic DNA can be divided into two major tasks. The first one is to locate potential protein-coding regions (called "gene finding") and the second task is to locate other functionally important regions like sequences providing the basis of transcriptional control (e.g., promoters, enhancers, locus control regions, and so on). The second type of analysis will be the focus of

From: *Methods in Molecular Biology, vol. 132: Bioinformatics Methods and Protocols*
Edited by: S. Misener and S. A. Krawetz © Humana Press Inc., Totowa, NJ

this article because the tools to be described were developed for the computer-assisted analysis of transcriptional control.

In general, the promoter is an integral part of the gene mediating and controlling initiation of transcription from that part of a gene that is located immediately downstream of the promoter (3′). Most protein-coding genes are controlled by RNA polymerase II promoters that are composed of several segments with different functions *(1)*. Among the most important segments are protein-binding sites like the TATA box or binding sites for other proteins called transcription factors (TFs, AP-1 is an example of such a factor, for review *see* **ref. 2**).

Although other promoter elements like intrinsically curved DNA, direct or inverted repeat elements, may also contribute to promoter function, protein-binding sites appear to be the most crucial elements. Therefore, detection of potential binding sites for transcription factors in DNA sequences is very important for any kind of promoter analysis. As will be detailed later, detection of organizational features of these binding sites within promoters is another important step towards a complete promoter analysis.

1.1. IUPAC Consensus Sequences and Weight Matrices

At present, there are two major approaches to TF-binding site recognition, the use of IUPAC consensus sequences and weight matrices for TF-binding sites. IUPAC consensus sequences are composed of the most frequent set of nucleotides describing that site and use special letters to indicate the potential presence of more than one nucleotide at a given position of a binding site, e.g., R may indicate either an A or a G. The weight matrix uses the complete composition of nucleotides for each position of the consensus derived alignment to achieve a more differentiated rating of a matching sequence. For example, a single position of an alignment of 12 sequences containing (T,T,T,T,T,T,T,A,A,A,C,C) (each letter representing one sequence at this position) would be assigned T in the IUPAC consensus. A new sequence with a T at this position would be considered a match, whereas an A at the same position would cause the whole sequence to be dismissed as no match. Even a simple nucleotide distribution matrix would assign a weight score (in this case proportional to the percentage of the nucleotide) of 0.58 to the T and still 0.25 to an A and 0.17 to C. Thus, weight scores represent the similarity of the tested sequence to all of the sequences in the alignment much better than IUPAC consensi. Most weight-matrix-based methods use more complex weighting functions, e.g., by comparison of the actual nucleotide distribution with random values or by other statistical measures (e.g., information content).

1.2. Is a Weight Matrix Better than a IUPAC String?

Programs for IUPAC string searches have been available for more than a decade (as part of the GCG package and in the program *SIGNAL SCAN* **(3)**. However, they are strongly influenced by selection of the sequences they are derived from, and will not discriminate between an insignificant mismatch, i.e., that can be tolerated by the binding protein, and one that will abolish binding. Weight matrices on the other hand are much less sensitive to sequence selection and provide a quantitative rating (score) suggesting likelihood of protein binding to the site analyzed. Even a single mismatch at a critical position will greatly reduce the score of the match and often correctly result in its rejection.

However, analysis of DNA sequences with weight matrices for TF-binding sites requires libraries of precompiled weight matrices because weight matrices are not simple to compile by the inexperienced user. Only weight matrix search programs providing precompiled libraries of weight matrices are of immediate use for molecular biologists. *MatInspector* **(4)** is a tool that provides such a library and other programs developed by the biocomputing and bioinformatics group, GSF-AG BIODV, like *FastM* **(5)** and *GenomeInspector* **(6,7)** take advantage of these libraries.

2. Required Resources

To analyze anonymous DNA sequences for transcription control regions, several software tools may be required. These can be separated into two categories. The tools of the first category are used to analyze single sequences, whereas the second category is geared towards the analysis of a set of several functionally related sequences.

The first category of GSF-AG BIODV programs includes programs for the detection of TF binding sites like *MatInspector* and *ConsInspector* **(8)**, as well as programs to model and analyze organizational sequence feature information. These programs are *FastM*, and *GenomeInspector*.

The second category consists of programs for matrix generation like *MatInd*, *ConsInd*, and *CoreSearch* **(9)**, a novel multiple alignment program, *DIALIGN* **(10,11)**, and *ModelGenerator* **(12)**, which develops organizational models from a set of training sequences. More detailed descriptions of the individual programs and further references can be found on the GSF-BIODV WWW server at `http://www.gsf.de/biodv`. *MatInspector* and *FastM* both feature a WWW interface that makes them very easy to use; the other programs require downloading and local installation usually on a UNIX workstation. A detailed discussion of the programs *MatInd*, *MatInspector*, *FastM*, *ModelInspector*, and *GenomeInspector* follows.

2.1. MatInspector

MatInspector is a program that utilizes a library of precompiled weight matrices for transcription factor binding sites to scan sequences for potential binding sites for the corresponding factors. The program determines an individual score for each sequence segment with all matrices selected for the analysis and reports all matches that score equal or above a user-defined threshold. (MatInspector supports matrix groups, fungi, insects, vertebrates, plants, miscellaneous as well as individual matrix selection.) The calculation of the matrix similarity includes the information content of individual positions within the matrix, which results in an enhanced sensitivity of the method with respect to any alteration at highly conserved positions (for a detailed description of the matrix similarity calculation *see* **ref. *4***). On one hand, if the matrix contains a position where only an A occurs in all training sequences and the test sequence features a T nucleotide at the corresponding position, the overall score of this region will be dramatically reduced. On the other hand, if a position in the matrix contains A, C, and G in about equal proportions but not T, the occurrence of a T at the corresponding position will only have a minor influence on the overall score.

The program *MatInspector* uses two different similarity thresholds, called core and matrix similarity. The core similarity refers to the four best conserved consecutive nucleotides in the matrix, which usually indicate the most critical part for protein binding. Mismatches within this region often impair protein binding even if the rest of the sequence appears to fit the binding site quite well. The core similarity is checked first, preventing further analysis of matches for matrix similarity that have a core sequence most likely unacceptable for protein binding. This reduces the number of spurious matches in most cases, whereas biologically meaningful matches are usually not affected by the core similarity threshold.

Experience showed that inclusion of the information content and the core similarity improved the correlation of the *MatInspector* similarity scores with the actual biological binding affinities. This has been tested and exemplified in a recent comparison of several matrix-based search programs for TF binding sites ***(13,14)***.

Analysis of sequences with the full library usually results in a rather long list of matches that often appear to cover the whole sequence. Many of these matches are not meaningful for several reasons. The most important reason is that the default thresholds are too low for some matrices (like AP-1), yet already exclude some meaningful matches from other matrices. For a number of factors more than one matrix was derived from independent experiments under different experimental conditions. It is not possible to use just one matrix without sacrificing useful information from the other matrices. Accordingly, the list of matches from *MatInspector* can be rather long. The user has to select those matches that appear to be most likely true matches according to the expertise of the user.

Much (but not all) of this task can be solved automatically by the commercial version of *MatInspector* from Genomatix (`http://genomatix.gsf.de`) that relies on an extended library of matrices (more matrices than the public domain version) all of which feature individually optimized thresholds for matrix similarity (no general default). In addition, similar and/or functionally related matrices are combined into matrix families that ensures full use of the information of all matrices without the burden of redundant matches of several related matrices. *MatInspector professional* only reports the best match for each matrix family. The combination of optimized thresholds with matrix families often reduces the output to less than half that of *MatInspector public* without a significant reduction of true positive matches. However, false-positive matches cannot be completely eliminated from the output by these enhanced features. The user has to decide whether these advantages justify the costs for this commercial tool.

It should be noted that Genomatix also offers an enhanced *MatInspector public* version on its server free of charge for academic scientists. It is capable of database searches. All of these tools can be accessed via a WWW interface (`http://genomatix.gsf.de/products`).

2.2. FastM/ModelInspector

A single TF binding site often does not encode a transcriptional function of its own (just binding of the protein) because biological function usually requires cooperation of two or more TF binding sites. For example, the TATA box binding protein (TBP) has a very relaxed sequence specificity. Therefore, it is principally impossible to define a highly specific matrix for a TATA box and every TATA box matrix produces many matches that are clearly not TATA boxes. However, the context of a potential TATA box (or other TF binding site) is often more important for functionality than the actual sequence of the binding site. The most simple form of context is represented by a pair of binding sites. The cooperative effects of many pairs are well known and a database of such functional pairs exists (*COMPEL*, **ref. *15***). Therefore, analysis of a potential promoter sequence for pairs of binding sites shared with other promoters is a very useful second step towards promoter recognition.

FastM is a program that allows the easy generation of models of such pairs that can then be used to scan sequences or *GenBank* (or *EMBL*) sections for the occurrence of these pairs in other sequences. That way *FastM* allows one to model hypotheses about potentially important pairs and provides immediate verification of the significance of such pairs. This can be used to scan the database for potential target genes for a signal to which the particular pair of binding sites may respond ***(5)***.

The actual search is carried out by the program *ModelInspector* ***(12)***, which locates matches to the individual TF binding sites and calculates a combined score for the complete model. Because *ModelInspector* will only report matches that fulfill the threshold for the complete model, the amount of false-positive matches usually is orders of magnitude lower than that of a search for individual matches. The details of the scoring calculations carried out by *ModelInspector* are described in **ref. *12***.

Both *FastM* and *ModelInspector* have also been developed into more powerful commercial products by Genomatix. Highlights of these products are that both programs are no longer confined to TF binding site matrices as elements of the models and models may contain up to 10 different individual elements instead of the two matrices in the public domain versions. *ModelInspector* professional also comes with a precompiled library of functional promoter modules than can be used to scan sequences without the necessity to first carry out the modeling.

2.3. GenomeInspector

The *FastM*-based approach can detect functional subunits that have been defined. *GenomeInspector* ***(6,7)*** can deduce such subunits without *a priori* knowledge solely from their repeated appearance in a long genomic sequence, which may actually be a whole genome like the yeast genome.

The basic principle underlying this analysis is very simple. There is only one way to arrange sequence elements on a linear molecule like DNA. They exhibit a sequential order (one necessarily comes before the other with respect to one strand) and they can be separated by a certain distance. This is exactly what *GenomeInspector* analyzes: the repeated occurrence of element pairs with the same sequential order within a given distance window.

GenomeInspector uses point and region sets for this correlation analysis. All matches to a certain transcription factor binding site matrix are represented by a point set and longer segments like reading frames are represented by region sets. To find a correlated pair of TF binding sites, *GenomeInspector* analyzes the corresponding point sets for the unusual accumulation of distance correlated pairs of sites, one from each set. For example, association of an element (which does not need to be a matrix) with yeast promoters can be assessed easily by correlating the point set of the binding site matrix with the regions upstream of the reading frames represented by a region set. That way, much information can be deduced from scratch. This is the most powerful feature of the method which is described in more detail in **refs.** ***6,7***.

GenomeInspector has an X-11-based graphical user interface but is not available via WWW. The program must be downloaded to a UNIX computer on which X-11 is installed. *GenomeInspector* is a large collection of individual analysis methods and could be envisaged as a tool box for correlation analysis. Therefore, the program is versatile but also requires some user training, because problems need to be translated into steps *GenomeInspector* can carry out.

3. Methods

3.1. Method 1: Generation of a Weight Matrix

This step can be omitted as long as the more than 250 precompiled weight matrices in the *MatInspector* library are considered sufficient for the analysis. However, if there is a need for a new matrix this is how to generate it:

1. Go to the BIODV WWW server at `http://www.gsf.de/biodv` (direct access to *MatInd/MatInspector* page `http://www.gsf.de/biodv/matinspector.html`). The download selection on the bottom half of this page will automatically provide you with both *MatInd* and *MatInspector*. Download the files and install them on your computer system.
2. Select at least four or five sequences containing the binding site for the particular protein. They have to be contained in a single file formatted in the IG format:
   ; comment lines (as many as desired)
   sequence_identifier (must be a single string with NO internal spaces)
   AGCTGACGTCGACGTCG1 (sequence in blocks, lines etc., must end with a 1)
   next sequence (comment lines, sequence identifier, sequence)
3. Start the *MatInd* program, indicate the sequence file, and let the program determine the weight matrix. If the matrix fulfills the minimum quality requirement (re value < 5), three files are generated. Two matrix files with extensions .sel and .mat as well as one file with extension .ali, which contains the actual sequence alignment. If the random expectation value (re) is larger than 5, i.e., more than five matches would be expected in 1000 nucleotides of a random DNA sequence, only the .mat and the .ali files will be generated indicating poor matrix quality. This may be caused by inclusion of sequences that do not contain the binding site or by selecting inappropriate sequence regions (e.g., truncating the binding site in some cases). If this happens, the selection of the training sequences should be corrected (if possible) and the *MatInd* analysis should be repeated.
4. The new matrix file can now be copied into the library directory of *MatInspector* and can be directly used by a local installation of *MatInspector*.

It is also possible to generate a *MatInd* weight matrix from a precompiled nucleotide distribution matrix (without sequences) which are often found in publications. Refer to the *MatInd* user guide (which comes as part of the download package) for further details of this method.

Note: *MatInd*-generated matrices defined by users cannot be used with the WWW version of *MatInspector* on the GSF server (`http://www.gsf.de/cgi-bin/matsearch.pl`). However, Genomatix offers a commercial version (*MatInspector professional*) on its server (`http://genomatix.gsf.de`) with this feature in addition to several other improvements over the public domain version.

3.2. Method 2: Analyzing Sequences with MatInspector

The program *MatInspector* is available both for download and online on the GSF WWW server. I will describe the procedure for the WWW version (`http://www.gsf.de/cgi-bin/matsearch.pl`) which includes principally the same steps as analysis with a local installation.

1. Paste or upload the sequence to be analyzed (uploading is the same as selecting a sequence file from a directory).
2. Set the search parameters (thresholds). *MatInspector* provides default settings for both core and matrix similarity. However, these thresholds can be adjusted by the user. Unless there is a specific reason for changing the defaults it is recommended to use the defaults. They work well for most applications.
3. Select the matrix group from the library to analyze the sequence. The basic selection (all or individual matrices) is made on the first page of the WWW version and can be refined later on. *MatInspector* offers several choices for matrix selection:
   a. Select the whole library (all matrices).
   b. Select a subset from the available matrix groups: vertebrates, insects, fungi, plants, miscellaneous matrices (all matrices from that subset).
   c select one matrix or several matrices individually from the list of all matrices in the library or a specified group (user-defined matrix set).
4. Select the output options. *MatInspector* allows one to arrange matches in the order of occurrence on the sequence, by matrix name, and by match quality. The last feature is especially useful if preference is given to high scoring sites without suppressing lower scoring sites from the output by raising the matrix threshold.
5. Submit the query. In case individual matrix selection was chosen, there will be a list of matrices from which individual matrices can be selected by clicking their check box. The next query submission will then start the analysis, which typically takes only seconds to complete. The results will be immediately displayed on the screen in the form of a table (*see* **Fig. 1**). The columns indicate the matrix name, the position in the sequence in which the match starts, the strand orientation (+ or –), the core and the matrix similarity, and the actual sequence in which the match was found. The matrix name usually is a hyperlink to a more detailed description of the matrix which is present in the *TRANSFAC* database ***(15)***. A

```
MatInspector Release 2.2   October 1997            Wed Sep  9 08:35:42 1998

   Solution parameters:
   ~~~~~~~~~~~~~~~~~~~~
   sequence file:    LTRtest_seq.seq
   core sim:         0.75
   matrix sim:       0.85

   Explanation for column output:
   ~~~~~~~~~~~~~~~~~~~~~~~~~~~~~~~
        ->       Matrix positions correspond to sense strand numbering, but all
                 sequences are given in 5'-3' direction.

        ->       n/a in column 'core simil.' indicates, that no core search
                 was conducted.

        ->       Capital letters within the sequence indicate the core string.
matrix                        | matrix         | core   | matrix | sequence
name                          | position(str)  | simil. | simil. |
----------------------------------------------------------------------------
  Inspecting sequence mmtpro [?] (1 - 1450):
  V$BARBIE_01                 |        2 (-)   | 1.000  | 0.850  | tttaAAAGaagcacg
  V$TATA_C                    |        9 (-)   | 0.928  | 0.885  | ttTTTAAAAg
  V$TATA_C                    |        9 (+)   | 0.928  | 0.879  | ctTTTAAAAa
  V$HNF3B_01                  |       18 (-)   | 0.848  | 0.852  | cccTTTTttctt
  V$IK2_01                    |       25 (+)   | 1.000  | 0.904  | aaggGGGAaatg
  V$IK1_01                    |       25 (+)   | 1.000  | 0.877  | aaggGGGAaatgc
  V$MZF1_01                   |       25 (+)   | 1.000  | 0.969  | aagGGGGa
  V$NFKB_Q6                   |       27 (+)   | 1.000  | 0.907  | ggGGGAaatgccgc
  .
  .
  .
  V$USF_C                     |     1430 (+)   | 0.817  | 0.857  | aCATGTgc
  V$USF_C                     |     1430 (-)   | 0.876  | 0.930  | gCACATgt
  V$CHOP_01                   |     1432 (+)   | 1.000  | 0.870  | atgTGCAatgagt
  V$AP1_C                     |     1439 (-)   | 0.848  | 0.871  | tTCACTCAt
  V$AP1_C                     |     1439 (+)   | 0.856  | 0.862  | aTGAGTGAa

A total of 1450 basepairs was scanned.

In   0 seq.   0 matches to the matrix V$AHRARNT_01 (re:    0.81) were found.
In   0 seq.   0 matches to the matrix V$AHRARNT_02 (re: <  0.01) were found.
In   0 seq.   0 matches to the matrix V$AHR_01      (re: <  0.01) were found.
In   1 seq.   7 matches to the matrix V$AP1FJ_Q2    (re:    2.45) were found.
In   1 seq.  12 matches to the matrix V$AP1_C       (re:    1.08) were found.
 .
 .
 .
```

Fig. 1. *MatInspector 2.2* output from the analysis of a 1450-nucleotide sequence containing a retroviral mammalian B-type LTR (long terminal repeat). The output has been shortened for display. Three dots oriented in vertical orientation indicate omitted lines in the output.

statistics summary is provided at the end of the list, which details how often each individual matrix was found in the sequence analyzed, including matrices that were not found.

3.3. Method 3: Analyzing Sequence Organization with FASTM

Once a list of potential binding sites has been established by scrutinizing the *MatInspector* output, it is recommended to analyze these sites for potential promoter modules as detailed in **Subheading 2.2.** with the program *FASTM*.

This program also features a WWW interface (`http://www.gsf.de/cgi-bin/fastm.pl`) that reduces the complex task of modeling to a series of selections from a WWW form. Here is how to check the significance of a potential pair of binding sites.

1. Paste or upload the sequence to be analyzed (uploading is the same as selecting a sequence file from a directory). Alternatively, a database section can be specified. All sequences within this section will be analyzed.
2. Select the matrix for the upstream TF binding site from the list (which is identical to the *MatInspector* library, so that every match found by that program can be specified in *FASTM* also). This includes specification of the strand orientation (sense or antisense). Unless the matrix is completely symmetric there are always two possible orientations. It is also possible to allow both orientations. The core and matrix similarity is individually set for each matrix in *FASTM* (same parameters as in a *MatInspector* search). sequence. If a matrix for a particular binding site is not available it is possible and may be useful to type a IUPAC consensus. However, IUPAC-based models are of much lower specificity than matrix-based models and should only be used with rather restrictive IUPAC sequences (**no Ns!**). As for matrices, the strand orientation of IUPAC sequences can be specified and in addition, *FASTM* tolerates one or two mismatches to the IUPAC sequence.
3. Select the matrix for the downstream TF binding site from the second matrix list or specify a IUPAC sequence instead. Selection of strand and mismatches is carried out in the same way as for the first matrix.
4. Specify the distance range between the two binding sites. Three common distance ranges are provided as ready-made selections (close, medium, and long) and there is the possibility to enter a user-defined distance range as well. The smaller the distance range is (maximum–minimum distance, independent of the absolute distance) the more specific the resulting model will be.
5. Finally, a correct e-mail address has to be specified to which the results will be mailed. This is necessary because analysis of a database section with a *FASTM* model may take some time.
6. Start the analysis of a sequence or a database section with this model. The result file will look similar to **Fig. 2**, which is an example for the search of a pair of TF binding sites. A summary of the search model is shown on top of the table, then all individual matches are detailed as shown below. For each match, positions and scores of the individual matrices are given and a combined score for the model is shown below these results. The maximum score two matrix models can reach is 2.0.

Biologically functional pairs of cooperating binding sites often produce less than 100 matches within the database, allowing inspection of the whole list. Usually, the number of matches alone is a good indicator if the selected pair of binding sites is specific or not. In this manner a number of potential pairs of

```
FastM / ModelInspector WWW Release 1.1                    Wed Sep  9 08:43:16 1998

Sequence model of "seq":


      0.75/0.80 0.75/0.80

      V$GRE_C   V$GRE_C

      (+)       (+)

------[me]------[me]------


Distances between elements:

V$GRE_C           -          V$GRE_C:   80 -  100 bp (+/- 10 bp)

Input parameter:

Inspecting both strands of sequences(s).
Maximum number of matches in output file: 100


Inspecting sequence EP011026 (1 - 600):

(-)              ------------------------------------------------------------
                    position    core sim.   mat. sim.

    V$GRE_C            418 (-)       1.00          0.83
    V$GRE_C            308 (-)       0.86          0.85

element score =   1.67




Inspecting sequence EP023010 (1 - 600):

(+)              ------------------------------------------------------------
                    position    core sim.   mat. sim.

    V$GRE_C            264 (+)       1.00          0.80
    V$GRE_C            346 (+)       0.92          0.81

element score =   1.61


     .
     .
     .

Sequences searched: 1306 (783600 bp); 7 matches found in 7 sequences.
```

Fig. 2. *FASTM* output from the analysis of the *Eukaryotic Promoter Database* with a model of two binding sites for the glucocorticoid receptor in a distance of 80–100 nucleotides. 0.75/.80 indicates the core/matrix similarity thresholds of the individual sites in the model, (+) indicates sense strand orientation, [me] indicates that the individual element is a *MatInspector* weight matrix. The matches are shown as individual matches to the matrix and the total element score is given below each match.

binding sites can be quickly tested for significance, adding further information about the internal structure of the putative promoter in the sequence analyzed.

A library of approx 50 precompiled promoter modules (most of them represented by pairs of binding sites) all of which have been experimentally verified as functional modules is available together with the program *ModelInspector* professional from Genomatix as a commercial product. This works in principle like the *MatInspector* program but modules are between one and three orders of magnitude more specific than individual matrices, minimizing output files to very few matches within a sequence of several 10 kb in length.

The *FASTM*-based approach can detect sequences that contain the same functional module even in the absence of overall sequence similarity. Therefore, this is an extension of the conventional alignment-based analysis tools like *FASTA* or *BLAST*, which strictly rely on significant overall similarity of two sequence regions. The results of a model search yield important information about functional promoter elements if other promoters can be found that belong to functionally related genes.

4. Concluding Remarks

Analysis of transcriptional control sequences with the above-described tools is a bit similar to the so-called automated sequencing that was introduced several years ago. The tools facilitate and automate several important steps of the analysis. However, as with the sequencers, full success is only reached by experienced users who are familiar with the basics. There are no methods available yet that can achieve excellent results in the hands of an unexperienced user. However, if the user has a good knowledge of the principles of transcription control, these tools can efficiently boost the analysis of sequence data and reveal a lot which would otherwise require extensive experimental analysis.

The development of the WWW and the browser-oriented, user-friendly interfaces has overcome most of the difficulties in using the software. It will be worthwhile for users to learn how to interpret the results of these analyses which at this time cannot be done satisfactorily by automated tools.

Acknowledgments

This work was supported in part by the German Federal Ministry for Education and Research (BMBF) research ring project FANGREB (Functional Annotation of Genomic Regulatory Regions) 0311641 and by European Union grant BI04-CT95-0226, Transcription Databases and Analysis Tools (TRADAT).

References

1. Smale, S. T. (1997) Transcription initiation from TATA-less promoters within eukaryotic protein-coding genes. *Biochim. Biophys. Acta—Gene Struct. Expression* **1351,** 73–88.
2. Sauer, F. and Tjian, R. (1997) Mechanisms of transcriptional activation: differences and similarities between yeast, *Drosophila*, and man. *Curr. Opin. Genet. Develop.* **7,** 176–181.

3. Prestridge, D. S. (1996) SIGNAL SCAN 4.0: additional databases and sequence formats. *Comp. Appl. Biosci.* **12,** 157–160.
4. Quandt, K., Frech, K., Karas, H., Wingender, E., and Werner, T. (1995) Matlnd and Matlnspector: new fast and versatile tools for detection of consensus matches in nucleotide sequence data. *Nucleic Acids Res.* **23,** 4878–4884.
5. Lavorgna, G., Bonicelli, E., Wagner, A., and Werner, T. (1998) Detection of potential target genes in silico? *Trends Genet.* **14,** 375–376.
6. Quandt, K., Grote, K., and Werner, T. (1996) Genomelnspector: basic software tools for analysis of spatial correlations between genomic structures within megabase sequences. *Genomics* **33,** 301–304.
7. Quandt, K., Grote, K., and Werner, T. (1996) Genomelnspector: a new approach to detect correlation patterns of elements on genomic sequences. *Comp. Appl. Biosci.* **12,** 405–413.
8. Frech, K., Herrmann, G., and Werner, T. (1993) Computer-assisted prediction, classification, and delimitation of protein binding sites in nucleic acids. *Nucleic Acids Res.* **21,** 1655–1664.
9. Wolfertstetter, F., Frech, K., Herrmann, G., and Werner, T. (1996) Identification of functional elements in unaligned nucleic acid sequences by a novel tuple search algorithm. *Comp. Appl. Biosci.* **12,** 71–80.
10. Morgenstern, B., Dress, A., and Werner, T. (1996) Multiple DNA and protein sequence alignment based on segment-to-segment comparison. *Proc. Natl. Acad. Sci. USA* **93,** 12,098–12,103.
11. Morgenstern, B., Frech, K., Dress, A., and Werner, T. (1998) DIALIGN: finding local similarities by multiple sequence alignment. *Bioinformatics* **14,** 290–294.
12. Frech, K., Danescu-Mayer, J., and Werner, T. (1997) A novel method to develop highly specific models for regulatory units detects a new LTR in GenBank which contains a functional promoter. *J. Mol. Biol.* **270,** 674–687.
13. Frech, K., Quandt, K., and Werner, T. (1997) Finding protein-binding sites in DNA sequences: the next generation. *Trends Biochem. Sci.* **22,** 103–104.
14. Frech, K., Quandt, K., and Werner, T. (1997) Software for the analysis of DNA sequence elements of transcription. *Comp. Appl. Biosci.* **13,** 89–97.
15. Heinemeyer, T., Wingender, E., Reuter, I., Hermjakob, H., Kel, A. E., Kel, O. V., Ignatieva, E. V., Ananko, E. A., Podkolodnaya, O. A., Kolpakov, F. A., Podkolodny, N. L., and Kolchanov, N. A. (1998) Databases on transcriptional regulation: TRANSFAC, TRRD and COMPEL. *Nucleic Acids Res.* **26,** 362–367.

19

Computational Approaches for Gene Identification

Gautam B. Singh

1. Introduction

Genetics is gaining increasing significance as the discovery of new genes continues to have considerable impact in the field of medical sciences. The Human Genome Project is a multidisciplinary endeavor that aims at learning the identity of every single base stored in the human genome has been ongoing for some time now. The genome stores the blueprints for the synthesis of a variety of proteins—the macromolecules that enable an organism to be structurally and functionally viable. The blueprint or the program for the synthesis of a single protein is called a gene, a unit of the DNA sequence that is generally between 1×10^3–1×10^6 bp in length based upon the complexity of the protein that it codes for. A higher level eukaryote contains as many as 30,000–40,000 genes. It has been estimated that gene coding region accounts only for 10–20% of the genome. The gene identification problem is to recognize these regions from an anonymous sequence of DNA.

The earlier phases of the genomic research focussed on the construction of physical maps. However, the current emphasis has been shifting continually towards intensive sequencing. This has enabled us to study the structure and function of eukaryotic genes that may span tens or hundreds of kilobases. Only a few percent of the total gene-span actually code for the protein. This renders the detection of eukaryotic genes using the traditional approaches, i.e., those based on cDNA selection, exon trapping, and the random cloning of cDNA, to be quite laborious for sequences that are larger than a few kilobases. Consequently, genome sequencing centers routinely use computational approaches for exon prediction in addition to other means for detecting genes.

From: *Methods in Molecular Biology, vol. 132: Bioinformatics Methods and Protocols*
Edited by: S. Misener and S. A. Krawetz © Humana Press Inc., Totowa, NJ

The gene-finding tools have been proposed by theoretical biology researchers since early 1980s. These programs analyzed the DNA sequence and labeled a region to be potentially coding based upon its local codon usage, presence of ancient conserved patterns, or its significant deviations from the composition of a random sequence. A summary of the gene prediction software can be found in **refs.** ***1**,**2***, in which strengths and weaknesses of some of these have been outlined.

As scientists learn about the genetic basis in the etiology of many diseases, the quest for discovering new genes using computational methods continues to be more significant than ever before. Newer programs that utilize a wide variety of decision theoretic and machine learning technologies are being developed. This chapter provides a comprehensive summary of some of these approaches and describes the methods that are being used today to establish a software's performance in gene-prediction accuracy.

2. Material—Gene Identification Systems

This section provides a compendium of the currently used software systems for identification of genes in anonymous segments of DNA.

2.1. AAT*:* Analysis and Annotation Tool

This tool is used for identifying genes in a DNA sequence by comparing the sequence against protein and cDNA sequence databases. The *Analysis and Annotation Tool* (*AAT*) includes two pairs of programs ***(3)***, with each pair comprising of a database search and an alignment program. The first program pair is designed to compare the query sequence to the protein database, whereas the second pair performs a similar comparison against cDNA databases. The alignment programs construct a consensus of all sequence database alignments into a multiple sequence alignment to enhance the predictions of splice junctions. The sequence alignments that score low are filtered out from the results and the final protein and cDNA alignments are combined and presented to the user.

The first program pair compares a query DNA sequence against a protein database using two programs called *DPS* and *NAP*. The *DPS* program is used for computing high-scoring chains of segment pairs between the query DNA sequence and a protein database. The global alignment program *NAP* finds the optimal alignment between a DNA and the matching protein sequence ***(4)***. The alignment model for *NAP* accommodates introns and frameshifts within codons, and is thus able to identify the exact locations of introns using the (GT) and (AG) consensus for splice-site identification. The second program pair, comprised of *DDS* and *GAP*, is used for comparing the query DNA sequence against a cDNA library. The *DDS* program is an improvement over the *BLASTN* program *(5)*. The *GAP* program is a global alignment program that is sufficiently powerful for aligning a DNA sequence containing introns to a cDNA sequence *(6)*.

One of the goals of the AAT is to help in an automatic annotation of DNA sequences. This task has been done manually traditionally, where the alignments between the coding regions of a DNA sequence and the existing proteins is established by *BLASTX* and linked to the sequence as an annotation in a post hoc manner. This helps in providing a clue for the functional significance of a given gene as is evident in the function of the related protein sequences. The *AAT* on the another hand, performs such an alignment and is able to display it as the basis for predicting genes. Furthermore, the alignment produced by *BLASTX* is prone to frameshift errors. This shortcoming is overcome by *AAT* by the development of a customized program for DNA–protein sequence alignment.

2.2. MZEF: Michael Zhang's Exon Finder

MZEF is an internal coding exon prediction program. It utilizes the method of quadratic discriminant analysis, or QDA, for the purpose of describing the distributions of exons and pseudoexons. This method is an extension of the statistical pattern recognition ideas earlier presented in linear discriminant analysis, or LDA, used in *HEXON (7)* (*FGENEH* in *GeneFinder* is an improvement on *HEXON*). In fitting a QDA, the surface that separates the distribution of exons and pseudoexons can be more accurately approximated.

An overview of the algorithm is as follows: Each potential exon that matches the template of AG → ORF → GT is analyzed. The exons that meet a minimum length criteria are next considered to be putative exons and must be separated from the pseudoexons. The putative exons are represented using a nine-value feature vector, comprised of parameters such as, exon length, branch score, and various differences between the hexamer frequency preferences on the two sides of the donor and acceptor splice sites. This nine dimensional feature vector χ is categorized to be an exon or a pseudoexon based on the following log ratio test derived from QDA:

$$\eta = \log\left(\frac{p_1}{p_2}\right) = \log\left(\frac{p_1^0}{p_2^0}\right) - \left(\frac{\delta_1 - \delta_2}{2}\right) - \left[\frac{1}{2} \quad \log \quad \frac{|\Sigma_1|}{|\Sigma_2|}\right] \tag{1}$$

The parameters used in the above equation are as follows: μ_i and Σ_i represent the group mean and covariance matrices obtained from the training sets (the training set was comprised of 1879 true exons and 184,217 pseudoexons). The quantities p_1^0 and p_2^0 denote the prior probability for a putative exon for membership into the group G_1 of true exons, or to the group G_2 of the pseudoexons.

The quantity, $\delta_i = (\chi - \mu_i)^T\Sigma_i^{-1} (\chi - \mu_i)$ is the squared Mahalanobis distance between the observed feature vector χ and μ_i; $|\Sigma_i|$ is the determinant of the covariance matrix Σ_i *(8)*.

2.3. GENSCAN

GENSCAN works by building a probabilistic model of the gene structure of human genomic sequences and applying this model to the problem of gene prediction. The probabilistic model of a gene includes the specific compositional and functional units of a eukaryotic gene, including exons, introns, splice sites, promoters, and the polyadenylation signals. The occurrence of a partial set of these units and the representation of a partial gene is supported by the implementation of the model search algorithm. Furthermore, the predictions made by the program are not a mere reflection of the types of genes that are found in the protein databanks, but rather an independent evaluation that provides information that complements our existing knowledge.

The modeling of a DNA sequence by *GENSCAN* is based on a generalized hidden Markov model (GHMM), that uses a double-stranded DNA and can find occurrence of multiple genes in a single sequence, on either one or both DNA strands. A simplistic representation of such a HMM is shown in **Fig. 1**, where the arcs represent a set of nucleotides belonging to a class and the nodes represent the DNA regions of transition from one class of nucleotides to another. The program's ability to model functional signals and their interrelationships in a natural manner using the maximal dependence decomposition method is instrumental in providing it the strength for a generalized gene detection task. The text output of the program is a list of one or more (or possibly zero) predicted genes and peptide sequences, whereas the graphical outputs provide a representation of the relative locations of the predicted exons. Versions of the program that are suitable for vertebrate, maize and *Arabidopsis* sequences are currently available *(9)*. (The vertebrate version works well on *Drosophila* sequences.)

2.4. Veil

VEIL (the *Viterbi Exon–Intron Locator*) is based on the observation that the hidden Markov models (HMMs), provide a precise probabilistic method for modeling sequences of discrete data. Consequently, it uses a custom-designed HMM to segment uncharacterized genomic DNA sequences into exons, introns, and intergenic regions. The exon–HMM module is designed to capture the regularities in codon usage and periodicity that appear in the exons as well as to rule out in-frame stop codons. A similar module represents the intron–HMM. The HMM models for the probabilistic representation of the splice sites resemble a pipeline, as these signals are of a well-defined length. For example,

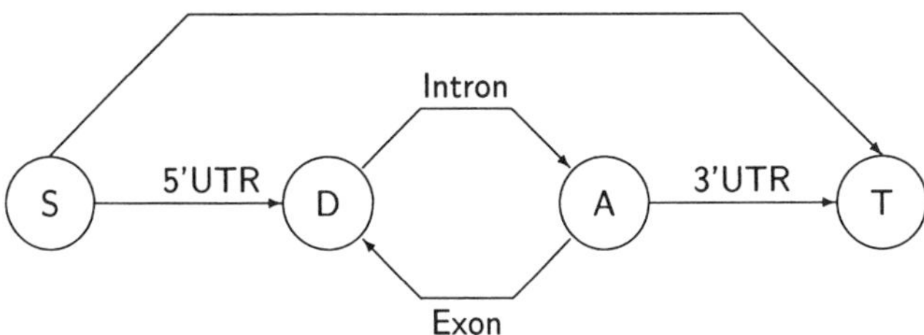

Fig. 1. A simple hidden Markov model for a multiple-exon gene. The arcs represent the DNA sequence belonging to a functional unit of the gene, whereas the nodes represent the occurrence of specific signals that transition the functional state or the sequence class that the model is scanning from 5′→3′. The nodes *S* and *T* represent the start and terminating codons, respectively, whereas the nodes *D* and *A* denote the donor and acceptor sites. The single arc from *S* to *T* enables the HMM to recognize a single exon gene. Such a model can learn the probabilities of the various state transitions using a set of DNA sequences representing genes.

the donor-acceptor site models in *VEIL* are comprised of 9 and 15 stages, respectively. Other HMMs included those for the start codon, and the polyadenylation signal AAATAA, and intergenic regions that are upstream of the start codon. These simpler models were put together into the overall gene model with the schematic similar to the one shown in **Fig. 1**. The final HMM is comprised of 241 states and 1003 edges. The representation of a gene in *GENSCAN* and *VEIL* and Genie are quite similar—with the differences being manifested in the representation of the individual segments of a gene ***(10)***.

After the determination of these regions, the Viterbi algorithm is used for parsing the query sequence into its component exons and introns. Such an algorithm is based on dynamic programming techniques in which the most likely set of states that a given sequence is expected to traverse are determined. In addition, the probability that the model will produce a given sequence is computed by the Viterbi algorithm. This represents the probability that a given DNA sequence contains a gene. Glimmer is a corresponding program designed with a similar idea of an interpolated hidden Markov model and is applicable for analysis of microbial genomes ***(11)***.

2.5. Morgan

The *MORGAN* (*Multiframe Optimal Rule-Based Gene Analyzer*) distinguishes itself from the other systems for finding gene by using a decision tree classifier, a technique that is derived from the statistics community. Decision trees are often utilized for a variety of classification tasks, such as cancer diagnosis, speech understanding, image understanding, and optical character recognition. Decision trees are often applied to objects represented in terms of their features. For example, the representation of an object in a *d*-dimensional feature space may be denoted as $f_1, f_2 \ldots, f_d$. Subsequently, the knowledge about the classification process is embedded into a tree-like structure and by performing a series of tests, generally of the form $f_k < T$, the identity of an unknown object is established. Thus, each question node in the decision tree corresponds to a linear discriminant, and helps in partitioning the search space into a set of compartments or leaves which individually represent an entity that we are interested in classifying. Similarly, there were two partitions created by *MORGAN* in gene classification. These are labeled as C for coding and N for noncoding.

The *MORGAN* system was built using a 19-feature set. These included features that measure the longest ORF, the dicodon usage, two hexamer frequencies, codon usage, position asymmetries for A,C,T,G and Fourier coefficients for periods 2–9. The decision tree was trained on 290,628 examples, and each example comprised of 54 bp of human DNA sequence. Using these examples, and knowing their classification *a priori*, the *MORGAN* decision tree was trained. It resulted in 20 leaf and 19 test or intermediate nodes. For a given DNA sequence that needs to be classified as coding or noncoding, the 19 features are first computed for its subsequences. The optimal segmentation is dependent on a separate scoring function that takes a subsequence and assigns to it a score reflecting the probability that the sequence is an exon. Each subsequence is scored by a decision tree and the individual scores are combined to give a probability estimate ***(12,13)***.

2.6. Genie

A Generalized Hidden Markov Model or GHMM is general enough to enable the generation of subsequences of DNA symbols at each state of a hidden Markov chain. The model required to produce a subsequence at each state of an GHMM (as shown in **Fig. 1**) can be quite complex and in turn be another HMM. In *Genie*, each component is designed and trained independently and combined into a modular system. The length distributions of introns and exons in the training set are used to learn the average length (and variance) for a string generated by a state in GHMM. The donor and acceptor sites are recognized by a neural network that uses a 15-bp window for donor and a 41-bp window for acceptor sites. Thus, the nodes representing the donor and acceptor sites in **Fig. 1** have a neural network embedded in them that returns the posterior probabilities for a given position to be a donor or acceptor site. After the construction of such a model, it is trained to learn the probabilities of transitions between states in the GHMM and for the generation of each nucleotide base given a particular state. Machine learning techniques are applied to optimize these probabilities using a standardized gene data set ***(14)***.

2.7. GeneFinder (FGENEH)

The *GeneFinder* is a suite of tools available from Baylor College of Medicine. The tools that are of interest from a standpoint of gene identification are, *fgeneh* for prediction of gene structure, *fexh* for 5′, internal, and 3′ exon prediction, *hspl* for splice-site prediction, and *hexon* for prediction of internal exons in human DNA.

The algorithm used in *fgeneh* begins by predicting all possible potential internal exons, and potential 5′ and 3′ exons. Each exon is described using a linear discriminant function that combines the contextual features of these exons. Next, the method of dynamic programming is used to search for the optimal combination of these exons that yields a gene model. The algorithm used in *hexon* is also based on the discriminant analysis of open reading frames flanked by GT and AG base pairs. Prediction is performed by linear discriminant function that combines characteristics of the donor and acceptor splice sites, 5′ and 3′ intron regions and the properties of the coding region of the open reading frame. This program can only predict internal exons with GT and AG conserved base pairs for donor and acceptor splice sites. However, this usually include more than 99% of the all authentic splice sites. It is expected that the future versions of this program will have option for other variants including an extension to other species ***(7,15)***.

2.8. GeneParser

The *GeneParser* program looks for gene specific features in the given sequence and subsequently applies dynamic programming (DP) to form their combinations to obtain a configuration that maximizes a likelihood function. Specifically, the anonymous DNA sequence is scanned for finding the locations of splice sites as well as the computation of content parameters such as codon usage, local compositional complexity, 6-tuple frequency, length distribution, and periodic asymmetry. The program scores all subintervals in a sequence for content statistics indicative of introns, exons, and their boundaries. The content statistics are fed into a neural network that provides a log-likelihood estimate that the given subinterval exactly represents an intron, first exon, internal exon, or the last exon. (Weights for the feed-forward neural network are optimized to maximize the number of correct predictions.) A dynamic programming algorithm is then applied to this classification of each subinterval so that the overall likelihood of a gene model is maximized. The DP algorithm operates under the constraints that introns and exons must be adjacent and nonoverlapping. The highest-scoring combination the groups of high-scoring combinations represents the maximally likely model of a gene ***(16,17)***.

2.9. GeneLang

GeneLang is a syntactic pattern recognition system, which uses the tools and techniques of computational linguistics to find genes and other higher-order features in biological sequence data. For example, formal language theory uses a set of rules, called the grammar, to define the valid sets of strings over a given alphabet. The motivation for building a gene grammar is the availability of parsers that can recognize whether the input string satisfies the rules specified by a gene grammar. Thus, in *GeneLang*, the patterns over the DNA alphabet are described using a set of rules and a general purpose parser, implemented in the logic programming language Prolog ***(18)***. The system treats components of a gene such as donor and acceptor sites, introns and exons, start and stop codons, etc., as words that are formed on the four-character DNA alphabet. The gene is next in the level of this syntactic hierarchy, i.e., the gene is a valid sentence formed by these words. Genes in a DNA sequence can be recognized in a manner analogous to the recognition of grammatically correct sentences in English language.

2.10. Grail

The *GRAIL* gene identification systems utilize a neural network for the recognition of genes. The neural networks in *GRAIL 1*, *GRAIL 1a*, and *GRAIL 2* are trained to combine the results from a number of coding predictors. *GRAIL 1* has been in place for more than six years. Using a neural network that recognizes coding potential within a fixed size window of 100 bp, the coding potential is computed without taking into consideration the information about additional features such as the splice junction ***(19,20)***. *GRAIL 1a* also uses fixed-length windows to first locate potential coding regions. It then evaluates a number of discrete candidates of different lengths around each potential coding region. The information from the two 60-base regions adjacent to that coding region are analyzed to find the correct boundaries for a coding region. *GRAIL 2* uses variable-length windows for each potential open reading frame bounded by a pair of start/donor, acceptor/donor, or acceptor/stop sites. Thus, *GRAIL 2* uses genomic context information to score the coding regions and is therefore not appropriate for sequences where the regions adjacent to an exon are absent ***(21,22)***. However, these changes have improved *GRAIL 2*'s overall performance, particularly for short exons. All three systems have been trained to recognize coding regions in human DNA sequences, although they also work well on a number of other organisms, particularly other mammals.

GRAIL is accessible by several methods, including an e-mail server at Oak Ridge National Laboratory (ORNL), which processes DNA sequence(s) contained in e-mail messages. An interactive graphical X-based client-server system called *Xgrail* is also available from ORNL. *Xgrail* supports a wide range of analysis tools, including tools for gene modeling and database search.

2.11. Other Tools for Gene Identification

A dicodon statistic is used for the prediction of splice signals and coding regions by the *GenView* system. This system aims at minimizing the number of nonreal exons in analyzed DNA *(23,24)*. *GeneBuilder* system is also based on prediction of splice signals and coding regions by dicodon statistics. Potential gene structure is constructed using dynamic programming approach *(25)*. The *XPound* system utilizes a probabilistic model for detecting coding regions in DNA sequences *(26)*. *GeneID* is a mail-server-based system for analyzing vertebrate genomic DNA and prediction of exons and gene structure by finding potential exons and using a rule-base to assemble these into genes *(27)*. The *PROCRUSTES* system is based on the spliced alignment algorithm and explores all possible exon assemblies and chooses the one that best fits a related protein *(28)*. The *GENMARK* system uses the HMM for exons and introns to find the locations of coding regions *(29)*.

3. Methods—Comparison of the Gene Identification Algorithms

To compare the performance of gene-finding systems, performance comparison metrics were defined and a dataset was created by Burset and Guigo *(17,30)*. This dataset is comprised of sequences from *GenBank* release 85.0 since January 1993. Thus, it was comprised of relatively new entries. The process by which this dataset was constructed is as follows.

First, Burset and Guigo collected vertebrate protein coding sequence from *GenBank*. Next, in a series of quality control steps, they removed all entries with pseudogenes, with in-frame stop codons, with no introns (essentially the cDNA sequences), and with nonstandard splice junctions (i.e., those that did not have the GT and AG at the beginning and end of an intron). They further removed the immunoglobulins and histocompatibility antigens, and were finally left with 570 complete sequences, each with exactly one gene and at least one intron. There are a total of 2649 exons and 2079 introns in this set. This data is public and is available from

```
ftp://www-hgc.lbl.gov/pub/genesets/OtherDataSets/
  GUIGO_96
```

3.1. Parameters for Performance Comparison

The parameters described below are utilized for the computation of accuracy statistics at the nucleotide level. Each nucleotide in the sequence being analyzed is classified as predicted positive (PP) if it is in a predicted coding region, predicted negative (PN) otherwise. The nucleotide's actual positive (AP) or actual negative (AN) value is also known according to the sequence annotation. These assignments are then compared to calculate the number of true positives (TP), false positives (FP), true negatives (TN), and false negatives (FN). The sensitivity and specificity of the prediction program is given by **Eq. 2** and **3**.

$$\text{Sensitivity: } Sn = \frac{\text{TP}}{\text{AP}} \tag{2}$$

$$\text{Specificity: } Sp = \frac{\text{TP}}{\text{PP}} \tag{3}$$

and Approximate Correlation, AC, is defined by **Eq. 4**.

$$AC = \frac{\left(\frac{TP}{TP+FN}\right) + \left(\frac{TP}{TP+FP}\right) + \left(\frac{TN}{TN+FP}\right) + \left(\frac{TN}{TN+FN}\right)}{2} - 1 \tag{4}$$

At the exon level, predicted exons (PE) are compared to annotated exons (AE). True exons (TE), is the number of predicted exons which are exactly identical to an annotated exon (i.e., both endpoints correct). The sensitivity and specificity at the exon level is given by **Eq. 5** and **6** respectively.

$$\text{Sensitivity: } Sn = \frac{TE}{AE} \tag{5}$$

$$\text{Specificity: } Sp = \frac{TE}{PE} \tag{6}$$

The average of Sn and Sp is typically used as an overall measure of accuracy at the exon level in lieu of a correlation measure. Two additional accuracy measures are also calculated at the exon level: missing exons (ME), the fraction of annotated exons not overlapped by any predicted exon; and wrong exons (WE), the fraction of predicted exons not overlapped by any true exon. Accuracy measures for a set of sequences are calculated by averaging the values obtained for each sequence separately, the average being taken over all sequences for which the measure is defined.

3.2. Performance

The performance of the various tools discussed in this review on the above dataset is shown in **Table 1**. The presentation was ordered in the tools performance as evident in the *Sn* parameter for the correctly predicted nucleotides. However, it may sometimes be more realistic to compare a tool's performance

Table 1
Accuracy of the Various Gene Identification Systems on Burset and Guigo *(30)* Data Set Comprised of 570 Vertebrate Sequences

	Predicted Nucleotides			Predicted Exons		
Method	Sn	Sp	AC	Sn	Sp	$\frac{Sn+Sp}{2}$
AAT	0.94	0.97	0.95	0.74	0.78	0.76
GENSCAN	0.93	0.93	0.91	0.78	0.81	0.80
MZEF	0.88	0.95	0.90	0.84	0.92	0.88
VEIL	0.83	0.72	0.73	0.53	0.49	0.51
MORGAN	0.82	0.80	0.78	0.58	0.54	0.56
Genie	0.78	0.84	0.77	0.61	0.64	0.63
GeneFinder	0.77	0.85	0.78	0.61	0.61	0.61
GeneID	0.63	0.81	0.67	0.44	0.45	0.45
GeneParser2	0.66	0.79	0.66	0.35	0.39	0.37
GeneLang	0.72	0.84	0.75	0.50	0.49	0.50
GRAIL-II	0.72	0.84	0.75	0.36	0.41	0.38
Xpound	0.61	0.82	0.68	0.15	0.17	0.16

based on $\frac{\text{Sn+Sp}}{2}$ for the predicted exons. In such a comparison, *MZEF*, *GENSCAN*, and *AAT* are optimum.

4. Conclusions

The prediction of human protein-coding genes in newly sequenced DNA becomes very important in large genome sequencing projects. Many of the systems continue to be developed in this field with the goal of increasing the reliability of identifying genes in DNA. Although the systems developed so far are not completely accurate, the results that they provide are vital to keep pace with the rapid analysis of sequence data in this age of high-throughput genomic sequencing. Some of the problems that are faced so far stem from the inadequacy to accurately predict the intron/exon boundaries, which in turn results in an inadequacy to identify the eukaryotic gene structure and a poor performance on most of the short exons. Furthermore, a large number of false splice site predictions eventually lower the reliability of the predicted exons. Because of the different types of processing performed by the various gene-identification systems, it therefore seems plausible to examine their results in a combined manner and look for a region of coding consensus produced by the various gene identification systems specified in **Table 2**. Furthermore, understanding the methodology adopted for a given method will be vital to explain the results produced by a specific method.

Table 2
A Listing of Gene Identification Resources on the WWW

Method	Site
AAT	`http://genome.cs.mtu.edu/aat.html`
GENSCAN	`http://gnomic.stanford.edu/GENSCANW.html`
MZEF	`http://sciclio.cshl.org/genefinder/`
VEIL	`http://www.cs.jhu.edu/labs/compbio/veil.html`
MORGAN	`http://www.cs.jhu.edu/labs/compbio/morgan.html`
Genie	`http://www-hgc.lbl.gov/inf/genie.html`
GeneFinder	`http://dot.imgen.bcm.tmc.edu:9331/gene-finder/gf.html`
GeneID	`http://www.imim.es/GeneIdentification/Geneid/geneid_input.html`
GeneParser2	`http://beagle.colorado.edu/~eesnyder/GeneParser.html`
GeneLang	`http://cbil.humgen.upenn.edu/~sdong/genlang.html`
GRAIL-II	`http://compbio.ornl.gov/Grail-bin/EmptyGrailForm`
SORFIND	`http://www.rabbithutch.com/`
PROCRUSTES	`http://www-hto.usc.edu/software/procrustes/index.html`
GenView	`http://www.itba.mi.cnr.it/webgene/`

References

1. Singh, G. B. and Krawetz, S. A. (1994) Computer based EXON detection: an evaluation metric for comparison. *Intl. J. Genome Res.* **1,** 321–338.
2. Fickett, J. (1996) Finding genes by the computer: the state of the art. *Trends Genet.* **12,** 316–320.
3. Huang, X., Adams, M. D., Zhou, H., and Kerlavage, A. R. (1997) A tool for analyzing and annotating genomic sequences. *Genomics* **46,** 37–45.
4. Huang, X. and Zhang, J. (1996) Methods for comparing a DNA sequence with a protein sequence. *Comput. Applic. Biosci.* **12,** 497–506.
5. Altschul, S., Gish, W., Miller, W., and Myers, E. (1990) A basic local alignment search tool. *J. Mol. Biol.* **215,** 403–410.
6. Huang, X. (1994) On global sequence alignment. *Comput. Applic. Biosci.* **10,** 227–235.
7. Solovyev, V., Salamov, A., and Lawrence, C. (1994) The prediction of human exons by oligonucleotide composition and discriminant analysis of spliceable open reading frames, in *Proc. 2nd Intl. Conf. on Intelligent Systems in Molecular*

Biology, Altman, R., Brutlag, D., Karp, R., Latrop, R., and Searls, D., eds.) AAAI Press, Menlo Park, CA, pp. 354–362.

8. Zhang, M. (1997) Identification of protein coding regions in the human genome based on quadrati discriminant analysis. *Proc. Natl. Acad. Sci. USA* **94,** 565–568.
9. Burge, C. and Karlin, S. (1997) Prediction of complete gene stuctures in human genomic DNA. *J. Mol. Biol.* **268,** 78–94.
10. Henderson, J., Salzberg, S., and Fasman, K. (1997) Finding genes in human DNA with a hidden Markov model. *J. Comp. Biol.* **4,** 127–141.
11. Salzberg, S., Delcher, A., Kasif, S., and White, O. (1998) Microbial gene identification using interpolated markov models. *Nucleic Acid Res.* **26,** 544–548.
12. Salzberg, S. (1995) Locating protein coding regions in human DNA using a decision tree algorithm. *J. Comp. Biol.* **2,** 473–485.
13. Salzberg, S., Delcher, A., Fasman, K., and Henderson, J. (1997) A decision tree system for finding genes in DNA. Technical Report 1997-03, Department of Computer Science, Johns Hopkins University, March 1997.
14. Kulp, D., Haussler, D., Reese, M. G., and Eeckman, F. H. (1996) A generalized hidden Markov model for the recognition of human genes in DNA, in *Proc. 4th Conf. on Intelligent Systems in Molecular Biology*, June 1996. St. Louis, MO (States, D., Agarwal, P., Gaasterland, T., Hunter, L., Smith, R., eds.), AAAI Press, Menlo Park, CA.
15. Solovyev, V., Salamov, A., and Lawrence, C. (1994) Predicting internal exons by oligonucleotide composition and discriminant analysis of spliceable open reading frames. *Nucleic Acid Res.* **22,** 5156–5163.
16. Snyder, E. E. and Stormo, G. D. (1993) Identification of coding regions in genomic DNA sequences: an application of dynamic programming and neural networks. *Nucleic Acid Res.* **21,** 607–613.
17. Snyder, E. E. and Stormo, G. D. (1995) Identification of coding regions in genomic DNA. *J. Mol. Biol.* **248,** 1–18.
18. Dong, S. and Searls, D. B. (1994) Gene structure prediction by linguistic methods. *Genomics* **23,** 540–551.
19. Uberbacher, E. and Mural, R. (1991) Locating protein coding regions in human DNA sequences using a multiple-sensor neural network approach. *Proc. Natl. Acad. Sci. USA* **88,** 11,261–11,265.
20. Uberbacher, E. and Mural, R. (1991) GRAIL seeks out genes buried in DNA sequence. *Science* **254,** 805.
21. Uberbacher, E., Xu, Y., and Mural, R. (1996) Discovering and understanding genes in human DNA sequence using GRAIL. *Comp. Meth. Macromol. Seq. Anal.* **266,** 259–281.
22. Xu, Y. and Uberbacher, E. (1997) Automated gene identification in large-scale genomic sequences. *J. Comp. Biol.* **4,** 325–338.
23. Milanesi, L., Kolchanov, N., Rogozin, I., Ischenko, I., Kel, A., Orlov, Y., Ponomarenko, M., and Vezzoni, P. (1993) GenView: a computing tool for protein-coding regions prediction in nucleotide sequences, in *Proc. 2nd. Intl. Conf. on Bioinformatics, Supercomput. and Complex Genome Analysis* (Lim, N.,

Fickett, J., Cantor, C., and Robbins, R. J., eds.) World Scientific Publishing, Singapore, pp. 573–588.

24. Milanesi, L., Kolchanov, N., Rogozin, I., Kel, A., and Titov, I. (1993) Sequence functional inference, in *Guide to Human Genome Computing* (Bishop, M. J., ed.) Academic, New York, pp. 249–312.
25. Rogozin, I. B., Milanesi, L., and Kolchanov, N. A. (1996) Gene structure prediction using information on homologous protein sequence. *Comput. Applic. Biosci.* **12,** 161–170.
26. Thomas, A. and Skolnick, M. (1994) A probabilistic model for detecting coding regions in DNA sequences. *IMA J. Math. Appl. Med. Biol.* **11,** 149–168.
27. Guigo, R., Knudsen, S., Drake, N., and Smith, T. (1992) Prediction of gene structure. *J. Mol. Biol.* **226,** 141–157.
28. Gelfand, M., Mironov, A., and Pevzner, P. (1996) Gene recognition via spliced sequence alignment. *Proc. Natl. Acad. Sci. USA* **93,** 9061–9066.
29. Borodovsky, M. and McIninch, J. (1993) GENMARK: parallel gene recognition for both DNA strands. *Comp. Chem.* **17,** 123–133.
30. Burset, M. and Guigo, R. (1996) Evalution of gene structure prediction programs. *Genomics* **34,** 353–367.

20

Primer3 on the WWW for General Users and for Biologist Programmers

Steve Rozen and Helen Skaletsky

1. Introduction

Designing PCR and sequencing primers are essential activities for molecular biologists around the world. This chapter assumes acquaintance with the principles and practice of PCR, as outlined in, for example, **refs.** ***1–4***.

Primer3 is a computer program that suggests PCR primers for a variety of applications, for example to create STSs (sequence tagged sites) for radiation hybrid mapping *(5)*, or to amplify sequences for single nucleotide polymorphism discovery *(6)*. *Primer3* can also select single primers for sequencing reactions and can design oligonucleotide hybridization probes.

In selecting oligos for primers or hybridization probes, *Primer3* can consider many factors. These include oligo melting temperature, length, GC content, 3′ stability, estimated secondary structure, the likelihood of annealing to or amplifying undesirable sequences (for example interspersed repeats), the likelihood of primer–dimer formation between two copies of the same primer, and the accuracy of the source sequence. In the design of primer pairs *Primer3* can consider product size and melting temperature, the likelihood of primer–dimer formation between the two primers in the pair, the difference between primer melting temperatures, and primer location relative to particular regions of interest or to be avoided.

1.1. Primer3 *Can Be Used Through its WWW Interface or as a Software Component*

Most casual users will prefer *Primer3*'s WWW interface (**Fig. 1**), which is suitable for selecting primers from a few sequences. **Subheading 2.** discusses this interface in detail.

From: *Methods in Molecular Biology, vol. 132: Bioinformatics Methods and Protocols*
Edited by: S. Misener and S. A. Krawetz © Humana Press Inc., Totowa, NJ

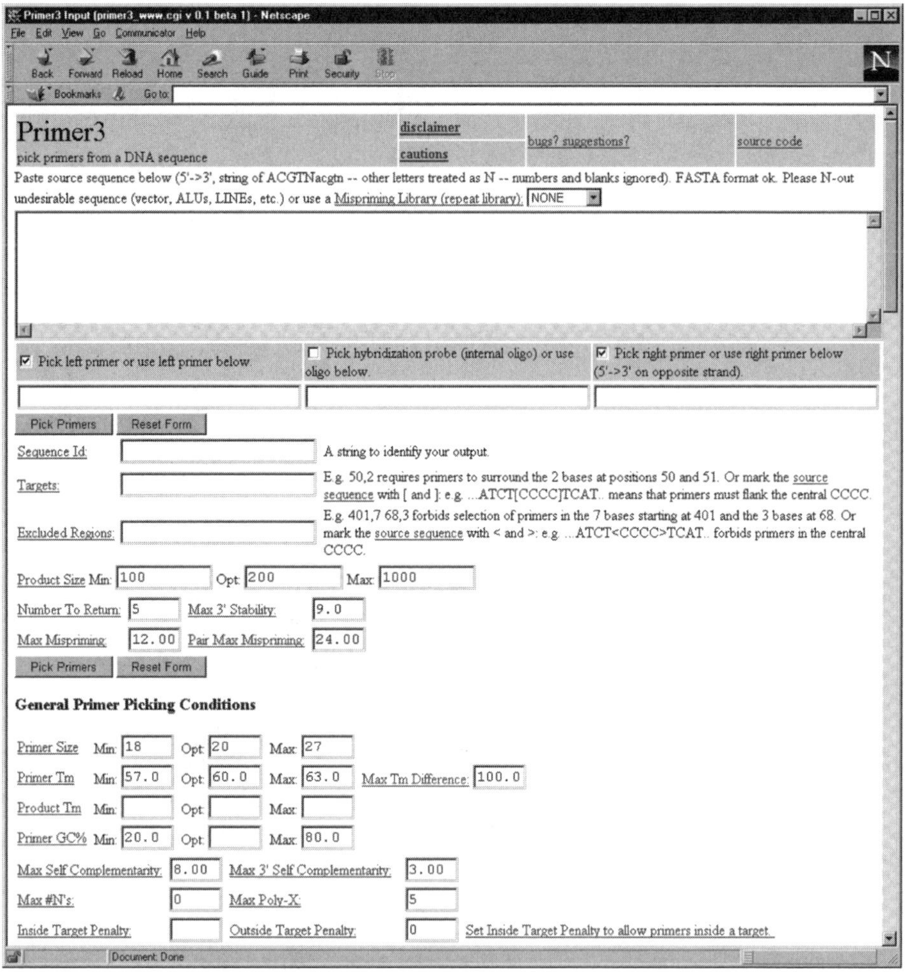

Fig. 1. Top of *Primer3*'s WWW input page without user input.

Scientists who must select primers for hundreds or thousands of sequences will prefer to use *Primer3* (specifically the *primer3_core* program) as a software component, which accepts input in a format convenient for other programs to produce and generates output in a format convenient for other programs to interpret. (We assume that no one would want to deal manually with primer picking results for hundreds or thousands of sequences.) We present examples of the use of *primer3_core* as a software component in **Subheading 3.** The underlying primer design process is identical for both the WWW interface and *primer3_core*, and in fact the WWW interface uses the WWW CGI protocol ***(9,10)*** layered on top of *primer3_core*.

In a few cases it will be preferable to modify *primer3_core* itself rather than simply use it as a software component. Therefore, for maximum portability and modifiability we wrote it in standard ANSI C *(7)* using standard POSIX calls *(8)* with simple and universally usable ASCII input and output. Furthermore, the distribution includes a thorough set of tests for *primer3_core*, which make it relatively easy to ensure that modifications do not introduce errors.

Although the WWW interface is adequately self-explanatory for many casual users, for others the background information we present here will be helpful. For potential high-volume users, customizers, and biologist programmers, this chapter introduces the use of *primer3_core* to streamline the particular primer-picking tasks at hand.

1.2. Where to Find Primer3

Public WWW interfaces for use by anyone with a Web browser (for example *Netscape*) are reachable from `http://www.genome.wwi.mit.edu/cgi-bin/primer/info.cgi`. You can also download the *Primer3* programs from this location. The program *primer3_core* is available in source form only, and generating an executable program requires a C compiler; *see* **Subheading 3.1.** for details. The source code for the WWW interface is also available and can be used on computers running a Web server. The WWW interface (like *primer3_core*) can be modified to meet the needs of particular sets of end-users. It is written in *perl*, the language of choice for this sort of application *(11)*.

1.3. What Primer3 *Does not Offer*

We regret that we do not have resources to distribute *Primer3* in ready-to-run executable form, with "native" front ends (e.g., for Microsoft Windows or Mac), or on tape, diskette, or CD. Other primer selection software is available in fully supported commercial form (though possibly not as a customizable software component). Examples include *OLIGO*®, available through Molecular Biology Insights, Inc., Cascade, Colorado, `http://www.mbinsights.com/`, (*see* also **ref.** ***12***), *DNA*Star's *PrimerSelect* module for *LaserGene* (`http://www.dnastar.com/`), and the *Prime* module in Genetics Computer Group's Wisconsin Package. Primer selection programs available from academic institutions include *Primer 0.5* (upon which *Primer3* was based, but available as a stand-alone program and as a ready-to-run executable for Macs and PCs) *(13)*, and *OSP–oligonucleotide selection program* ***(14)***.

The following tasks are *not* built in to *Primer3*:

- Automatically adding standard 5′ tails to each primer.
- Selecting nested primer pairs.

- Selecting primer pairs for multiplex amplification.
- Designing a tiling of amplicons across a sequence.
- Picking primers from a reverse-translated amino acid sequence.

(However, we have used *primer3_core* as a software component in conjunction with other codes to accomplish each of these tasks but the last.) The packages mentioned above perform some of these tasks.

1.4. PCR and Primer Design Applications Are Diverse

Primer design is really many different problems. Sometimes one wishes to design primers for a large number of sequences, and if for some reason it is difficult to find good primers for a particular source sequence one simply discards that source sequence. An example would be high-throughput whole genome mapping (the application for which *Primer3* and its predecessors were originally designed). In this case one designs STSs from tens of thousands of sequences and then uses these STSs in hundreds of amplifications. In this application no one sequence is particularly valuable compared to the cost of primers and subsequent amplifications, so it is not worth proceeding with a sequence for which there are only dubious primer pairs.

In other applications one wishes to design primers to amplify a *particular* sequence if at all possible; if there are no obvious good primers one will choose several possibilities in the hope that at least one will work. Examples include designing primers to distinguish two very similar sequences or to amplify a particular exon flanked by a CpG island in which it is hard to find a good primer. In this situation the precious resource is the particular sequence to amplify, and the scientist is willing to spend considerable effort getting a clean amplicon.

There are many other variables in primer-design goals. Sometimes one wants large amplicons (for example to amplify as much of a cDNA as possible), and sometimes one wants very short amplicons (for example to flank a single nucleotide polymorphism as closely as possible). Sometimes the amplification template is complex (for example a mammalian genome), and sometimes it is simple (for example a single bacterial artificial chromosome). Some *Taq* formulations are less likely than others to produce primer dimers or self-priming hairpins.

Because primer design is really many different problems *Primer3* gives users numerous options to specify which primers are acceptable and which primers are better than others. The number of these options can be overwhelming to new and experienced users alike, but typically for any particular application only a few need changing from default values.

This chapter cannot discuss all of *Primer3*'s options, but it covers those you are most likely to change. The WWW interface and the **README** distributed with the program document the more esoteric options.

Primer3 Input (primer3_www.cgi v 0.1 beta 1) - Netscape

File Edit View Go Communicator Help

Back Forward Reload Home Search Guide Print Security Stop

Bookmarks Go to:

Primer3
pick primers from a DNA sequence

disclaimer | cautions | bugs? suggestions? | source code

Paste source sequence below (5'->3', string of ACGTNacgtn -- other letters treated as N -- numbers and blanks ignored). FASTA format ok. Please N-out undesirable sequence (vector, ALUs, LINEs, etc.) or use a Mispriming Library (repeat library): RODENT

```
gcaacagtga agttttcttt catctggttc ccttcctccc taggcattgg caagcataag
tcatgtggcc catgtcacta ttacaaccaa aacagctgat gggaaatgtg cgtacaggta
tgtgataatg acctctgtgt ccaccctaaa catacgtgtt attccccttg acccccgcta
```

☑ Pick left primer or use left primer below.
☐ Pick hybridization probe (internal oligo) or use oligo below.
☑ Pick right primer or use right primer below (5'->3' on opposite strand).

Pick Primers | Reset Form

Sequence Id: Example Sequence 1 — A string to identify your output.

Targets: 40, 78 — E.g. 50,2 requires primers to surround the 2 bases at positions 50 and 51. Or mark the source sequence with [and]: e.g. ...ATCT[CCCC]TCAT.. means that primers must flank the central CCCC.

Excluded Regions: — E.g. 401,7 68,3 forbids selection of primers in the 7 bases starting at 401 and the 3 bases at 68. Or mark the source sequence with < and >: e.g. ...ATCT<CCCC>TCAT.. forbids primers in the central CCCC.

Product Size Min: 100 Opt: 200 Max: 1000

Number To Return: 5 Max 3' Stability: 9.0

Max Mispriming: 12.00 Pair Max Mispriming: 24.00

Pick Primers | Reset Form

General Primer Picking Conditions

Primer Size Min: 18 Opt: 20 Max: 27

Primer Tm Min: 57.0 Opt: 60.0 Max: 63.0 Max Tm Difference: 100.0

Product Tm Min: Opt: Max:

Primer GC% Min: 20.0 Opt: Max: 80.0

Max Self Complementarity: 8.00 Max 3' Self Complementarity: 3.00

Max #N's: 0 Max Poly-X: 5

Inside Target Penalty: Outside Target Penalty: 0 Set Inside Target Penalty to allow primers inside a target.

Fig. 2. *Primer3*'s WWW input page after the user has entered sequence, a "Sequence Id," and a "Target."

2. *Primer3* from the End-User Perspective

This section refers primarily to the WWW interface, which calls upon *primer3_core* to perform almost all of the work of selecting primers.

Primer3 takes as input a sequence and selects single primers or PCR primer pairs. **Figure 2** shows example input in the WWW interface. The user has pasted the source sequence into the large data-entry field near the top of the page, selected the **RODENT Mispriming Library** and entered a **Sequence Id** ("Example Sequence 1") and a **Target** ("40, 78").

Below the field for the source sequence are three check boxes labeled **Pick left or use left primer below**, **Pick hybridization probe (internal oligo) or use oligo below**, and **Pick right primer or use primer below. . . .** These check boxes govern whether *Primer3* tries to design a primer pair, a primer pair plus hybridization probe, or an individual primer (e.g., for sequencing) or hybridization probe. In **Fig. 2** the **Pick left...** and **Pick right...** boxes are checked, so *Primer3* will select PCR primer pairs.

Below each of these three check boxes is an input box. Placing an oligo in one of these boxes instructs *Primer3* to evaluate that oligo and choose matching oligos (depending on the check boxes).

After the user clicks on any of the **Pick Primers** buttons in the input page (**Fig. 2**) *Primer3* returns suggested primers as shown in **Fig. 3**. **Subheading 2.2.** discusses the interpretation of this output in detail; **Subheading 2.3.** suggests some strategies for proceeding when *Primer3* is unable to find any acceptable primers or primer pairs.

The label for each input option is a link to documentation on the meaning of the option and how *Primer3* uses it. For example, clicking on **Max End Stability** takes one to the following documentation:

> **Max End Stability**
> The maximum stability for the five 3' bases of a left or right primer. Bigger numbers mean more stable 3' ends. The value is the maximum delta G for duplex disruption for the five 3' bases as calculated using the nearest neighbor parameters published in Breslauer, Frank, Bloeker and Marky, Proc. Natl. Acad. Sci. USA, vol 83, pp 3746-3750. Rychlik recommends a maximum value of 9 (Wojciech Rychlik, "Selection of Primers for Polymerase Chain Reaction" in BA White, Ed., "Methods in Molecular Biology, Vol. 15: PCR Protocols: Current Methods and Applications", 1993, pp 31-40, Humana Press, Totowa NJ).

The **Max Mispriming** and **Pair Max Mispriming** input fields are important in many situations because the source sequence might contain one of the interspersed repeats (ALUs, LINEs, and others) that make up more than 35% of the human genome ***(15)***. The user should either replace these sequences by Ns before picking primers or should select a **Mispriming library**. (Not all mispriming is strictly speaking caused by repeats; it could be any sequence that one does not wish to inadvertently amplify.) However, the WWW interface at the `www.genome.wi.mit.edu` web site only offers repeat libraries for human and for mouse and rat (**RODENT**).

The maximum primer length is restricted because the nearest neighbor melting temperature model agrees well with reality only for relatively short sequences ***(16,17)***.

2.1. How Primer3 *Picks Primers*

Primer3 accepts many options that specify which primers are acceptable and which primers are better than others. In the WWW interface the user selects

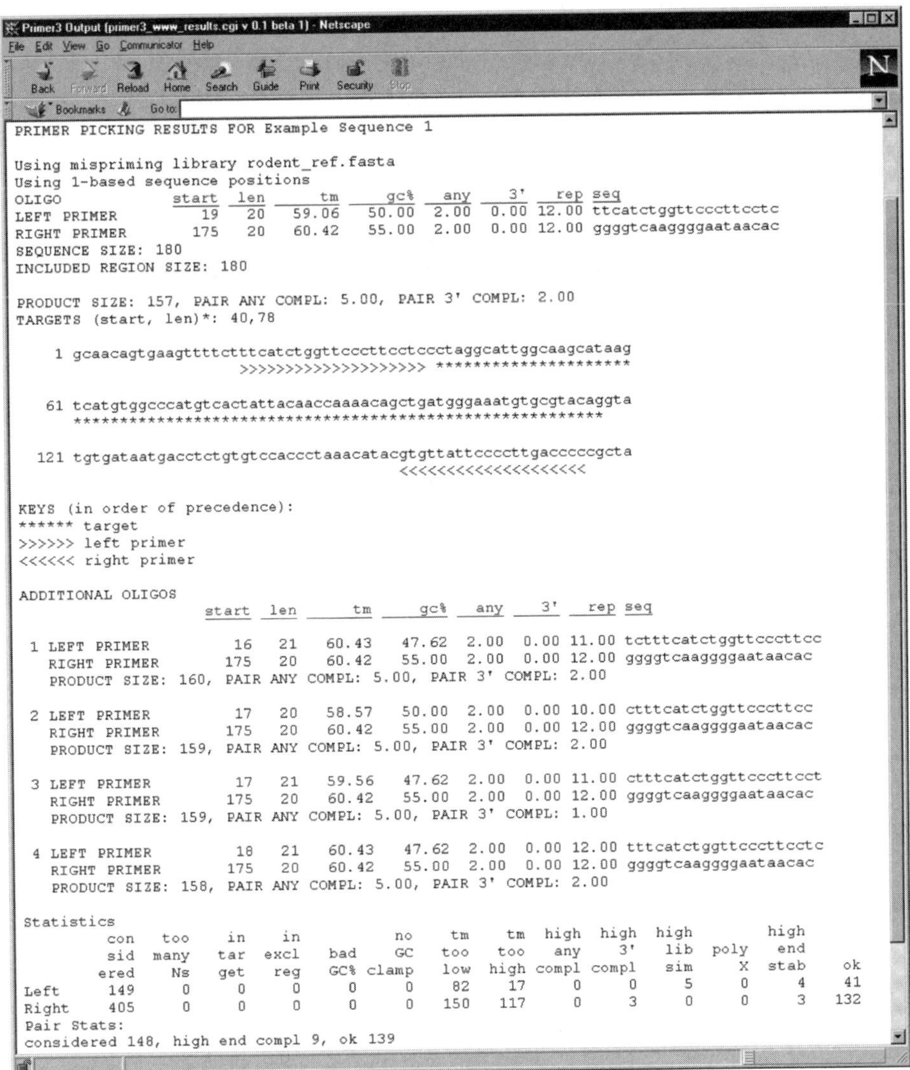

```
PRIMER PICKING RESULTS FOR Example Sequence 1

Using mispriming library rodent_ref.fasta
Using 1-based sequence positions
OLIGO            start  len     tm     gc%   any    3'   rep seq
LEFT PRIMER         19   20  59.06   50.00  2.00  0.00 12.00 ttcatctggttcccttcctc
RIGHT PRIMER       175   20  60.42   55.00  2.00  0.00 12.00 ggggtcaaggggaataacac
SEQUENCE SIZE: 180
INCLUDED REGION SIZE: 180

PRODUCT SIZE: 157, PAIR ANY COMPL: 5.00, PAIR 3' COMPL: 2.00
TARGETS (start, len)*: 40,78

    1 gcaacagtgaagttttctttcatctggttcccttcctccctaggcattggcaagcataag
                        >>>>>>>>>>>>>>>>>>>> *********************

   61 tcatgtggcccatgtcactattacaaccaaaacagctgatgggaaatgtgcgtacaggta
      *********************************************************

  121 tgtgataatgacctctgtgtccaccctaaacatacgtgttattccccttgacccccgcta
                                    <<<<<<<<<<<<<<<<<<<<

KEYS (in order of precedence):
****** target
>>>>>> left primer
<<<<<< right primer

ADDITIONAL OLIGOS
                     start  len     tm     gc%   any    3'   rep seq

 1 LEFT PRIMER          16   21  60.43   47.62  2.00  0.00 11.00 tctttcatctggttcccttcc
   RIGHT PRIMER        175   20  60.42   55.00  2.00  0.00 12.00 ggggtcaaggggaataacac
   PRODUCT SIZE: 160, PAIR ANY COMPL: 5.00, PAIR 3' COMPL: 2.00

 2 LEFT PRIMER          17   20  58.57   50.00  2.00  0.00 10.00 ctttcatctggttcccttcc
   RIGHT PRIMER        175   20  60.42   55.00  2.00  0.00 12.00 ggggtcaaggggaataacac
   PRODUCT SIZE: 159, PAIR ANY COMPL: 5.00, PAIR 3' COMPL: 2.00

 3 LEFT PRIMER          17   21  59.56   47.62  2.00  0.00 11.00 ctttcatctggttcccttcct
   RIGHT PRIMER        175   20  60.42   55.00  2.00  0.00 12.00 ggggtcaaggggaataacac
   PRODUCT SIZE: 159, PAIR ANY COMPL: 5.00, PAIR 3' COMPL: 1.00

 4 LEFT PRIMER          18   21  60.43   47.62  2.00  0.00 12.00 tttcatctggttcccttcctc
   RIGHT PRIMER        175   20  60.42   55.00  2.00  0.00 12.00 ggggtcaaggggaataacac
   PRODUCT SIZE: 158, PAIR ANY COMPL: 5.00, PAIR 3' COMPL: 2.00

Statistics
        con  too   in   in        no   tm   tm high high high      high
        sid many  tar excl  bad   GC  too  too  any   3'  lib poly  end
      ered   Ns  get  reg  GC% clamp low high compl compl sim    X stab   ok
Left   149    0    0    0    0     0   82   17    0    0    5    0    4   41
Right  405    0    0    0    0     0  150  117    0    3    0    0    3  132
Pair Stats:
considered 148, high end compl 9, ok 139
```

Fig. 3. Output from *Primer3's* WWW interface when primers have been found.

these options through text boxes, check boxes, and pull-down menus. For example in **Fig. 2** these include the **Mispriming Library** pull down (above the sequence input field), **Product Size Min**, **Opt**, and **Max** input fields, and all the input fields beneath the **General Primer Picking Conditions** heading.

Some options specify which primers are *acceptable*. For example to the left of **Product Length** the **Min** and **Max** options set lower and upper bounds on the length of products. Such options are called *constraints* because they constrain the set of acceptable primer pairs. Other options that specify constraints include **Primer Tm Min** and **Max**, **Max End Stability**, and **Max Mispriming**. (Tm is an abbreviation for "melting temperature".)

Other options specify characteristics of the optimum output primers or primer pairs (beyond specifying those that are simply acceptable). Examples include the **Product Length Opt** (Optimum) and the **Primer Tm Opt** inputs. Roughly speaking, if **Product Length Opt** is specified *Primer3* tries to pick a pair of primers that produce an amplicon of approximately the specified length. For some options the user need not specify the optimum value because *Primer3* can take it for granted; for example there is no input field for the optimum mispriming library similarity, which *Primer3* assumes to be 0.

Primer3 examines all primer pairs that satisfy the constraints and finds pairs that are closest to the optimum. How does *Primer3* calculate how close a primer pair is to the optimum? By default the WWW interface tries to balance equally primer length, primer melting temperature, and product length. (For compatibility with earlier versions *primer3_core* by default uses only primer length and primer melting temperature.)

However, to accommodate the diversity of primer picking applications *Primer3* is flexible in the formula it uses to calculate how close a primer or primer pair is to the optimum. The technical term for this formula is *objective function*. Thus, suppose you deem the difference in melting temperature between the two primers to be more important than their lengths, melting temperatures, and the product size. Then you can use the **Objective Function Weights...** sections of the input page (as partially shown in **Fig. 4**) to tell *Primer3* to use these considerations in calculating optimality. In **Fig. 5** this effect is accomplished by the values in the fields labeled **Product Size Lt** and **Gt**, **Tm Difference**, and **Primer Penalty Weight**. (**Primer Penalty Weight** is an adjustment factor for the entire per-oligo contribution to the objective function. More details are available in the online documentation.)

2.2. Interpreting the Output when Primers are Found

Please refer to **Fig. 3**. The top of the output displays the sequence id (**Example Sequence 1**) and a number of informational notes. The next part of the output displays the best left and right primers, and their characteristics (starting position, length, melting temperature, and so forth). Then the output displays information specific to the input sequence and the selected pair.

The next information is a quasi-graphical representation of the location of the left (>>>>> ...) and right (<<<<< ...) primers in the source sequence, as well as any important features of the source sequence, in this example only the position of the target (marked by asterisks***** ...). Following the sequence is information for some number of additional primer pairs. (The user can control

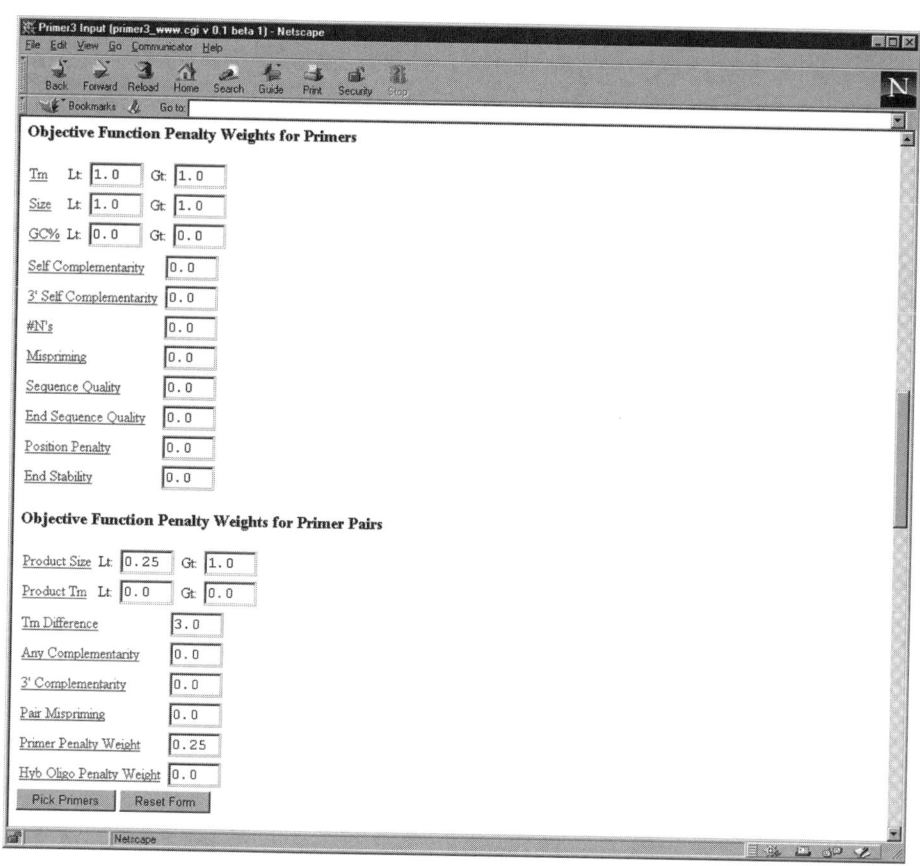

Fig. 4. The part of *Primer3*'s WWW interface that allows modification of the objective function. In this example "Product Size Lt" and "Gt," "Tm Difference," and "Primer Penatly Weight" have been changed from the defaults.

the number returned by entering a different value in the **Number to Return** input field.)

Finally the output contains a section headed **Statistics**, which we discuss in detail below.

2.3. What if There Are no Acceptable Primers?

Recall from **Subheading 1.3.** that in some situations one wants only good primers for uniform conditions and would rather discard some source sequence than deal with dubious primers. Most of *Primer3*'s default option values are tuned to suit these situations: constraints are strict. Given strict constraints the specifics of the objective function are not critical because any acceptable

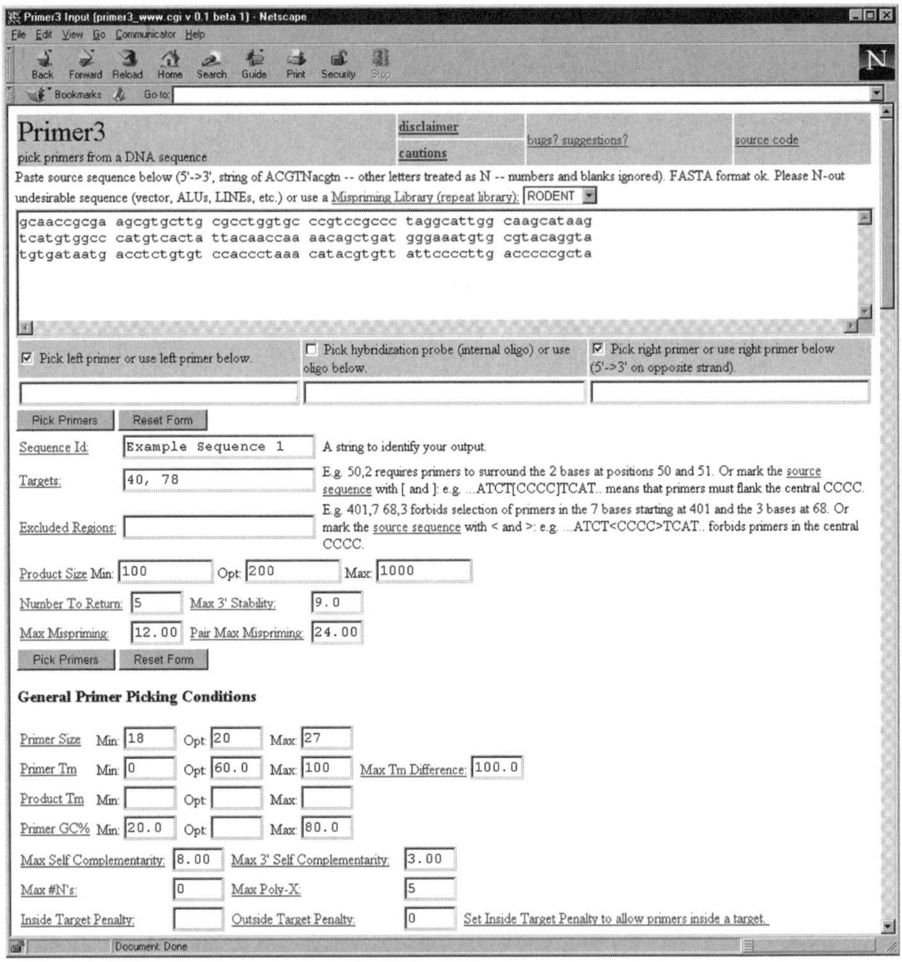

Fig. 5. *Primer3*'s WWW interface in which the "Primer Tm Min" and "Primer Tm Max" constraints have been made completely liberal.

primer or primer pair will be reasonably good, but potentially usable primers might be rejected as unacceptable.

In other situations, however, one must design primers for a given sequence at almost any cost. Suppose you are faced with such a situation, and *Primer3* cannot find acceptable primers given the default constraints. In this case *Primer3* will return a screen similar to that shown in **Fig. 6**.

Then what? The most intuitive course is to relax the constraints that you think are least important in your particular situation and that are most likely preventing any primers or primer pairs from being acceptable. The **Statistics**

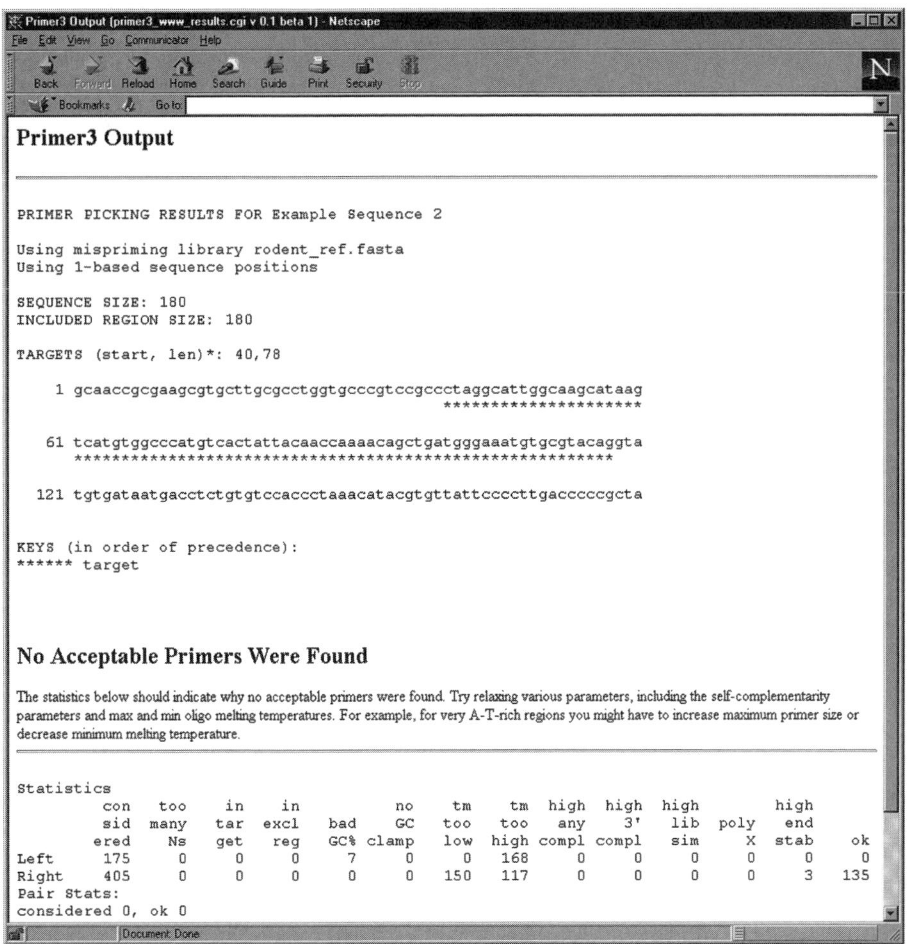

Primer3 Output (primer3_www_results.cgi v 0.1 beta 1) - Netscape

Primer3 Output

```
PRIMER PICKING RESULTS FOR Example Sequence 2

Using mispriming library rodent_ref.fasta
Using 1-based sequence positions

SEQUENCE SIZE: 180
INCLUDED REGION SIZE: 180

TARGETS (start, len)*: 40,78

    1 gcaaccgcgaagcgtgcttgcgcctggtgcccgtccgccctaggcattggcaagcataag
                                             *********************

   61 tcatgtggcccatgtcactattacaaccaaaacagctgatgggaaatgtgcgtacaggta
      *********************************************************

  121 tgtgataatgacctctgtgtccaccctaaacatacgtgttattccccttgacccccgcta


KEYS (in order of precedence):
****** target
```

No Acceptable Primers Were Found

The statistics below should indicate why no acceptable primers were found. Try relaxing various parameters, including the self-complementarity parameters and max and min oligo melting temperatures. For example, for very A-T-rich regions you might have to increase maximum primer size or decrease minimum melting temperature.

Statistics

	con sid ered	too many Ns	in tar get	in excl reg	bad GC%	no GC clamp	tm too low	tm too high	high any compl	high 3' compl	high lib sim	poly X	high end stab	ok
Left	175	0	0	0	7	0	0	168	0	0	0	0	0	0
Right	405	0	0	0	0	0	150	117	0	0	0	0	3	135

Pair Stats:
considered 0, ok 0

Document: Done

Fig. 6. Output from *Primer3*'s WWW interface when no primers were found.

section at the bottom of the output indicates reasons that individual primers are unacceptable. In the example above it is clear that the main problem is that all acceptable left primers have too high a melting temperature (as indicated by the column headed **tm too high**).

A word of caution: *Primer3* never considers a primer that is unacceptable because of its position. Thus, if a primer is outside of the included region or can never be acceptable given the length of the sequence, the position of any specified "included region" and targets and the range of allowable product sizes then it is *not* counted in the **considered** column. If it seems as though very few primers are even being considered, you might want to modify your maximum and minimum product size options, or expand the included region. An example of such a situation is the following, in which no left primers are considered:

```
Statistics
          con   too     in    in          no    tm     tm  high  high          high
          sid  many    tar  excl   bad    GC   too    too   any    3'  poly    end
         ered    Ns    get   reg   GC% clamp   low   high compl compl     X   stab     ok
Left        0     0      0     0     0     0     0      0     0     0     0      0      0
Right    1401    25      0     0     9     0   884    191     0     0     0      0    292
```

The **Pair Stats:** section indicates reasons that pairs of primers (as opposed to single primers or oligos) are rejected. For example, there might only be a few acceptable primers, all of which when paired would create a product with too low or too high a melting temperature.

Examining the **Statistics** (and less commonly) the **Pair Stats:** sections should suggest constraints that if relaxed would allow primers to be chosen.

In some cases (especially when relaxing several constraints at once) it might be desirable to also modify the objective function to reflect specific primer design objectives. In the sequence in **Fig. 6**, the temperatures of all possible left primers are too high. One way to proceed is to incrementally relax what seem to be the limiting constraints, for example increasing the **Primer Tm Max** option until an acceptable left primer is found. Alternatively, it can be more expeditious to simply relax the limiting constraint totally, as in **Fig. 5**, in which the **Primer Tm Min** and **Max** are set to 0 and 100° C, respectively. The primer pair selected with these relaxed constraints is:

```
OLIGO                start  len      tm     gc%   any    3'   rep seq
LEFT PRIMER             10   19   68.72   63.16  6.00  1.00 10.00 aagcgtgcttgcgcctggt
RIGHT PRIMER           175   20   60.42   55.00  2.00  0.00 12.00 ggggtcaaggggaataacac
```

And the Statistics are

```
          con   too     in    in          no    tm     tm  high  high  high          high
          sid  many    tar  excl   bad    GC   too    too   any    3'   lib  poly    end
         ered    Ns    get   reg   GC% clamp   low   high compl compl   sim     X   stab     ok
Left       53     0      0     0     7     0     0      0     0     9     0     0     16     21
Right     387     0      0     0     0     0     0      0     0     4     0     0      5    378
```

Now there are acceptable primer pairs but a large difference in melting temperatures between the left and right primers. To reduce this difference one can include it as part of the objective function, as shown in **Fig. 4**. After this adjustment, *Primer3* selects the following primer pair:

```
OLIGO               start  len      tm     gc%   any    3'   rep seq
LEFT PRIMER            11   18   67.83   66.67  4.00  1.00 10.00 agcgtgcttgcgcctggt
RIGHT PRIMER          180   20   66.95   60.00  2.00  2.00 10.00 tagcgggggtcaaggggaat
```

3. *Primer3* for Biologist Programmers

3.1. Installation Instructions

The source distribution is available as a UNIX "tar" archive, which can be managed by the UNIX tar utility, by the Windows / Windows NT *WinZip* utility (Nico Mak Computing; `http://www.winzip.com/winzip.htm`) or by the Mac *DropStuff with Expander Enhancer* utility `www.aladdinsys.com`. To run *primer3_core* you will first need to compile it using an ANSI C compiler with *POSIX* libraries and run the tests supplied with the distribution as documented in the **README**.

3.2. Examples of How to Use primer3_core as a Software Component

In this section we present two examples of using *primer3_core* as a software component. The code for these examples is available in the *Primer3* distribution.

3.2.1. *Using* primer3_core *with* UNIX *Pipes*

The first example is a relatively lightweight application of the kind that requires only a minimum of *perl* scripting experience.

This example shows how to postprocess *primer3_core*'s output to complete an oligo design task. The example task is the specification of "overgos" (John D. McPherson, pers. comm.), in which a 36-mer double-stranded hybridization probe is constructed from annealing overlapping 22-mers and filling in the singled stranded tails:

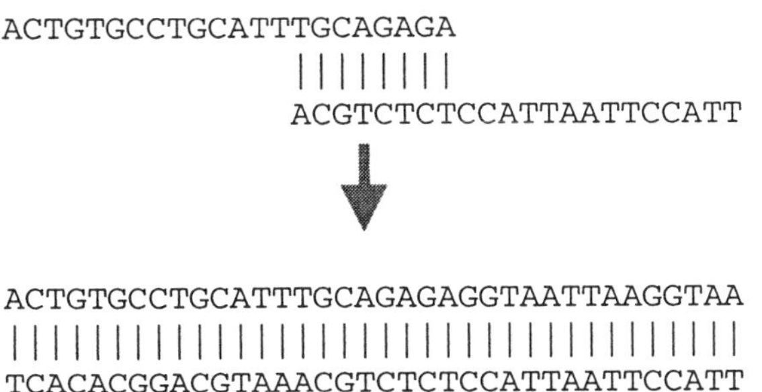

Specifically, we will show code that takes an existing primer pair and then designs an overgo that will hybridize to the site amplified by that primer pair. Here is the UNIX command line one would use:

```
prompt> ./primer3_core < input | ./overgo.pl
```

In this command line *primer3_core* runs first, taking its input from the file "input", whereas its output is sent directly to the *perl* program *overgo.pl* via a

UNIX pipe (specified by the vertical bar, '|', on the command line). The input could be prepared by hand in a text editor or (more likely) produced by another program. It has the form of *tag*=value pairs, a format dubbed *Boulder-IO* ***(18)***:

```
PRIMER_SEQUENCE_ID=Overgo Example
PRIMER_PICK_INTERNAL_OLIGO=1
PRIMER_INTERNAL_OLIGO_MAX_MISHYB=36
PRIMER_INTERNAL_OLIGO_MIN_SIZE=36
PRIMER_INTERNAL_OLIGO_MAX_SIZE=36
PRIMER_INTERNAL_OLIGO_OPT_SIZE=36
PRIMER_INTERNAL_OLIGO_MIN_TM=10
PRIMER_INTERNAL_OLIGO_MAX_TM=90
PRIMER_INTERNAL_OLIGO_OPT_TM=70
PRIMER_INTERNAL_OLIGO_SELF_ANY=30
PRIMER_INTERNAL_OLIGO_SELF_END=30
PRIMER_INTERNAL_OLIGO_MISHYB_LIBRARY=humrep
PRIMER_PRODUCT_SIZE_RANGE=70-1000
PRIMER_EXPLAIN_FLAG=1
PRIMER_PAIR_WT_IO_QUALITY=1
PRIMER_PAIR_WT_PR_QUALITY=0
PRIMER_IO_WT_REP_SIM=1
PRIMER_IO_WT_TM_GT=0
PRIMER_IO_WT_TM_LT=0
PRIMER_IO_WT_SIZE_GT=0
PRIMER_IO_WT_SIZE_LT=0
PRIMER_NUM_RETURN=1
PRIMER_LEFT_INPUT=GAAATGTGTCCTTCCCCAGA
PRIMER_RIGHT_INPUT=GAGTTCACCCATACGACCTCA
SEQUENCE=GGATCACAACGTTTTTGACACACCCTATAATGATGTATT . . .
=
```

Boulder-IO is a format for moving semistructured data between programs. *Primer3* receives its input and (by default) produces its output in a simple subset of *Boulder-IO*. The **README** in the *Primer3* distribution describes the meanings of all these *tag*=value pairs in the input, as well as those in the output. The output from *primer3_core* given the input above is:

```
PRIMER_SEQUENCE_ID=Overgo Example
PRIMER_PICK_INTERNAL_OLIGO=1
PRIMER_INTERNAL_OLIGO_MAX_MISHYB=36
```

```
PRIMER_INTERNAL_OLIGO_MIN_SIZE=36
PRIMER_INTERNAL_OLIGO_MAX_SIZE=36
PRIMER_INTERNAL_OLIGO_OPT_SIZE=36
PRIMER_INTERNAL_OLIGO_MIN_TM=10
PRIMER_INTERNAL_OLIGO_MAX_TM=90
PRIMER_INTERNAL_OLIGO_OPT_TM=70
PRIMER_INTERNAL_OLIGO_SELF_ANY=30
PRIMER_INTERNAL_OLIGO_SELF_END=30
PRIMER_INTERNAL_OLIGO_MISHYB_LIBRARY=humrep
PRIMER_PRODUCT_SIZE_RANGE=70-1000
PRIMER_EXPLAIN_FLAG=1
PRIMER_PAIR_WT_IO_QUALITY=1
PRIMER_PAIR_WT_PR_QUALITY=0
PRIMER_IO_WT_REP_SIM=1
PRIMER_IO_WT_TM_GT=0
PRIMER_IO_WT_TM_LT=0
PRIMER_IO_WT_SIZE_GT=0
PRIMER_IO_WT_SIZE_LT=0
PRIMER_NUM_RETURN=1
PRIMER_LEFT_INPUT=GAAATGTGTCCTTCCCCAGA
PRIMER_RIGHT_INPUT=GAGTTCACCCATACGACCTCA
SEQUENCE=GGATCACAACGTTTTTGACACACCCTATAATGATGTATT . . .
PRIMER_LEFT_EXPLAIN=considered 1, ok 1
PRIMER_RIGHT_EXPLAIN=considered 1, ok 1
PRIMER_INTERNAL_OLIGO_EXPLAIN=considered 224,
long poly-x seq 12, ok 212
PRIMER_PAIR_EXPLAIN=considered 1, ok 1
PRIMER_PAIR_QUALITY=15.0000
PRIMER_LEFT_SEQUENCE=GAAATGTGTCCTTCCCCAGA
PRIMER_RIGHT_SEQUENCE=GAGTTCACCCATACGACCTCA
PRIMER_INTERNAL_OLIGO_SEQUENCE=ACTGTGCCTGCATTTGCA . . .
PRIMER_LEFT=99,20
PRIMER_RIGHT=198,21
PRIMER_INTERNAL_OLIGO=140,36
PRIMER_LEFT_TM=59.903
PRIMER_RIGHT_TM=59.981
PRIMER_INTERNAL_OLIGO_TM=72.885
PRIMER_LEFT_SELF_ANY=3.00
PRIMER_RIGHT_SELF_ANY=4.00
PRIMER_INTERNAL_OLIGO_SELF_ANY=8.00
```

```
PRIMER_LEFT_SELF_END=0.00
PRIMER_RIGHT_SELF_END=1.00
PRIMER_INTERNAL_OLIGO_SELF_END=3.00
PRIMER_INTERNAL_OLIGO_MISHYB_SCORE=15.00, MLT1b
(MLT1b subfamily) - consensus sequence
PRIMER_LEFT_END_STABILITY=8.2000
PRIMER_RIGHT_END_STABILITY=8.2000
PRIMER_PAIR_COMPL_ANY=4.00
PRIMER_PAIR_COMPL_END=1.00
PRIMER_PRODUCT_SIZE=100
=
```

The second program, *overgo.pl*, takes the sequence of the 36-mer hybridization probe from this output and produces the overlapping 22-mers that constitute the overgo:

```
#!/usr/local/bin/perl5 -w
$/ = "\n=\n"; # Set the record terminator.
while (<>) {
  %rec = split /[=\n]/;  # A DANGEROUS approach
                         # to parsing the sequence.
  for (keys %rec) { $rec{$_} =~ s/\n// }
  $seq = $rec{'PRIMER_INTERNAL_OLIGO_SEQUENCE'};
  next unless $seq;
  print "MARKER\t\t$rec{'PRIMER_SEQUENCE_ID'}\n";
  $left  = substr($seq,0,22);          # Get left oligo.
  $r     = substr($seq, 14);           # Get the right
                                       # oligo,
  $right = reverse($r);                # reverse it, and
  $right =~ tr/GATC/CTAG/;             # complement it.
  print "LEFT_MID_OLIGO\t$left\n";
  print "RIGHT_MID_OLIGO\t$right\n";
  print "MAX_SCORE\t
  $rec{'PRIMER_INTERNAL_OLIGO_MISHYB_SCORE'}\n";
  $gc    = ($seq =~ tr/GC/GC/);
  printf "GC_content\t%d%%\n\n", $gc * 100 / 36;
}
```

The statement `$/ = "\n=\n";` tells *perl* that each record is terminated by an "=" sign on a line by itself (which is the standard record terminator for

Boulder-IO). The statement `%rec = split /[=\n]/;` parses *Boulder-IO* record into the perl hash `%rec`. This method of parsing the output requires that we know that "=" will not appear in the *value* part of any *Boulder-IO tag*=value pair. For situations in which more robustness is required, use Lincoln Stein's *perl Boulder* module (available at `http://www.genome.wi.mit.edu/genome_software/other/boulder.html`). Using this module *overgo.pl* would be rewritten as:

```
#!/usr/local/bin/perl5 -w

use Boulder::Stream;
$in = new Boulder::Stream;
while ($rec = $in->read_record()) {
  $seq
    = $rec->get('PRIMER_INTERNAL_OLIGO_SEQUENCE');
  next unless $seq;
  print "MARKER\t\t",
  $rec->get('PRIMER_SEQUENCE_ID'), "\n";
  $left  = substr($seq,0,22);       # Get the left
                                    # oligo.
  $r     = substr($seq, 14);        # Get the right
                                    # oligo,
  $right = reverse($r);             # reverse it, and
  $right =~ tr/GATC/CTAG/;          # complement it.
  print "LEFT_MID_OLIGO\t$left\n";
  print "RIGHT_MID_OLIGO\t$right\n";
  print "MAX_SCORE\t",
  $rec->get('PRIMER_INTERNAL_OLIGO_MISHYB_SCORE'),
  "\n";
  $gc    = ($seq =~ tr/GC/GC/);
  printf "GC_content\t%d%%\n\n", $gc * 100 / 36;
}
```

Using the *Boulder* module is preferable because it is more robust. It will run correctly even if someone puts an "=" in, for example, the value for **PRIMER_SEQUENCE_ID**. The only disadvantage is that you need to get the *Boulder* module before you can try it. The output for the input above is

```
MARKER            Overgo Example
LEFT_MID_OLIGO    ACTGTGCCTGCATTTGCAGAGA
RIGHT_MID_OLIGO   TTACCTTAATTACCTCTCTGCA
MAX_SCORE  15.00, MLT1b (MLT1b subfamily) . . .
  consensus sequence
```

3.2.2 *Calling* primer3_core *from* perl

The second example is *Primer3's* WWW interface itself. This code fragment is adapted from the CGI script, *primer3_www_results.cgi*, which implements part of that interface. (The CGI module is available from `http://www.genome.wi.mit.edu/ftp/distribution/software/WWW/`.) *Primer3_www.cgi* calls *primer3_core*, with a flag requesting formatted output (-format_output), and then grabs *primer3_core*'s output and tweaks it a bit:

```
#!/usr/local/bin/perl5 -w

...

use FileHandle; # Standard part of perl distribution
use IPC::Open3; # Standard part of perl distribution
use CGI;

    ...

    $query = new CGI;
                    # $query now contains the parameters
                    # to the cgi script

    ...

    my @names = $query->param;

    ...

    for (@names) {
       next if ... # Some cgi parameters do not get
                   # sent to primer3_core

       ...

       $line = "$_=$v\n";
       push @input, $line;  # Save a Boulder-IO line for
                            # primer3_core's eventual consumption.

    }
    my $cmd = "./primer3_core -format_output -strict_tags"
    my $primer3_pid;
    my ($childin, $childout) = (FileHandle->new, FileHandle->new);
    {
       local $^W = 0;
       $primer3_pid = open3($childin, $childout, $childout, $cmd);
    }
```

```
if (!$primer3_pid) {
      print "Cannot excecure $cmd:<br>$!\n$wrapup\n";
     exit;
   }
   print "<pre>\n";
   print $childin @input;
   $childin->close;
   my $cline;
   while ($cline = $childout->getline) {
     if ($cline =~ /(.*)start  len    tm   gc%  any   3\' seq/)
     {
       # Grap a particular line and
       # add hyperlinks to it:
       $cline = $1
               . "<a href=\"$DOC_URL#PRIMER_START\">start</a>"
               . "<a href=\"$DOC_URL#PRIMER_LEN\">len</a>"
               . "<a href=\"$DOC_URL#PRIMER_TM\">tm</a>"
               . "gc%   any    3\' seq\n"
     }
     print $cline;
   }
   print "</pre>\n";
   waitpid $primer3_pid, 0;
   if ($? != 0 && $? != 64512) { # 64512 == -4
       ... # primer3_core exited with
           # an error code; alert the browser.
   }
```

Of course the -formated_output flag in $cmd is not an essential part of the paradigm at work in this example. The script could have parsed *Boulder-IO* output and then formatted or processed the information in some other way.

3.2.3 Other Uses of primer3_core *as a Software Component*

The two examples above show how to use *primer3_core* as a component in a lightweight Unix pipeline (the overgo design example) and how to use *perl*'s **open3** command to start an execution of *primer3_core* and then grab its output for further processing. An intermediate approach that is simpler to program than using **open3** is to simply use the *perl* **open** command and then return *primer3_core*'s output unmodified, e.g.

```
if (!open(PRIMER,"| $cmd")) {
  print "Cannot execute <pre>$cmd\n</pre>\n$wrapup";
  return;
}
print PRIMER @input;
close PRIMER;    # primer3_core's output is the same
                 # as this script's output.
if ($? != 0 && $? != 64512) { # 64512 == -4
    . . . # $cmd exited with an error code.
}
```

At the Whitehead institute we have used *primer3_core* as part of an industrial-strength primer design pipeline that includes vector clipping (identification and electronic "removal" of vector arms), microsatellite repeat identification, and automatic screening for vector contaminants. We have also used it in pipelines that add constant 5' tails to each primer and in pipelines that find a tiling of amplicons across a sequence. For this last application we set **PRIMER_FILE_FLAG=1** in *primer3_core*'s input, which directs *primer3_core* to create files containing all acceptable left and right primers. A different program then selects primers from these lists to produce the tiling.

3.3. Efficiency Considerations

The running time of *Primer3* is seldom an issue for users of the WWW interface. However, users of *primer3_core* for high volume applications should be aware of the factors that determine running time. The most expensive operation in selecting individual primers is a check against a mispriming or mishyb library (the actual time needed for each oligo is a linear function of the size of the library). The next most expensive operations are checks for oligo self-complementarity, and, if *Primer3* examines a large number of primer pairs, checking oligo pairs for self-complementarity.

Primer3's running time depends also on the size of the sequence in which to select primers. Selecting a single primer pair anywhere within a 10-kb sequence will take approx 10 times as long as selecting a single primer pair anywhere within a 1-kb sequence (all other options being equal).

The following are additional determinants of *Primer3*'s running time:

- Strict as opposed to liberal constraints on oligos. *Primer3* excludes primers based on cheap computations (e.g., oligo melting temperature) before examining more expensive-to-compute characteristics (e.g., similarity to mispriming library entries) so relaxing cheap-to-compute constraints entails evaluation of a larger number of expensive-to-compute constraints.

- Acceptable locations for primers (considering also constraints on product size). This item is similar to the preceding one. *Primer3* does not perform expensive operations to characterize primers which, because of their location, can never be part of an acceptable primer pair.
- The **PRIMER_FILE** input flag. This flag causes *Primer3* to compute every characteristic, including mispriming similarity and self-complementarity, of every possibly acceptable primer.
- Cost of computing the objective function. There are two subcases.
  - The objective function depends on expensive-to-compute characteristics of oligos or primers, such as similarity to mispriming or mishyb libraries or complementarity between primers in a pair. In this case *Primer3* must perform these expensive computations on essentially all acceptable primers.
  - The objective function depends on characteristics of primer pairs *per se*, such as product melting temperature or product size. In this case *Primer3* must calculate whether each individual primer is acceptable, which usually requires some expensive computation to determine acceptability.

(When the objective function depends neither on expensive characteristics of individual primers nor on characteristics of primer pairs then *Primer3* organizes its search so that it only checks expensive constraints on the best primers.)

Acknowledgments

The development of *Primer3* and the *Primer3* WWW interface were funded by Howard Hughes Medical Institute and by the National Institutes of Health, National Human Genome Research Institute, under grants R01-HG00257 (to David C. Page) and P50-HG00098 (to Eric S. Lander).

We gratefully acknowledge the support of Digital Equipment Corporation, which provided the Alphas that we used for much of the development of *Primer3*, and of Centerline Software, Inc., whose TestCenter memory error, memory leak, and test coverage checker helped us discover and correct a number of otherwise latent errors in *Primer3*.

Primer3 is the most recent of a number of primer-picking programs implemented at Whitehead Institute, starting with primer 0.5 ***(13)***. *Primer3* started as a reimplementation of *Primer 0.5* as software component; the design of *Primer3* draws heavily on the design of *Primer 0.5* and *Primer v2* (Richard Resnick) and the WWW interface designed by Richard Resnick for *Primer v2*.

Thanks to Alex Bortvin, Mark Daly, Nathan Siemers, and William J. Van Etten for reviewing drafts of this chapter.

References

1. Dieffenbach, C. W. and Dveksler, G. S. (1995) *PCR Primer A Laboratory Manual.* Cold Spring Harbor Laboratory Press, Cold spring Harbor, NY.
2. Innis, M. A., Gelfand, D. H., Sninsky, J. J., and White, T. J., eds. (1990) *PCR Protocols A Guide to Methods and Applications.* Academic Press, San Diego, CA.
3. Rychlik, W. (1993) Selection of primers for polymerase chain reaction, in *Methods in Molecular Biology, vol. 15: PCR Protocols: Current Methods and Applications* (White, B. A., ed.) Humana, Totowa, NJ, pp. 31–40.
4. Wetmur, J. G. (1991) DNA probes: applications of the principles of nucleic acid hybridization. *Crit. Rev. Biochem. Mol. Biol.* **26,** 227–259.
5. Schuler, G. D. et al. (1996) A gene map of the human genome. *Science* **274,** 540–546.
6. Wang, D. G. et al. (1998) Large-scale identification, mapping and genotyping of single-nucleotide polymorphisms in human genome. *Science* **280,** 1077–1082.
7. Harbison, S. and Steele, G. (1995) *C A Reference Manual*, 4th ed. Prentice Hall, Englewood Cliffs, NJ.
8. Dougherty, D. (1991) *POSIX Programmer's Guide.* O'Reilly, Cambridge, MA.
9. Gundavaram, S. (1997) *CGI Programming with Perl.* O'Reilly, Cambridge MA.
10. Stein, L. D. (1997) *How to Set Up and Maintain a Web Site*, 2nd ed. Addison-Wesley, Reading, MA.
11. Wall, L., Christiansen, T., and Schwartz, R. L. (1996) *Programming Perl*, 2nd ed. O'Reilly, Cambridge, MA.
12. Rychlik, W. and Rhoads, R. E. (1989) A computer program for choosing optimal oligonucleotides for filter hybridization, sequencing and in vitro amplification of DNA. *Nucleic Acids Res.* **17,** 8543–8551.
13. Daly, M. J., Lincoln S. E., and Lander E. S. (1991). *"PRIMER"*, unpublished software, Whitehead Institute/MIT Center for Genome Research. Available at `http://www.genome.wi.mit.edu/ftp/pub/software/primer.0.5`, and via anonymous ftp to `genome.wi.mit.edu, directory /pub/software/primer.0.5`.
14. Hillier, L. and Green, P. (1991) OSP: an oligonucleotide selection program. *PCR Meth. Appl.* **1,** 124–128. Documentation available at `http://genome.wustl.edu/gsc/manual/protocols/ospdocs.html`. OSP is available from the author on request.
15. Smit, A. F. A. (1996) The origin of interspersed repeats in the human genome. *Curr. Opin. Genet. Devel.* **6,** 743–748.
16. Breslauer, K. J., Frank, R., Bloeker, H., and Marky L. A. (1986) Predicting DNA duplex stability from the base sequence. *Proc. Natl. Acad. Sci. USA* **83,** 3746–3750
17. Rychlik, W., Spencer, W. J., and Rhoads, R. E. (1990) Optimization of the annealing temperature for DNA amplification in vitro. *Nucleic Acids Res.* **18,** 6409–6412.
18. Stein, L. (1997) How perl saved the human genome project. *Dr Dobb's Journal* Spring 1997 Special Report on Software Careers. Available at `http://www.ddj.com/ddj/1997/careers1/stei.htm`.

21

Using the WWW to Supply the Molecular Biology Lab

MaryAnn Labant and Roger Anderson

1. Introduction

The worldwide embrace of the Internet has led to the development of an electronic marketplace for purchasing research laboratory supplies. The exponential growth of the World Wide Web has given suppliers cost-effective tools for continually providing up-to-date information on products, pricing, availability, and order status. Acceptance of the use of credit cards, typically referred to as P-cards or procurement cards was one of the final tools necessary to open up a new era in procurement practices.

"Internet commerce holds the promise of increasing efficiencies, lowering costs and furthering globalization of trade" (*Cyber Commerce—Internet Tsunami*, Goldman Sachs Technology Report, Aug. 4, 1997). Today, organizations are steadily improving the way they process information, particularly in procurement and financial systems. Suppliers are steadily improving the way they communicate information, particularly through the use of electronic media. These systematic improvements of buying and selling processes directly affect the laboratory researchers because researchers are the final "customers of the process." In return, laboratory researchers directly impact successful implementation of these new processes and systems. Customers, competition, and change are forcing suppliers to provide greater cooperation and support and the winner, in all of this, will be the researcher.

This new era of procurement is still evolving. Many options still exist for organizations and researchers to purchase goods and services. The purpose of this chapter is to provide some insight into the options and restrictions that the molecular biology laboratory researcher currently faces.

From: *Methods in Molecular Biology, vol. 132: Bioinformatics Methods and Protocols*
Edited by: S. Misener and S. A. Krawetz © Humana Press Inc., Totowa, NJ

2. Change is a Constant on the Web

In today's rapidly changing world of acquisitions, mergers, area codes, and addresses, out-of-date information is rampant on the Web. Many companies or individuals who had developed independent, informational sites have discontinued operations. These sites may or may not remain on the Web resulting in invalid links or links to outdated information. In a quick sample of 25 "informational sites," we found over 30% had not been updated within the last 6 mo to a year. Discarded sites are often not removed by their developers, as the companies run out of funding or individuals maintaining the sites graduate, change positions, or lose interest. Before you begin searching a site, always check for update records to ensure up-to-date information.

One of the benefits of Web publishing is the dynamic nature of the content. The disadvantage is that design, context, and content changes occur frequently as technology providers develop new tools. There are many website development tools and even more website developers. The matrix of this combination yields a tremendous diversity in site design and functionality. No two suppliers, groups, companies, associations or journals have similar sites. Menus, link placement, layout, and content, are all different. Companies also change their sites on a regular basis to build customer traffic and to keep abreast of demands as users' tastes and needs change. Expect the unexpected, including having to relearn site maps for your regular haunts.

3. Finding Product Information on the Web

An abundance of information is available on the Web today. There is no hard and fast rule for what you will find at a website so the only course of action is to go there and if you like what you find, bookmark it so you can return easily and quickly. Suppliers, who originally established their websites with static pages are taking advantage of more sophisticated applications and databases to populate their pages with real-time information to meet customers' needs and to ensure accuracy. As an example, at the time of this writing, Millipore Corporation, a market leader in filtration products, provided in-depth, searchable technical information, multilingual catalogs, on-line ordering, and order status tracking capabilities at their site.

Companies are utilizing search engines more and more, adding power to their sites. These search routines have some common features. For example, words enclosed in quotations are searched for together; words listed separately, without quotations, are searched for individually and when found together yield a higher match score. A lot of effort is being put forth to produce better search routines. Of course, none of this matters if the site you visit does not have the bulk of their data available in a searchable format. It is always useful to read the search routine instructions for a few sites. These give you a feel for the various methods in use.

Online product breadth can range from very extensive to very limited. Some suppliers only list their new or most frequently purchased items. These suppliers may perceive a lack of interest in their websites for information on their entire product line or they may not have their product information in an electronic format that can be easily updated. Therefore, even if you know the company sells the product, your search may be unsuccessful. Public pressure may change this as users reinforce their requests for product information by purchasing from an alternative source. Users need to let suppliers know what information they are interested in and why they visit a particular company's website. Feedback is taken seriously, especially if it is presented in a professional manner with justification. Let them know how you feel.

Sophistication levels and capabilities of suppliers' websites also vary greatly. Most suppliers do offer the ability to request information and literature. Some suppliers allow on-line ordering from anyone with a procurement or credit card, whereas others restrict their privileges to customers with existing accounts. Suppliers who sell direct are beginning to team up with the express package carriers for online order status tracking.

In earlier days, suppliers had "useful and interesting links" sections. Most of these sections have been discontinued as of this writing. Instead suppliers continue to maintain and expand their proprietary online technical support.

In addition to suppliers' sites, alternate avenues exist for locating technical and product information. Good starting points include trade publications and association sites (**Table 1**). Whereas some sites, such as the site for the journal *Science*, only lists information from their advertisers, others, such as the site for the publication *Biomedical Product News*, and the site for the publishing company International Scientific Communications, have their entire Buyer's Guides online for quick reference. Buyer's Guides are helpful for identification of suppliers servicing specific product areas.

A number of other independent sites, in addition to the trade publications and associations, have been developed to consolidate product and supplier information for the scientific marketplace. These sites differ in their approaches, from the amount of product information provided, the types of supplies listed, categorization methods, and search mechanisms to the ability to communicate information requests or product orders directly to suppliers. These sites are most helpful when looking for initial product information or suppliers.

In the mid 1990s, there were many independently developed websites with directories of links to other websites (**Table 2**). This was particularly important because powerful search engines and web spiders were not yet available.

Table 1
Some Examples of Journal and Association Sites

Publication, Company, or Association	Directory Name	Site URL
Biomedical Products News	Life Science Lab Reference Supplier Directory	www.biomedprod.com
International Scientific Communications	Lab Crawler - ISC Buyer's Guides	www.iscpubs.com
The Scientist	Lab Consumer	www.the-scientist.com
American Chemical Society	LabGuide	pubs.acs.org
	General Information and links	www.acs.org
	Products	www.cas.org
	Product Information Network (PIN)	www.chemcenter.org
BioTechniques	Buyer's Guide and BioMall	www.biotechniques.com
Nature	Buyer's Guide and Biotechnology Directory	www.nature.com
Science	Electronic Marketplace	www.sciencemag.org

This list is a general sampling and is not meant to provide a comprehensive listing of publication and association sites.

Many of these original sites still exist but they have not been updated in years. One favorite was "Pedro's Tools." As of this writing, on many pages, the last recorded update is June 1996.

A cautionary note about independent sites is the determination of information validity. It is not a requirement to disclose relationships, such as sponsorship, to suppliers or other for-profit entities. This can be misleading. Any information found on the Internet should not be taken as the absolute truth.

We encourage you to become an active participant in the growth of scientific resources on the web. Many sites rely on user input to find, identify and critique resources.

4. You Found It—Can You Buy It On The Web?

You finally located the product that you have been searching for on a website. And this particular site allows online ordering. Now that you found the product that you want, can you buy it online? The ability to purchase from sites depends not only on the existing purchasing rules at your organization but also on the payment methods accepted by the particular company.

4.1. The Organizational Purchasing Process

In research environments, researchers may be given extensive purchasing authority. However, they also need to adhere to negotiated contracts, organiza-

Table 2
Some Examples of Independent Sites that Can Be Used for Finding Suppliers and Product Information

URL	Description
`www.atcg.com`	Lists detailed product information from over 3000 suppliers on 600,000 life science, MRO, and office supplies for biological and analytical research.
`www.biosupplynet.com`	Lists product information on 15,000 products from 2700 suppliers.
`www.bio.com`	Lists software for chemical and biological research.
`www.chemconnect.com`	Lists chemicals, suppliers, and reference materials.
`www.biolinks.com`	Lists scientific suppliers and other useful information.
`www.sciquest.com`	Lists product information on over 300,000 products and services for the analytical, life science, and clinical industries.
`www.bio.net`	BioSci is a set of electronic communication forums. The moderated methods and reagents discussion group often talks about new products.
`www.antibodyresource.com`	Lists antibody suppliers, databases, and other resources.
`www.cato.com`	Lists a directory of products and services for the biotechnology and pharmaceutical industries.

Information on these websites was taken from published literature at the date of this writing.

tional rules, and account for their purchases. Often, a group assigns someone the responsibility for purchasing supplies, leaving this person to devise their own procurement method. Unfortunately, these people most likely duplicate the efforts of someone in the purchasing department and may inadvertently overpay for some of their supplies.

Although researchers may object, control is a critical feature in any system that allows purchasing, whether it is a supplier's website, an independent site, or an organization-wide procurement system. Administrators must oversee spending, negotiate, and check adherence to contracts, verify compliance with government regulations, and manage supplier performance.

For their protection, suppliers must be able to verify that the person placing the order is who they say they are, and authorized to make purchases. Suppliers also need assurances that they will receive payment for goods shipped. Both buyers and sellers need, and should require, security and privacy protection. Organization-wide electronic catalogs and commerce systems offer flexibility, privacy, and open communication for users yet allow the organization to maintain control and adapt to changing procurement requirements and simultaneously reduce the occurrence of fraud.

There are additional factors that influence the determination of where you can and cannot purchase goods. The *Year 2000* computer issues offer many challenges to organizations. Financial applications are being repaired or replaced. To complicate matters further, many organizations have decided that a major infrastructure application update warrants re-engineering efforts and process redesign. Depending on when the last thorough process analysis was completed, a totally new approach to doing business at your organization may be underway. This directly impacts everyone, researchers, administration, and suppliers.

A current trend is to place more responsibility for product selection, and to put controlled order placement, into the hands of the researchers and other end-users. Several ways for organizations to accomplish decentralization of the purchasing process include the use of electronic catalogs, use of suppliers' sites, an electronic commerce or web-based ordering system, and procurement cards. All of these systems should interact with the financial applications to ensure that sufficient funds are in place and encumbered before an order is sent to the supplier for fulfillment. Although it may sound cumbersome, bureaucratic, and time consuming, in reality, today, information flows can be efficient, effective and user-friendly.

If a system is designed correctly, the person placing the order should not notice the seamless data flows taking place behind the scenes. They will, however, recognize the benefits—and one of the major benefits is the ability to access accurate real-time account balances. At the end of a grant season or fiscal year, everyone saves time because the reconciliation of funds has been ongoing, eliminating "use it or lose it" shopping sprees or over-spending clean-up hassles.

4.2. Marketplaces vs Individual Supplier Sites

Today, there are a variety of available choices for online ordering, ranging from collective marketplaces to individual suppliers' sites. Some marketplaces and large suppliers use standard Electronic Data Interface (EDI) data sets as the basic catalog building blocks. The advantage of using EDI is that data can be exchanged real-time allowing for the display of availability and correct prices. The disadvantage is that descriptions of products may be indecipherable because of limited data field sizes, such as 30 characters for an item name, so that other information sources must be utilized to make a product decision.

Other marketplace approaches use a mall scenario, either taking individual

Table 3
Some of the Product Categories

• Antibodies	• Clinical Supplies	• Lab Organisms	• Office Supplies
• Apparel	• Columns	• Labware and Equipment	• Photographic Materials
• Books	• Filters and Membranes	• Libraries	• Plasticware
• Broths, Media and Sera	• Gels and Gel Materials	• MRO Supplies	• Proteins and Peptides
• Chemicals	• Glassware	• Modifying Enzymes	• Restriction Enzymes
• Chromatography Supplies	• Kits	• Nucleic Acids	• Vectors

supplier's sites and making them accessible through one common gateway, often a frame around the target site, or by taking individual supplier's product information and providing each supplier a store within the mall. In the first approach, the researcher is forced to learn the layout of each supplier's website. This is similar to learning the layout of each supplier's paper catalog. The latter mall scenario organizes each store similarly although the user must hop from store to store within the mall. Product comparison, in the mall model, requires the use of multiple windows or browsers. Given the increasing demands of browsers, this can be a burden on computer resources.

For the rest of this discussion we will concentrate on how we have constructed the marketplace resource at `www.atcg.com`. This resource is comprised of two components:

- The *ATCG* catalog, a consolidated, source of information
- *ProductWindow*™, an electronic commerce service.

The *ATCG* catalog is a comparative marketplace. The catalog is a searchable information source that currently contains close to a million items from thousands of suppliers and manufacturers in a single product database. We have endeavored to make it an aggregated source of relevant product information allowing users to quickly find and compare product offerings and then order products from many suppliers at one time.

The open construction of the resource allows suppliers and scientific data editors to maintain the product data in the catalog **(Table 3)**.

The goal is to provide enough relevant technical information in each of the product listings, in the catalog, so that users can make purchasing decisions based on the information they find there. Links to additional information, such as buffer compositions, that aid in product decision making are also available. Each product listing includes the supplier's name with a link to a supplier information page that contains supplier contact information and links to the supplier's online site and e-mail address (*see* **Fig. 1**). Suppliers can elect to link graphics or to link catalog numbers back to their site for additional technical or marketing information.

Our marketplace is grouped by product category, not by supplier, creating one large searchable, comparative information source. This allows users to compare like products from multiple suppliers, on one page, and to select the appropriate product for their needs. Information can be sorted by price, unit, price/unit or supplier to facilitate final selection.

Products can be found in four ways: by product category; by catalog number; by text; and by supplier (*see* **Fig. 2**). Product categories use a **Selection Criteria** page to help the user fine-tune the search. Selection criteria are customized to each product category. Multiple items in any criteria list, including suppliers, can be selected.

As a service to the scientific community, the product resource at `www.atcg.com` is displayed publicly for access by all researchers to look up product information. Registered U.S. and Canadian users can also access list pricing. Anderson Unicom Group also provides custom electronic catalog development and implementation of *ProductWindow*™, our electronic commerce service. During a custom organization-wide implementation, organization-specific suppliers are added to the marketplace as is negotiated pricing with availability restricted to authorized users at that particular organization.

4.3. Ordering Online

Web-base ordering systems, such as the one offered by the Anderson Unicom Group or those available at individual suppliers' sites, allow users to assemble and process orders anytime, 24 hours a day, 7 days a week.

A general scenario of an online ordering procedure is as follows (**Fig. 3**):

1. Locate the site which has the desired products. This can be a supplier's site, a marketplace, or your organization's web-based catalog.
2. At some point, regardless of where you wish to purchase online, you will be required to register and provide contact, ship-to, and bill-to information. This information may be stored on your computer in the form of a "cookie." If you change or delete this file or access the site from another location you may have to re-register or log-in again. If you have already registered at the desired site, then log-in using the assigned, or selected, username and password to gain access to ordering utilities. Pre-entering information allows forms to auto-fill with information from your profile, eliminating errors and the hassles of looking up information.

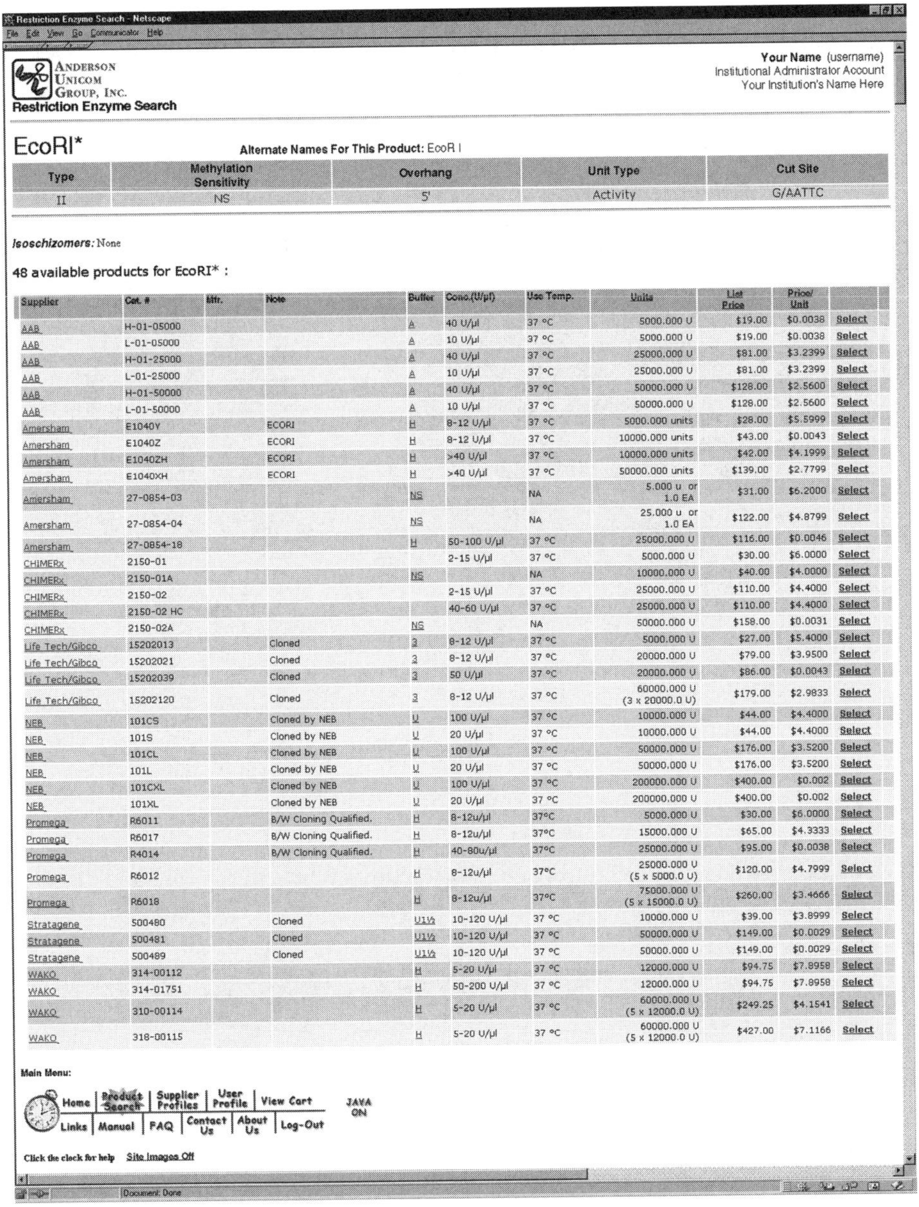

Restriction Enzyme Search - Netscape

File Edit View Go Communicator Help

ANDERSON UNICOM GROUP, INC.

Restriction Enzyme Search

Your Name (username)
Institutional Administrator Account
Your Institution's Name Here

EcoRI*

Alternate Names For This Product: EcoR I

Type	Methylation Sensitivity	Overhang	Unit Type	Cut Site
II	NS	5'	Activity	G/AATTC

Isoschizomers: None

48 available products for EcoRI* :

Supplier	Cat. #	Mfr.	Note	Buffer	Conc.(U/µl)	Use Temp.	Units	List Price	Price/ Unit	
AAB	H-01-05000			A	40 U/µl	37 °C	5000.000 U	$19.00	$0.0038	Select
AAB	L-01-05000			A	10 U/µl	37 °C	5000.000 U	$19.00	$0.0038	Select
AAB	H-01-25000			A	40 U/µl	37 °C	25000.000 U	$81.00	$3.2399	Select
AAB	L-01-25000			A	10 U/µl	37 °C	25000.000 U	$81.00	$3.2399	Select
AAB	H-01-50000			A	40 U/µl	37 °C	50000.000 U	$128.00	$2.5600	Select
AAB	L-01-50000			A	10 U/µl	37 °C	50000.000 U	$128.00	$2.5600	Select
Amersham	E1040Y		ECORI	H	8-12 U/µl	37 °C	5000.000 units	$28.00	$5.5999	Select
Amersham	E1040Z		ECORI	H	8-12 U/µl	37 °C	10000.000 units	$43.00	$0.0043	Select
Amersham	E1040ZH		ECORI	H	>40 U/µl	37 °C	10000.000 units	$42.00	$4.1999	Select
Amersham	E1040XH		ECORI	H	>40 U/µl	37 °C	50000.000 units	$139.00	$2.7799	Select
Amersham	27-0854-03			NS		NA	5.000 u or 1.0 EA	$31.00	$6.2000	Select
Amersham	27-0854-04			NS		NA	25.000 u or 1.0 EA	$122.00	$4.8799	Select
Amersham	27-0854-18			H	50-100 U/µl	37 °C	25000.000 U	$116.00	$0.0046	Select
CHIMERx	2150-01				2-15 U/µl	37 °C	5000.000 U	$30.00	$6.0000	Select
CHIMERx	2150-01A			NS		NA	10000.000 U	$40.00	$4.0000	Select
CHIMERx	2150-02				2-15 U/µl	37 °C	25000.000 U	$110.00	$4.4000	Select
CHIMERx	2150-02 HC				40-60 U/µl	37 °C	25000.000 U	$110.00	$4.4000	Select
CHIMERx	2150-02A			NS		NA	50000.000 U	$158.00	$0.0031	Select
Life Tech/Gibco	15202013		Cloned	3	8-12 U/µl	37 °C	5000.000 U	$27.00	$5.4000	Select
Life Tech/Gibco	15202021		Cloned	3	8-12 U/µl	37 °C	20000.000 U	$79.00	$3.9500	Select
Life Tech/Gibco	15202039		Cloned	3	50 U/µl	37 °C	20000.000 U	$86.00	$0.0043	Select
Life Tech/Gibco	15202120		Cloned	3	8-12 U/µl	37 °C	60000.000 U (3 x 20000.0 U)	$179.00	$2.9833	Select
NEB	101CS		Cloned by NEB	U	100 U/µl	37 °C	10000.000 U	$44.00	$4.4000	Select
NEB	101S		Cloned by NEB	U	20 U/µl	37 °C	10000.000 U	$44.00	$4.4000	Select
NEB	101CL		Cloned by NEB	U	100 U/µl	37 °C	50000.000 U	$176.00	$3.5200	Select
NEB	101L		Cloned by NEB	U	20 U/µl	37 °C	50000.000 U	$176.00	$3.5200	Select
NEB	101CXL		Cloned by NEB	U	100 U/µl	37 °C	200000.000 U	$400.00	$0.002	Select
NEB	101XL		Cloned by NEB	U	20 U/µl	37 °C	200000.000 U	$400.00	$0.002	Select
Promega	R6011		B/W Cloning Qualified.	H	8-12u/µl	37°C	5000.000 U	$30.00	$6.0000	Select
Promega	R6017		B/W Cloning Qualified.	H	8-12u/µl	37°C	15000.000 U	$65.00	$4.3333	Select
Promega	R4014		B/W Cloning Qualified.	H	40-80u/µl	37°C	25000.000 U	$95.00	$0.0038	Select
Promega	R6012			H	8-12u/µl	37°C	25000.000 U (5 x 5000.0 U)	$120.00	$4.7999	Select
Promega	R6018			H	8-12u/µl	37°C	75000.000 U (5 x 15000.0 U)	$260.00	$3.4666	Select
Stratagene	500480		Cloned	U1½	10-120 U/µl	37 °C	10000.000 U	$39.00	$3.8999	Select
Stratagene	500481		Cloned	U1½	10-120 U/µl	37 °C	50000.000 U	$149.00	$0.0029	Select
Stratagene	500489		Cloned	U1½	10-120 U/µl	37 °C	50000.000 U	$149.00	$0.0029	Select
WAKO	314-00112			H	5-20 U/µl	37 °C	12000.000 U	$94.75	$7.8958	Select
WAKO	314-01751			H	50-200 U/µl	37 °C	12000.000 U	$94.75	$7.8958	Select
WAKO	310-00114			H	5-20 U/µl	37 °C	60000.000 U (5 x 12000.0 U)	$249.25	$4.1541	Select
WAKO	318-00115			H	5-20 U/µl	37 °C	60000.000 U (5 x 12000.0 U)	$427.00	$7.1166	Select

Main Menu:

Home | Product Search | Supplier Profiles | User Profile | View Cart
Links | Manual | FAQ | Contact Us | About Us | Log-Out

JAVA ON

Click the clock for help   Site Images Off

Document: Done

Fig. 1. A search for the restriction enzyme *Eco*R1 results in a listing of 48 products from a range of suppliers.

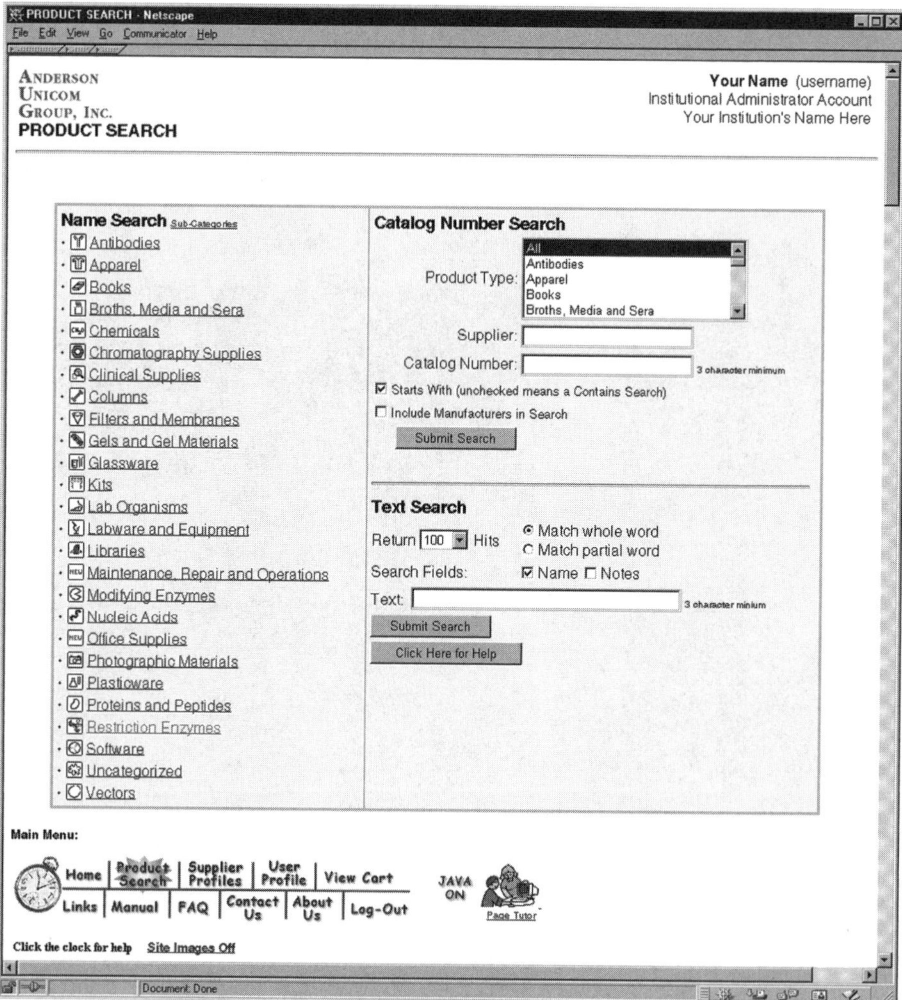

Fig. 2. The search page in the ATCG catalog provides a choice of four methods to find products: by product category, by catalog number, by supplier, and by text.

3. Next, search or browse the site to locate the desired products and select these items to be ordered. Sites typically either use a "shopping cart" or online order form. Be aware that not all sites use carts and many delete the cart if the user leaves the site and comes back later.
4. Once items are in the shopping cart, or on the order form, adjust the quantity, add additional items, or delete items.

5a. If the site uses a cart, then the next step is to submit the cart. This step assembles the order. Depending on the site there may be an additional review step in place to allow a last opportunity to change the information in the cart. This is not always

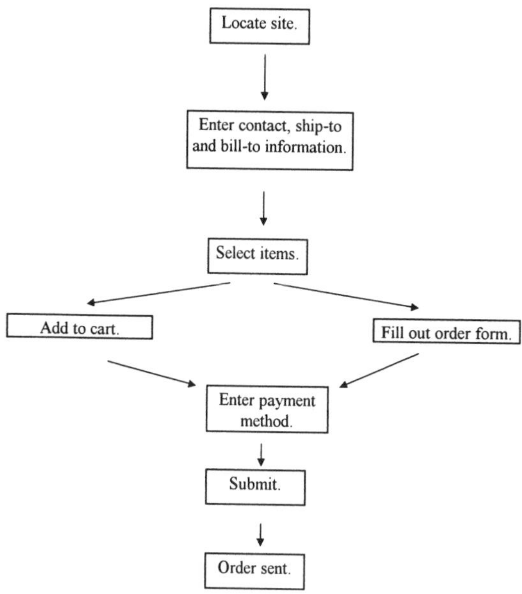

Fig. 3. Ordering online.

the case so don't click too fast. If you are at a marketplace like the *ATCG* catalog, items can be added to a shopping cart from multiple suppliers. In these types of systems, once the cart is submitted, items selected for order are broken into orders for individual suppliers and populate supplier-specific product order forms.

5b. If the site uses an order form, then submit the order form.

6. Either prior to, or after, the cart or order form has been submitted, the ordering system will ask for the method of payment, usually a purchase order or procurement card number.

7. After you enter and submit this information, the order is sent. The cart, or order form, is now empty to begin again.

If the system in use is a local program or contracted service that is set up to feed information into your organization's financial application, then when a shopping cart is submitted for order, the order is electronically routed along a predefined approval path. Administrators set the approval paths and the dollar limits under which no approval is needed. E-mail notices are typically sent to the individuals approving orders informing them that there are orders waiting for their approval. This information can usually be accessed online so users are always aware of an order's status. After order approval, the order is forwarded to the supplier for fulfillment. Order confirmations are sent directly to the user, or the organization, by the supplier.

In organization-wide systems, once orders are placed, information is archived and viewable using online functions to generate custom reports. The researcher has just saved hours of time.

5. Summary

The World Wide Web is changing the face of business today and will continue to reshape it in the coming years. Nonvalue-added processes are being eliminated as organizations reengineer their internal ways of doing business in a continual effort to reduce costs and remain competitive.

Current electronic market followers claim that gateways, or portals, of information will be the wave of the future. This does not mean that individual suppliers' websites will become extinct. It means that content will continue to change and adapt so that sites will continue to be more useful. If gateways do turn out to be the user's choice for product information presentation in the future, expect more consolidated sites devoted to specific product categories, such as chemicals, molecular biology supplies, and the like.

4

Computers and Molecular Biology: *Issues and Constraints*

22

Network Computing

Restructuring How Scientists Use Computers and What We Get Out of Them

Brian Fristensky

1. Introduction

This article describes the network computer (NC), an alternative to the stand-alone PC. By shifting data storage and most processing to the server, any user can do any task from any NC. NCs provide a reliable, consistent interface for all users, and make it easy to provide group access to resources such as laboratory databases. NCs are intrinsically insulated from obsolescence, and offer economies of scale through shared hardware, software, and administration. In the future, stand-alone PC-driven laboratory equipment could be superseded by Java-based NC robots, controlled and monitored across the network.

1.1. The Problem: The Fat Client

The standalone PC is referred to as a "fat client" in that it must have all the hardware and software necessary for every task. The rapid evolution of processing capabilities drives software development to utilize those capabilities. Consequently, most PCs, and their accompanying hardware and software, must be replaced every 3–5 yr.

System administration is the largest hidden cost of PCs. Because each PC is typically configured by different users for different purposes, each PC is a special case. PCs are becoming increasingly complex, particularly because of their increasing integration in networks. Keeping everything working on all machines in a department, and upgrading smoothly, is becoming an increasingly unrealistic goal even for professional PC/LAN administrators.

From: *Methods in Molecular Biology, vol. 132: Bioinformatics Methods and Protocols*
Edited by: S. Misener and S. A. Krawetz © Humana Press Inc., Totowa, NJ

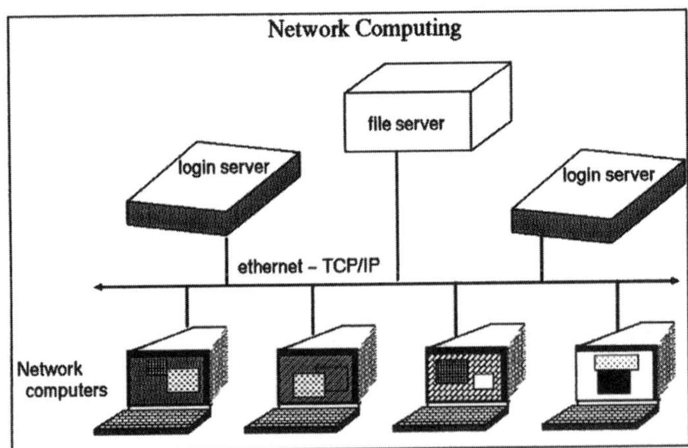

Fig. 1. Network Computing: data and software reside on a central file server, mounted remotely to all servers. Any user can log into any server from any NC, and do any task.

Everyone is familiar with the inconvenience of having to wait to use a program because the only machine on which it is installed is occupied. Again, each PC is a special case. In another example, to put a sequence alignment into a paper, you might first do the alignment on a lab computer that has sequence software, then upload the alignment to a LAN directory, go to your office, download the file, and import it into a word processor. Then you decide you want to present the alignment in a slightly different way. You must then repeat all of these steps.

Fragmentation of data and software among PCs wastes time, causes frustration, and makes it difficult to remember where a specific file is, or where the most recent version of the file is. Also, since shared PCs are seldom backed up, and PC-based operating systems almost entirely lack security features, other users can, through mishap or malice, destroy valuable data.

1.2. The Solution: The Thin Client

Early computers were centrally administered, serving numerous users via remote text-only terminals. Today, systems like UNIX can provide dozens or hundreds of users with a point-and-click desktop. Several variations of network computing are being developed. The *X11*, or *X-Windows* system, pioneered at MIT in the mid 1980s and now developed by The Open Group *(1)*, is the most stable and widely used protocol. As illustrated in **Fig. 1**, a user connects to a login server, which sends *X11* commands to the NC. The *X11* commands specify the content and location of each window as it appears on the

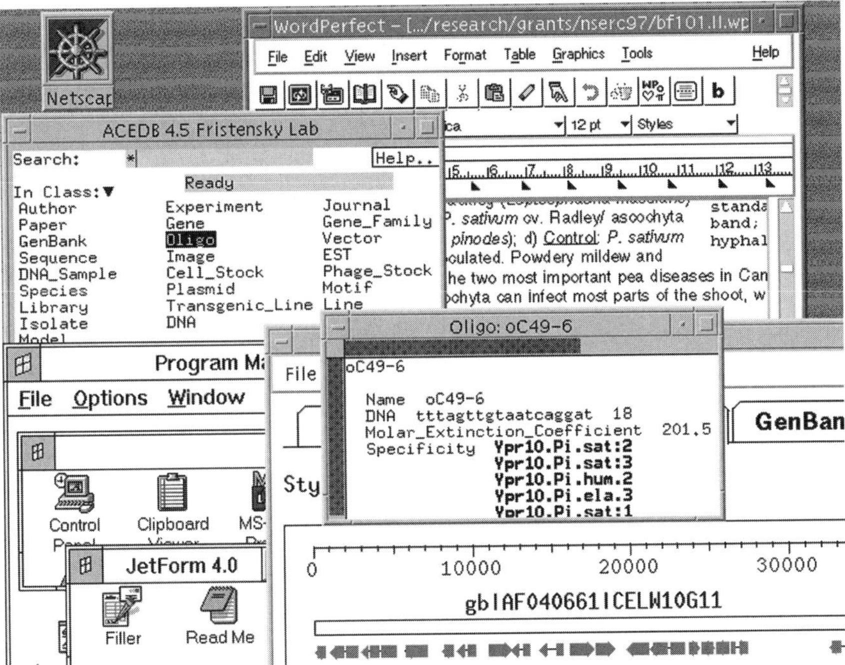

Fig. 2. Screen shot of a typical UNIX session, illustrating the wide range of desktop and scientific software available.

screen. Typically, most of the work done by a program involves redrawing of windows as you scroll, cut and paste, or open or close windows. *X11* offloads these tasks to the NC, reducing load on the server. Because all other tasks are done at the server, the NC is referred to as a "thin client." *X*-terminals are diskless, and do not themselves run application programs.

Where multiple servers are available, all software and files are centralized on a file server, whose file systems are remotely mounted across the network to all login servers. Regardless of which server you log into, you will have the same desktop and the same home directory. Put simply, the central concept of network computing is that any user can do any task at any NC.

Figure 2 illustrates the capabilities available to users on our Sun/UNIX system at the University of Manitoba. At top, a grant proposal is being written using *WordPerfect* for UNIX *(2)*. Background information on oligonucleotides is obtained from our *AceDB* **(3)**-driven lab database. Information on genes to be used in the project is accessed from *GenBank* using the *NEntrez* network client *(4)*, in the background at bottom right. At lower left, the *MS-Windows Program Manager* runs via Sun's *Wabi* **(5)**, making it possible to fill out the spread-sheet part of the grant using *JetForm*. Finally, a Web-browser can be opened at the click of an icon when information is needed from the Web.

On such a system, software is usually installed by the system administrators, but there is nothing to stop the user from installing or writing their own software. To illustrate, all development and management of our university-wide sequence analysis facility, *BIRCH*, was done on a regular user account without root privileges ***(6)***.

The main point I wish to make is that NCs allow you to do all tasks commonly done on PCs. I have used UNIX exclusively for all my computing since 1990 and my lab does not even have a PC. Our first *X*-terminal, purchased in 1993, still lets us run the latest software on the server. Upgrades to the operating system and applications software, and even an upgrade to a 64-bit server, were entirely transparent, requiring no reconfiguration of our terminals.

2. Hardware, Software, and Know-How

2.1. Hardware

You will need a UNIX-based workstation, such as a Sun/Solaris, IBM/AIX, or Intel/Linux. Avoid buying a 32-bit workstation. Most versions of UNIX are either completely adapted to 64-bit architecture, or will be in the next few years. If your institution provides centralized UNIX servers, you may not even need to buy a server. You will also need an *X*-terminal, or a PC running *X-Windows* emulation software (*see* **Subheading 3.5.**). The most important single factor to ensuring smooth performance is RAM. The more memory the server has, the less frequently it will need to swap programs and data to disk. Additional memory can often be purchased cheaply from third party vendors who routinely advertise on Internet newsgroups. Additionally, most versions of UNIX can support two or more CPUs.

2.2. Software

Because almost everything related to the Internet was invented under UNIX, typical configurations come standard with full Internet services, including telnet, ftp, e-mail programs, and a Web browser such as *Netscape*. UNIX systems typically also include the *X11 Display Manager* (xdm) that serves *X*-terminals. If the computer center at your institution already operates networked UNIX servers, they may be willing (perhaps for a small fee) to administer your server as a clone of their other machines. This has the advantage that if your server needs to be down, you can log into any other server. To cite an example, in 1996 I moved my home directory from my personal workstation to our campus system. During a 3-month interim prior to purchasing a new server for several lab groups in our department, I used the publicly-available dual-

CPU Sun Sparc20 servers, which typically had 30–60 simultaneous users. With few exceptions, I seldom noticed any degradation in performance.

2.3. Know-How

If your computer center administers your server as described above, you reap the greatest benefit of the NC model: Let the experts do the work. This is particularly important with respect to security. Most professionally administered systems have someone who keeps up on the latest security holes, and installs the appropriate patches in a timely fashion. Alternatively, someone in your lab or department will need a working knowledge of UNIX, some previous experience in programming (preferably C or Java) and an ability to write HTML. All other users will benefit from an introduction to UNIX *(7)*, but by and large, the skills and concepts learned on Windows-based systems will easily transfer to UNIX desktops.

3. Practical Matters

Network computing is still evolving, and there will probably be many different implementations of the thin client model. As the UNIX/*X-windows* approach to network computing has been fairly stable for almost a decade, and is likely to continue to be one of the major network computing models, I will discuss some of the things you need to know to do all your computing on an *X-windows* platform.

3.1. Make a Complete Switch

When you move to the NC platform, do it all the way. If you divide your computing and datafiles between NCs and PCs, you will actually make things more complicated. Also, you will learn your way around server-based computing faster if you use it for everything.

3.2. The Third Party Software Problem

The chief impediment to the growth of network computing is the perceived lack of third party software for server-based platforms. At one level it is true that the majority of desktop software vendors write specifically for PCs. Whereas use of Java may result in platform independence in the future, at present it is often more difficult to find applications for server-based systems such as UNIX, VMS, or AS400. On the other hand it is surprising how much software is available for these platforms. In many cases, server-based versions of programs such as *WordPerfect (2)* or *Adobe PhotoShop (8)* are available. In other cases, comparable applications exist that are specifically targeted to client/server platforms.

3.3. Using Windows Applications under UNIX

Generally, it is best to use native UNIX applications wherever possible. When no UNIX version is available, Windows applications can be displayed from a *WindowsNT* server to an *X-Windows* desktop via NCD's *Wincentre (9)*. This approach has the advantages that applications run on native *Intel* architecture, and some tasks are offloaded to the NT server. The disadvantages are that a dedicated NT server is needed, and the NT server must be configured to work with the UNIX file server. On Sun systems, *Wabi (5)* can run *Windows 3.1* in software emulation.

3.4. X11 Programs from Remote Servers

Sometimes it is desirable to run an *X-Windows* application on a server other than the one you are logged into currently. For example, an application might be licensed to run on only a few servers. This problem can be solved easily by logging into the licensed server and setting the *X11* display to your terminal or workstation.

If a user named raven is logged into marigold.uofm.ca, but wants to run *SAS,* which is installed on petunia.uofm.ca. Log into petunia using telnet:

```
{marigold:/home/raven}telnet petunia
Trying 130.122.36.48 . . .
Connected to petunia.uofm.ca.

UNIX (r) System V Release 4.0 (petunia) (pts/18)

login: raven

Password:

{petunia:/home/raven}
```

Next, set the environment variable **DISPLAY** to the name of your terminal or workstation.

```
{petunia:/home/raven}setenv DISPLAY ncd12.uofm.ca:0.0
```

This command will cause all subsequent *X11* programs launched in this shell to display on the screen of the *X*-terminal called ncd12.

Finally, launch *SAS*:

```
{petunia:/home/raven}sas &
```

Note the ampersand (&) after the command. Putting an ampersand after any UNIX command causes it to run in the background. This does two things. First, it frees up the command window to let you do other things as the application is running. Secondly, if you wanted to, you could log out from petunia without causing *SAS* to terminate.

For applications that frequently need to be run from remote servers, those steps could be automated by putting these commands into a shell script, which could be launched from the workspace menu.

3.5. Turning PCs into NCs

Many labs have made large investments in PC hardware. As well, most departments are full of older PCs that can no longer run the latest software (e.g., 486s that can not run *Windows'95*). Finally, it is usually not feasible to purchase a large number of NCs at one time. For all these reasons, a number of *MS Windows* and/or Mac-based programs allow a PC to act as an *X*-terminal.

Some of the pluses of *X11* emulation software include:

- Inexpensive.
- The software has been around long enough to be reasonably reliable and easy to use and install.
- Even 486s will often perform almost as quickly as an *X*-terminal. Also, the task of drawing windows on a screen always remains about the same, so once an old 486 works, it should always work.
- Typically includes network transport protocols such as *SLIP* and *PPP*, making it possible to run an *X-Windows* session over a fast modem from home.

There are some very good reasons why *X* emulation software is an interim fix, rather than a permanent solution:

- You still have to keep *MS Windows* working. Any time you upgrade *Windows*, or install software, or alter the PC networking software, on the PC or on the LAN server, you risk affecting the *X* emulation program.
- *X* emulators are not perfect. Because the PC and Windows add a layer of complexity, there is never a guarantee that the *X11* software will do everything that an *X*-terminal is supposed to do.
- If you are using a PC with an *X* emulator, there is a temptation to do some things on the *X* desktop, and some things on the PC. Thus, you fragment your files and necessitate uploading and downloading of information, and have less incentive to really learn how to use the *X* desktop. Things are simpler if you stick to one system.

The strategy should therefore be to upgrade existing PCs to *X* emulation in the short term, and in the long term, buy new *X*-terminals, rather than new PCs. PC *X* emulation programs include Hummingbird's *eXceed* ***(10)***, White Pine Software's *eXodous* ***(11)***, and NCD's *PC-Xware* ***(12)***. Many of these vendors offer free downloads of a trial copy of the program.

4. The Future: How Network Computers Will Change the Way We Work

The purpose of this section is to show the kinds of developments that become possible once network computing becomes commonplace. I will stick to things that are already in development or are already being implemented, with the exception of **Subheading 4.4.**, which is a synthesis of these developments.

4.1. Work the Same Way Anywhere

Today, each researcher and student is wedded to a single PC, or worse, to several PCs specialized for different tasks. We have to copy files to diskette to work at home, or carry a laptop when we travel.

Network computers are even now beginning to appear in hotels, airports, and university computer centers and libraries. In the near future, you will be able to use a network computer at home to work on papers, run resource-intensive computations, or even check on an experiment in progress (*see* **Subheading 4.4.**). Because NCs are cheap, they will be everywhere. During airport layovers, in your hotel room, or even on sabbatical half way around the world, you can do anything you can do in your lab or office, *in exactly the same way*. You will no longer be limited to the files, software, and computing power that you can carry with you.

To simplify the use of NCs by the traveling public, companies such as Network Computer Inc. ***(13)*** have developed smart cards that carry the information required to find your home server across the Internet and to connect you to your account.

4.2. Electronic Seminars, Presentations, Teaching

Classroom and symposium presentations are already being transformed by Web-based presentations. However, Web browsers are still limited compared to NCs. For example, I present all lectures for my cytogenetics course using an *X* terminal from which the screen output is sent to a 1024 × 768 projector. Whereas most of the lecture is on the course Web site ***(14)***, I often use graphics applications for simple demos. The server-based nature of the terminal guarantees that the demo will work in class exactly the way it worked in my office.

As conference centers adapt to network computing, symposium speakers will have all figures and programs from their home server available at the podium. Even unanticipated data or figures could be presented in response to questions.

4.3. Java: Write Once, Run Anywhere

Java ***(15)***, the new programming language from Sun Microsystems, was specifically designed to run on any computer platform. This is how it works: Java programs are run within a shell called the Java virtual machine. Because the Java virtual machine has been ported to virtually all computer platforms (e.g., UNIX, Windows, OpenVMS, IBM mainframes, Macintosh, and so on) all Java programs should run on all platforms. All you need is the Java virtual machine. Software written in Java is therefore platform independent. Software developers need only write and maintain one Java version of the software, rather than many versions for many platforms.

The Java *Molecular Biology Workbench* ***(16)*** is an example of a suite of Java programs. In this case, the programs are run as applets, which are downloaded from the remote server at runtime. Java programs can also be downloaded and run as standalone applications on one's local server, workstation or PC.

One of the main advantages of Java is that it is modular. Applications written in traditional languages are single entities taking up many megabytes of memory. Java applications are packages of small objects, each carrying out a single function. When a NC runs a Java application, only the objects needed at a given time are downloaded from the server. Consequently, Java-based NCs don't need large amounts of memory and processing power, which provides some protection from obsolescence.

4.4. Clean integration of Computing Platform, Network, Lab Notebook, and Lab Equipment

Java was conceived originally as a hardware-independent language to allow electronic devices to be programmed, rather than having their capabilities hardwired. That aspect of Java may lead to networked laboratory equipment that is far more versatile and upgradeable than the machines currently in our labs. Today, laboratory devices such as fluorescent imagers, DNA sequencers, and even plant growth chambers each require a PC to operate them. Because software for analysis and data acquisition both usually reside on the dedicated PC, you cannot analyze your data if the device is in use by someone else. In the future, many devices will be networked Java NCs. One obvious result is that it will no longer be necessary to purchase and configure a PC for each device. Dedicated hardware such as monitors, printers, or LCD displays will also be unnecessary, making Java devices smaller and cheaper. A Java chip ***(17)*** resident in each device will perform all operations. Each device can be controlled and monitored from any NC. At the completion of the experiment, data can be uploaded directly to the user's directory for analysis.

More exciting is the concept of the virtual robot. In principle, virtual robots could be created on the desktop by linking Java devices together in an equipment control program. In this example, the equipment control program represents each real device by a screen icon. Note that the pipetting robot, which is called by each device, does not need to be represented in the control program. (This is analogous to class inheritance in object-oriented languages like Java). In **Fig. 3**, a DNA sequence analysis program is used to design primers, which are sent to the equipment control program. A DNA sequencing robot is created by linking a DNA synthesizer, thermal cycler, and DNA sequencer in succession. Results are relayed from the sequencer via the control program, into a DNA sequence analysis program. The electronic lab notebook is also a Java device, and can be used to tell the program which DNA samples to use for sequencing, or where to store samples generated by the thermal cycler.

Whereas multicomponent robots could be created using existing PCs and operating systems, each virtual device is a special case requiring extensive programming to implement. A software-based control program would make it possible to tailor virtual robots for each task. For example, if you wanted to quantitate your PCR product before loading onto the sequencer, a fluorescence detector might be linked between the thermal cycler and the sequencer, and only samples that successfully amplified would be sequenced.

5. Looking Back at Today from Tomorrow's Perspective

When we look back at the PC era in five or ten years, we will be amazed at the things we took for granted. Most ridiculous will be the notion that every user was a system administrator. In the NC era, servers will be professionally administered. By their nature, there is essentially nothing to administer on an NC. All the user will do is use them.

The economics of computing will be changed, breaking the obsolescence cycle. Users will still spend money on computing, but rather than personally buying RAM, disk drives, coprocessors, and software, we will pay our service providers a monthly fee to provide these things for us. Professionally maintained centralized resources will be more stable and reliable.

Today, most users are locked into the *MS Windows* operating system. In the NC era, even desktops and operating systems will be subject to competition, resulting in more choices and competitive pricing. This article has focused on UNIX, largely because UNIX is particularly well suited to network computing. However, the open nature of the NC model means that other systems such as OpenVMS, AS/400, and possibly WindowsNT (if scalability and security issues can be resolved) could play a role in providing NC services. One possible outcome is that NC-service providers will use a mixture of different servers and operating systems to deliver a complete range of applications to a single desktop, in a fashion that is transparent to the user. Regardless of the changes at the server end, the user's investment in NC hardware will be protected, because the thinner the client, the less there is that can become obsolete.

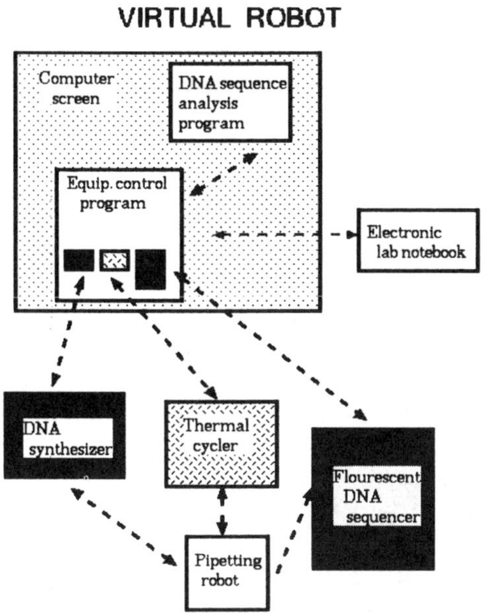

Fig. 3. The virtual robot: Networked devices programmed in Java can be linked together via software to create virtual robots tailored for a specific task, and controlled from the NC.

Our anticipatory retrospective on the PC era was best summarized in 1977 by Ken Olsen, Chairman and founder of DEC: "There is no reason anyone would want a computer in their home" (or lab, B.F.).

For more information on Network Computing, and additional Internet links, *see* `http://home.cc.umanitoba.ca/~psgendb/nc`.

References

1. The Open Group (1998) *X Window System*, `http://www.camb.opengroup.org/`.
2. Corel *WordPerfect* for UNIX, `http://www.corel.com/products/cwp7unix.htm`.
3. Durbin, R. and Thierry-Meig, J. (1991) A C. elegans Database. Code and data available from anonymous FTP servers at `lirmm.lirmm.fr`, `cele.mrc-lmb.cam.ac.uk` and `ncbi.nlm.nih.gov`.
4. Natl. Center for Biotech. Info. Nentrez, `http://www3.ncbi.nlm.nih.gov/Entrez/Network/nentrez/overview.html`.
5. Sun Microsystems, *Wabi*, `http://www.sun.com/solaris/wabi/`.
6. Fristensky, B. (1999) Building a multiuser sequence analysis facility using freeware, this volume, pp. 131–198.

7. Sobell, Mark G. (1995) A Practical Guide to the UNIX System. Addison-Wesley Publishers, Reading, MA.
8. Adobe Corp. *PhotoShop*. http://www.adobe.com.
9. Network Computing Devices Wincenter. http://www.ncd.com/pwin/pwin.html.
10. Hummingbird Ltd., http://www.hummingbird.com/products/exceed/.
11. White Pine Software, http://www.wpine.com/exodus/.
12. Network Computing Devices, http://www.ncd.com/ppcx/ppcx.html.
13. Network Computer Inc. http://www.nc.com/prodcard.html.
14. Fristensky, B. *Introductory Cytogenetics*, Univ. of Manitoba. http://www.umanitoba.ca/afs/plant_science/COURSES/CYTO/.
15. Sun Microsystems. *The Java Platform*. http://java.sun.com/aboutJava/.
16. Toldo, L. (1997) *JaMBW 1.1: Java-based molecular biologists' workbench. Comput. Appl. Biosci.* **13,** 475–476. http://www.embl-heidelberg.de/~toldo/JaMBW.html.
17. Sun Microsystems. *Java Computing*. http://www.sun.com/java/.

23

Computing with DNA

Lila Kari and Laura F. Landweber

1. A New Player in the History of Computation

A brief look at the history of humanity shows that since the earliest days people needed to count and compute, either for measuring the months and the seasons or for commerce and construction. The means used for performing calculations were whatever was available, and thus progressed gradually from manual (digits) to mechanical (abacus, mechanical adding engine), and from there on to electronic devices. Electronic computers are only the latest in a long chain of human efforts to use the best technology available for performing computations. Although it is true that their appearance, some 50 years ago, has revolutionized computing, electronic computers mark neither the beginning nor the end of the history of computation. Indeed, even electronic computers have their limitations: There is a limit to the amount of data they can store, and physical laws dictate the speed thresholds they will soon reach. The most recent attempt to break down these barriers is to replace, once more, the tools for performing computations with biological ones instead of electrical ones.

DNA computing (also sometimes referred to as biomolecular computing or molecular computing) is a new computational paradigm that employs (bio)molecule manipulation to solve computational problems, at the same time exploring natural processes as computational models. Research in this area began with an experiment by Leonard Adleman, who surprised the scientific community in 1994 *(1)* by using the tools of molecular biology to solve a difficult computational problem. Adleman's experiment solved an instance of the Directed Hamiltonian Path Problem solely by manipulating DNA strands. This marked the first solution of a mathematical problem by use of biology.

From: *Methods in Molecular Biology, vol. 132: Bioinformatics Methods and Protocols*
Edited by: S. Misener and S. A. Krawetz © Humana Press Inc., Totowa, NJ

Computing with biomolecules (mainly DNA) generated a tremendous amount of excitement by offering a brand new paradigm for performing and viewing computations. The main idea was the encoding of data in DNA strands and the use of tools from molecular biology to execute computational operations *(1a)*. Besides the novelty of this approach, molecular computing has the potential to outperform electronic computers. For example, DNA computations may use a billion times less energy than an electronic computer, while storing data in a trillion times less space *(2)*. Moreover, computing with DNA is highly parallel: In principle there could be billions upon trillions of DNA molecules undergoing chemical reactions, that is, performing computations, simultaneously *(3)*.

Despite the complexity of this technology, the idea behind DNA computing follows from a simple analogy between the following two processes, one biological and one mathematical:

a. The complex structure of a living organism ultimately derives from applying a set of simple operations (copying, splicing, inserting, deleting, and so on) to initial information encoded in a DNA sequence.
b. Any computation, no matter how complex, is the result of combining very simple basic arithmetical and logical operations.

Adleman realized that the two processes are not only similar but that advances in molecular biology allow one to use biology to do mathematics. More precisely, DNA strings can encode information while various molecular biology laboratory techniques perform simple operations. (The reader is referred to **ref.** *4* for further molecular biology notions.) These practical possibilities of encoding information in a DNA sequence and performing simple DNA strand manipulations led Adleman *(1)* to solve a seven node instance of the Directed Hamiltonian Path Problem.

A directed graph G with designated nodes v_{in} and v_{out} is said to have a Hamiltonian path if and only if there exists a sequence of compatible one-way edges $e_1, e_2, ..., e_z$ (that is, a path) that begins at v_{in}, ends at v_{out} and enters every other node exactly once. A simplified version of this problem, known as the traveling salesman problem, poses the following question: given an arbitrary collection of cities through which a salesman must travel, such as the graph in **Fig. 1**, what is the shortest route linking those cities? Adleman's version limited the number of connecting routes between the cities by specifying the origin and final destination cities of his journey. Because not all cities are connected, the challenge was to discover a continous path to link them all, if one exists.

The following (nondeterministic) algorithm solves the problem:

1. Generate random paths through the graph.
2. Keep only those paths that begin with v_{in} and end with v_{out}.

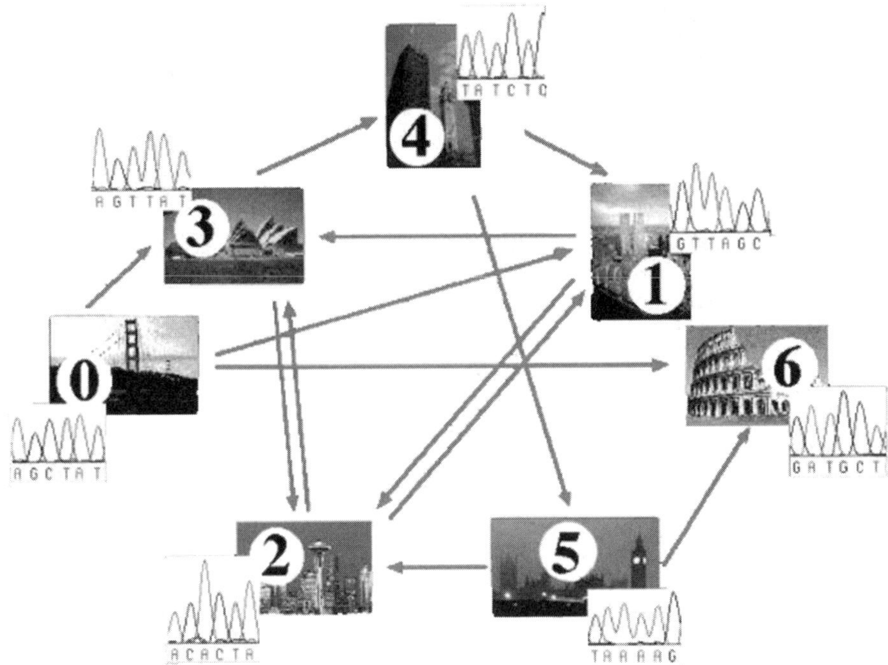

Fig. 1. An example of the graph used in Adleman's experiment *(1)*. Cities, or nodes, are represented as arbitrary DNA sequences. The traveling salesman must find the simplest path which takes him through all seven cities shown, in this case departing from San Francisco (city 0) and arriving in Rome (city 6) as the final destination.

3. If the graph has n nodes, then keep only those paths that enter exactly n nodes.
4. Keep only those paths that enter all of the nodes of the graph at least once.
5. If any paths remain, say "yes"; otherwise say "no".

To implement **step 1**, each node of the graph was encoded as a random 20-base strand of DNA, or oligonucleotide. Then, for each (oriented) edge of the graph, a different 20-base oligonucleotide was generated that contains sequences complementary to the second half of the source node plus the first half of the target node. By using these complementary DNA oligonucleotides as splints, all DNA sequences corresponding to compatible edges would self-assemble and be ligated, or linked together, by the enzyme T4 DNA ligase. Hence, annealing and ligation reactions generated DNA molecules encoding random paths through the graph (**Fig. 2**).

To implement **step 2**, the product of **step 1** was amplified by polymerase chain reaction (PCR) using oligonucleotide primers representing v_{in} and v_{out}.

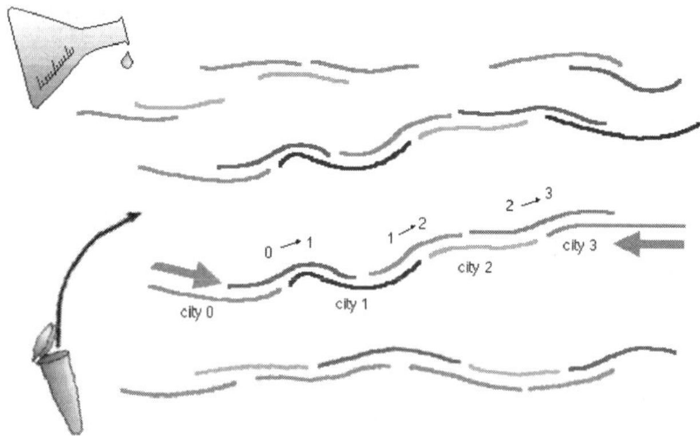

Fig. 2. Self assembly of DNA molecules representing paths through a graph. PCR primers marking origin and final destination oligonucleotides (cities 0 and 3 here) are shown as arrows. Complementary overlap exists between the second half of the sequence representing city i and the first half of a sequence representing edge $i \rightarrow j$, and also between the second half of the sequence representing $i \rightarrow j$ and the first half of the sequence representing city j.

This amplified and thus retained only those molecules encoding paths that begin with v_{in} and end with v_{out}.

For implementing **step 3**, agarose gel electrophoresis allowed separation and recovery of DNA strands of the correct length. The desired path, if it exists, would pass through all seven nodes, each of which was assigned a length of 20 bases. Thus PCR products encoding the desired path would have to be $7 \times 20 = 140$ bp.

Step 4 was accomplished by successive use of affinity purification for each node other than the start and end nodes. This process permits the separation and recovery of single strands encoding a given node from a heterogeneous pool of strands. DNA strands complementary to the node sequence were attached to magnetic beads. The heterogeneous solution containing single-stranded DNA was then passed over the beads and those strands containing the node sequence were selectively retained. Strands that lack one of the required node sequences generally do not survive **step 4**, because they pass through at least one of the columns without being retained.

To implement **step 5**, the presence of a molecule encoding a Hamiltonian path was checked by PCR. The first PCR amplified the results of **step 4** and checked for the presence of a product, as in **step 2**. If a product was present, then a second PCR confirmed the presence of each internal node by using the DNA oligonucleotides complementary to each node as PCR primers. This step also elegantly allowed mapping and readout of connected nodes in the graph, without need for DNA sequencing.

A remarkable observation about Adleman's experimental result is that it not only solved a mathematical problem, but that it was also a difficult computational problem in the sense explained below (see **refs.** ***5*** and ***6***).

Problems can be ranked in difficulty according to the length of time the best algorithm will require to solve the problem on a single computer. Algorithms whose time complexity function is bounded by a polynomial function, in terms of the size of the input describing the problem, are in the polynomial time class *P*. Such algorithms are generally considered efficient. Any algorithm whose time complexity function cannot be so bounded belongs to the inefficient exponential class EXP. A problem is called *intractable* if it is so hard that no polynomial time algorithm can possibly solve it.

A special class of problems, apparently intractable, including *P* and included in EXP is the "nondeterministic polynomial time" class, or *NP*. The following chain of inclusions between problem classes holds:

$$P \subseteq NP \subseteq \text{EXP} \subseteq \text{Universal}$$

NP contains the problems for which no polynomial time algorithm to solve them is known, but that can be solved in polynomial time on a nondeterministic computer (a computer that has the ability to pursue an unbounded number of independent computational searches in parallel). The Directed Hamiltonian Path problem is a special kind of problem in *NP* known as "*NP*-complete." An *NP*-complete problem has the property that every other problem in *NP* can be reduced to it in polynomial time. Thus, in a sense, *NP*-complete problems are the "hardest" problems in *NP*.

The question of whether or not the *NP*-complete problems are intractable, mathematically formulated as "Does *P* equal *NP*?", is now considered one of the foremost open problems of contemporary mathematics and computer science. Because the Directed Hamiltonian Path problem has been shown to be *NP*-complete, it seems likely that no efficient (that is, polynomial time) algorithm exists for solving it with an electronic computer.

Other experiments have followed Adleman's to tackle mathematical problems using DNA manipulation. Kaplan et al. ***(7)*** repeated Adleman's experiment; Guarnieri, Fliss and Bancroft used a horizontal chain reaction for DNA-based addition ***(8)***; a Wisconsin team of computer scientists and biochemists made partial progress towards solving a five-variable instance of the Satisfiability (SAT) problem using surface chemistry ***(9)***; Quyang et al. ***(10)*** solved a six-variable *NP*-complete problem (the Maximal Clique Problem) using restriction enzymes; and most recently one of our laboratories ***(11)*** has solved a nine-variable SAT problem using RNA.

At the same time, numerous experiments have investigated a variety of aspects of the feasibility of DNA computing: They have addressed the effect of good encodings on solutions to Adleman's problem *(12)*; studied the complications raised by PCR *(13)*; investigated the use of self-assembly of DNA *(14)*; pointed out the experimental gap between design and assembly of unusual DNA structures *(15)*; reported joining and rotating data with molecules *(16)*; studied concatenation with PCR *(16,17)*; made progress towards evaluating simple Boolean formulas *(18)*; conducted ligation experiments in computing with DNA *(19)*; implemented an expert "Inference Engine" based on molecular computing *(20)*; and obtained a partial solution to the Shortest Common Superstring Problem *(21)*.

Theoretical studies have supplemented experimental research of DNA algorithms by suggesting potential strategies for solving various problems by means of DNA manipulation. Descriptions of such proposed experiments include the SAT Problem *(22)*, breaking the Data Encryption Standard *(23,24)*, expansions of symbolic determinants *(25)*, matrix multiplication *(26)*, graph connectivity and the knapsack problem using dynamic programming *(27)*, the road coloring problem *(28)*, exascale computer algebra problems *(29)*, the Bounded Post Correspondence Problem *(30)*, and simple Horn clause computation *(31)*.

2. Towards a DNA Computer

The experiments mentioned so far are singular experiments that construct algorithms to solve particular problems. This immediately leads to two fundamental problems, posed already in **refs.** ***1*** and ***6***: What classes of problems can be efficiently solved by DNA algorithms? and Is it possible, at least in principle, to design a programmable DNA computer? Even though the models of DNA computation that have been proposed to answer these questions all differ from each other, they have a number of common features.

Indeed, any kind of computer, whether mechanical, electronic, or biological, needs two basic capacities to function: storage of information and the ability to perform operations on stored data. In the following we address both issues: how can information be stored in DNA strands, and what molecular biology techniques are potentially useful for computation. To distinguish between ordinary mathematical operations and biomolecular procedures performed on DNA strands, we use the term bio-operations to refer to the latter.

A single strand of DNA can be described as a string composed of a combination of four different symbols, A, G, C, T. Mathematically, this means we have at our disposal a four-letter alphabet $\Sigma = \{A, G, C, T\}$ to encode information. Incidentally, this is more than enough, considering that an electronic computer needs only two digits, 0 and 1, for the same purpose.

Concerning the operations performed on DNA strands, the proposed models of DNA computation generally use various combinations of the following "primitive" bio-operations:

— *Synthesizing* a desired polynomial-length strand, used in all models.
— *Mixing*: Combine the contents of two test tubes to achieve union ***(1,32–36)***.
— *Melting*: Dissociate a double-stranded DNA into its single-stranded complementary components by heating the solution ***(35–39)***.
— *Annealing*: Bond together two single-stranded complementary DNA sequences upon cooling the solution ***(35–39)***.
— *Amplifying* (copying): Make copies of DNA strands by using the polymerase chain reaction (PCR) ***(1,25,32–38,40)***.
— *Separating* the strands by length using gel electrophoresis or other size fractionating methods ***(1,32,33,36,37,40)***.
— *Extracting*: Capture strands that contain a given pattern as a substring by affinity purification ***(1,32,34,40)***.
— *Cutting* DNA double strands at specific sites by using commercially available restriction enzymes. ***(37,38,40–42)***.
— *Ligating*: Join DNA strands with compatible sticky ends by using DNA ligases ***(37–42)***.
— *Substituting*: Substitute, insert, or delete DNA sequences by using PCR site-specific oligonucleotide mutagenesis (*see* **refs. *40,43***).
— *Marking* single strands by hybridization: Complementary sequences are attached to the strands, making them double stranded. The reverse operation is unmarking of the double-strands by denaturing ***(9,33,35)***.
— *Destroying* the marked strands by using a variety of nucleases, ***(9,11)***. or by cutting marked strands with a restriction enzyme and purifying intact strands by gel electrophoresis ***(10,33)***.
— *Detecting and reading*: Given the contents of a tube, say "yes" if it contains at least one DNA strand that meets the requirements of the applied operations, and then interpret the sequence; say "no" otherwise, ***(1,32–34,36)***.

A biocomputation consists of a sequence of bio-operations performed on tubes containing DNA strands. The bio-operations listed above, and possibly others, may then be used to write "programs." A program receives a tube containing DNA strands encoding information as input, and returns as output either statements "yes" or "no" or a new collection of tubes.

Various models of DNA computing, based on combinations of the above bio-operations, have been proposed and studied from the point of view of their computational power plus feasibility (*see*, for example, ***1,32,37,39,43–51)***. There are advantages and disadvantages for each of the proposed models but, overall, the existence of different models with complementary features shows the versatility of DNA computing and increases the likelihood of practical construction of a DNA computing device.

Many substantial engineering challenges to constructing a DNA computer remain at almost every stage. These arise primarily from difficulties in dealing with large-scale systems and in coping with ensuing errors ***(52)***. However, we remark that the issues such as active monitoring and adjusting the concentrations of biological molecules, as well as fault tolerance, are all addressed in biological systems by nature: Cells must adjust the concentrations of various compounds, to promote reactions of rare molecules, and they also cope with undesirable byproducts of their own activity. Because cells can successfully manage these problems *in vivo*, this may ultimately suggest strategies we can mimic *in vitro*. Taking a theoretical step in this direction, ***(53)*** suggests the use of membranes to separate volumes (vesicles) and active transport systems to shuttle selected chemicals across these borders ***(53)***. Moreover, familiar computer design principles for electronic computers could be exploited to build biomolecular computers ***(3,54,55)***.

3. A Formal Model for DNA Computing and its Computational Power

One aspect of theoretical research on DNA computing is the search for a suitable formal model to describe molecular computations. This approach often compares the computational power of such a model to the power of a Turing machine, which is the formal model of today's electronic computers.

We illustrate this type of research by *contextual insertion/deletion systems* ***(43,51)*** a formal language model of DNA computing. We show that this model of DNA computation, besides being feasible in the laboratory, has the full power of a Turing machine.

Before formally stating the model, we summarize its terminology ***(56)***. For a set Σ, card(Σ) denotes its cardinality, that is, the number of elements in Σ. An *alphabet* is a finite nonempty set. Its elements are called *letters* or *symbols*. The letters will usually be denoted by the first letters of the alphabet, with or without indices, i.e., a, b, C, D, a_i, b_j, and so on. (In the case of DNA computing, the alphabet at our disposal is $\Sigma = \{A, C, G, T\}$.) If $\Sigma = \{a_1, a_2, \ldots, a_n\}$ is an alphabet, then any sequence $w = a_{i1}a_{i2} \ldots a_{ik}$, $k \geq 0$, $a_{ij} \in \Sigma$, $1 \leq j \leq k$ is called a *string* (*word*) over Σ. The length of the word w is denoted by $|w|$ and, by definition, equals k. The words over Σ will usually be denoted by the last letters of the alphabet, with or without indices, for example x, y, w_j, u_i, and so on. The set of all words consisting of letters from Σ will be denoted by Σ^*.

As a formal language operation, the *contextual insertion* is a generalization of catenation and insertion of strings and languages, ***(57)***: Words can be inserted into a string only if certain *contexts* are present. More precisely, given a set of contexts we add the condition that insertion of a word can be performed only between a pair of words in the context set. Analogously, contextual deletion allows erasing of a word only if the word is situated between a pair of words in the context set.

Besides being theoretically interesting, one of the motivations for studying insertions and deletions is their relevance to laboratory manipulation. Indeed, by using synthetic oligonucleotides and the technique of PCR site-directed mutagenesis ***(58)***, one can insert and delete oligonucleotide sequences in a variety of given contexts.

Kari et al. ***(43,51)*** investigated the mathematical properties of contextual insertions and deletions (below we refer to them as simply insertions and deletions): One of their results is that the actions of every Turing machine can be simulated entirely by insertion and deletion rules. Beaver ***(40)*** proposed that a similar operation, base substitution, simulates a universal Turing machine.

Using insertion-deletion systems, we briefly present several characterizations of recursively enumerable (RE) languages (the equivalents of the Turing machine model of computation). Such a system generates the elements of a language by inserting and deleting words, according to their contexts. Grammars based on insertion rules were already considered ***(59)*** with linguistic motivation. Insertion/deletion operations are also basic to DNA and RNA processing, particular RNA splicing and editing reactions ***(60)***. Our results show that these operations, even with strong restrictions on the length of the contexts and/or on the length of the inserted/deleted words, are computationally complete, that is, they can simulate the work of any Turing machine.

An insertion-deletion (in/del) system, ***(43)***, is a construct

$$\gamma = (V, T, A, I, D)$$

where V is an alphabet, $T \subseteq V$, A is a finite subset of V^*, and I, D are finite subsets of $V^* \times V^* \times V^*$.

The alphabet T is the terminal alphabet of γ, A is the set of axioms, I is the set of insertion rules, and D is the set of deletion rules. An insertion (deletion) rule is written as a triple (u, z, v), which means that z can be inserted in (deleted from) the context (u, v), where u represents the left context and v represents the right context.

For $x, y \in V^*$ we say that x derives y and we write $x \Rightarrow y$ if one of the following two cases holds:

1. $x = x_1uvx_2$, $y = x_1uzvx_2$, for some $x_1, x_2 \in V^*$ and $(u, z, v) \in I$ (insertion)
2. $x = x_1uzvx_2$, $y = x_1uvx_2$, for some $x_1, x_2 \in V^*$ and $(u, z, v) \in D$ (deletion).

Denoting by $\Rightarrow^*$ the reflexive and transitive closure of the relation $\Rightarrow$, the language generated by γ is defined by

$$L(\gamma) = \{w \in T^* | \, x \Rightarrow^* w, \text{ for some } x \in A\}.$$

Informally, $L(\gamma)$ is the set of strings obtained from the initial axiom set A by repeated application of insertion and deletion rules.

An in/del system $\gamma = (V, T, A, I, D)$ is said to be of weight (n, m, p, q) if

$$\max \{|z| \mid (u, z, v) \in I\} = n,$$

$$\max \{|u| \mid (u, z, v) \in I \text{ or } (v, z, u) \in I\} = m,$$

$$\max \{|z| \mid (u, z, v) \in D\} = p,$$

$$\max \{|u| \mid (u, z, v) \in D \text{ or } (v, z, u) \in D\} = q.$$

Thus n (respectively p) represents the maximum length of the inserted (deleted) sequences, whereas m (respectively q) represent the maximum length of the right/left contexts of an insertion (respectively deletion).

We denote by $INS_n^m DEL_p^q$, $n, m, p, q \geq 0$, the family of languages $L(\gamma)$ generated by in/del systems of weight (n', m', p', q') such that $n' \leq n$, $m' \leq m$, $p' \leq p$, $q' \leq q$. When one of the parameters n, m, p, q is not bounded, we replace it by ∞. Thus, the family of all in/del languages is $INS_\infty^\infty DEL_\infty^\infty$.

The main results obtained regarding insertion and deletion systems are:

Theorem 1 *(34)* $RE = INS_3^6 del_2^7$.
Theorem 2 *(35)* $RE = INS_1^2 DEL_1^1$.
Theorem 3 *(35)* $RE = INS_2^1 DEL_2^0$.
Theorem 4 *(35)* $RE = INS_1^2 DEL_2^0$.

The interpretation of Theorem 1 is that the actions of every Turing machine can be simulated by an insertion/deletion system with finitely many rules, where the length of inserted strings is at most 3, and the length of the right and left contexts of insertion is at most 6, whereas the length of deleted strings is at most 2 and the length of the right and left contexts of deletion is bounded by 7. This suggests the possibility of using PCR site directed mutagenesis to simulate a Turing machine. Theorems 2–4 show that the same computational power can be obtained even with shorter contexts and inserted/deleted strings. These results point to yet another possible way of implementing biocomputations, namely by using RNA editing ***(60)*** which consists of insertions and deletions

```
DNA    G    G       GTTTTGG    AGA          G ATTTGG    A
RNA  uGuuuuGuuuuuGUUUUGGuuuAGAuuuuuuuGuAU**GGuuAuuu
```

Fig. 3. RNA editing by *u* insertion/deletion. Comparison of an edited RNA sequence encoding *H. mariadeanei* cytochrome oxidase subunit III (bottom) with its genomic DNA copy (top) ***(60)***. DNA sequences in upper case; uridines in mRNA that are added by RNA editing in lowercase (boldface); two encoded thymidines deleted from the mRNA indicated by asterisks.

of a single nucleotide. The most recent result, ***(51)***, proves that faithful restricted in/del systems have universal Turing machine power, where a faithful restricted in/del system has insertions and deletions of one letter only, but the length of contexts and inserted/deleted sequences is not bounded.

Overall, the general result that contextual insertions and deletions, by either site-directed mutagenesis or RNA editing, are sufficient to simulate the actions of a Turing machine suggests the existence of many platforms for biomolecular computing.

4. Nature's Solutions to Computational Problems

Research in molecular computing will undoubtedly have a great impact on many aspects of science and technology. In particular, molecular computing sheds new light on the very nature of computation, while it also introduces the prospect of designing computing devices that differ radically from today's computers. Probing the limits of biomolecular computation both *in vitro* and *in vivo* may provide new insights into the informational capacity of DNA in cellular organisms and the range of computational processes that exist in nature.

4.1. RNA Editing

Already, we have shown that computational processes exist in a variety of single- and multicellular organisms whose RNA molecules undergo RNA editing ***(61)***. Found in a wide variety of eukaryotes, from parasitic protozoa to humans, RNA editing by addition, deletion, or substitution of nucleotides alters the sequence of a messenger RNA before translation into protein. For example, **Fig. 3** shows a gene with an enormous number of uridine (*U*) insertions. (Sequence information in RNA is encoded in *A*, *C*, *G*, or *U*, with *U* replacing *T*.) In organisms such as trypanosomes RNA editing adds and deletes literally hundreds of uridine residues. These create initiation and termination codons, alter the structural features of transcripts, and construct over 90% of the coding capacity of this gene. On average, *U* insertions and deletions contribute to more than 60% of the nucleotides contained many genes. The other bases–A, C, and G–are completely conserved between the DNA and the RNA sequence ***(60)***.

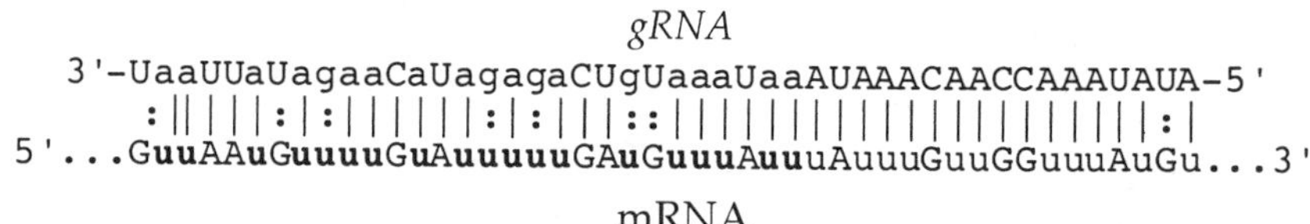

Fig. 4. Guide RNA–messenger RNA base-pairing interactions direct RNA editing. Lowercase *a*'s and *g*'s in the top gRNA sequence base pair with and guide the insertion of boldface lowercase **u**'s in this portion of the bottom messenger RNA sequence.

RNA editing restores coding messenger RNA (mRNA) sequences from encrypted pieces of the genome. Base-pairing interactions between small "guide RNA" (gRNA) molecules and the "pre-edited mRNA molecule" provide the context for determining these insertions and deletions (**Fig. 4**). Astonishingly, this process can create a single conserved protein coding sequence from over a dozen or so RNA molecules, each encoded in a unique circular DNA molecule (with the gene itself located on a *maxicircle* and the genes for each guide RNA usually found on one of the thousands of *minicircles*).

For every inserted U in the messenger RNA sequence, a corresponding A or G in the gRNA pairs with the fully edited product (**Fig. 4**). Complete editing proceeds 3′ to 5′ on the mRNA and requires a full set of overlapping gRNAs. Editing by each guide RNA creates an anchor sequence for binding the next guide RNA, leading to an ordered cascade of insertions and deletions—a genuinely RNA-based computer (***61***).

4.2. Gene Unscrambling

Ciliated protozoa possess two types of nuclei: an active macronucleus (soma) and a functionally inert micronucleus (germline) that contributes only to sexual reproduction. The macronucleus develops from the micronucleus after sexual reproduction. The micronuclear copies of some protein-coding genes in hypotrichous ciliates are obscured by intervening nonprotein-coding DNA sequences (internally eliminated sequences, or IESs) which must be removed before the assembly of a functional macronuclear DNA copy. Furthermore, the protein-coding DNA segments (macronuclear destined sequences, or MDSs) in *Oxytricha* and *Stylonychia* are sometimes present in a permuted order relative to their final position in the macronuclear copy. For example, we have found that the gene encoding DNA polymerase α in *S. lemnae* is scrambled in several dozens of pieces in the micronucleus. Destined to unscramble its micronuclear genes by putting the pieces together again, *O. trifallax* impressively solves a potentially complicated computational

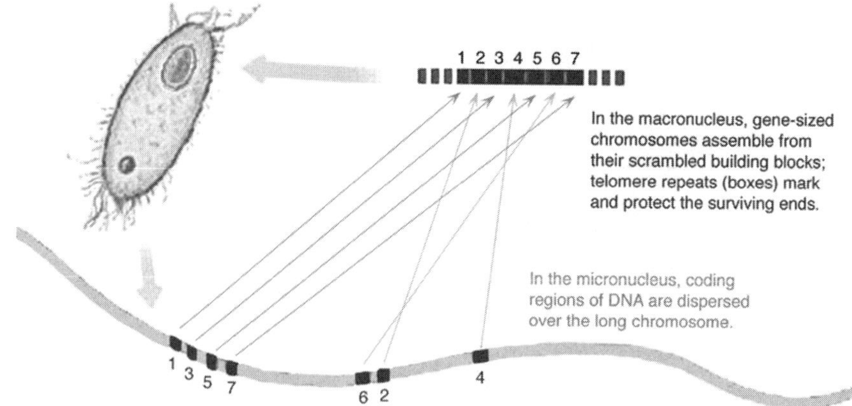

Fig. 5. Gene unscrambling as a computational problem. Dispersed coding MDSs, such as 1–3–5–7–6–2–4 (bottom), reassemble during macronuclear development to form the functional gene copy (top). Telomere addition marks and protects the ends of the gene, replacing the role of PCR primers in Step 2 of Adleman's experimental computation, because only those strands that have telomeres at both ends survive (*61*).

problem when assembling its functional sequences from their smaller constituents *(61)*.

The process of unscrambling bears a striking resemblance to the DNA algorithm Adleman *(1)* used to solve a seven-city instance of the Directed Hamiltonian Path problem. The developing ciliate macronuclear 'computer' (**Fig. 5**) makes use of the information contained in short 2–14 nucleotide direct repeats. These act as guides in a series of homologous recombination events. For example, the DNA sequence present at the junction between MDS n and the downstream IES is generally the same as the sequence between MDS $n+1$ and its upstream IES, leading to correct ligation of MDS n to MDS $n + 1$. By providing the splints analogous to edges in Adleman's graph, this mechanism assembles protein-encoding segments (MDSs, or 'cities' or nodes in this graph) in the correct order in which they belong in the final protein coding sequence ("Hamiltonian Path"), though the details of this mechanism are still unknown. As such, the unscrambling of gene sequences accomplishes an astounding feat of cellular computation, especially as Hamiltonian Path Problems of this size (approx 50 nodes) present a challenge to any computer.

Together RNA editing and gene unscrambling provide a unique array of potentially usable paradigms for biological computation. Furthermore, these processes underscore the diversity of computational paradigms that exist in biological systems and suggest a plethora of models for importing biology into mathematics.

References

1. Adleman, L. (1994) Molecular computation of solutions to combinatorial problems. *Science* **266,** 1021–1024.

1a. Kari, L. (1997) DNA computing: arrival of biological mathematics. *The Mathematical Intelligencer,* **19,** 2, 9–22.

2. Baum, E. (1995) Building an associative memory vastly larger than the brain. *Science* **268,** 583–585.
3. Reif, J. (1995) Parallel molecular computation: models and simulations, in *Proceedings of the 7th Annual ACM Symposium on Parallel Algorithms and Architectures*, Santa Barbara, CA, pp. 213–223.
4. Kendrew, J. (eic) (1994) *The Encyclopedia of Molecular Biology*, Blackwell Science, Oxford.
5. Garey, M. and Johnson, D. (1979) *Computers and Intractability. A Guide to the Theory of NP-completeness.* W. H. Freeman and Company, San Francisco.
6. Gifford, D. K. (1994) On the path to computation with DNA. *Science* **266,** 993–994.
7. Kaplan, P., Cecchi, G., and Libchaber, A. (1995) *Molecular computation: Adleman's experiment repeated.* NEC Technical Report.
8. Guarnieri, F., Fliss, M., and Bancroft, C. (1996) Making DNA add. *Science,* **273,** 220–223.
9. Liu, Q., Guo, Z., Condon, A., Corn, R., Lagally, M., and Smith, L. (1999) A surface-based approach to DNA computation, in *DNA Based Computers II* (L. F. Landweber and E. B. Baum, eds.), DIMACS Series in Discrete Mathematics and Theoretical Computer Science, vol. 44, American Mathematical Society, Providence, RI, pp. 123–132.
10. Ouyang, Q., Kaplan, P. D., Liu, S., and Libchaber, A. (1997) DNA solution of the maximal clique problem. *Science*, **278,** 446–9.
11. Cukras, A., Faulhammer, D., Lipton, R., and Landweber, L. F. (1998). Chess games: a model for RNA-based computation, in *Proceedings of the Fourth International Meeting on DNA Based Computers* (Kari, L., Rubin, H., and Wood, D. H., eds.), University of Pennsylvania, Philadelphia, PA, pp. 27–37.
12. Deaton, R., Murphy, R., Rose, J., Garzon, M., Franceschetti, D., and Stevens, S. (1997) A DNA based implementation of an evolutionary search for good encodings for DNA computation, in *Proceedings of the IEEE International Conference on Evolutionary Computation,* Indianapolis, IN, IEEE, Piscataway, NJ, pp. 267–271.
13. Kaplan, P., Cecchi, G., and Libchaber, A. (1999) DNA-based molecular computation: template-template interactions in PCR, in *DNA Based Computers II*

(Landweber, L. F. and Baum, E. B., eds.), DIMACS Series in Discrete Mathematics and Theoretical Computer Science, vol. 44, American Mathematical Society, Providence, RI, pp. 97–104.

14. Winfree, E., Yang, X., and Seeman, N. (1999) Universal computation via self-assembly of DNA: some theory and experiments, in *DNA Based Computers II* (Landweber, L. F. and Baum, E. B., eds.), DIMACS Series in Discrete Mathematics and Theoretical Computer Science, vol. 44, American Mathematical Society, Providence, RI, pp. 191–213.
15. Seeman, N., Wang, H., Liu, B., Qi, J., Li, X., Yang, X., Liu, F., Sun, W., Shen, Z., Sha, R., Mao, C., Wang, Y., Zhang, S., Fu, T.-J., Du, S., Mueller, J. E., Zhang, Y., and Chen, J. (1999) The perils of polynucleotides: the experimental gap between the design and assembly of unusual DNA structures, in *DNA Based Computers II* (Landweber, L. F. and Baum, E. B., eds.), DIMACS Series in Discrete Mathematics and Theoretical Computer Science, vol. 44, American Mathematical Society, Providence, RI, pp. 215–233.
16. Arita, M., Hagiya, M., and Suyama A., (1997) Joining and rotating data with molecules, in *Proceedings of the IEEE International Conference on Evolutionary Computation*, Institute of Electrical and Electronics Engineers (IEEE), pp. 243–248.
17. Arita, M., Suyama, A., and Hagiya, M., (1997) A heuristic approach for Hamiltonian Path Problem with molecules, in *Genetic Programming 1997: Proceedings of the Second Annual Conference* (Koza, J. R., Deb, K., Dorigo, M., Fogel, D. B., Garzon, M., Iba, H., and Riolo, R. L., eds.), Stanford University, Palo Alto, CA, Morgan Kaufmann, pp. 457–462.
18. Hagiya, M. and Arita M. (1999) Towards parallel evaluation and learning of Boolean μ-formulas with molecules, in *DNA Based Computers III* (D. H. Wood, ed.), DIMACS Series in Discrete Mathematics and Theoretical Computer Science, American Mathematical Society, Providence, RI, in press
19. Jonoska, N. and Karl, S. (1997) Ligation experiments in computing with DNA. *Proceedings of the IEEE International Conference on Evolutionary Computation*, Indianapolis, IN, IEEE, Piscataway, NJ, pp. 261–266.
20. Mulawka, J., Weglenski, P., and Borsuk, P. (1998). Implementation of the Inference Engine based on Molecular Computing Technique, in press.
21. Gloor, G., Kari, L., Gaasenbeek, M., and Yu, S. (1998) Towards a DNA solution to the Shortest Common Superstring Problem, in *Proceedings of the IEEE International Joint Symposia on Intelligence and Systems*, Rockville, MD, IEEE Computer Society Press, Los Alamitos, CA, pp. 140–145.
22. Lipton, R. (1995) DNA solution of hard computational problems. *Science,* **268,** 542–545.
23. Boneh, D., Dunworth, C., and Lipton, R. J. (1996). Breaking DES using a molecular computer, in *DNA Based Computers: Proceedings of a DIMACS Workshop* (Lipton, R. J. and Baum, E. B., eds.), American Mathematical Society, Providence, RI, 27, pp. 37–65.

24. Adleman, L., Rothemund, P., Roweis, S., and Winfree, E. (1999) On applying molecular computation to the Data Encryption Standard, in *DNA Based Computers II* (Landweber, L. F. and Baum, E. B., eds.), DIMACS Series in Discrete Mathematics and Theoretical Computer Science, vol. 44, American Mathematical Society, Providence, RI, pp. 31–44.
25. Leete, T., Schwartz, M., Williams, R., Wood, W., Salem, J., and Rubin, H. (1999) Massively parallel DNA computation: expansion of symbolic determinants, in *DNA Based Computers II* (Landweber, L. F. and Baum, E. B., eds.), DIMACS Series in Discrete Mathematics and Theoretical Computer Science, vol. 44, American Mathematical Society, Providence, RI, pp. 45–58.
26. Oliver, J. (1997) Matrix multiplication with DNA. *Journal of Molecular Evolution*, **45,** 161–7.
27. Baum, E. and Boneh, D. (1999) Running dynamic programming algorithms on a DNA computer, in *DNA Based Computers II* (Landweber. F. and Baum, E. B., eds.), DIMACS Series in Discrete Mathematics and Theoretical Computer Science, vol. 44, American Mathematical Society, Providence, RI, pp. 77–85.
28. Jonoska, N. and Karl, S. (1999) A molecular computation of the road coloring problem, in *DNA Based Computers II* (Landweber, L. F. and Baum, E. B., eds.), DIMACS Series in Discrete Mathematics and Theoretical Computer Science, vol. 44, American Mathematical Society, Providence, RI, pp. 87–96.
29. Williams, R. and Wood, D. (1999) Exascale computer algebra problems interconnect with molecular reactions and complexity theory, in *DNA Based Computers II* (Landweber, L. F. and Baum, E. B., eds.), DIMACS Series in Discrete Mathematics and Theoretical Computer Science, vol. 44, American Mathematical Society, Providence, RI, pp. 267–275.
30. Kari, L., Gloor, G., and Yu, S. (1999) Using DNA to solve the Bounded Post Correspondence Problem. *Theoretical Computer Science*, in press.
31. Kobayashi, S., Yokomori, T., Sampei, G., and Mizobuchi, K. (1997) DNA implementation of simple Horn clause computation, in *Proceedings of the IEEE International Conference on Evolutionary Computation*, Indianapolis, IN, IEEE, Piscataway, NJ, pp. 213–217.
32. Adleman, L. (1996) On constructing a molecular computer, in *DNA Based Computers: Proceedings of a DIMACS Workshop* (Lipton, R. J. and Baum, E. B., eds.), American Mathematical Society, Providence, RI, pp. 1–21.
33. Amos, M., Gibbons, A., and Hodgson, D. (1999) Error-resistant implementation of DNA computation, in *DNA Based Computers II* (Landweber, L. F. and Baum, E. B., eds.), DIMACS Series in Discrete Mathematics and Theoretical Computer Science, vol. 44, American Mathematical Society, Providence, RI, pp. 151–161.
34. Lipton, R. (1996) Speeding up computations via molecular biology, in *DNA Based Computers* (Lipton, R. J. and Baum, E. B., eds.), American Mathematical Society, Providence, RI, pp. 67–74.
35. Roweis, S., Winfree, E., Burgoyne, R., Chelyapov, N., Goodman, M., Rothemund, P., and Adleman, L. (1999) A sticker based architecture for DNA computation, in

DNA Based Computers II (Landweber, L. F. and Baum, E. B., eds.), DIMACS Series in Discrete Mathematics and Theoretical Computer Science, vol. 44, American Mathematical Society, Providence, RI, pp. 1–29.

36. Ogihara, M. and Ray, A. (1998). The minimum DNA computation model and its computational power. University of Rochester, Technical report TR-672.
37. Beaver, D. (1995) Computing with DNA. *J. Comput. Biol.*, **2,** 1–7.
38. Smith, W. (1996) DNA computers *in vitro* and *in vivo*, in *DNA Based Computers: Proceedings of a DIMACS Workshop* (Lipton, R. J. and Baum, E. B., eds.), American Mathematical Society, Providence, RI, DIMACS series, 27, pp. 121–185.
39. Winfree, E. (1996) On the computational power of DNA annealing and ligation, in *DNA Based Computers: Proceedings of a DIMACS Workshop* (Lipton, R. J. and Baum, E. B., eds.), American Mathematical Society, Providence, RI, DIMACS series, vol. 27, pp. 199–221
40. Beaver, D. (1996) A universal molecular computer, in *DNA Based Computers: Proceedings of a DIMACS Workshop* (Lipton, R. J. and Baum, E. B., eds.), American Mathematical Society, Providence, RI, pp. 29–36.
41. Head, T. (1987) Formal language theory and DNA: an analysis of the generative capacity of recombinant behaviors. Bulletin of Mathematical Biology, **49,** 737–759.
42. Rothemund, P. (1996) A DNA and restriction enzyme implementation of Turing machines, in *DNA Based Computers: Proceedings of a DIMACS Workshop* (Lipton, R. J. and Baum, E. B., eds.), American Mathematical Society, Providence, RI, pp. 75–119.
43. Kari, L., and Thierrin, G. (1996) Contextual insertions/deletions and computability. *Information and Computation,* **131,** 47–61.
44. Yokomori, T. and Kobayashi, S. (1999) DNA-EC: a model of DNA computing based on equality checking, in *DNA Based Computers III* (in Wood, D. H., ed.), DIMACS Series in Discrete Mathematics and Theoretical Computer Science, American Mathematical Society, Providence, RI. , in press.
45. Head, T., Paun, G., and Pixton, D. (1996) Language theory and molecular genetics, in *Handbook of Formal Languages* (Rozenberg, G. and Salomaa, A., eds.), Springer Verlag, Berlin, **2,** 295–358.
46. Paun, G. and Salomaa, A. (1996) DNA computing based on the splicing operation. *Mathematica Japonica,* **43,** 3, 607–632.
47. Paun, G. (1995) On the power of the splicing operation. *International Journal of Computer Mathematics,* **59,** 27–35.
48. Freund, R., Kari, L., and Paun, G. (1999) DNA computing based on splicing: the existence of universal computers. *Theory of Computing Systems* **32,** 69–112.
49 Csuhaj-Varju, E., Freund, R., Kari, L., and Paun, G. (1996). DNA computing based on splicing: universality results, in *Proceedings of 1st Annual Pacific Symposium on Biocomputing*, Hawaii (Hunter, L. and Klein, T., eds.), World Scientific Publ., Singapore, pp. 179–190.
50. Yokomori, T., Kobayashi, S., and Ferretti, C. (1997) On the power of circular

splicing systems and DNA computability, in *Proceedings of the IEEE International Conference on Evolutionary Computation*, Indianapolis, IN, IEEE, pp. 219–224.

51. Kari, L., Paun, G., Thierrin, G., and Yu, S. (1999) At the crossroads of DNA computing and formal languages: characterizing recursively enumerable languages using insertion/deletion systems, in *DNA Based Computers III* (Wood, D. H., ed.), DIMACS Series in Discrete Mathematics and Theoretical Computer Science, American Mathematical Society, Providence, RI, in press.
52. Hartmanis, J. (1995) On the weight of computations. *Bulletin European Association of Theoretical Computer Science,* **55,** 136–138.
53. Kurtz, S., Mahaney, S., Royer, J., and Simon, J. (1999) Active transport in biological computing, in *DNA Based Computers II* (Landweber, L. F. and Baum, E. B., eds.), DIMACS Series in Discrete Mathematics and Theoretical Computer Science, vol. 44, American Mathematical Society, Providence, RI, pp. 171–179.
54. Amenyo, J. (1999) Mesoscopic computer engineering: automating DNA-based molecular computing via traditional practices of parallel computer architecture design, in *DNA Based Computers II* (Landweber, L. F. and Baum, E. B., eds.), DIMACS Series in Discrete Mathematics and Theoretical Computer Science, vol. 44, American Mathematical Society, Providence, RI, pp. 133–150.
55. Mihalache, V. (1997) Prolog approach to DNA computing, in *Proceedings of the IEEE International Conference on Evolutionary Computation*, Indianapolis, IN, IEEE, pp. 249–254.
56. Salomaa, A. (1973) *Formal Languages.* Academic Press, New York.
57. Kari, L. (1991) *On insertions and deletions in formal languages.* Ph.D. thesis, University of Turku, Finland.
58. Dieffenbach, C. W. and Dveksler, G. S., (eds.), (1995) *PCR primer: a laboratory manual,* Cold Spring Harbor, NY, Cold Spring Harbor Laboratory Press, pp. 581–621.
59. Galiukschov, B. S. (1981) Semicontextual grammars (in Russian). *Mat. logica i mat. ling.*, Kalinin Univ., 38–50.
60. Landweber, L. F. and Gilbert, W. (1993). RNA editing as a source of genetic variation. *Nature* **363,** 179–182.
61. Landweber, L. F. and Kari, L. (1998) The Evolution of DNA Computing: Nature's Solution to a Combinatorial Problem, in *Genetic Programming 1998: Proceedings of the Third Annual Conference, July 22–25, 1998*, (Koza, J. R., Banzhaf, W., Chellapilla, K., Deb, K., Dorigo, M., Fogel, D. B., Garzon, M. H., Goldberg, D. E., Iba, H., and Riolo, R. L., eds), University of Wisconsin, Madison, WI, San Francisco, CA, Morgan Kaufmann, pp. 700–708.

24

Detecting Biological Patterns

The Integration of Databases, Models, and Algorithms

Gautam B. Singh

1. Introduction

Every cell in an organism essentially contains the same genome. However, the subset of genes expressed in one cell type is different from another and is determined by the functional role of that cell. It is also known that the protein *coding* regions in an eukaryotic genome only account for 10–20% of the DNA. Research has shown that the majority of nuclear DNA is *noncoding* and is important for the regulation of gene expression. Furthermore, the mechanisms of gene regulation are coordinated by the presence of biologically significant DNA sequence patterns that are observed in the noncoding regions within the neighborhood of genes ***(1)***.

Special sequences of regulatory importance such as introns, promoters, enhancers, matrix association regions (MARs), and repeats are found in the noncoding DNA. Many of these regions contain patterns that represent functional control points for cell specific or *differential* gene expression ***(2,3)***, whereas others such as the *repetitive* DNA patterns can serve as a biological clock ***(4)***. These and numerous other examples indicate that *patterns* in the eukaryotic DNA may play a vital role. Other examples of these patterns include the AT or GC rich regions, telomeric repeats of sequence AGGGTT (in human DNA), rare occurrence or absence of dinucleotides TA and GC, and tetranucleotide CTAG, and the GNN periodicity in the gene-coding regions. There is evidence that suggests that some deviations from patterns are deleterious to the viability of the organism.

From: *Methods in Molecular Biology, vol. 132: Bioinformatics Methods and Protocols*
Edited by: S. Misener and S. A. Krawetz © Humana Press Inc., Totowa, NJ

Thus, DNA is not a homogeneous string of characters, but is comprised of a mosaic of sequence-level motifs that come together in a synergistic manner to coordinate and regulate in synthesis of proteins. The following four events are needed for the successful transcription or the synthesis of RNA from the DNA template:

1. Potentiation or the structural DNA conformation: This step is a prerequisite for expression. Essentially, the potentiated genes are those that are located on the 10-nm fiber and can be transcribed.
2. Initiation or binding of RNA polymerase to double-stranded DNA: This step involves a transition to single-strandedness in the region of binding in which RNA polymerase binds on the region of the DNA called the promoter.
3. Elongation or the covalent addition of nucleotides to the 3'-end of the growing polynucleotide chain.
4. Termination or the recognition of the transcription termination sequence and the release of RNA polymerase.

It is believed that the structural properties of DNA are responsible for maintaining the potentiated conformation of a gene within a cell. Although the actual transcription is performed by RNA polymerase, other proteins, known as transcription factors are needed to initiate this process. These factors are either associated directly with RNA polymerase or help in building the actual transcription apparatus. That is, the transcription factors facilitate transcription by binding to the RNA polymerase, to other transcription factors, and by binding to the *cis*-acting DNA sequences.

The transcription factors are further categorized into three major groups, namely, basal, upstream, and inducible. The basal transcription factors are needed by all genes and are essential for the very process of transcription. The TATA-box and CAAT-box are examples of basal factors. The upstream transcription factors affect the efficiency of transcription. The nuclei of the cells that contain these factors are able to synthesize the gene in larger numbers. The inducible transcription factors regulate the expression of a gene in response to various stimuli, and during growth and development.

The basal transcription apparatus required for the transcription of the protein-coding genes (i.e., those transcribed by RNA polymerase II), exhibits similarity in the prokaryotic as well as eukaryotic genes, albeit their locations in the two cell types are different. In general, the prokaryote promoters are turned *on*, and control is exerted in a negative manner. In contrast, a typical eukaryotic promoter will carry out basal level transcription, but requires positive activators for efficient transcription. This upstream binding of transcription factors generally enhance the levels of transcription and are often unique to a promoter. They assist in transcription initiation by contacting members of the basal

A

Cell Type	Location	Sequence
Prokaryote	~-10 bp	TATAAT
	~-35 bp	TTGACA
Eukaryote	~-25 bp	TATA
	~-80 bp	CAAT

Basal

B

Promoter Name	Sequence	TF Name
CAAT box	GGCCAATCC	CTF
GC box	GGGCGG	SP1
Octomer box	ATTTGCAT	Oct1

Upstream

Fig. 1. Promoters bind with the DNA and help in the formation of transcription apparatus necessary for gene expression.

apparatus as well as by binding coactivators to the basal apparatus. **Figure 1** shows examples of basal and upstream elements and their corresponding DNA-binding sites.

The group known as upstream promoters is diverse, and the combinations of their co-occurrences is quite large. Because of the potentially large number of sequence combinations possible, their control can be quite sophisticated. The three factors shown in **Fig. 1B** are general, i.e., they are found in all human tissues. In the presence of these factors, the rate of transcription initiation is increased. Sometimes these sites are even necessary for *in vivo* transcription. Additionally, a set of distant elements that effect transcription are called enhancers. In contrast to the upstream promoter elements, which influence gene expression only when present in a narrow region, enhancer binding can promote transcription when they occur anywhere around a gene.

Finally, some factors are turned on in response to the environmental stimuli, and thus provide the final control point in regulating gene expression. These are termed inducible factors. In addition to containing binding sites for constitutive transcription factors like TFIID, SP1, and so on, the eukaryotic promoters also contain sites for inducible transcription factors. These factors will activate transcription of genes in response to growth factors, nutrient levels, heat shock, or other types of signals. There are consensus binding sites that are found toward the 5′ end of the transcription initiation sites of the genes that are regulated by these factors. These transcription activators usually control expression by oligomerization. That is, the binding of a inducible TF to its corresponding promoter region on the DNA results in the recruitment of other transcription factors necessary to initiate transcription. Examples of the inducible transcription factors are shown in **Fig. 2**.

Transcription factor classification has also been performed based on the

Stimuli	Site	Size	TF Name
Heat Shock	HSE	27 bp	HSF or HSTF
Glucocorticoid	GRE	20 bp	Glu. Receptor
Serum	SRE	20 bp	SRF

Fig. 2. Examples of inducible transcription factors.

structural features of DNA-binding domains. Such a structural classification results in their categorization into a set of superfamilies. Some of these are, b/ZIP, bHLH, homeodomain proteins, ETS, REL, zinc finger nuclear proteins, the HMG-family, the winged helix, serum response factors, and CTF/NF-1. Each of these has been further subdivided into several families, and the proteins belonging to a specific family recognize *cis*-elements with distinct structure. For example, the families AP-1, CREB/ATF, and C/EBP belong to the b/ZIP superfamily. The EGR and Sp-1 families that are members of C_2H_2 zinc finger superfamily. The *cis*-elements recognized by these two superfamilies have different structures. Such a diversity of transcription factors is necessary for cell-specific differential gene regulation.

2. Materials

This section provides a brief background of the various databases that store biological patterns. Most of these databases contain evolving information, and have thus gone through several revisions since they were first introduced in mid-1980s. From their earlier emphasis on the specification of motif consensus, their present focus is towards an integration of functional information.

2.1. TFD: *The* Transcription Factor Database

This database was created to manage the growing body of sequence information that is being generated as we understand the processes of eukaryotic gene regulation. A relational database model was adopted for organizing this information as it enables the efficient management of relationships between sequence motifs, the factors that bind, the domains that they belong to, and the gene/organism specificity. This information is recorded using five relations or tables called **SITES**, **DOMAINS**, **FACTORS**, **cDNAs**, and **ELEMENTS**, with the **SITES** table providing the specific information about the DNA sequence recognized by a specific transcription factor *(5)*. From a functional standpoint, the *Transcription Factor Database* (*TFD*) is thought to be consisting of two parts. The first part contains the set of protein sequences that are the amino acids that constitute the transcription-regulating factors. The second part is a pattern database with nucleotide patterns that are recognized by transcription factors. This component of TFD is also available within the DNA analysis programs available from Genetic Computer Group, Milwaukee, WI. TFD is updated four times every year. Currently, the GCG version of TFD consists approx 5000 patterns. TFD entries pertaining to the sequence binding sites provide the following information ***(6)***:

```
UAS(G)-pMH100    CGGAGTACTGTCCTCCG GAL4   J. Mol.
  Biol. 209, 423-432 (1989)

TFIIIC-Xls-5S.1 TGGATGGGAG        TFIIIC EMBO. J
  6, 3057-3063 (1987)

GCN4-his3-189    ATGACTCAT         GCN4   Science
  234, 451-457 (1986)
```

ooTFD (*object-oriented Transcription Factors Database*) is a recent successor to *TFD*, and uses object-oriented database technology to represent several relationships between transcription factors and related proteins, cDNA as well as links to appropriate literature *(7)*. Specifically, *ooTFD* uses this new technology to represent containment and composition—the relationships that are often cumbersome to map to a relational schema. In this manner, *ooTFD* can represent information about all transcription factors, including both eukaryotic and prokaryotic, basal and regulatory factors formed by multiprotein complexes or monomeric proteins. *ooTFD* and associated tools are available at `http://www.isbi.net`.

2.2. TRANSFAC

The development path for the *TRANSFAC* database was to provide a biological context for understanding the function of regulatory signals found in genomic sequences. The aim of this compilation of signals was meant to provide all relevant data about the regulating proteins and allow researchers to trace back transcriptional control cascades to their origin ***(8,9)***. The *TRANSFAC* database contains information about regulatory DNA sequences and the transcription factors binding to and acting through them. The *TRANSFAC* database was used to describe these elements, to define consensi and matrices for elements of certain function, and thus to provide a means to identify regulatory signals within anonymous genomic sequences ***(10–12)***.

Currently, three databases have evolved from *TRANSFAC*. These are the *TRANSFAC*, *TRRD* (*Transcription Regulatory Region Database*), and *COMPEL* (*Composite Elements* database). *TRANSFAC* is a database on transcription factors and their DNA-binding sites and is organized as a relational schema. A representation of these relations can be obtained in terms of six files, namely **SITE**, **FACTOR**, **CLASS**, **MATRIX**, **CELL**, and **GENE**, with the semantics closely resembling those of TFD. The additional MATRIX file provides a profile for the binding site signals and is useful for assigning an information theoretic significance to the signal search process. **SITES** and **FACTORS** maintain links to external databases, including *EPD*, *SwissProt*, *EMBL*, and *PIR*. **TRRD** provides a hierarchical representation of the structure of the eukaryotic transcription-regulatory region, and specific patterns of gene expression. It provides information on factors that work together as well as organization of promoter and enhancer regions that span several hundred bases. **COMPEL** is a database on composite regulatory elements of vertebrate genes. Such elements are located in the transcription-control region of a gene and contain two closely situated binding sites for different transcription factors. The starting point for accessing these databases are `http://transfac.gbf.de/TRANSFAC`, `http://www.bionet.nsc.ru/TRRD`, and `http://www.bionet.nsc.ru/COMPEL` ***(13)***.

2.3. PROSITE *and* EPD *(*Eukaryotic Promoter Database*)*

PROSITE is a compilation of sites and patterns found in protein sequences. Its development was motivated by its applications in the determination of functions of uncharacterized proteins that are generated from the translation of genomic or cDNA sequences ***(14,15)***. The *PROSITE* database consists of biologically significant patterns and profiles that can enable one to establish the family of protein (if any) to which a new sequence belongs, or which known domain(s) it contains ***(16)***.

The *Eukaryotic Promoter Database* (*EPD*) is an annotated nonredundant collection of experimentally characterized eukaryotic pol II promoters. All information presented in *EPD* is based on experimental evidence as reported in the biological literature. The promoters in *EPD* are annotated by specifying the *EMBL* pointer of the flanking sequences, the promoter-defining evidence, cross-references to other databases, and bibliographic references. *EPD* is designed primarily as a resource for comparative sequence analysis. Consequently, *EPD* facilitates the dynamic extraction of biologically meaningful promoter subsets. The database is available through the web site at `http://cmpteam4.unil.ch` ***(17)***.

2.4. The MAR *Database*

The matrix or scaffold attachment regions are relatively short (100–1000 bp long) sequences that anchor the chromatin loops to the nuclear matrix. MARs often include the origins of replication (ORI) and can possess a concentrated area of transcription factor binding sites *(18)*. Approximately 100,000 matrix attachment sites are believed to exist in the mammalian nucleus, of which approx 30,000–40,000 serve as ORIs *(19)*. MARs have been observed to flank the ends of genic domains encompassing various transcriptional units. It has also been shown that MARs bring together the transcriptionally active regions of chromatin such that the transcription is initiated in the region of the chromosome that coincides with the surface of the nuclear matrix *(19,20)*.

It is expected that indicators such as MARs will be significant during the next phase of the Human Genome Project that will focus on completing the transcript map. In light of the key role of MARs in genetic processes, and their localization to functional chromatin domains, a means to model these markers so that they could be placed on the map from sequencing data is significant. There is no known consensus sequence for a MAR. However, MARs have been experimentally defined for several gene loci, including, the chicken *lysozyme* gene *(21)*, human *interferon-β* gene *(22)*, human *β-globin* gene *(23)*, chicken *α-globin* gene *(24)*, *p53* *(25)*, and the human protamine gene cluster *(26)*. Several motifs that characterize MARs have emerged and have been extensively studied by researchers. These motifs are functionally categorized and represented as AND-OR patterns as described below.

The AND-OR pattern specification is a disjunction (**OR**) of the conjunctions (**AND**) on motifs detected in the sequence. The sequence level motifs serve as the lowest level predicates used to detect the presence of a higher level pattern. In general the following operations may be applied to the lower level motifs:

- Motif consensus sequence, m, represented as a regular expression or a profile, or
- The logical **OR** of two motifs m_i and m_j, represented as $m_i \vee m_j$, or
- The augmented logical **AND** of two motifs m_i and m_j, represented as a $m_i \wedge_a^b m_j$ (in this augmented **AND** operator, the parameters a and b specify the acceptable separation between the co-occurrence of the two motifs), or
- The logical negation of a motif, m, represented as $\overline{m}$, specifying the absence of a given motif.

In such a general framework, the pattern description language is defined that is powerful enough to represent the variety of patterns likely to be discovered as our understanding about the DNA–protein interactions and the control of genetic machinery reaches a higher level of maturity. Also associated with each motif (and a pattern) is the probability of its random occurrence. This value can be derived using the base composition of the sequence being analyzed *(27)*. Using the **AND-OR** methodology, the following MAR rules were developed:

1. The origin of replication (ORI) rule: It has been established that replication is

associated with the nuclear matrix, and the origins of replication share the ATTA, ATTTA, and ATTTTA motifs.

2. Curved DNA: Curved DNA has been identified at or near several matrix attachment sites and has been involved with DNA–protein interactions, such as recombination, replication, and transcription ***(18,28)***. Optimal curvature is expected for sequences with repeats of the motif, AAAAn_7AAAn_7AAAA as well as the motif TTTAA.
3. Kinked DNA: Kinked DNA is typified by the presence of copies of the dinucleotides TG, CA, or TA that are separated by 2–4 or 9–12 nucleotides. For example, kinked DNA is recognized by the motif TAn_3TGn_3CA, with TA, TG, and CA occurring in any order.
4. Topoisomerase II sites: It has been shown that topoisomerase II binding and cleavage sites are also present near the sites of nuclear attachment. Vertebrate and *Drosophila* topoisomerase II consensus sequence motifs can be used for finding regions of matrix attachment ***(29,30)***.
5. AT-rich sequences: It is quite typical for many MARs to contain stretches of AT-rich sequences. Furthermore, the occurrence of AT-rich sequences must be spaced regularly in a periodic manner.
6. TG-rich sequences: Some TG rich spans are indicative of MARs. These regions are abundant in the 3′ UTR of a number of genes, and may act as signals at the recombination sites ***(18)***.

2.4.1. Pattern Specification

We next discuss the issue of assimilating the motif variability into a representation of patterns. Such a variability may be captured using the previously discussed **AND-OR** rules. As an example, consider the rule to define the origin of DNA replication. This can be based on an **OR**, i.e., the $\vee$ operator applied to the three motifs m_1 = ATTA, m_2 = ATTTA, and m_3 = ATTTTA. The motif detectors bypass the **AND** layer in this case.

$$R_1 = m_1 \vee m_2 \vee m_3 \tag{1}$$

Similarly, the requirement for multiple motif occurrences can be specified using the **AND**, i.e., the $\wedge$ operator. In an **AND** rule, an additional parameter is incorporated to constrain the allowable gap between the two co-occurring motifs. For example, the AT-richness rule can been formulated as the occurrence of two hexanucleotide strings, m_4=WWWWWW (Note: The IUPAC code W denotes an ambiguous base A or T), that are separated by distance of 8–12 nucleotides, using the augmented **AND** operator using $\wedge_{\text{low}}^{\text{high}}$ to define the acceptable distance between the two motifs:

$$R_2 = m_4 \underset{8}{\overset{12}{\wedge}} m_4 \tag{2}$$

The significance of the occurrence of a pattern in a DNA sequence is inversely related to the probability that the pattern will occur purely by chance. The probabilities of random occurrences of the underlying predicates are combined mathematically to evaluate the probability of the random occurrence of a pattern specified by a given rule. As a simplistic case, the random occurrence probabilities for the patterns described by the two rules above can be computed. This value for the set of acceptable patterns described by rule R_2 is based on the occurrence of at least one motif within an acceptable distance from the reference motif. In this manner, the random occurrence probabilities for rules R_1 and R_2 in the above **Eqs. 1** and **2** are given by:

$$Pr(R_1) = Pr(m_1) + Pr(m_2) + Pr(m_3)$$

$$Pr(R_2) = Pr(m_4) \cdot \{1 - \exp[-5 \cdot Pr(m_4)]\} \tag{3}$$

In a similar manner, the random occurrence probability rules constructed on underlying predicates that are defined as profiles can be computed using generating functions *(27)*. As described in the sections below, the rule probabilities are used to estimate the statistical significance of the set of patterns that are detected in a given region of the DNA sequence. The database of higher level patterns used in detection of the matrix association regions is described in *(31)*.

3. Methods

3.1. Pattern Detection Software

The generation of "signal search data" represents a general method of describing the common properties of a set of DNA sequences presumed to be functionally analogous *(32)*. Besides the detailed description of this method, two computer programs that use signal search data as input data are presented: One that processes them to a "constraint profile" and another one which lists overrepresented "signals" of potential functional relevance.

3.2. Web Tools for Pattern Detection

1. *Signal Scan* (`http://bimas.dcrt.nih.gov/molbio/signal`): This web search tool resulted from the integration of a variety of tools described in **ref.** *33*. The tool finds homologies between the sequences published in *TRANSFAC* and *TFD* and the sequence being analyzed. The tools however do not provide an interpretation of the locations of the signals found. The significance of a signal detected is assumed to be related to the signal length. For example, there will be many matches for short signal like the CP1 and thus has a high probability of random occurrence. On the other hand, there will be fewer, and possibly more significant matches for longer signal sequences such as the

binding sites for glucocorticoid elements. Also, if a signal does not follow the reported consensus sequence, it will be missed by this search tool.

The results produced by *Signal Scan* show the name of the signal, the published signal sequence, and the location of the first base pair matching in the sequence being analyzed. The factor binding name, if known, is also provided. Each signal found in the query sequence also shows the corresponding site number for the *TFD* or *TRANSFAC* database. The output produced by the software may be sorted by location or by the type of transcription factor binding sites found.

2. *IMD Matrix Search* (`http://bimas.dcrt.nih.gov/molbio/matrixs`): This really represents an extension of the *Signal Search* program—however, the transcription factor binding sites are represented as matrices that are derived using information theory. Accordingly, this search program looks for high scoring regions where the footprint of bases match the distribution specified in the matrix ***(34)***. The starting position of patterns with scores above the cutoffs of each matrix are indicated—the cut-offs being determined by the level of stringency needed to reduce the false positive rates. The software reports a p value that is proportional to the strength of each match above the cut-off. In the case of overlapping sites for the same factor, only the one with the highest information score is selected.
3. *Tfsearch* (`http://www.rwcp.or.jp/papia`): This program was written by Yutaka Akiyama, Koyoto University, Japan, and utilizes correlation calculations to identify transcription factor binding sites. The correlation of DNA sequences with the known binding site profiles are used to search out the signals in the query sequence. The transcription profiles are obtained from *TFMATRIX* component of *TRANSFAC*.
4. *MatInd* and *MatInspector* (`http://www.gsf.de/biodv/matinspector.html`): The *MatInd* tool enables the development of a profile from a series of short sequences provided. Such a profile description of 280 entries exists for the sequences described in *TRANSFAC v3.4*, and is utilized by the *MatInspector* tool for analysis of query sequences. The analysis tool allows the user to specify a large number of sequences for a single run as well as provides the option to select a subset of profile matrices that may be used for search. The tool *FastM* provides the ability to generate models for regulatory regions in DNA sequences by enabling users to search for the transcription factor binding sites that are separated by a specified distance ***(35)***. A complete description of this tool is given in Chapter 18 of this volume.
5. *MarFinder* (`http://www.ncgr.org/MarFinder`): The search for MARs results from defining a group of patterns that are bonded together because of similarity of their function. After such a grouping, a search for the patterns in a given group can be performed to discover regions in the query DNA sequence. If a large subset of members of a functionally related group of patterns is found in a specific region of the uncharacterized DNA sequence, one can begin to learn

about its function. This process is called a Functional Pattern Search and is demonstrated by the *MarFinder* system that performs a search for the group of patterns that are associated with MARS ***(36)***.

It is quite intuitive to think of the pattern-cluster density as a property defined along the span of a sequence. Thus a sliding window algorithm can be applied for measuring this value, where the measurements are characterized by the two parameters, W and δ. The cluster-density is measured in a window of size W centered at location x along that sequence. Successive window measurements are done by sliding this window in the increments of δ nucleotides. If δ is small, linear interpolation can be used to join the individual window estimates that are gathered at x, $x + \delta$, . . . $x + k\delta$. In this manner, a continuous distribution of the cluster-density is obtained as a function of x.

The task of estimating the density of pattern clusters in each window can be statistically defined as some inverse of the probability of rejecting the null hypothesis, that states that the frequency of the patterns observed in a given window is not significantly different from the expected frequencies from a random W nucleotide sequence of the same composition as the sequence being analyzed. The inverse function chosen as, $\rho = -\log(\alpha)$, where the parameter α is the probability of erroneously rejecting H_0. In other words, α represents the probability that the set of patterns observed in a window occurred purely by chance. The value of ρ is computed for both the forward and the reverse DNA strands and the average of the two values is considered to be the density estimate at a given location.

To compute ρ, assume that we are searching for k distinct types of patterns within a given window of the sequence. In general, these patterns are defined as rules $R_1, R_2, \ldots, R_k$. By using the probability formulation similar to those defined in **Eq. 3**, the probability of random occurrence of the various k patterns is calculated. Let these values be $p_1, p_2, \ldots, p_k$. Next, a random vector of pattern frequencies, F, is constructed. F is a k-dimensional vector with components, $F = \{x_1, x_2, \ldots, x_k\}$, where each component x_i is a random variable representing the frequency of the pattern R_i in the W-bp window. The component random variables x_i are assumed to be independently distributed Poisson processes, each with the parameter $\lambda_i = p_i \cdot W$. Thus, the joint probability of observing a frequency vector $F_{obs} = \{f_1, f_2, \ldots, f_k\}$ purely by chance is given by:

$$P(F_{obs}) = \prod_{i=1}^{k} \left[\frac{(e^{-\lambda_i} \lambda^{f_i})}{f_i!}\right] \text{ where } \lambda_i = p_i \cdot W \tag{4}$$

The steps required for computation of α, the cumulative probability that pattern freqeuncies equal to or greater than the vector F_{obs} occurs purely by chance is given by **Eq. 5**. This corresponds to the one-sided integral of the multivariate Poisson distribution, and represents the probability that the H_0 is erroneously rejected.

$$\alpha = Pr(x_1 \geq f_1, x_2 \geq f_2, \ldots, x_k \geq f_k)$$

$$= Pr(x_1 \geq f_1) \cdot Pr(x_2 \geq f_2) \cdot \ldots \cdot Pr(x_k \geq f_k)$$

$$\left\{\sum_{x_1=f_1}^{\infty}\left[\frac{(\exp^{-\lambda 1}\lambda_1^{x1})}{x_1!}\right]\right\}\cdot\left\{\sum_{x_2=f_2}^{\infty}\left[\frac{(\exp^{-\lambda 2}\lambda_2^{x2})}{x_2!}\right]\right\}\cdots\left\{\sum_{x_k=f_k}^{\infty}\left[\frac{(\exp^{-\lambda k}\lambda_1^{xk})}{x_k!}\right]\right\} \quad (5)$$

The p value, α, in **Eq. 5** is utilized to compute the value of ρ or the cluster density as specified in **Eq. 6** below:

$$\rho = \ln\left(\frac{1.0}{\alpha}\right) = -\ln(\alpha)$$

$$= \sum_{i=1}^{k} \lambda_i + \sum_{i=1}^{k} \ln f_I! - \sum_{i=1}^{k} f_i \ln \lambda_i$$

$$-\sum_{i=1}^{k} \ln\left[1 + \left(\frac{\lambda_i}{f_i+1}\right) + \left(\frac{\lambda_i^2}{(f_i+1)(f_i+2)}\right) + \ldots + \left(\frac{\lambda_i^t}{(f_i+1)(f_i+2)\ldots(f_i+t)}\right)\right] \quad (6)$$

The infinite summation term in **Eq. 6** quickly converges and thus can be adaptively calculated to the precision desired. For small values of λ_1, the series may be truncated such that the last term is smaller than an arbitrarily small constant, ε. **Figure 3** presents a snapshot of the output generated from the analysis of the human *β-globin* gene sequence. The areas of strong associations between the patterns are shown as the areas with high MAR potential.

4. Conclusions

A summary of the various types of patterns that are present in DNA sequences was provided. It is demonstrated that the task of detecting these patterns is possible due to the synergies between the availability of pattern database, a model for representing the patterns (such as profiles, or rules) and applying the a search algorithm for their detection.

References

1. Kliensmith, L. and Kish, V. (1995) *Principles of Cell and Molecular Biology*, 2nd. ed., HarperCollins, New York, NY, pp. 400–468.
2. Kadonaga, J. (1998) Eukayotic transcription: an interlaced network of transcription factors and chromatin-modifying machines. *Cell* **92,** 307–313.
3. Roeder, R. (1996) The role of general initiation factors in transcription by RNA polymerase II. *Trends Biochem. Sci.* **21,** 327–335.
4. Hartwell, L. and Kasten, M. (1994) Cell cycle control and cancer. *Science* **266,** 1821–1828.
5. Ghosh, D. (1990) A relational database of transcription factors. *Nucleic Acid Res.* **18,** 1749–1756.

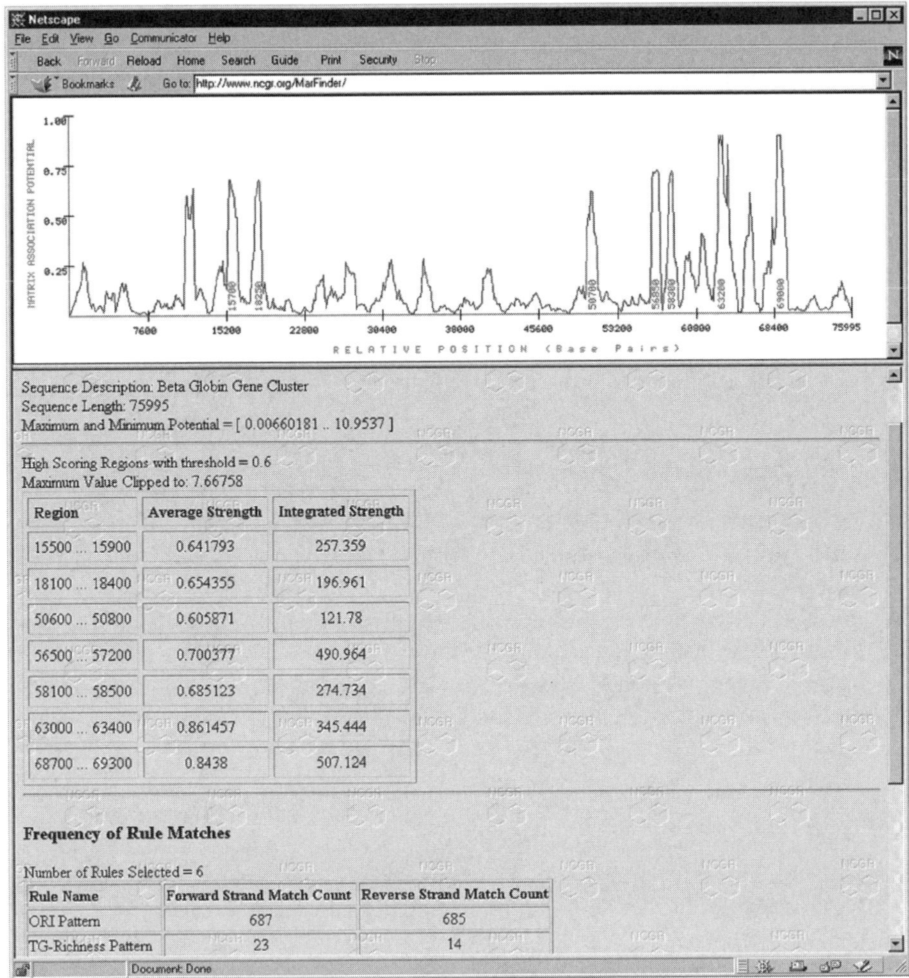

Region	Average Strength	Integrated Strength
15500 ... 15900	0.641793	257.359
18100 ... 18400	0.654355	196.961
50600 ... 50800	0.605871	121.78
56500 ... 57200	0.700377	490.964
58100 ... 58500	0.685123	274.734
63000 ... 63400	0.861457	345.444
68700 ... 69300	0.8438	507.124

Rule Name	Forward Strand Match Count	Reverse Strand Match Count
ORI Pattern	687	685
TG-Richness Pattern	23	14

Fig. 3. The results of analyzing the beta globin gene cluster using the *MarFinder* pattern detection tool. The areas of high pattern cluster density are also functionally significant for matrix association.

6. Ghosh, D. (1992) TFD: the transcription factors database. *Nucleic Acid Res.* **20(suppl),** 2091–2093.
7. Ghosh, D. (1998) OOTFD (Object-Oriented Transcription Factors Database): an object-oriented successor to TFD. *Nucleic Acid Res.* **26,** 360–362.
8. Wingender, E. (1998) Compilation of transcription regulating proteins. *Nucleic Acid Res.* **16,** 1879–1902.
9. Wingender, E. (1990) Transcription regulating proteins and their recognition sequences. *Crit. Rev. Eukaryot. Gene Expr.* **1,** 11–48.

10. Wingender, E. (1994) Recognition of regulatory regions in genomic sequences. *J. Biotechnol.* **35,** 273–280.
11. Wingender, E., Karas, H., and Knuppel, R. (1996) TRANSFAC: a database on transcription factors and their DNA binding sites. *Nucleic Acid Res.* **24,** 238–241.
12. Wingender, E., Karas, H., and Knuppel, R. (1997) TRANSFAC database as a bridge between sequence data libraries and biological function. *Pac. Symp. Biocomput.* 477–485.
13. Wingender, E., Kel, A., Kel, O., Karas, H., Heinemeyer, T., Dietze, P., Knuppel, R., Romaschenko, A., and Kolchanov, N. (1997) TRANSFAC, TRRD and COMPEL: towards a federated database system on transcriptional regulation. *Nucleic Acid Res.* **25,** 265–268.
14. Bairoch, A. and Bucher, P. (1994) PROSITE: recent developments. *Nucleic Acid Res.* **22,** 3583–3589.
15. Bairoch, A., Bucher, P., and Hofmann, K. (1995) The PROSITE database, its status in 1995. *Nucleic Acid Res.* **24,** 189–196.
16. Bairoch, A., Bucher, P., and Hofmann, K. (1997) The PROSITE database, its status in 1997. *Nucleic Acid Res.* **25,** 217–221.
17. Perier, R., Junier, T., and Bucher, P. (1998) The Eukaryotic Promoter Database. *Nucleic Acid Res.* **26,** 353–357.
18. Boulikas, T. (1993) Nature of DNA sequences at the attachment regions of genes to the nuclear matrix. *J. Cell. Bioch.* **52,** 14–22.
19. Bode, J., Stengert-Ibert, M., Kay, V., Schlake, T., and Dietz-Pfeilstetter, A. (1996) Scaffold/matrix attchment regions: topological switches with multiple regulatory functions. *Crit. Rev. in Eukaryot. Gene Expr.* **6,** 115–138.
20. Nikolaev, L., Tsevegiyn, T., Akopov, S., Ashworth, L., and Sverdlov, E. (1996) Construction of a chromosome specific library of MARs and mapping of matrix attachment regions on human chromosome 19. *Nucleic Acid Res.* **24,** 1330–1336.
21. Phi-Van, L. and Strätling. (1988) The matrix attachment regions of the chicken lysozyme gene co-map with the boundaries of chromatin domain. *EMBO J.* **7,** 655–664.
22. Jade, J., Rios-Ramirez, M., Mielke, C., Stengert, M., Kay, V., and Klehr-Wirth, D. (1995) Scaffold matrix attachment regions: structural properties creating transcriptionally active loci. *Intl. Rev. Cytol.* **162A,** 389–454.
23. Jarman, A. and Higgs, D. (1988) Nuclear scaffold attachment sites in the human globin gene complexes. *EMBO J.* **7,** 3337–3344.
24. Farache, G., Razin, S., Targa, F., and Scherrer, K. (1990) Organization of the 3′-boundary of the chicken alpha globin gene domain and characterization of a CR 1-specific protein binding site. *Nucleic Acid Res.* **18,** 401–409.
25. Deppert, W. (1996) Binding of MAR-DNA elements by mutant p53: possible implications for oncogenic function. *J. Cell. Biochem.* **62,** 172–180.
26. Kramer, J. A. and Krawetz, S. A. (1996) Nuclear matrix interactions within the sperm genome. *J. Biol. Chem.* **271,** 11,619–11,622.
27. Staden, R. A. (1988) Methods for calculating the probabilities of finding patterns in sequences. *Comput. Applic. Biosci.* **5,** 89–96.

28. von Kries, J., Phi-Van, L., Diekmann, S., and Strätling, W. (1990) A non-curved chicken lysozyme 5′ matrix attachment site is 3' followed by a strongly curved DNA sequence. *Nucleic Acid Res.* **18,** 3881–3885.
29. Spitzner, J. and Muller, M. (1988) A consensus sequence for cleavage by vertebrate DNA topoisomerase II. *Nucleic Acid Res.* **16,** 5533–5556.
30. Sander, M. and Hsieh, T. (1985) Drosophila topoisomerase II double stranded DNA cleavage: analysis of DNA sequence homology at the cleavage site. *Nucleic Acid Res.* **13,** 1057–1067.
31. Singh, G. B., Kramer, J. A., and Krawetz, S. A. (1997) Mathematical model to predict regions of chromatin attachment to the nuclear matrix. *Nucleic Acid Res.* **25,** 1419–1425.
32. Bucher, P. and Bryan, B. (1984) Signal search analysis: a new method to localize and characterize functionally important DNA sequences. *Nucleic Acid Res.* **12,** 287–305.
33. Prestridge, D. (1991) Signal Scan—a computer program that scans DNA sequences for eukaryotic transcriptional elements. *Comput. Applic. Biosci.* **7,** 203–206.
34. Chen, Q., Hertz, J., and Stormo, G. (1995) MATRIX SEARCH 1.0: a computer program that scand dna sequences for transcriptional elements using a database of weight matrices. *Comput. Applic. Biosci.*, **11,** 563–566.
35. Quandt, K., Frech, K., Karas, H., Wingender, E., and Werner, T. (1995) Matind and matinspector–new fast and versatile tools for detection of consensus matches in nucleotide sequence data. *Nucleic Acid Res.* **23,** 4878–4884.
36. Singh, G. B., Kramer, J. A., and Krawetz, S. A. (1997) Mathematical model to predict regions of chromatin attachment to the nuclear matrix. *Nucleic Acid Res.* **25,** 1419–1425.

5

Teaching Bioinformatics and Keeping Up-to-Date with the Literature

25

Design and Implementation of an Introductory Course for Computer Applications in Molecular Biology and Genetics

Stephen A. Krawetz

1. Introduction

A computational biology course is becoming an integral component of many graduate programs. It is clear that the students who are interested in pursuing training in computational biology possess substantially different levels of expertise. How then, does one best meet the needs of all participants? This can be accomplished by providing two separate courses that are specifically designed for either the basic or advanced student. The first basic level course should be tailored to those students who are just acquiring computer skills in computational biology. The second advanced course would then serve senior graduate students, postdoctoral fellows, and principal investigators. Advanced courses tacitly assume that the participants will have a basic working knowledge of sequence analysis and computer systems. There are several advanced courses offered at various institutions around the globe. A partial listing is available at the *Frontiers in Bioscience* web site:

`http://www.bioscience.org/events.htm`

Of particular note are the courses offered at Cold Spring Harbor Laboratory:

`http://nucleus.cshl.org/meetings/`

The Genome Mapping Project Resource Centre (U.K.):

`http://www.hgmp.mrc.ac.uk/About/Docs/Courses/`

From: *Methods in Molecular Biology, vol. 132: Bioinformatics Methods and Protocols*
Edited by: S. Misener and S. A. Krawetz © Humana Press Inc., Totowa, NJ

and the cosponsored EMBNet courses (Europe):

```
http://www.icgeb.trieste.it/net/netcourse.html
```

Whereas each of these advanced courses is excellent in its own right, they are not intended to accommodate the needs of those students just entering the field of molecular medicine and genetics. The following chapter describes the introductory course that has been developed to serve this student base.

1.1. Course History and Design Considerations

Initial interest in acquiring computational skills for the individual molecular biology project was spawned with the advent of rapid and efficient DNA sequencing technologies. It was soon realized that only computers could provide the means to address the various data management and analyses issues. These skills were usually passed down from one member of the laboratory to the next. However, interest in this technology was accruing at such a rapid rate that the demands placed on that knowledgable individual increased exponentially. It was realized that it would be more efficient to develop a course to meet the needs of the junior graduate student population rather than to provide individual training on an as-required basis. To address this need, an entry level introductory course entitled Computer Applications in Molecular Biology was initiated in 1990 at Wayne State University. It has now evolved to become a required core course for new graduate students entering the Molecular Medicine and Genetics graduate program. The course has always been well received because of the inherent appreciation for the powerful tools that are provided when this technology is appropriately used. The success of the course relies on three integral components. First, the ability to provide hands-on training within a defined learning center. During the early years, the teaching/training resource center was sponsored with the temporary loan of computer software and hardware from several manufacturers. As such, the center was transitory and only available to the students for the duration of the course. The teaching/training resource center is now provided as part of the Medical School Library system. Second, each session utilizes both a lecturer and facilitator. The facilitator has been integral to the success of this course, providing immediate assistance to each student at his or her work station when the need arose during the lecture/demonstration. At least one facilitator per group of five students is required, however a ratio of one facilitator per group of three students has been found to be optimal. The lecturer/facilitator concept has proven a successful means to meet the daily course objectives. It has also provided the opportunity to directly interact with each student on a daily basis because all lecturers also participate as facilitators. In this manner the faculty became familiar with each student's strengths and the course was modified accordingly. Third, the course is structured as a single intensive 2-week session. It was expected that during this period the student's primary responsibility was to this course. This requirement has proven essential, given the amount of time each student must invest to fulfill the course requirements of a daily 3-hour lecture demonstration and take-home problem assignment.

1.2. Learning Objectives

Four learning objectives have been defined as requisite components to a successful course outcome. Attaining each objective permits the student to build upon a series of fundamental skills. In doing so, they develop self confidence and the ability to explore and utilize independent resources in a productive manner. The learning objectives are:

1. To introduce and broaden familiarity to basic computer operating systems, desktop environments, and functions that include file management, transfer, and e-mail.
2. To develop fundamental skills for accessing and searching the information resources provided by the virtual library.
3. To develop Web-based application skills to solve a range of problems in computational biology.
4. To develop essential and complementary sequence analysis skills using the institutional core analysis suite.

Upon completion, it is expected that each participant will be able to independently identify the appropriate analytical tool and/or resource to solve a related problem.

1.3. Course Description

At present, the course is given for single credit and is taught at the introductory level. Enrollment is restricted to 10 students to accomodate the number of lecturers and facilitators that can devote the substantial amount of time that is required to provide the "individualized" instruction. The course evolves from year to year to reflect the advances in the field and interests of the participating faculty. The 1998 course description is as follows.

> "This course will provide the opportunity for students to become familiar with the basic use and concepts of the Center for Molecular Medicine and Genetics core computer facility, the virtual library, and various Internet resources. The students will be introduced to the UNIX operating system, the motif interface, and electronic mail. Access to resources on the Internet will utilize *Netscape*. Topics that will be covered include the GCG analysis suite, the virtual library, NCBI, sequence- and text-based database searching, identifying repetitive elements, multiple sequence alignment, computer prediction of biological meaningful sequence segments, and program retrieval and installation. The use of e-mail in various computer environments will be demonstrated and participants

will be given the opportunity to develop a set of limited basic skills in each environment. Effective use of the virtual library will also be demonstrated. This will include the retrieval of the latest published paper of interest, the use of MEDLINE, and other journal indexing services to provide a series of related journal articles. Grant acquisition resources will also be reviewed. This intense laboratory course will provide extensive hands-on training. Each session will begin with an overview, followed by a hands-on demonstration, and an example problem. The students will then be expected to solve a related problem. Course evaluation will be based on daily assignments and the solution to a more complex problem by the end of the course. Evaluation will also include an oral presentation and a written report. Enrollment is by permission of the course faculty only."

A single presentation has also been used to provide an overview to the faculty and the uninitiated. This has also given the students the opportunity to develop some useful rudimentary analysis skills. The outline of the 2-hour overview follows.

"This introductory session will provide the means for the uninitiated to become familiar with the basic use and concepts of e-mail, the virtual library, and biological resources on the Internet. Each concept will begin with an overview, followed by a hands-on demonstration, and an example problem. The use of e-mail in various computer environments will be demonstrated and participants will be given the opportunity to develop a set of limited basic skills in these environments. Effective use of the virtual library will be demonstrated. This will include the retrieval of the latest published paper of interest, the use of MEDLINE, and other journal indexing services to provide a series of related journal articles. Grant acquisition resources will also be reviewed. Some of the various Internet resources that are available in support of research will be demonstrated. Other Internet resources including access to newsgroups, NCBI, and sequence- and text-based database searching will be discussed. This intense laboratory course will provide hands-on training. It is expected that upon completion the participant will have sufficient familiarity to surf the net, use some of the tools that are available, and electronically communicate with colleagues. Class size is restricted to the number of computers available for "hands-on" training."

Participation in this overview session is not a prerequisite to enroll in the full course. However, attending this overview does raise the level of enthusiasm for additional training. This is evidenced by increased enrollment in the two-week intense course if the opportunity is provided.

1.4. Course Outline and Sample Assignments

The course outline shown in **Fig. 1** provides a synopsis of each 1998 class session. This overview is provided as a guide to the group of sessions that form the core of the program. These sessions include computer basics, the review of key molecular genetic Internet sites, searching for similar sequences, multiple sequence alignment, the virtual library, higher-order sequence analysis searching, and an introduction to the GCG analysis suite. The course can then be tailored to best meet the needs of that student population and the expertise of

Fig. 1. Sessions.

Session 1: Computer Basics

Course introduction & welcome
MS Windows basics
UNIX basics
PC *X-Windows* (NCD PCXWARE)
File management
E-Mail (*PINE*, *EUDORA*, *NETSCAPE MAIL*)
File transfer (ftp, WSftp)

Session 2: Review of key molecular genetic internet sites & searching for similar sequences & multiple sequence alignment

Internet World Wide Web resources (a list and description is provided in **Fig. 2**)
Simimlarity searching *BLAST/FASTA*
Retrieving and installing a program (*TreeTool*)
Multiple sequence alignment (*CLUSTAL W* and *GeneBee*)

Session 3: The virtual library

Searching *MEDLINE* on the *PubMed* system from the National Center for Biotechnology Information.
Searching the *Science Citation Index* and *Current Contents Connect* from the Institute for Scientific Information.
Using bibliographic databases and tables of content services to stay current of the biomedical literature.
Accessing full-text journals on the Internet and printing articles.
Finding grant and funding resources on the Internet.

Session 4: Higher-order sequence analysis searching for simple repeat sequences restriction site analysis

MARFinder
Identifying Repetitive Elements
Identifying Transfactor Binding site candidates

Sessions 5 & 6: GCG sequence analysis or other comparable suite

Introduction to GCG: sequence analysis
GCG Manual: `http://cmmg.biosci.wayne.edu/gcg/gcgmanual.html`
SeqLab: the *X* interface to GCG
SeqWeb: the Web interface to GCG
Basic sequence analyses
Multiple Sequence analysis

Session 7–9: Final assignment

Session 10: Presentations and final report

Fig. 2. Some useful sites on the Internet.

DATABASES AND SEARCH TOOLS

NCBI
*http://www.ncbi.nlm.nih.gov/

EMBL SERVER
*http://www2.ebi.ac.uk/services.html

Genome Navigator: Saccharomyces cerevisiae Genome Index
http://www.mpimg-berlin-dahlem.mpg.de/~andy/GN/S.cerevisiae/

SEQUENCE ALIGNMENT

GENEBEE MULTIPLE SEQUENCE ALIGNMENT
*http://www.genebee.msu.su/

TREEVIEW
*http://taxonomy.zoology.gla.ac.uk/rod/treeview.html

CLUSTAL W
*http://www2.ebi.ac.uk/clustalw/

GENEDOC: Multiple Sequence Alignment Editor, Analyser and Shading Utility for Windows.
http://www.cris.com/~ketchup/genedoc.shtml

SEQUENCE ANALYSIS

Restriction Enzyme Site Digestion

Webcutter 2.0: Analyze your sequence also a direct reference to REBASE.
*http://www.firstmarket.com/cutter/cut2.html

Search for potential transcription factor binding sites with MatInspector V2.2
*http://transfac.gbf-braunschweig.de/

MAR-Finder: Deduce the presence of matrix association regions, or MARs, in DNA sequences.
*http://www.ncgr.org/MarFinder/

Computational Genomics Group of the Sanger Centre Informatics Division
http://genomic.sanger.ac/uk/

BCM Search Launcher
http://kiwi.bcm.tmc.edu:8088/search-launcher/launcher.html

REPETITIVE ELEMENTS

RepeatMasker2 Web Server
*http://ccr-081.mit.edu/Repeats.html

CENSOR Web Server
*http://charon.girinst.org/~server/censor.html

IMAGE ANALYSIS, EXPERIMENTAL PROTOCOLS, AND COMPUTER COURSES

ANALYSIS-NIH IMAGE PROGRAM MAC & PC (FREE)
http://www.scioncorp.com/

PCR and multiplex PCR: guide and troubleshooting guide
http://info.med.yale.edu/genetics/ward/tavi/PCR.html

Welcome to the VSNS BioComputing Division
http://www.techfak.uni-bielefeld.de/bcd/welcome.html

OTHER USEFUL SITES
The Really Quite Useful MolBioPage
`http://www.lars.bbsrc.ac.uk/plantsci/molbiol/molbiol.html`
Alex's Cyber-Science Jumpstation
`http://www.flnet.nl/~bossers//`
On line analysis tools
`http://www-biol.univ-mrs.fr/english/logligne.html`
Welcome to the Globin Gene Server
`http://globin.cse.psu.edu/`
ExPASy Molecular Biology Server: Swiss Institute of Bioinformatics (SIB) 2-D PAGE.
`http://expasy.hcuge.ch/`

*Internet sites used for data analysis.

the lecturers using any chapter from this book as a additional or supplemental session. The programs from the internet sites that have been used for data analysis for each session are indicated by the stared entries given in **Fig. 2**. Examples of each problem assignment for each session are presented in **Fig. 3**, and can also be used as a guide. As indicated in **Fig. 4** (Format section) greater than 1 h of each class is set aside for questions and individualized assistance to those in need. This has proven an effective means to ensure that the students understand the material that was presented.

1.5. Lecture Outline

The lecture begins with the presentation of the background theory to the program that is the subject of presentation. This is followed by a hands-on demonstration in which the class follows along with the lecturer at their own workstations. Subsequent to the execution of the program, the resulting analysis is interpreted as part of a class discussion. If time permits, the parameters of the program are adjusted to demonstrate their effect on the output and the resulting interpretation.

Although the students should be familiar with the basic concepts of gene structure, they are often not familiar with the concept that a functional gene can also be defined by an ordered array of sequence elements along a nucleotide sequence string. It is usually helpful to provide a brief review of sequence structure, signals, and motifs; their location, and how these elements can be manually identified. This is usually accomplished by annotating a line drawing representation of the sequence file header information that can be obtained from any one of the sequence files from a nucleic acid database. Furthermore, and most importantly, it is never too early to stress that the computer output is a prediction that must always be placed in a biological context. The

Fig. 3. Assignments.

ASSIGNMENT 1:

a. Send an e-mail message to everyone in the class, including the instructors, introducing yourself and your research interests and why you are taking this course.
b. Register as a user at `http://www.ncgr.org/MarFinder/`
c. Draw a graphical representation of a *GenBank* sequence file using the annotations as a guide.

ASSIGNMENT 2:

a. Recover and install a sequence editor from one of the sites on the Internet. Be prepared to demonstrate its functionality to the class.
b. There are three genes in the gene cluster of U15422. Use at least two different WWW-based resources to determine the optimal sequence alignment of the genes of this cluster? Are any of these alignments optimal? Could you improve on the alignment.

ASSIGNMENT 3:

Retrieve a complete bibliographic listing of papers on the sequences that are known to be bound to the nuclear matrix. Retrieve the corresponding sequences from the list.

ASSIGNMENT 4:

a. Identify candidate transcription factor binding sites within the sequence U15422. What is their relationship to the structure of this region? Which sites are reasonable biologically active candidates and why?
b. Compare and contrast the repetitive elements that are contained within U15422 and identified using at least two different WWW-based resource.

ASSIGNMENT 5:

Choose DNA sequence that encodes a protein of interest to you, such as a gene that you are working on in your lab (or any other gene). Please prepare a restriction map of that DNA sequence. Then, do a database search at NCBI to find other proteins related to the protein product of your gene. Send the results of your analyses by e-mail before next class. Please include a description of the DNA sequence you have chosen and the reason you chose it, describe the procedures you use to obtain your analyses, and explain your results.

ASSIGNMENT 6:

Use the GCG Multiple Sequence Analysis tools to align the amino acid sequences of a group your group of relateded proteins of interest in order to compare their sequence similarities and phylogenetic relatedness. Save the dendogram and phylogram and transfer the files to your lab or library computer, and print out copies for the next class using both a Postscript printer and a Hewlett Packard Graphics printer.

FINAL ASSIGNMENT

You will be given a list of sequences from human chromosome 16. Within these sequences identify the regions that are reasonable candidates to be bound to the nuclear matrix. What is the relationship among these sites?

Fig. 4. Daily session format.

0.5 hour—Questions and individualized assistance.
1.0 hour—Instruction.
0.5 hour—Break/Questions and individualized assistance.
1.0 hour—Instruction.
Post class—Questions and individualized assistance.

output can only be used to guide experiments. The predictions require biological verification.

The philosophy that many different tools and strategies can be used to solve the problem has been adopted throughout the course. For example, when discussing sequence similarity searching, both the *BLAST* and *FASTA* programs ***(1,2)*** are presented and the various program options discussed. Differences among various search strategies and their impact upon analysis are addressed. Discussion of how the adjustment of program parameters, e.g., word size, window size, and threshold, may be used to optimize the output to answer the question being posed is encouraged. The concept of a scoring matrix, how a matrix is derived and how it is used is also presented in the context of the various programs like *BLASTP* (protein vs protein database), *BLASTN* (nucleic acid vs nucleic acid data base), and *BLASTX* (six frame nucleotide translation vs protein data base). Subsequent to the search, the interpretation of the similarity scores is discussed. In addition, the students are shown that simple statistical concepts like E, expectation (the number of times that pattern or event is expected to occur at random) can also be used as a quality measure of the resulting aligned structures. When database-searching strategies are demonstrated it is quite valuable to review a list of biologically related sequences in which some of the members are quite distant. This emphasizes that care needs to be taken when interpreting the results and that one must be clear when posing a question to a computer.

As above, data interpretation is always emphasized throughout the course. One of the best examples is searching for transcription factor binding sites ***(3)***. In this case, the student is always left with numerous candidates that must be manually sorted for likely biological candidates. This exercise is often revealing of those students who understand basic gene structure, i.e., the location of the promoter in relation to the start codon. Typically these students use this information to limit the search to these regions in which these biologically relevant elements are likely to reside. In addition, the basic biological concept of tissue distribution and specificity of the identified factor is emphasized by asking the following simple question. Does the candidate transacting factor have a similar tissue distribution as the gene of study?

1.5. Other Suggested Topics

There are three elements that are not covered in this course. They are the use of a sequence editor, sequence project management, and phylogenetic analysis. First, whereas file format structure and conversion among file formats is demonstrated, the students are expected to become familiar with a sequence editor as the necessity arises. Most students initially solve this problem on their own with the use of a text-based editor. However, once they embark upon the more complex assignments the necessity of a sequence editor becomes clear. It has been our experience that even though a sequence editor is a necessity, most students dread this session as much as the first session on fundamentals in which many are lacking. If this skill is considered essential, the students could begin with a simple text-based editor using a wordprocessor to emphasize the file format constraints imposed by the various programs. They could then proceed to more complex editing functions using a sequence file editor. Examples of fully functional editors can be found within the GCG suite (Chapters 1 and 2) and *Staden Suite* ***(4,5***; Chapter 7). In addition, an editor like *Sequin* ***(6)***, that is used for standard database submissions may also be appropriate. The latter would further familiarize the student with the wealth of information that is contained within the header of a *GenBank* file.

Second, sequence project management presents distinctive challenges. The uniqueness is not usually appreciated until the individual embarks upon a sequencing project. For this reason it is recommended that sequence project management be addressed as a separate course. However, upon becoming familiar with the fundamentals most students can easily acquaint themselves with sequence management software like the *Staden Suite* (Chapter 7).

Third, the practical aspects of phylogenetic analysis are discussed using *CLUSTAL W* and tree tool but a comprehensive analysis as given in Chapter 12 of this volume is not presented. This topic could be treated as an entire course, once the students have mastered the fundamentals.

1.6. Using Help Guides to Prepare Handouts

Most of the handouts are prepared with the aid of the help files that are immediately available at the program's Web site. These program documentation files provide an excellent resource for the operation of the program. They often detail the use of the analysis program, input and output parameters, data interpretation, as well as addressing file format issues. Where appropriate, additional information can be gathered from the current literature. Interestingly, past experience has indicated that students prefer receiving a package of handouts at the beginning of the course along with the assigned reading list, rather than retrieving similar documents from the web.

Fig. 5. Daily course evaluation.

Lecturer: Date:

Please complete the following evaluation.
Circle your response for each question.

1) Did you prepare for this lecture?
Yes No

2) Did you read all of the assigned material for this lecture?
Yes No

3) Did you find the material covered in this lecture useful?
Yes No

4) Was the presentation clear?
Yes No

5) Was adequate time allotted for discussion?
Yes No

6) Did you understand the material covered?
Yes No

7) If you answered no to any of the questions, please clarify?

____

____

____

____

8) Please indicate lecturer(s) for today's session:

9) Please indicate facilitator(s) for today's session:

1.7. Course Evaluation

The daily interaction with the facilitator and the daily assignments that each student submits by e-mail are used to assess the students progress and identify any areas that require remediation. It is also very helpful for the students to provide a third and anonymous means of almost immediate feedback for each session. This can be revealing to both the lecturers and facilitators, as it provides the students' perspective of how the class is proceeding and if, in their own mind, they understood the material that was presented. These three criteria can then be used as a guide to appropriately modify the subsequent session or reiterate a concept in which it was clear that the majority of students require assistance. A typical survey that is given after each lecture is shown in **Fig. 5**.

1.8. Conclusions

This course provides an overview to the field of computational molecular biology and is not an end in itself. The strengths of the course lie in familiarizing the participant with the many resources that are available, yet instilling the concept that many different methods can be used to solve the same problem. Whereas the mathematical foundation for each of the programs utilized is defined, emphasis is placed on the biology of the system. In this manner the computer is treated as an aid to the design of the next experiment and not as a means of proof of concept. The essential outcome of this course is that students become armed with a set of computational tools and the confidence that a computer program is not an unmanagable ghoul. They should possess the skills to access and utilize new programs or systems as required for their research program.

Acknowledgments

I would like to thank D. Womble, Wayne State University, Center for Molecular Medicine and Genetics for permitting the use of his questions in this chapter and for his continued help in making the course the success that it is. I would also like to thank my wife for her continued support and encouragement. This chapter is dedicated in my father's memory.

References

1. Altschul, S. F., Madden, T. L., Schäffer, A. A., Zhang, J., Zhang, Z., Miller, W., and Lipman, D. J. (1997) Gapped BLAST and PSI-BLAST: a new generation of protein database search programs. *Nucleic Acids Res.* **25,** 3389–3402.
2. Pearson, W. R. (1990) Rapid and sensitive sequence comparison with FASTP and FASTA. *Meth. Enzymol.* **183,** 63–98.
3. Quandt, K., Frech, K., Karas, H., Wingender, E., and Werner, T. (1995) MatInd and MatInspector—new fast and versatile tools for detection of consensus matches in nucleotide sequence data. *Nucleic Acids Res.* **23,** 4878–4884.
4. *Wisconsin Package Version 9.0.* Genetics Computer Group (GCG), Madison, WI.
5. Staden, R. (1996) The Staden Sequence Analysis Package. *Mol. Biotechnol.* **5,** 233–241.
6. Benson D. A., Boguski, M. S., Lipman, D. J., Ostell. J., and Ouellette, B. F. F. (1998) GenBank. *Nucleic Acids Res.* **26,** 1–7.

26

The Virtual Library I

Searching MEDLINE

Keir Reavie

1. Introduction

Staying current with the molecular biology literature has eased appreciably over the last few years, as information resource producers provide more of their information via the World Wide Web (WWW). The ability to access these resources using one access point (WWW) and single piece of software (a Web browser), enables the researcher and clinician to easily stay current of the published literature and research in molecular biology and genetics. In addition, current developments now allow the researcher to quickly link across data resources to retrieve full articles and related numeric information on DNA and protein sequences.

The next three chapters discuss the resources available, how they can be used to access information at the point of need, and to provide regular updates of the latest research. This chapter focuses on the *MEDLINE* database and how it can be searched using the National Library of Medicine's (NLM) *PubMed* system. Chapter 27 looks at the *Science Citation Index*, an important resource for access to the basic sciences literature; *Current Contents* and alerting services, to obtain timely access to the literature as it is published; and the use of bibliographic management software to enhance and simplify the research process. Chapter 28 looks at access to electronic journals, and the major Internet resources for staying current of funding opportunities. Throughout all these chapters the current state of information access in the electronic world is discussed, particularly within the context of where it is incomplete, the

From: *Methods in Molecular Biology, vol. 132: Bioinformatics Methods and Protocols*
Edited by: S. Misener and S. A. Krawetz © Humana Press Inc., Totowa, NJ

difficulties in getting access to electronic information resources, both technical and otherwise, and what we can expect to see in the future.

2. *MEDLINE*

The most accessible and used source for searching the biomedical literature is *MEDLINE*, produced by NLM, located at the National Institutes of Health (NIH). *MEDLINE* is the online version of NLM's *Index Medicus*, with additional citations from the nursing and dentistry literature. It contains bibliographic citations with abstracts from approx 3900 biomedical journals published in the United States and 70 other countries. Criteria for inclusion of a journal title in the database is available from NLM *(1)*. There are approximately nine million records in the database, going back to 1966, when *MEDLINE* first began its conversion from the print *Index Medicus* into its electronic format. *MEDLINE* is updated weekly. The majority of records in the database are from English language publications or have English abstracts.

In August, 1996, NLM introduced a supplement to *MEDLINE*—*PREMEDLINE*. *PREMEDLINE* provides access to basic citation information and abstracts before the full records are entered into *MEDLINE*. This enables access to the journal literature in the biomedical sciences within days of publication. Records are entered into *PREMEDLINE* daily, and each record is assigned a *MEDLINE* Unique Identifier (UI). In time, the record will be fully processed to include complete *MEDLINE* citation elements and be entered into the database. NLM identifies priority journal titles, which are added to *PREMEDLINE* and processed for addition to *MEDLINE* faster than others. Priority indexing is based on a journal's relative importance to biomedical researchers and clinicians.

NLM sells the *MEDLINE* data to a variety of third parties, who in turn make the data available through a variety of different search interfaces and in a variety of formats, such as CD-ROMs. *MEDLINE* databases made available through an institution's library, either on standalone workstations, or via computer networks are normally purchased from a third party vender—most often Ovid Technologies or SilverPlatter. The vendors also provide search software for access to the *MEDLINE* data. In 1997, NLM made *MEDLINE* and *PREMEDLINE* freely available via the World Wide Web (WWW) through two different interfaces, the *Internet Grateful Med* and *PubMED*. *PubMED* is the system with which we will be concerned here, as it was developed at the National Center for Biotechnology Information (NCBI) at NLM, in conjunction with NCBI's *Entrez* database service.

Entrez was designed by NCBI to integrate access to DNA and protein sequence databases along with taxonomy, genome, and protein structure information. *Entrez* also contains direct access to *MEDLINE* articles describing sequences. The natural progression of this development was to include the entire *MEDLINE* database, along with *PREMEDLINE*, and provide direct links from *MEDLINE* records to sequence information in *Entrez*. NLM also negotiates with publishers for direct links from *MEDLINE* records to the full text of journal articles. Further information on the production of *PubMed* and all its features is located at the NCBI Web site *(2)*.

2.1. *Searching* MEDLINE *with* PubMed

The following information on using *PubMed* to search *MEDLINE* is meant as a basic practical guide. Detailed information on using the *PubMed* system is available by clicking on the **Help** links on the *PubMed* search pages.

The *PubMed MEDLINE* search system (`http://www.ncbi.nlm.nih.gov/PubMed/`) offers two search options, a **Simple Search**, to which you are connected on the main *PubMed* page, and an **Advanced Search** option. It is advisable to use the **Advanced Search** option, as it enables you to take full advantage of the more advanced *PubMed MEDLINE* search features and improve search retrieval. For our purposes we will discuss mainly the advanced search options in *PubMed*.

The main *PubMed* page provides a number of links in a column to the left of the screen. The **Help** link provides detailed information on searching the system. The other two links of concern to us are **Advanced Search** and **MeSH Browser**. It is sufficient at this point to go directly into the **MeSH Browser**, which we will use to build our search strategy. *PubMed* can then execute the strategy to retrieve information from the *MEDLINE* database. When using the **MeSH Browser** to search *PubMed*, the results are displayed with additional advanced search options, enabling revision of the search strategy. We will discuss some of the advanced field specific search options later in this section.

2.1.1. *Using the* **MeSH Browser**

Click on the **MeSH Browser** link to enter into the browser. The purpose of the browser is to assist searchers in using the *Medical Subject Headings* (*MeSH*) to search *MEDLINE*. The *Medical Subject Headings* are a controlled vocabulary used to index articles entered into the database. Subject experts at NLM read incoming articles for inclusion in *MEDLINE*, identify the topics of those articles, and then go to *MeSH* to select the most appropriate terminology to define those topics. The terms are then attached to the article as its index terminology, or *Medical Subject Headings*, and entered into *MEDLINE* along with the article's citation. It is important to try and use this same terminology to extract information from *MEDLINE*, as it makes the search process more accurate and comprehensive. Among other things, use of *MeSH* enables us to avoid having to worry about the use of synonymous terminology or acronyms, because all synonyms and acronyms are collated into one selected *MeSH* term. Use of the *MeSH* also allows us to broaden or narrow our search to easily include articles on related or more specific subject areas in our retrieval.

To illustrate the use of *PubMed*'s **MeSH Browser**, we will use an example query:

Locate recent articles on using resources on the Internet for sequence analysis.

There are two main concepts in this example to include in the search strategy—*the Internet* and *sequence analysis*. It is important before doing any search to analyze the query's main concepts. A search strategy can then be developed to embody all the concepts in a way that retrieves relevant information, and answers the question. In this example the search will include the concepts of *the Internet* and *sequence analysis*, which will be combined to retrieve articles that discuss both of these topics together.

By entering the term *Internet* in the **MeSH Browser** and clicking the **Browse** button, *PubMed* will identify the appropriate *MeSH* terminology for searching this topic. The Browser explains that *Interne*t is not a *MeSH* term, but is closely associated with the *MeSH* term *Computer Information Networks*—all articles in *MEDLINE* that discuss the Internet will be indexed using this term. Hence, we use this *MeSH* term to retrieve articles about the Internet. The **MeSH Browser** also provides us with the term's definition and a **MeSH Tree Location**. The Tree Location identifies where a *MeSH* term is located within a hierarchy of terminology. We can see that *Computer Information Networks* is listed beneath *Computer Systems*, and that *Local Area Networks* is listed as a more specific type of *Computer Information Network*. If we click on the **Detailed Display** link, next to the *MeSH* term, we retrieve further information.

Near the middle of the screen in the **Detailed Display** is a list of **Subheadings**, any combination of which can be selected as qualifiers to the *MeSH* term and assist in narrowing a search. Subheadings are general terms that can be applied to a broad range of subjects and help to make subject searching in *MEDLINE* more specific. Their use will be demonstrated later in this section.

Below the subheadings in the **Detailed Display** there are two additional options: **Restrict Search to Major Topic headings only** and **Do Not Explode this term**. The first option enables us to retrieve only those documents where our subject of interest is considered to be a major point of discussion in the article, rather than a minor topic. This will narrow the search retrieval. One needs to be careful in selecting this option, as articles are indexed by humans and the process is not completely objective. What one person may think is a major topic of an article, another may not. Narrowing a search to retrieve major topics only can sometimes eliminate relevant articles from the retrieval.

While selecting **Major Topic headings only** narrows a search, selecting the **Explode** option broadens the search. When a *MeSH* term can be exploded, *PubMed* automatically uses the **Explode** option. Explode enables us to search the selected *MeSH* term and all the more specific terms listed beneath it, thereby incorporating additional articles, on more specific subjects in the results. The rule at NLM is to index the articles in *MEDLINE* using the most specific *MeSH* terms available. If an article discusses sequence analysis on local area networks, it will be indexed using *Local Area Networks*, and not *Computer Communication Networks*, the broader term. Such an article will not be retrieved in the search if we do not explode *Computer Communication Networks*. We need to decide whether articles on local area networks need to be included in the search retrieval. If not, we can turn off the **Explode** feature with the **Do Not Explode this term** option.

Having selected all of our options for this term, we can click the **Add** button on the left of the screen to include the term in our query. The term will be added to our search. To add additional terms to our search, select **Enter another MeSH term to browse** from the top of the page and enter the next topic—*sequence analysis*. Because we had previously used the **Detailed Display** option, *PubMed* automatically connects to the detailed display for the *MeSH* term *Sequence Analysis*. After making the appropriate selections for narrowing to major *MeSH* term and exploding, we can click on the **Add** button, to include this term to our search.

When adding additional terms to a *PubMed* query, the system automatically assumes the inclusion of the **AND** operator between terms. There are however three options: **AND**, **OR**, and **BUTNOT**, which can be viewed and selected by clicking on the small window next to the **Add** button. These are Boolean operators, named for the Irish mathematician Boole. **AND** will retrieve articles indexed under both topics together; **OR** will retrieve articles that are indexed using either one or the other topic, or both; and **BUTNOT** will remove any articles that are indexed using that terminology from the retrieved set. In this example we will use the **AND** operator. Assuming we chose **Major MeSH terms** for both topics and left the **Explode** option turned on, the search will appear as follows:

```
Computer Communication Networks [MAJR] AND
Sequence Analysis [MAJR]
```

Click on the **Return to PubMed** button next to the search strategy to now have PubMed execute the search against the *MEDLINE* database.

Before discussing the PubMed display screens, let's look at another example: *Find articles on the mapping of chromosome X.*

The two main concepts in this search query will be *mapping* and *chromosome X*. Again, we can begin with the **MeSH Browser**. However, if we enter *mapping* into the browser, it acts differently than in the last example. *PubMed* is unable to identify specific *MeSH* terms for the concept of *mapping*, and consequently, it provides us with a screen that says: **No exact match for your term was found**, along with a list of terms from which we are able to select the most appropriate for this subject. In this example we are likely to select *Mapping, Gene*, and then click on the **Browse This Term** button. *PubMed* then directs us to the correct *MeSH* term—*Chromosome Mapping*. At this point the *PubMed* query can be constructed as in the previous example.

2.1.2. MeSH *Subheadings*

Subheadings have already been discussed as a method of narrowing search retrieval by qualifying *MeSH* terms. A complete list of subheadings can be viewed by clicking on *PubMed*'s **Detailed Help** link. It is hoped that future versions of *PubMed* will provide direct links from subheadings to notes on how specific headings are defined and can be used. Subheadings of relevance for searches in the areas of molecular biology and genetics include *abnormalities, congenital*, and *genetics*. For example, if we were interested in locating articles discussing the genetic aspects of a disease such as phenylketonuria, we would first enter *phenylketonuria* into the **MeSH Browser**. We would see that *Phenylketonuria* is a *MeSH* term, and from the **Detailed Display** screen, be provided the option to qualify the term with specified subheadings. As phenylketonuria is a disease, the subheadings displayed will be relevant to qualifying a disease term, such as *diagnosis*, *pathology*, or *therapy*. We can select any combination of subheadings for our search. In this example we will simply select *genetics* and ask *PubMed* to add this term to our query. Assuming we also chose to have *PubMed* retrieve only articles where this is a major *MeSH* topic, the search query should look as follows:

```
Phenylketonuria/genetics [MAJR]
```

2.1.3. Modifying and Limiting Search Results

Once *PubMed* has executed a search query against the *MEDLINE* database, it will display the **PubMed Query** screen. At the top is the search query, followed by a button showing the number of documents retrieved. We can click on this button to begin viewing the retrieved documents with the default of 20 citations at a time. *PubMed* automatically displays the number of available articles dating back to 1966. We can limit the number of articles viewed by changing the **Entrez Date Limit** option. The options range from **30 days** to **No Limit** (back to 1966).

The next section of the query result screen is titled **Add Term(s) to Query**. This section can be used to add additional terms to our search query, to make it more specific. By clicking the arrow in the **Search Field** window, you can see that we are able to search specific fields in the *MEDLINE* database, including the *MeSH* and major *MeSH* fields. It is unfortunate that we cannot return to the **MeSH Browser** at this stage to add additional *MeSH* terms. However, we can enter *MeSH* terms into the **Add Term(s) to Query** section and select *MeSH* in the **Search Field** window. This presumes that searchers know the *MeSH* term that they wish to search. If we click on **Search**, *PubMed* will automatically **AND** the term into the current search query. We might also select the **List Terms** option from the **Search Mode** window. This tells *PubMed* to first provide a list of appropriate terms, from which the searcher can select, before running the search. This is a useful feature if the searcher is not sure of the exact usage for a *MeSH* term they want to search.

There are some standard things that we might want to do before browsing our retrieval to limit the search results. Some of these limits include narrowing the retrieval to English language articles, studies done only on humans, or review articles. To ensure that this feature works properly, we need to search these terms in the correct *MEDLINE* database fields. English should be searched in the **Language** field, and review in the **Publication Type** field. A complete list of publication types is available by clicking the *PubMed* **Detailed Help** link. Enter the term and then select the appropriate field from the **Search Field** window. Searching for human studies is a little more complicated. *Human* is a *MeSH* term, but is used mainly as a **Check Tag** in indexing. If the article is a human study, it is tagged with the *MeSH* term *Human*. Other kinds of **Check Tags** *(3)* are used to identify geographical regions, or age groups studied. Tags should never be searched as major *MeSH*, because they are rarely, if ever, used as major indexing terms (there may be exceptions). To narrow the retrieval to human studies only, search *human* in the *MeSH* field. It is hoped that later versions of the *PubMed* search system will include additional search options that provide easier methods of using the standard search limits discussed here.

It is important to be careful with the use of a limit like English. If your search needs to be comprehensive, remember that many important articles are published in languages other than English and may be missed when narrowing to English articles only. Articles published in other languages may, however, have English abstracts in *MEDLINE*.

2.1.4. Displaying Retrieval Results in PubMed

To display the results of the search retrieval, from the *PubMed* search retrieval screen, select an option for the **Entrez Date Limit** and click on the button that states the number of documents retrieved in the search. The next screen displays the first 20 citations from the retrieval. Detailed information for a citation is available by clicking on the linked author's name. Each citation also provides a link to **Related Articles**. Use of this link will retrieve closely related articles that use similar terminology in titles, abstracts and the *MeSH* fields. This feature is particularly useful once we have found an article that matches our query closely. Similar articles are then easily retrieved.

The standard display option is **Abstract Report**, which provides basic bibliographic information and an abstract, when available. We can select a collection of citations for simultaneous display by clicking on the boxes to the left of each. Once selections have been made, select a display option from the **Display** window and click on the **Display** button. In addition to **Abstract Report**, *MEDLINE* **Report** formats the citations for downloading and subsequent importing into a variety of bibliographic management software packages (*see* Chapter 27). *PubMed* also provides options to display links for related information from other *Entrez* databases. In the full display mode each citation will display additional **Links** buttons at the top, enabling direct linkage to Entrez databases and in some instances, the full text of the article. These link buttons include: **Protein**—Protein sequences from *Swiss-Prot*, *PIR*, *PRF*, *PDB*, and translated protein sequences from the DNA sequences databases; **DNA**—DNA sequences from *GenBank*, *EMBL*, and *DDBJ*; **OMIM**—information from the *Online Mendelian Inheritance in Man*; and links to publisher sites or the full text. Full-text access is available on a case-by-case basis, depending on the agreement that NLM has made with individual publishers, and whether the *PubMed* user has authorization to access the full text. Many of the conditions for full text access are discussed in Chapter 28.

Citations can be easily printed using your Web browser's print option, or downloaded in text or HTML formats by using the **Save** options at the bottom of the *PubMed* full display screen. If you are intending to import your results into bibliographic management software, citations should be displayed in the *MEDLINE* format and downloaded as a text file.

2.1.5. Saving Searches

PubMed searches can be easily executed and then saved for future use from the **Simple Search** screen. For this to work, the searcher must enter the search manually, without the assistance of the **MeSH Browser**. If we wanted to save and rerun the first search we created in this section, we would enter:

```
Computer Communication Networks [MAJR] AND
Sequence Analysis [MAJR]
```

into the **Simple Search** screen, and click on the **Search** button. The search can be saved for future use by bookmarking the search retrieval page in your Web browser. To re-execute the search at any time, simply click on the bookmark. Unfortunately, when the search is re-executed, it will search the entire *MEDLINE* database. If you want to retrieve only current information—entered into *MEDLINE* since you last searched—you will have to include date limits in your search strategy statement.

Developing a search strategy without using the **MeSH Browser** requires knowledge of the *PubMed* search command language. Information on using this language can be found in the online manual, accessed by clicking **Help** from any of the *PubMed* screens.

2.1.6. Loansome Doc

Loansome Doc is a service provided by NLM that enables *PubMed* searchers to route search results to a library of their choice, so that the library in turn can retrieve and deliver the full articles to the user. It should be noted that you will need to make arrangements with a local library to provide this service, and depending on the library, there may be document delivery charges. Further information on using *Loansome Doc* is located at NLM's Web site *(4)*. A list of libraries in the United States that will receive and fill *Loansome Doc* can be found at the National Network of Libraries of Medicine (NN/LM) Web site (`http://www.nnlm.nlm.nih.gov/`).

In the *PubMed* display screens, the **Order** button is located directly below the **Display** button. Articles can be selected for ordering by clicking on the box to the left of the citation and then clicking on **Order**. *Loansome Doc* requires users to login with an ID and password. If you are a first-time *Loansome Doc* user, you will be asked to register for the service, and assign yourself an ID and password for future use. You will also need a library ID to complete the registration. The library ID tells *Loansome Doc* where to route the order requests. Library IDs are available from the NN/LM site, or by contacting your local library.

Appendix

The following is a list of the key WWW sites discussed in this chapter.

National Institutes of Health: `http://www.nih.gov/`
National Library of Medicine: `http://www.nlm.nih.gov/`
PubMed MEDLINE: `http://www.ncbi.nlm.nih.gov/PubMed/`

References

1. National Library of Medicine (August 27, 1998), *Journal Selection for Index Medicus/MEDLINE*, National Library of Medicine [3 pages], http://www.nlm.nih.gov/pubs/factsheets/jsel.html.
2. National Center for Biotechnology Information (January 9, 1998), *The NLM PubMed Project*, National Library of Medicine [3 pages], http://www.ncbi.nlm.nih.gov/PubMed/overview.html.
3. National Library of Medicine, Medical Subject Headings Section (1998), *Medical Subject Headings, Annotated Alphabetic List, 1999*, National Library of Medicine, Bethesda, MD.
4. National Library of Medicine (June 2, 1998), *Loansome Doc*, National Library of Medicine [2 pages], http://www.nlm.nih.gov/pubs/factsheets/loansome_doc.html.

27

The Virtual Library II

Science Citation Index and Current Awareness Services

Keir Reavie

1. Introduction

This chapter continues the discussion on searching the biomedical literature on the Internet started in Chapter 26. We will discuss searching the *Science Citation Index* as an adjunct to the information available in *MEDLINE*, and the use of current awareness services such as the *Current Contents* database and tables-of-contents alerting services, to obtain timely updates of the literature as it is published. The chapter concludes with a discussion of bibliographic management software and how it can be incorporated into the research process.

2. *Science Citation Index*

The *Science Citation Index* (*SCI*) is a multidisciplinary database of bibliographic information produced by the Institute for Scientific Information (ISI). It provides an important search enhancement to the literature that can be obtained from the *MEDLINE* database, particularly for basic science research publications. It will include many biomedical sciences journals not indexed in *MEDLINE*. *SCI* online corresponds to the print version of *Science Citation Index*. As an enhancement to the data in *SCI,* each record includes the article's cited references, allowing the researcher to search the database for publications that cite a particular author or article. Cited reference searching enables the retrieval of more recent articles on similar topics, which may not always be retrieved by subject searches. Using this search feature, the researcher can track the historical development of particular research activities, and analyze the impact of specific published research.

From: *Methods in Molecular Biology, vol. 132: Bioinformatics Methods and Protocols*
Edited by: S. Misener and S. A. Krawetz © Humana Press Inc., Totowa, NJ

Subject coverage in *SCI* includes all scientific and technical disciplines. Approximately 3500 of the world's leading scientific and technical journals are included in *SCI. SCI Expanded*, available on ISI's *Web of Science*, indexes 5600 journal titles, with retrospective data going back to 1974. The electronic version of *SCI* is available via a number of methods, including CD-ROMs, published once a month, and magnetic tape editions, updated weekly. ISI's *Web of Science* is available via the WWW. Weekly updates to the database include 17,000 new articles and approx 300,000 new cited references. Approximately 70% of the articles contain English language author-generated abstracts. Additional information on *SCI*, its subject coverage and accessibility are available from ISI's Web pages *(1)*.

2.1. Searching SCI *on the* Web of Science

The following discussion looks at a few techniques for searching *SCI* on the *Web of Science*. SCI on the *Web of Science* does not have the sophisticated search capabilities of *PubMed MEDLINE*. This is mainly because of *SCI*'s lack of controlled vocabulary, like *MeSH*, for the indexing of articles. Nor does *SCI* on the *Web of Science* have the convenient links to molecular and genetic information from the *Entrez* databases like *PubMed*. Detailed information on searching the system can be found in the online manual, accessible via the **Help** button that appears on each of the *Web of Science* search screens. Use of *SCI* on the *Web of Science* is not free, as is *PubMed MEDLINE*. Access will be available only to users in institutions that have subscribed to the service.

The initial screen for the Web of Science search interface (`http://webofscience.com/`) presents two options for searching *SCI*, **Quick Search** and **Full Search**. This discussion will deal only with the **Full Search** options. If we click on **Full Search**, we are then taken to a database selection screen, from which we need to select the databases for searching. The contents of this screen will depend on the level of the subscription held by an institution. We can select the entire database, **This week's update**, the **Latest 2 Weeks**, **Latest 4 Weeks**, or we may opt to search specific years of *SCI*. We also have two options for searching, a **General Search** or a **Cited Ref Search**. **General Search** will allow subject searching, while **Cited Ref Search** searches for articles that cite a specific author or publication.

2.1.1. Subject Searching

Selecting the **General Search** option enters us into the *Web of Science* search screen. In the middle of the screen there are options for searching **Topics**, **Authors**, **Source Title** (to search specific journal titles), and **Address** (to search for publications from specific institutions). The last option is useful for retrieving publications from a single institution and measuring that institution's research impact by analyzing the frequency of citations to its publications.

As mentioned, *SCI* does not have controlled vocabulary, so subject searching needs to be performed carefully, to ensure that no important articles are missed. *SCI* does however include keywords supplied by authors, which are used to represent the content of publications. In addition, ISI generates **KeyWords Plus** for many articles. **KeyWords Plus** are words or phrases that frequently appear in the titles of an article's references, but do not appear in the title of the article itself. A subject search automatically searches these keywords (when available), as well as the titles and abstracts of actual articles. The lack of controlled vocabulary requires that search queries use synonyms and acronyms as needed, because different authors will use different keywords to represent similar subjects. Also, truncating search terms is important. The "*" symbol is used to retrieve variant word endings. *Genetic** will retrieve *Genetic*, *Genetics*, *Genetically*, and so on.

A weakness of the *Web of Science* interface is that it does not provide for easy modification of searches once they have been executed. The searcher must attempt to do a complete and accurate search in a single step. As an example, let's execute a similar search to the one we did in Chapter 26 on *MEDLINE–the genetic aspects of phenylketonuria*. The search strategy might look something like this:

```
phenylketonuria AND (genetic* OR molecular biology)
```

This is not necessarily the ideal search strategy for this topic, but it will do for demonstration purposes. *Web of Science* supports the use of the Boolean operators **AND**, **OR**, and **NOT**. The use of parentheses around *genetic* OR molecular biology* ensures that we retrieve articles on phenylketonuria and genetics or phenylketonuria and molecular biology, and not phenylketonuria and genetics and then all articles in the database on molecular biology, regardless of whether they have anything to do with phenylketonuria. Because the **AND** operation is performed before the **OR** operation in Boolean statements, the parentheses are important.

At the bottom of the **General Search** search screen are **limit** and **sort** options to help narrow the search. The only common limit that we might want to use here is English, to retrieve only English articles. Remember, if you need to be comprehensive, do not use the English limit, which may remove important publications in languages other than English from the retrieval. If not specified, *Web of Science* automatically sorts the retrieval by **Latest Date**. The search can now be executed by clicking on the **Search** button.

2.1.2. Displaying Search Results

Once a search has been executed, *Web of Science* immediately displays a listing of the first 10 retrieved citations. The retrieval display screen is similar to the *PubMed* display screen. Detailed citations, including the abstract and cited references, can be viewed by clicking on the article's title. We can also select articles for full display, printing and downloading by clicking on the box to the left of the citation. When selecting citations, before proceeding to the next page and the next 10 citations, click the **Submit** button to add the selected citations to a **Marked List**. The selected items will be deleted if you proceed to the next page before doing this. Once all relevant articles have been selected, click on the **Marked List** button to view the full list of selected citations, along with the available printing and downloading options.

The **Marked List** screen provides options for selecting fields, as well as a window for the selection of specific sort options for printing and downloading. Three buttons are then used for the different output options; **Format for Printing**, which formats a page within your Web browser for printing the citations; **Save to File**, to download the citations; and **Export**, to automatically export the citations to a bibliographic management software package (*see* **Subheading 5.**).

2.1.3. Cited Reference Searching

Cited reference searching in *SCI* provides a means of locating more recent publications on specific subjects by searching for articles that have cited a particular author or publication. To enter the cited reference search screen click on the **Cited Ref Search** button on the top of any *Web of Science* screen.

The cited reference search screen provides for searching a **Cited Author**, as well as a particular cited author's work, by entering the journal title abbreviation in the **Cited Work** section of the search screen, and the year of publication in the **Cited Year** section. Note that the system requires you to enter the journal title abbreviation. A list of abbreviated titles is available by clicking in the **list** link located above the **Cited Work** line. An author's name needs to be entered as last name, a space, and then first and middle initial. If the middle initial is not known, the name can be truncated by placing an "*" after the first initial. If the cited reference search is not further qualified with a cited work and year, the retrieval for a truncated author search may include unwanted results if there are several authors with the same last name and the same first initial. To search for articles citing publications by Stephen Krawetz, we would enter `krawetz s*` into the **Cited Author** line. If we know that the author's middle initial is A, we can then enter `krawetz sa`. We can now retrieve a list of this author's cited works by clicking on the **Lookup** button.

2.1.3.1. Displaying Cited Reference Information

In the above search, the cited reference search display screen will retrieve and list all cited works by the author, 10 at a time, identifying the frequency of citations for each individual publication to the left of the list, under the heading **Hits**. To retrieve all articles that cite a specific publication, click on the box to the left of the citation and click the **Search** button. There are also options to define limits and sorting preferences at the bottom of the display screen, as in the **General Search** screen, which can be modified before executing the search.

2.1.4. Saving Searches

Web of Science enables searchers to save and then execute a search at a later date. This can be done from either the **General Search** or the **Cited Ref Search** screens. First enter the search in the search screen. To save it, click on the **Save Query** button. *Web of Science* will prompt you to save the search to disk. The search should be saved to a disk location where it can be easily retrieved, as well as saved with a recognizable name. The search is saved in a file with the extension "cgi." To execute a saved search in *Web of Science*, first retrieve it from the disk to which it was saved. This can be done by selecting **Open File** from the **File** menu in a Web browser. The file can be opened from any location within *Web of Science*, but you do need to be logged into *Web of Science* for it to execute properly. The saved search will open in the web browser, and can then be executed the same as any other search in the *Web of Science*.

3. *Current Contents*

Current Contents (*CC*) is a current awareness service produced by the ISI. It provides the tables of contents from more than 7000 journals and 2000 books and conference proceedings in science, social science, technology, and arts and humanities, shortly before or after publication dates. Whereas the print version of *CC* provides just the tables of contents, electronic versions of *CC* include full bibliographic information for each article, as well as English author-generated abstracts when available. Detailed author information is also included to assist researchers in requesting reprints.

CC is published weekly in seven subject editions:

1. *Agriculture, Biology, and Environmental Sciences*
2. *Arts and Humanities*
3. *Clinical Medicine*
4. *Engineering, Computing, and Technology*
5. *Life Sciences*
6. *Physical, Chemical, and Earth Sciences*
7. *Social and Behavioral Sciences*

Each edition is not mutually exclusive of any others. Comprehensive searching in molecular biology requires the use of 1, 3, 4, 5, and 6. Because *CC* is a current awareness service, retrospective files are not maintained for long periods of time. It should be used only to retrieve the very latest journal publication information. Retrospective searching needs be carried out on *MEDLINE* and *SCI*.

The database is available in a variety of formats: ftp deliveries, floppy disks, and CD-ROMs, through ISI and a variety of third party vendors. ISI now provides access to *CC* on the World Wide Web from their *Current Contents Connect* system. Additional information on *CC* content and availability can be found at ISI's Web site *(2)*.

3.1. *Searching* Current Contents Connect

The following discussion looks at only a few of the search capabilities for *CC* on *Current Contents Connect* (*CCC*). Detailed information on searching the system is available in the online manual by clicking on the **Help** button found on each *CCC* page. Use of *CC* on *CCC* is not free. Access will be available only to users in institutions that have subscribed to the service.

Once connected to *CCC* on the WWW (`http://www.isicc.com/`), click on the **Start** button. *CCC* starts with a **Search Limits** screen, from which certain search limits can be selected before searching. To the left of the screen are the **Current Content Editions**, all of which are turned on. We can turn off any editions that are not needed for a search by clicking on the checked boxes to the left of the editions. When searching molecular biology topics, it is likely that we will not need the Social and Behavioral Sciences or Arts and Humanities editions. To the right of the screen are the **File Depth** options, from which are selected how much of the database is to be searched: the **Latest week**; **Latest four weeks**; **Latest six months**; or **Extended file**, which covers the last two years. Once the desired limits have been selected, click on the **Submit Limit Changes** button to enter the search system.

Searching *CCC* is similar to searching *SCI* on the *Web of Science*. At the top right of each *CCC* screen is information that tells the searcher which *CC* editions and depth of the file are being searched. Toward the middle and to the left of the screen is a window that says **Topic/Subject** (TS). Clicking on the arrow pulls down a menu that reveals the different fields that can be searched in *CCC*. To enter a search strategy, select a field, normally **Topic/Subject**, and enter the search strategy in the line to the right. As an example, let's execute the same search that we discussed in searching *SCI*. Select **Topic/Subject** and enter:

```
phenylketonuria and (genetic* or molecular
biology)
```

The search can be executed by clicking on the **Search** button. Like *SCI*, *CC* does not have a controlled vocabulary for indexing, but uses author keywords and **Keywords Plus** (*see* **Subheading 2.1.1.**).

After executing a search *CCC* returns to the same **Search** screen, unlike the *Web of Science*, which immediately displays retrieval results. The search strategy appears at the bottom of the screen with an indication of how many **Hits**, or documents, were retrieved. Clicking on the icon (the one with the eye) to the left of the strategy retrieves a list of the first 10 citations. New searches can be executed from the **Search** screen and the additional strategies are stacked at the bottom, along with the first search. Up to 10 lines of strategy can be accommodated. If more than 10 strategies are produced, the oldest will be removed from the page. Search strategy lines can also be combined from the **Search** screen. For example, we could select **Language** (LA) from the **Search Field** menu and enter `English` into the search line. This will create a set of all English articles in the database. This set can then be combined with our first search set by selecting **Combine Sets** from the **Search Field** menu and entering `1 AND 2` in the search line. Remember to select **Combine Sets** when doing this, otherwise *CCC* will retrieve all articles where the numbers 1 and 2 appear.

3.1.1. Displaying Search Results

CCC provides a similar citation display screen as does the *Web of Science*. However, the only option from the citation list screen is to **Mark All**, which saves all citations in the list, for printing or downloading at a later time. To mark and save individual citations, we need to first view the full citation by clicking on the title, and then click on the box to the top left of the full citation display. Hopefully, future versions of *CCC* will improve this feature, enabling the searcher to mark citations from the citation display screen.

Once citations are marked for printing and downloading, we can click on the **List** button to view all selected citations. The **List** screen also provides **Sort**, **Print**, and **Download**, or **Export**, options. Of note are the **Export Format** option, which includes **Request-a-Print File** format. **Request-a-Print File** formats the output for printing onto reprint request cards, which can be purchased from ISI, and then mailed to authors to request a reprint. We can also export to **Procite** or **Reference Manager**, two bibliographic management software packages (*see* **Subheading 5.**), by clicking on the **Export to Procite/ Reference Manager** button.

ISI also offers the option of ordering documents directly from their document delivery service. Be warned, the service can be expensive once copyright fees have been added to the delivery charges for a document.

3.1.2. Saving Searches

Because *CC* is a current awareness service, it is important to be able to save searches, which can be executed against the weekly updates. *CCC* saves searches in the same way as does the *Web of Science*. Complete the search and from the **Search** screen click on the **Save Session** button. *CCC* opens a screen called the **Current Contents Connect Profile**. The complete search will be listed in the profile. To save the search, click on the **Save Profile** button, and save the search to disk as you would with the *Web of Science* (*see* **Subheading 2.1.4.**). To execute the search at a later date, open the file in your web browser and click the **Run Profile** button. Before *CCC* runs a saved profile it will ask whether we want to change any of the limits set when the search was first executed. This is done from the **Search Limits** screen. We might select different *CC* editions, or most likely, we will select the **File Depth** as **Latest Week**, to search only the latest week's update.

4. Current Awareness Services

A more recent development in services enabling researchers to stay current of the literature are alerting services. Alerting services have the advantage of informing the user when new information is available, removing the burden of having to remember and regularly log into literature search services and run updates.

4.1. ISI Table of Contents Corporate Alerting Service

The *ISI Tables of Contents Corporate Alerting Service* (*ISI TOC*) is available via electronic mail. The user subscribes, selects from approx 8000 journal titles in the ISI database, and receives the tables of contents with author abstracts (when available), all through electronic mail. Further information can be found at ISI's *Alerting Services* site *(3)*.

The *ISI TOC* service is easy to use. A message requesting registration to the service is sent to a specific e-mail account set up at ISI. Within half an hour the requester is sent a list of over 200 subject areas. Subjects of interest to the requester are selected by placing an "X" in the space provided to the left of each subject. The list is then forwarded back to ISI. Again, within half an hour, lists of journal titles for each of the selected subject areas is sent to the requester. The messages are again forwarded back to ISI with selected journal titles. All selected titles are recorded in the requester's profile. As new issues of selected journals are published, the ISI TOC user receives their table-of-contents in his or her e-mail account.

An additional feature of this service is that it can be set up to enable users to select articles from the tables-of-content and have those selections forwarded to their local library, or back to ISI for document delivery.

5. Bibliographic Management Software

Bibliographic management software (BMS) is mentioned several times in **Subheading 2.** BMS has two major functions: 1) to enable researchers to create and maintain personal citation databases on specific subjects; and 2) to assist in the formatting of endnotes, footnotes, and references in papers written for publication.

Three commonly used BMS packages are *Reference Manager (RM)*, *ProCite*, and *Endnote*. They all have essentially the same features and costs. RM, which was originally designed for the biomedical researcher, will properly format reference information for specific journals, depending on where the paper is to be submitted for publication.

RM and *ProCite* are produced by Research Information Systems (RIS; `http://www.risinc.com/`), and *Endnote* is produced by Niles and Associates (`http://www.niles.com/`). RIS and Niles and Associates are now both subsidiaries of The Institute for Scientific Information (ISI; Carlsbad, CA). At the time of writing, they were in the process of combining to form a new company, ResearchSoft. RIS has stated that technical support and product development will continue for all three packages, although in the long term it seems unlikely that a single company will continue to market three very similar products.

RM, *Procite*, and *Endnote* also include the useful feature of being able to automatically import downloaded citation information from *MEDLINE*, *SCI*, and *CC*. In fact, *SCI* on the *Web of Science* and *CC* on *Current Contents Connect*, have an option that enables the searcher to directly load citation output from the WWW into *RM* and *Procite*. This requires a piece of software that can be downloaded from RIS. Software to perform the same operation with *Endnote* was imminent at the time this chapter was written.

The main features of all three BMS packages are discussed and compared by Stigleman *(4)*. This review is slightly out-of-date, but it does provide a good overview of the packages' capabilities and valuable information on how to evaluate BMS software for purchase.

Appendix

The following is a list of key WWW sites discussed in this chapter.

Institutes for Scientific Information
`http://www.isinet.com/`
Science Citation Index on the *Web of Science*
`http://webofscience.com/`
Current Contents Connect
`http://www.isicc.com/`
Research Information Systems
`http://www.risinc.com/`
Niles and Associates
`http://www.niles.com/`

References

1. Institute for Scientific Information (March 30, 1999) *Science Citation Index Database*, Institute for Scientific Information [6 pages] http://www.isinet.com/prodserv/citation/citsci.html.
2. Institute for Scientific Information (January 27, 1999) *Current Contents General Information*, Institute for Scientific Information [5 pages] http://www.isinet.com/prodserv/cc/cchp.html.
3. Institute for Scientific Information (March 30, 1999) *ISI Alerting Services*, Institute for Scientific Information [3 pages] http://www.isinet.com/prodserv/ias/ca.html.
4. Stigleman, S. (1996) Bibliography programs do Windows, *Database* **19,** 57–66.

28

The Virtual Library III

Electronic Journals, Grants, and Funding Information

Keir Reavie

1. Introduction

This chapter discusses the current state of access to electronic journals on the Internet, some of the difficulties in getting access, and what we might expect to see in the future for these resources. The last section of the chapter discusses the use of the major Internet resources for obtaining information on current biomedical funding opportunities.

2. Electronic Journals

Access to the full text of journals on the Internet is still a relatively new phenomena. Many of the details for providing easy and consistent access, issues regarding access rights, pricing, and technological problems related to access still need to be resolved. Journals tend to come from two sources, commercial publishers, and titles published by associations and societies. Access to titles from either source reveal similar characteristics and policies in the approach to making full text available on the Internet. There are a variety of access levels, title availability, and pricing mechanisms. If a title is available online, it may provide tables of contents, tables of contents with abstracts, or the full text of the journal. When available in full-text form, a variety of formats may be encountered: the two main ones being HTML (hypertext markup language, used for formatting standard documents on the WWW), or in PDF (portable document format). In many instances the information is available in both formats. PDF is the preferred format for a variety of reasons, which we will discuss in detail later in this section.

From: *Methods in Molecular Biology, vol. 132: Bioinformatics Methods and Protocols*
Edited by: S. Misener and S. A. Krawetz © Humana Press Inc., Totowa, NJ

Pricing for full-text access is a major issue, and it varies greatly from title to title, or publisher to publisher. Electronic journals published by societies or associations may be freely available to their members. They may also be freely available to subscribers of the journal's print counterparts, or for a small fee in addition to the print subscription price. Others will be available for a much larger cost in addition to the paper subscription. Currently many titles are free to subscribers of the paper copy, and some may remain that way. But all indications point to the future implementation of additional pricing, on top of paper subscriptions, for access to the online versions of a journal. Some larger publishers only provide access to their titles as a collection—you can purchase online access for a percentage cost above all the paper subscriptions that you purchase from the publisher. This normally provides access only to those titles for which the paper version is owned. However, in some instances, there may be a provision, at additional cost, to obtain access to more titles, or at least to individual articles from those titles as needed. Larger publishers may also charge an annual maintenance fee for access to the titles via their WWW interface. Purchasing access to these larger publisher sites is normally only affordable and feasible for large organizations such as academic institutions, who over the next few years will be purchasing these services for their faculty and students.

The situation with regard to access to the full text of journals electronically is currently rather chaotic, although many of the initial problems and issues involved should be rectified over time. Currently, the easiest way to deal with electronic journal access is to look at individual titles—presumably those that are most important to the user—and to investigate how access can be achieved for all the titles needed. Also, how much will it cost, in addition to print subscriptions, for access? In some cases a print subscription is required to allow access to the online version of the title, which seems contradictory to making the full text available electronically in the first place.

Aside from cost, other issues to consider when purchasing access to full-text journals electronically are: how access to the full text will be controlled, and will this in any way affect accessibility by the prime users of the resource; and will access to the old journal issues still be available if a subscription is cancelled? When a print subscription is cancelled, you retain access to the old print copies. Will this also be true of the older electronic versions? The access problem involves both policy and technical issues, on the part of the subscriber and the publisher. Access to electronic journals on the Internet is essentially available via two methods: a login ID and password, or through Internet addresses, which are assigned over a range of computer addresses, normally defined within a single entity, organization, or institution. The login ID requires administration through the distribution of a single ID and password to all users, or the assigning of IDs and passwords to each individual user. Access through Internet addresses may be a problem if some of the prime users of the resource work in locations that do not fall into the specified Internet addressing range for access. Publishers may be reluctant to spread the access too thinly without charging additional fees for access by what they may consider additional user groups. As well, users that have to dial-in to access these resources may be restricted if their Internet access service falls within a different Internet address range than that of the primary organization providing access to the electronic journals. From the policy aspect, if a user is part of the institution or organization, why should they be denied access simply because they may have an office external to the institution's main location? This is a common problem in large academic research facilities, where faculty may be spread throughout numerous buildings or geographic locations, and it is for these researchers that the ability to access journals electronically is most important.

Users of electronic journals are currently being forced to access titles from a variety of locations on the WWW. An important step in making access to electronic journals efficient will be the consolidation of access points and user interfaces. This is currently taking place on four fronts; the publisher, at the institutions providing access, at literature retrieval sites like *PubMed*, and through third party organizations. This latter group includes projects like Highwire Press at Stanford University. We will discuss this project in more detail in the next section as we look at some examples of electronic publishing.

2.1. Examples of Electronic Journal Titles

To illustrate some of the issues involved in accessing electronic journals, we will examine some specific online titles. These examples are not comprehensive and you may encounter other problems with titles not discussed here.

2.1.1. Nature

Nature, *Nature Genetics*, and *Nature Medicine* (`http://www.nature.com/`) became available in full text on the Internet in September, 1998. *Nature Neuroscience*, *Nature Biotechnology*, and *Nature Structural Biology* have since been added to this initial collection. At this time all are available freely to individual subscribers. Institutional subscribers do not have access. Researchers who want access to these titles electronically must maintain personal paper subscriptions.

2.1.2. Science

Science (`http://www.sciencemag.org/`) is available full text on the Internet to individual subscribers for an additional $12. This permits the user to access the full text on a single computer workstation, with access being controlled by the Internet address of that computer. Institutional subscribers can provide access to *Science* on individual computers within the institution's library for $25 a computer. Site-wide access is available for institutions using a pricing structure based on full-time employees (and the number of students for educational institutions) and the number of affiliated sites.

2.1.3. Proceedings of the National Academy of Science

The *Proceedings of the National Academy of Science* (*PNAS*) (`http://www.pnas.org/`) requires a subscription fee on top of the paper subscription. *PNAS* is one of many basic medical science journals that have been made available through the Highwire Press initiative at Stanford. Highwire Press (`http://highwire.stanford.edu`) has negotiated with publishers (mainly scholarly societies and university presses) to make available important biomedical journals in both HTML and PDF formats. Highwire provides a single site on the Internet where these titles can then be accessed. Policies for access to the titles available from Highwire, such as *PNAS*, will vary from title to title, and any subscription costs for access need to be negotiated between the individual or institution and the actual publisher of each title.

Access to *PNAS* is controlled for individual subscribers through an ID and password. Access for institutions is controlled via Internet addresses, which the institution needs to submit to the publisher of *PNAS*. The same system is used for all titles in Highwire, but access needs to be negotiated with each publisher separately. Highwire does, however, offer a service in which they will act as the intermediary for negotiating subscriptions to collections of titles available from their site.

2.1.4. EMBO Journal

EMBO Journal is also accessible from the Highwire site and is currently freely available to individuals and institutions with paper subscriptions.

2.1.5 Journal of Biological Chemistry

The *Journal of Biological Chemistry* (*JBC*) (`http://www.jbc.org/`) was one of the first journals to be available full text on the Internet. It is accessible through the Highwire site. *JBC* online is free for American Society for Biochemistry and Molecular Biology members, and available to institutions for an additional cost on top of a print subscription.

2.1.6. Cell

An individual paper subscription to *Cell* (`http://www.cell.com/`) provides free access to the electronic version. Institutions can obtain access for an additional fee on top of the cost for a print subscription. This additional fee is based on the size of the institution and the number of potential users of *Cell* online. The institutional pricing does, however, include access to the other Cell Press journals: *Immunity*, *Neuron*, and *Molecular Cell*.

2.1.7. Molecular Medicine Today

Molecular Medicine Today (`http://www.elsevier.com/locate/molmed/`) is published by Elsevier. Elsevier publications are available electronically via their *Science Direct* service. This is normally only available to institutions as it requires a fee for initial access to the *Science Direct* system, and then additional fees above the print subscriptions for Elsevier publications. This can become quite expensive.

2.1.8. New England Journal of Medicine

The *New England Journal of Medicine (NEJM)* (`http://www.nejm.org/`) is freely available to anyone with a paper subscription, both individual and institutional. An ID and password for access can be assigned through the *NEJM* web site. Institutional subscribers will have difficulty with this access method, since they also need to use an ID and password. Institutional access to electronic journals is preferable via Internet address controls, so that all users can access the resource from all institutional computers without having to worry about remembering another ID and password. Things can start to become complicated if one needs an ID and password for each electronic journal they use.

2.2. Reading PDF Files with the Adobe Acrobat Reader

The majority of electronic journal sites provide full text access to articles in both an HTML and PDF formats. HTML is problematic for this kind of publication, as the initial full-text HTML document is mainly text with thumbnails, or small images, for graphics, tables, and charts. To view graphics, tables, and charts, one normally has to click on the thumbnail link, which then loads the full image onto a separate Web page. When printing, one first needs to print the text, then expand each image to be printed separately. In addition, HTML documents will not have the correct pagination when printed. The PDF format however, displays the article as it would appear in the print journal, with the full images in the correct position on the pages and with correct pagination.

To view documents in the PDF format requires the use of the *Adobe Acrobat Reader*, available from the Adobe Corporation (`http://www.adobe.com/`). The *Acrobat Reader* can be downloaded freely from Adobe and set up for use with a Web browser. Most Internet sites that make information available in PDF format have direct links to the Adobe download site and the instructions for loading the software. Once loaded, your Web browser will automatically activate the *Acrobat Reader* software when retrieving files with a "PDF" extension from the Internet. The *Acrobat Reader* allows only the viewing and printing of PDF documents. There are no editing capabilities. It is essential when printing PDF documents that contain images to have a good laser printer.

3. Grants and Funding Information

Two primary Internet sites providing grants and funding information for molecular biology research support are the National Institutes of Health (NIH) Grants and Funding pages (`http://www.nih.gov/grants/`) and the National Science Foundation (NSF) Grants and Awards pages (`http://www.nsf.gov/home/grants.htm`). Both sites provide browse and search options that enable researchers to locate funding opportunities, information on recent awards, and to obtain requisite forms for proposal submission.

3.1. National Institutes of Health

NIH funding opportunities can be located in two ways, by searching for opportunities through the NIH Office of Extramural Research (OER) (`http://www.nih.gov/grants/oer.html`) and information from the NIH Guide to Grants and Contracts. Award information for all recent awards from the Department of Health and Human Services, which includes NIH, can be searched in the CRISP (Computer Retrieval of Information on Scientific Projects) database. Also, each Institute at the National Institutes of Health will provide grant and funding information at their individual web sites. A list with links to all Institutes' web sites is available from NIH's Home Page (`http://www.nih.gov/`).

3.1.1. Searching the Office of Extramural Research

The OER Grants Web site provides a basic search system for locating funding opportunities at NIH. The Search Site (`http://www.nih.gov/grants/search.htm`) has a single line, into which search terms or phrases can be entered. There is no controlled vocabulary akin to *MeSH*, so careful use of terminology, synonyms, and acronyms is important. The system does, however, use an operator that can be entered as `<thesaurus>`. The use of this operator in front of any term will retrieve items that contain synonyms for that term. Search tips are provided at the bottom of the page. Detailed searching help is available by clicking on the **Help** link above the search line. The use of Boolean operators **AND**, **OR**, and **NOT,** as well as truncation using the "*" symbol, is permissible. To look for funding opportunities for research on the genetic aspects of phenylketonuria, we could enter the search:

```
phenylketonuria AND genetic*
```

Click on the **Search** button to execute the search. Once the search has been executed a results screen is displayed. The list of retrieved documents contains the title and the first couple of lines of the document's summary. Click on the title to retrieve the full document. Note that the OER search system does not search Requests For Applications and Program Announcements in the *NIH Guide to Grants and Contracts*.

3.1.2. The NIH Guide to Grants and Contracts

Information on requests for applications and program announcements are released weekly in the *NIH Guide for Grants and Contracts*, both in print and an online version (`http://www.nih.gov/grants/guide/index.html`). The *NIH Guide* web page provides links for browsing the weekly *Guide* by year, as well as a basic search system to search the contents of the *Guide*'s archives back to 1992. A space for entering a search strategy is available in the middle of the screen. The *Guide* uses the same search system as the OER and, consequently, the same search features. Search help is available by clicking on the **Search Help** link to the right of the search line.

3.1.2.1. THE *NIH GUIDE* LISTSERV

The *NIH Guide* can also be received as a weekly e-mail message by joining the *NIH Guide* Listserv. Simply send an e-mail message to:

```
listserv@list.nih.gov
```

In the body of the message type the following information:

```
subscribe NIHTOC-L your name
```

The NIH Listserv system will add you to the *NIH Guide* list. Within a few minutes you should receive a message that asks you to confirm your request to added to the list. Follow the instructions in the message and return it to confirm your subscription. You will then receive a further confirmation of your subscription, as well as instructions for use of the list and how to unsubscribe. Once you have been added to the list, the *NIH Guide* will be sent to your e-mail account each week as it is published. For information regarding the NIH, Guide Listserv is available online ***(1)***.

3.1.3. CRISP

The *CRISP* database (`http://www-commons.cit.nih.gov/crisp`) provides access to information on research projects and programs supported by the Department of Health and Human Services. The majority of the funding information is from NIH, but it includes data from the Centers for Disease Control and Prevention, the Food and Drug Administration, the Health Resources and Services Administration, and the Agency for Health Care Policy and Research. *CRISP* provides a means by which we can see who received awards for specific research and which agencies support the award. The WWW version of the *CRISP* search system is new and has many deficiencies that prevent us from doing quick and efficient searches. Some of these will be mentioned below.

3.1.3.1. Searching *CRISP*

There are two databases available at the *CRISP* WWW site: *Current Award Information*, and *Historical Award Information*. To locate the most recent award select the *Current Award Information* site. *CRISP* provides a **Basic Query Form** and **Advanced Query Form** for searching. **Help** in searching *CRISP* from either form can be obtained by clicking on the question mark icon at the top right of the screen. The **Advanced Query Form** should be used for searches with Boolean operators. If we again wanted to search for awards given for research on the genetic aspects of phenylketonuria, go first to the **Advance Query Form**, then enter the phrase:

```
phenylketonuria AND genetic
```

into the **Enter Search Terms** section at the top of the form. Click in the space next to **and** in the **Global Logic** section of the form, to indicate an **AND** Boolean operator is to be used between the terms in the search. The current WWW version of *CRISP* does not provide for the use of more than one type of Boolean operator in the search phrase.

We can truncate terms in our search by selecting the **Stem** option in the **Expansion Logic:** line. In the above example, this will retrieve items that use the terms *genetic*, *genetics*, genetically, and so on.

Click on **Submit Query** to execute the search.

The *CRISP* database has a controlled vocabulary. However, the new WWW *CRISP* search system does not yet provide access to this vocabulary. To use the controlled vocabulary we could execute a search strategy, locate an item that matches our strategy closely, and select terminology from the **Thesaurus Terms** section of the record. We can then return to the search form, and re-execute the search using the selected terminology, and, hopefully, retrieve more accurate results.

The *CRISP* results page provides a list of award titles, along with the award number and the last name of the principal investigator. Click on an award title to retrieve the detailed information about the award, including an abstract.

It is also possible to search *CRISP* by principal investigator to obtain details on awards received by individuals, as well as by institute and to locate awards received by specific institutions.

3.2. National Science Foundation

The NSF Web site provides search systems for searching awards and funding opportunities using the *Verity* search language. Information on using the *Verity* search language is available by clicking on the **Verity search language** link from any of NSF's advanced search screens.

3.2.1. Documents Online

The *Documents Online* service at NSF (`http://www.nsf.gov/cgi-bin/pubsys/browser/odbrowse.pl`) enables one to search all NSF documents online, including NSF *Program Announcements and Information*. On the main **Documents Online** screen users can browse documents by type or execute a basic search on the full text of NSF documents. There is also a **Fielded Search** form, or advanced search form, for searching documents, available by clicking on the **Fielded Search** link (**Text Search** in the main screen).

On the fielded search screen one can specifically search *Program Announcements and Information* by clicking on the arrow in the **Document Type** and selecting *Program Announcements and Information* from the **Document Type** menu. The search strategy can then be entered into the **Full Text** window. Use of the Boolean operators **AND**, **OR**, and **NOT** is permissible, as is truncation using the "*" symbol. As an example, one would enter `genetic*` into the **Full Text** window. Click on the box next to **Current** in the **Document Status** line, at the bottom of the form, to retrieve only current documents in the search. Click on the **Search** button to execute the search. The retrieval screen will list a number recent program announcements dealing with genetics. One can then link to any of the items in the retrieved list to obtain detailed information.

3.2.2. Search Awards

The *NSF Search Awards* page (`http://www.nsf.gov/verity/srchawd.htm`) enables one to search for awards received from NSF. It provides a basic search screen, as well as the **Fielded Search** screen (`http://www.nsf.gov/verity/srchawdf.htm`). The **Fielded Search** screen for *NSF Awards* provides for more detailed searching than that for *Documents Online*. Boolean searches can be performed from the **Full Text** window at the bottom of the form, using the same search rules as the *Documents Online* fielded search. One can again execute the search:

```
phenylketonuria AND genetic*
```

by entering it into the **Full Text** window. Click on the **Search** button to execute the search. The search retrieval screen lists titles of awards, on which one can link to obtain the detailed information about the award.

The **Fielded Search** form enables the researcher to search several different fields, including investigator. When entering search information into more than one field in the form be sure to use the **AND** option for **Boolean operator used to combine field expressions**, at the bottom of the search form. This ensures that the **AND** operator is used to combine the search results from the different fields used.

4. Appendix

Some key WWW sites discussed in this chapter:

Highwire Press: `http://highwire.stanford.edu/`
Adobe Corporation (*Adobe Acrobat Reader* for reading PDF files): `http://www.adobe.com/`
National Institutes of Health Grants and Funding Resources: `http://www.nih.gov/grants/`
National Science Foundation Grants and Awards Resources: `http://www.nsf.gov/home/grants.htm`

References

1. National Institutes of Health, Office of Extramural Research (December 3, 1997). NIH Guide: Using the TOC Notification LISTSERV Service, National Institutes of Health [1 page]. `http://www.nih.gov/grants/guide/listserv.htm`.

Index

G

H

I

X

Y